The Ecology of Agricultural Pests

The Systematics Association Special Volume Series

Series Editor

David M. John

Department of Botany, The Natural History Museum, Cromwell Road, London SW7 5BD, UK.

The Systematics Association provides a forum for discussing systematic problems and integrating new information from cytogenetics, ecology and other scientific fields into taxonomic concepts and activities. It has achieved great success since the Association was founded in 1937 by promoting major meetings covering all areas of biology and palaeontology, supporting systematic research and training courses through the award of grants, production of a membership newsletter and publication of review volumes by its publisher Chapman & Hall. Its membership is open to both amateurs and professional scientists in all branches of biology who are entitled to purchase its volumes at a discounted price.

The first of the Systematics Association's publications, *The New Systematics*, edited by its then president Sir Julian Huxley, was a classic work. Over 50 volumes have now been published in the Association's 'Special Volume' series often in rapidly expanding areas of science where a modern synthesis is required. Its *modus operandi* is to encourage leading exponents to organise a symposium with a view to publishing a multi-authored volume in its series based upon the meeting. The Association also publishes volumes that are not linked to meetings in its 'Volume' series.

Anyone wishing to know about more about the Systematics Association and its volume series are invited to contact the editor.

Forthcoming titles in the series:

The Units of Biodiversity: Species in Practice
Edited by M.F. Claridge, A.H. Dawah and M. Wilson

Relationships of the Major Arthropod Groups
Edited by R.A. Fortey and R.H. Thomas

Other Systematics Association publications are listed after the index for this volume.

The Systematics Association Special Volume Series 53

The Ecology of Agricultural Pests

Biochemical Approaches

Edited by

W.O.C. Symondson

School of Pure and Applied Biology
University of Wales, Cardiff, UK

J.E. Liddell

School of Molecular and Medical Biosciences,
University of Wales, Cardiff, UK

THE Systematics ASSOCIATION

CHAPMAN & HALL
London · Weinheim · New York · Tokyo · Melbourne · Madras

Published by Chapman & Hall, 2–6 Boundary Row, London SE1 8HN, UK

Chapman & Hall, 2–6 Boundary Row, London SE1 8HN, UK

Chapman & Hall GmbH, Papelallee 3, 69469 Weinheim, Germany

Chapman & Hall USA, 115 Fifth Avenue, New York, NY 10003, USA

Chapman & Hall Japan, ITP-Japan, Kyowa Building, 3F, 2-2-1 Hirakawacho, Chiyoda-ku, Tokyo 102, Japan

Chapman & Hall Australia, 102 Dodds Street, South Melbourne, Victoria 3205, Australia

Chapman & Hall India, R. Seshadri, 32 Second Main Road, CIT East, Madras 600 035, India

First edition 1996

Typeset in 10/12pt Palatino by Acorn Bookwork, Salisbury
Printed in Great Britain by Cambridge University Press

ISBN 0 412 62190 8

A catalogue record for this book is available from the British Library

Library of Congress Catalog Card Number: 95-83581

∞ Printed on permanent acid-free text paper, manufactured in accordance with ANSI/NISO Z39.48-1992 and ANSI/NISO Z39.48-1984 (Permanence of Paper). [Paper = Magnum, 70 gsm]*

Contents

List of contributors ix

Preface xiii

Introduction – Biochemical approaches to understanding agricultural pests 1
M.F. Claridge

1. Biochemical systematics: principles and perspectives for pest management 7
S.B.J. Menken and L.E.L. Raijmann

2. Combined use of biochemical, immunological and molecular assays for infection, species identification and resistance detection in field populations of *Anopheles* (Diptera: Culicidae) 31
J. Hemingway, G.J. Small, S.W. Lindsay and F.H. Collins

3. Studies of interactions between alfalfa weevil strains, *Wolbachia* endosymbionts and parasitoids 51
T.H. Hsaio

4. Biochemical approaches to the study of ecological genetics: the role of selection and gene flow in the evolution of insecticide resistance 73
J.C. Daly and S. Trowell

5. A molecular and ecological investigation of the large arionid slugs of North-West Europe: the potential for new pests 93
L.R. Noble and C.S. Jones

6. Systematics of brown planthopper and related species using nuclear and mitochondrial DNA 133
P.L. Jones, P. Gacesa and R.K. Butlin

7. Application of novel molecular markers (DNA) in agricultural entomology 149
H.D. Loxdale, C.P. Brookes and P.J. de Barro

8. Location of resistance to *Brevicoryne brassicae* in wild brassica species 199
R.A. Cole and T. Fenning

9. The use of DNA markers in population genetics and ecological studies of the desert locust *Schistocerca gregaria* (Orthoptera: Acrididae) 213
D-X. Zhang and G.M. Hewitt

10. The use of DNA analysis and the polymerase chain reaction in the study of introduced pests in New Zealand 231
K.F. Armstrong and S.D. Wratten

11. Serological analysis of arthropod predation: past, present and future 265
M.H. Greenstone

12. Serological diagnosis of parasitism: a monoclonal antibody-based immunodot assay for *Microplitis croceipes* (Hymenoptera: Braconidae) 301
M.K. Stuart and M.H. Greenstone

13. Polyclonal, monoclonal and engineered antibodies to investigate the role of predation in slug population dynamics 323
W.O.C. Symondson and J.E. Liddell

14. The potential of combinatorial antibody libraries in pest–predator relationship studies 347
J.E. Liddell and W.O.C. Symondson

15. Serological analysis of predators of *Helicoverpa armigera* Hübner (Lepidoptera: Noctuidae) eggs in sorghum–pigeonpea intercropping at ICRISAT, India: a preliminary field study 367
L. Sigsgaard

16. Using gut content immunoassays to evaluate predaceous biological control agents: a case study 383
J.R. Hagler and S.E. Naranjo

17. An environmental risk assessment for release of an exotic microsporidium for European corn borer control in North America 401
D.W. Ragsdale and C.T. Oien

18. Progress in quantifying predation using antibody techniques 419
K.D. Sunderland

19. Electrophoretic approaches to predator–prey interactions 457
M.G. Solomon, J.D. Fitzgerald and R.A. Murray

20. Genetics of esterases in laboratory and feral cotton boll weevils, *Anthonomus grandis* Boheman 469
C.J. Biggers, B.R. Jones, H.R. Bancroft and L.A. Dickey

21. Practical applications for techniques to determine the nutritional state and age of field-caught tsetse flies 479
P.A. Langley

Index 499

List of Systematics Association Publications 514

Contributors

K.F. Armstrong
Department of Entomology and Animal Ecology, Lincoln University, PO Box 84, Canterbury, New Zealand.

H.R. Bancroft
Department of Biology, University of Memphis, Memphis, Tennessee 38152, USA.

C.J. Biggers
Department of Biology, University of Memphis, Memphis, Tennessee 38152, USA.

C.P. Brookes
Entomology and Nematology Department, Rothamsted Experimental Station, Harpenden, Hertfordshire, AL5 2JQ, UK.

R.K. Butlin
Department of Genetics, The University, Leeds, LS2 9JT, UK.

M.F. Claridge
School of Pure and Applied Biology, University of Wales Cardiff, PO Box 915, Cardiff, CF1 3TL, UK.

R.A. Cole
Horticulture Research International, Wellesbourne, Warwick, CV35 9EF, UK.

F.H. Collins
Department of Health and Human Services, Centers for Disease Control, Atlanta, Georgia 30333, USA.

J.C. Daly
CSIRO Division of Entomology, Black Mountain, GPO Box 1700, Canberra, ACT, Australia.

P.J. De Barro
Department of Biology, University of Southampton, Bassett Crescent East, Southampton, SO9 3TU, UK.

L.A. Dickey
Department of Biology, University of Memphis, Memphis, Tennessee 38152, USA.

T. Fenning
Horticulture Research International, Wellesbourne, Warwick, CV35 9EF, UK.

J.D. Fitzgerald
Horticulture Research International, East Malling, West Malling, Kent, ME19 6BJ, UK.

P. Gacesa
School of Molecular and Medical Biosciences, University of Wales Cardiff, PO Box 911, Cardiff, CF1 3US, UK.

M.H. Greenstone
Plant Science & Water Conservation Laboratory, 1301 N. Western Street, Stillwater, Oklahoma 74075, USA.

J.R. Hagler
Western Cotton Research Laboratory, USDA-ARS, 4135 East Broadway Road, Phoenix, Arizona 85040, USA.

J. Hemingway
School of Pure and Applied Biology, University of Wales Cardiff, PO Box 915, Cardiff, CF1 3TL, UK.

G.M. Hewitt
Population Biology Sector, School of Biological Sciences, University of East Anglia, Norwich, NR4 7TJ, UK.

T.H. Hsiao
Department of Biology, Utah State University, Logan, Utah 84322-5305, USA.

B.R. Jones
Department of Biology, Rhodes College, Memphis, Tennessee 38112, USA.

C.S. Jones
Department of Zoology, University of Aberdeen, Tillydrone Avenue, Aberdeen, AB9 2TN, UK.

P.L. Jones
School of Molecular and Medical Biosciences, University of Wales Cardiff, PO Box 911, Cardiff, CF1 3US, UK.

P.A. Langley
School of Pure and Applied Biology, University of Wales Cardiff, PO Box 915, Cardiff, CF1 3TL, UK.

J.E. Liddell
School of Molecular and Medical Biosciences, University of Wales Cardiff, PO Box 911, Cardiff, CF1 3US, UK.

S.W. Lindsay
Danish Bilharziasis Laboratory, Jaegersborg Alle 1D, DK 2920, Charlottelund, Denmark.

H.D. Loxdale
Entomology and Nematology Department, Rothamsted Experimental Station, Harpenden, Hertfordshire, AL5 2JQ, UK.

S.B.J. Menken
Institute for Systematics and Population Biology, University of Amsterdam, PO Box 94766, 1090 GT Amsterdam, The Netherlands.

R.A. Murray
Horticulture Research International, East Malling, West Malling, Kent, ME19 6BJ, UK.

S.E. Naranjo
Western Cotton Research Laboratory, USDA-ARS, 4135 East Broadway Road, Phoenix, Arizona 85040, USA.

L.R. Noble
Department of Zoology, University of Aberdeen, Tillydrone Avenue, Aberdeen, AB9 2TN, UK.

C.T. Oien
Department of Entomology, University of Minnesota, 1980 Folwell Avenue, St. Paul, Minnesota 55108, USA.

D.W. Ragsdale
Department of Entomology, University of Minnesota, 1980 Folwell Avenue, St. Paul, Minnesota 55108, USA.

L.E.L. Raijmann
Institute for Systematics and Population Biology, University of Amsterdam, PO Box 94766, 1090 GT Amsterdam, The Netherlands.

L. Sigsgaard
Department of Plant Pathology and Pest Management, Danish Institute of Plant and Soil Science, Ministry of Agriculture and Fisheries, Lottenborgvej 2, DK-2800 Lyngby, Denmark.

G.J. Small
School of Pure and Applied Biology, University of Wales Cardiff, PO Box 915, Cardiff, CF1 3TL, UK.

M.G. Solomon
Horticulture Research International, East Malling, West Malling, Kent, ME19 6BJ, UK.

M.K. Stuart
Department of Microbiology and Immunology, Kirksville College of Osteopathic Medicine, 800 West Jefferson Street, Kirksville, Missouri 63501, USA.

K.D. Sunderland
Horticulture Research International, Wellesbourne, Warwick, CV35 9EF, UK.

W.O.C. Symondson
School of Pure and Applied Biology, University of Wales Cardiff, PO Box 915, Cardiff, CF1 3TL, UK.

S. Trowell
CSIRO Division of Entomology, Black Mountain, GPO Box 1700, Canberra, ACT, Australia.

S.D. Wratten
Department of Entomology and Animal Ecology, Lincoln University, PO Box 84, Canterbury, New Zealand.

D-X. Zhang
Population Biology Sector, School of Biological Sciences, University of East Anglia, Norwich, NR4 7TJ, UK.

Preface

This book was produced following an international symposium held at the University of Wales Cardiff from 21st to 23rd September, 1994, also entitled *The Ecology of Agricultural Pests: Biochemical Approaches*, and organized by the editors. However, it represents much more than a simple collection of conference proceedings. The meeting was designed to be a unique event, bringing together for the first time leading researchers in three primary and very broad areas of technical expertise: electrophoresis; molecular biology; and immunological detection systems. The symposium was intentionally techniques-based, and designed to compare experience with a range of biochemical and molecular approaches to understanding and measuring interactions between pests, their hosts, their predators and other organisms, as well as their population genetics, dynamics and systematics. Many of the contributors were using a range of techniques to answer different ecological questions, while several different methodologies were being applied to study similar problems.

The symposium, therefore, was a forum for discussing and comparing the benefits and drawbacks of these techniques, within the superimposed framework of studies involving pests, giving the meeting a clear, applied emphasis. By inviting a high percentage of the speakers to the meeting, we were able to ensure that all the major modern technologies being used for such research were well represented by leaders in their respective fields. All the speakers were invited to provide chapters for this book, and all contributions were reviewed by at least two expert referees in appropriate fields. We believe, therefore, that, like the symposium itself, this book provides a unique resource.

The meeting was truly international, in that it attracted delegates from 14 different countries including England, Scotland, Wales, the Republic of Ireland, the USA, New Zealand, Germany, Spain, Norway, Sri Lanka, Australia, the Netherlands, Denmark, Iraq and the Philippines.

We would like to thank the Systematics Association, the Techniques

Group of the Biochemical Society, and both the School of Molecular and Medical Biosciences and the School of Pure and Applied Biology within the University of Wales Cardiff, for sponsoring this symposium. We are particularly grateful to the Systematics Association for agreeing to publish these papers, with Chapman & Hall, in the Special Volume Series. In many ways this book provides an update on the electrophoretic techniques described in Special Volume No. 39, *Electrophoretic Studies on Agricultural Pests*, and we think that the Systematics Association should be congratulated for their foresight in expanding their remit to such applied research. Biochemical and molecular differences between species, populations and individuals are fundamental to an understanding of both pest ecology and systematics, and therefore this close association between taxonomists and applied biologists is likely to be highly productive. It is therefore appropriate that the Symposium, and this book, begin with an opening address by Professor Mike Claridge, President of the Systematics Association, and an applied biologist in his own right.

We also wish to thank the session chairs, Professor Ifor Bowen, Dr David Glen, Dr Gordon Port, Professor Peter Langley and Dr Peter Haskell, for skilfully steering the meeting over the three days. Also Mr Nigel Pompeus, who took on much of the administrative load. We are particularly grateful to Dr Mike Wilson and Mr John Deeming from the Entomology section of the Department of Zoology at the National Museum of Wales, who gave the delegates a guided tour of their extensive collection of British and overseas insects, and then of the rest of the museum. We thank everyone involved with the catering and accommodation at the University of Wales Conference Centre and also all those involved in providing the excellent banquet, which was held in the magnificent baroque splendour of Cardiff Castle. Finally, our greatest thanks go to all the scientists who attended the Symposium, in particular those who presented their work, on whom the success of the meeting, this book and the future development of work on pest ecology, is so dependent.

W.O.C.S. and J.E.L.

Introduction

Biochemical approaches to understanding agricultural pests

M.F. Claridge

ABSTRACT

The ecological view of modern agricultural systems requires a multidisciplinary approach to the development of pest management strategies. Biochemical techniques have particularly enabled an understanding of: (i) problems of feeding interrelationships in the study of biocontrol agents; and (ii) the nature of variation within and between populations of crop plants, their pests and natural enemies.

INTRODUCTION

I guess that 20 or 30 years ago a symposium under the title of *'The Ecology of Agricultural Pests: Biochemical Approaches'* would have been inconceivable. Indeed, agricultural pests – and even the crop plants themselves – would hardly have been thought to have any ecology and would certainly not generally have been thought of as appropriate subjects for ecological study. Primary pest control strategies would have depended on the use of suitable pesticides and/or herbicides in an attempt to eliminate offending pests and weeds from the crop system. The history of agriculture over the past 30 years or so has been one of a growing realization that such simplistic approaches do not work, and certainly are not consistent with the agreed modern requirements of sustainable land use and conservation of biodiversity.

Of course, concepts of agroecosystems are now generally accepted and pests most certainly are seen not only to have ecology, but also genetics and therefore evolution. As a consequence it has become

The Ecology of Agricultural Pests, Edited by W.O.C. Symondson and J.E. Liddell.
Published in 1996 by Chapman & Hall, London. ISBN 0 412 62190 8

obvious that we should expect, for example, the evolution of resistance to pesticides as a result of inevitable field selection pressures. This phenomenon has often become such a problem that the increasingly higher levels of pesticide application needed for adequate control have become uneconomic. Also, the consequent increased levels of pesticide-induced pollution are equally inevitable with such a strategy.

A further consequence of the intensive use of pesticides is that they may cause pest outbreaks or even create new pests. For example, there is now little doubt that the brown planthopper, *Nilaparvata lugens* (Stål) – which has emerged as a major pest of tropical rice throughout tropical Asia only since the 1970s – is such a case. This has been a direct result of the differential reduction of natural enemies, both predators and parasitoids, following the introduction of newer high-yielding rice varieties with the consequent need for high levels of fertilizer and pesticide application (Kenmore *et al.*, 1984). Thus, population, and indeed also community, ecology are central to our understanding of pests and the possibility of manipulating them in any system of true biological control.

A further consequence of the more ecological approach to pest problems is the need to recognize variation and understand its nature both within and between populations of crops and their pests and natural enemies. Thus, there is still a need for biosystematic work even on some of the most important pests of world agriculture. For example, the nature of the so-called biotypes of the brown planthopper have caused major problems in developing appropriate control strategies in different parts of Asia (Claridge, 1990). Further emphasis on the importance of sustainability and the conservation of biodiversity has been provided by the signing of the Convention on Biodiversity by world leaders at Rio de Janeiro in 1993. Now, all signatory governments are committed to developing their own strategies. The UK Government has for example recently (Anon, 1994a,b) published strategy documents including action plans for both Sustainable Development and Biodiversity.

Thus, with the now obvious needs to understand pest ecology and diversity, all novel techniques that may contribute to an obviously multidisciplinary problem must be welcome. Biochemical and chemical technologies in pest management have a special role to play and have developed rapidly over the past 30 years. They will obviously continue to expand in importance into the forseeable future.

The contributors to this volume show that two major aspects of pest biology have so far particularly been illuminated by the application of chemical and biochemical techniques:

1. the nature of complex assemblages of natural enemies; and
2. variation within and between pest and natural enemy species.

NATURAL ENEMIES

Procedures and practices of integrated pest management require that populations of natural enemies of pests – predators, parasitoids, parasites and pathogens – must be conserved and promoted where possible in agroecosystems. Even selective introductions and releases of exotic natural enemies – as in what is often termed classical biological control – may be required.

Whatever the precise problem there is always one vitally important question that needs to be answered – what is eating what? Where the organisms involved are small to very small arthropods or microorganisms this apparently simple question is usually very difficult to answer with confidence. The classical development of gross serological tests, as for example by Dempster (1960), to identify the arthropod predators of particular insect herbivores, was in its day a great breakthrough. In more recent years these techniques have been refined and developed by the elegant use of both polyclonal and monoclonal antibodies. Major examples of different aspects of these techniques are given here by, among others, Greenstone; Stuart and Greenstone; Symondson and Liddell; Sigsgaard; Hagler and Naranjo; and Sunderland. There can be no doubt that this rapidly developing area of technology will continue to provide new tools for use in pest ecology.

VARIATION AND BIOSYSTEMATICS

The need to understand the nature of variation within and between populations of pests, their natural enemies and indeed crop plants themselves, and sometimes even assemblages of associated weeds, has been demonstrated time and again. This is a direct consequence of an ecological approach to agricultural systems. Thus, the importance of biosystematics – and the ability to recognize and identify biological species – is central to pest management. Traditionally, gross morphological markers have been used to recognize genetically and ecologically distinct biological species. However, these are often not adequate where groups of sibling species occur. Such species are genetically quite distinct with different biological characteristics, including such ecologically important ones as disease vector capability, host preference, etc. (Claridge, 1991).

Biochemical techniques have been and continue to be developed to provide adequate markers. The earliest really useful systems were those based on enzyme gel electrophoresis. Indeed, in 1988 a previous Systematics Association meeting at Rothamsted and the subsequent published volume (Loxdale and den Hollander, 1989) were wholly devoted to the use of electrophoretic techniques in many aspects of agricultural pest research. A major point was the study of biological systematics and genetic variation. Several of the papers in the present volume show how these techniques continue to have very great importance not only in systematics (e.g. Menken), but also in population ecology, predation studies and monitoring resistance (Biggins *et al.*, and Solomon).

In recent years a completely new and rapidly expanding series of molecular biology techniques have become available for use by the ecologist. Most excitingly these techniques provide access to information on variation in DNA – the ultimate material of heredity and genetic variation. A great practical advantage of these techniques is that nucleic acids are relatively stable molecules which can be extracted from dead – often long-dead – field samples, and even, for example, from museum-preserved specimens. Thus, the often very great problems of field preservation for protein electrophoresis techniques are usually much less for DNA systems. However, even more significantly these techniques provide us with a completely new set of highly refined markers. In the present volume Loxdale *et al.* give a review of the molecular markers currently available to agricultural entomology, Jones *et al.* show how molecular methods may help in understanding relationships among closely related groups of pests, and Hsiao shows how these techniques may allow us to understand the little known, but very important, world of the endosymbionts of pests.

There is often a danger in science of relying wholly on one particular technique – often the newest – to the exclusion of others. Several authors here (e.g. Hemingway *et al.* and Daly and Trowell) show how multidisciplinary methods are essential to understand pest problems. We must always beware the temptation to throw out older tried and tested techniques when new ones become available. Biological field problems are always complex and require all of the techniques that are available to us.

Acknowledgements

I am deeply indebted to the organizers of the Cardiff meeting and editors of this volume, Drs Eryl Liddell and Bill Symondson, for asking me to contribute an introductory address and this brief introductory text.

REFERENCES

Anon. (1994a) *Sustainable Development the UK Stratgey*, HMSO, London.
Anon. (1994b) *Biodiversity the UK Action Plan*, HMSO, London.
Claridge, M.F. (1990) Variation in pest and natural enemy populations – relevance to brown planthopper control strategies, in *Pest Management in Rice*,

(eds B.T. Grayson, M.B. Green and L.G. Copping), Society of Chemical Industry, Elsevier, London, pp. 143–154.

Claridge, M.F. (1991) Genetic and biological diversity of insect pests and their natural enemies, in *The Biodiversity of Microorganisms and Invertebrates: its Role in Sustainable Agriculture*, (ed. D.L. Hawksworth), CAB International, Wallingford, pp. 183–194.

Dempster, J.P. (1960) A quantitative study of the predators on the eggs and larvae of the brown beetle, *Phytodecta olivacea* Forster, using the precipitin test. *Journal of Animal Ecology*, **29**, 149–67.

Kenmore, P.E., Carino, F.O., Perez, C.A., Dyck, V.A. and Gutierrez, A.P. (1984) Population regulation of the rice brown planthopper (*Nilaparvata lugens* Stal) within rice fields in the Philippines. *Journal of Plant Protection in the Tropics*, **1**, 19–37.

Loxdale, H. and den Hollander, J. (1989) *Electrophoretic Studies on Agricultural Pests*, Clarendon Press, Oxford.

1

Biochemical systematics: principles and perspectives for pest management

S.B.J. Menken and L.E.L. Raijmann

ABSTRACT

Despite the increasing application of molecular techniques such as RFLPs, RAPDs, fingerprinting, and DNA sequencing, enzyme electrophoresis is still the simplest and most widely used technique and among the most cost-efficient in agricultural entomology.

This review will focus on: (i) species identification in groups of pest insects that are notoriously rich in sibling species or where species identification of immature stages on morphological grounds is impossible; and (ii) the analysis of population structure and intraspecific differentiation. In many pests, populations and sometimes even individuals within populations have been found to possess distinctive biological characteristics relevant to their pest status. The systematic status of such so-called biotypes appears to range from distinct sibling species to genetic variants in panmictic populations. Biochemical systematic studies aid much in understanding the complex population structures of pest species (and of their parasitoids and host plants). Details of analyses of the spatial and temporal population structure will be presented for *Yponomeuta padellus* (L.) (Lepidoptera: Yponomeutidae), an oligophagous defoliator of various Rosaceae species, which may form genetically differentiated taxa.

1.1 INTRODUCTION

The application of enzymatic proteins as markers of genetic variation has proved to be very useful for the identification of individuals, the analysis

The Ecology of Agricultural Pests, Edited by W.O.C. Symondson and J.E. Liddell.
Published in 1996 by Chapman & Hall, London. ISBN 0 412 62190 8

of population structures, the delimitation of species boundaries, and phylogenetic reconstruction. Since the Systematics Association Symposium at Rothamsted (Loxdale and Den Hollander, 1989) we have witnessed an enormous increase in such studies using DNA markers. Still, in many cases allozyme analyses provide the most cost-efficient approach with a good signal-to-noise ratio for many questions asked in agricultural entomology (Menken and Ulenberg, 1987). In fact, there is an increasing need for simple, relatively cheap and effective techniques for pest species identification, and the analysis of intraspecific variability in agricultural entomology. Which technique to apply depends much on the questions being asked and the organisms being studied. No single technique is a panacea (Loxdale *et al.*, 1996, this volume) and the complementarity of the various techniques cannot be emphasized enough.

The present review focuses on the delineation of sibling complexes and the analysis of spatial and temporal population structures of phytophagous pest insects. Firstly, however, we will briefly outline some aspects of the differences in sensitivity between starch and cellulose acetate gel electrophoresis; the latter system is becoming very popular because it is easy to handle and does not need extensive laboratory facilities, and because of its short running time (less than 30 minutes) and the small amounts of homogenate needed (Richardson *et al.*, 1986).

1.2 A NOTE ON ELECTROPHORETIC TECHNIQUES AND SENSITIVITY

A major drawback of gel electrophoresis is the fact that the majority of single amino acid substitutions are undetected under standard electrophoretic conditions. Thus, many allozyme bands are effectively heterogeneous collections of identically charged proteins (so-called electromorphs; King and Ohta, 1975). Such electromorphs can be separated when subject to a series of different electrophoretic conditions (so-called sequential electrophoresis; Johnson, 1977; McLellan, 1984). This even appears to be feasible for the group of truly non-polar amino acid substitutions, for which Righetti and co-workers developed the technique of isoelectric focusing with immobilized pH gradients, allowing the separation of proteins differing in their isoelectric point by only 0.003 of a pH unit (Cossu and Righetti, 1987 and references therein).

Conventional electrophoretic techniques such as starch and polyacrylamide gel electrophoresis are sometimes sensitive enough to detect allozyme variation in small insects or first instars (Menken, 1991). However, in minute insects and mites conventional techniques usually allow the detection of only some (if any) of the most active enzymes (Ward *et al.*, 1982; Hung and Huo, 1985; Gotoh *et al.*, 1993), thus impeding the application of electrophoretic analyses in important pest species (such as spider mites) or parasitoids that are used in biological control programmes (e.g., *Trichogramma* Westwood species [Hymenoptera: Trichogrammatidae]).

Recently, Kazmar (1991) developed two procedures for the electrophoretic analysis at a number of loci in single individuals of minute arthropods. Both include isoelectric focusing but differ in the type of supportive medium (viz., ultrathin agarose gels versus single or multiple layers of prefabricated cellulose acetate (CA) membranes [details of CA gel electrophoresis can be found in Richardson *et al.*, 1986]). Interpretable banding patterns could be obtained at some 20 loci in *Trichogramma* species (Kazmar, 1991; see also Pinto *et al.*, 1993). In *Anagrus epos* Girault (Hymenoptera: Mymaridae) and *Tetranychus pacificus* McGregor (Acari: Tetranychidae) fewer loci were tested but similar results are expected (Kazmar, 1991). CA gels are preferred to agarose gels because they allow for multiple layered electrophoresis followed by separate stainings for individual enzyme systems, they are more easy to handle, prefabricated, and of constant quality. In our laboratory we obtained good results with the multi-embryonic endoparasitoid *Ageniaspis fuscicollis* (Dalman) (Hymenoptera: Encyrtidae), *Liriomyza* Mik species (Diptera: Agromyzidae; Oudman, 1992; Oudman *et al.*, in press), and various thrips and phytophagous mite species.

There is, however, a general drawback of CA electrophoresis: its sensitivity is inferior to starch gels. A comparison between starch and CA gels shows that at highly polymorphic loci a considerable number of alleles is missing on the latter system (e.g. phosphoglucomutase: Figure 1.1). Out of 24 individuals 21 appeared to be heterozygous for a total of nine alleles on starch versus 16 heterozygotes for only four alleles on CA. Also, loci monomorphic on CA gels often differentiate into slightly polymorphic loci with one or more rare alleles on starch gels.

Another potential problem, particularly in Hymenoptera and Homoptera, is the lack of abundant allozyme polymorphisms (Lester and Selander, 1979; Packer and Owen, 1992; Loxdale *et al.*, 1996, this volume). Screening a large number of histochemical stains (for an exhaustive enumeration and description of recipes, see Murphy *et al.*, 1990) in combination with systematically varying experimental conditions, and different life stages or parts of the body, can result in a substantial increase in suitable markers (Menken, 1987; Choudhary *et al.*, 1992; Shoemaker *et al.*, 1992). Yet, students working with Hymenoptera or Homoptera will readily be inclined to use DNA analyses (e.g. Loxdale *et al.*, 1996, this volume).

1.3 DISCRIMINATION OF SIBLING SPECIES

Apart from being of concern to taxonomists, the presence of sibling species in groups of arthropod pests may necessitate drastic changes in

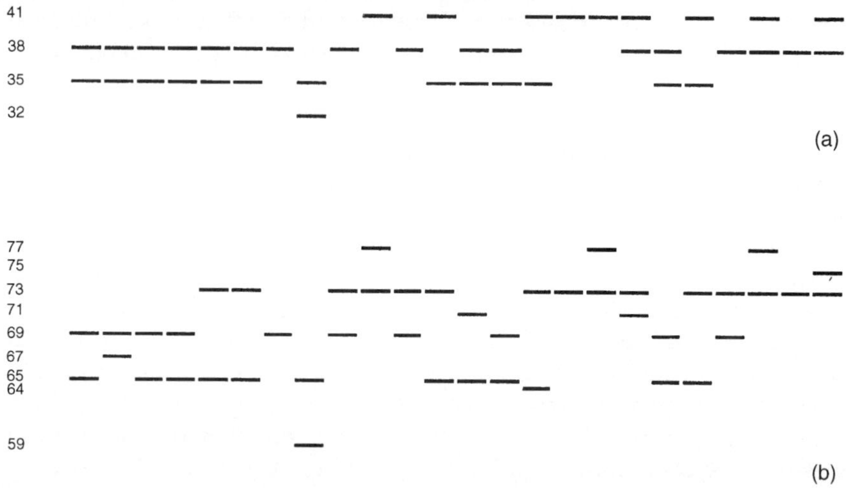

Figure 1.1 Comparison of sensitivity between cellulose acetate (a) and starch (b) gels. Electrophoretic patterns of 24 individuals of *Yponomeuta padellus* at the phosphoglucomutase locus are shown. Numbers in the left margin indicate absolute mobility in mm under standard conditions. Both gel systems were optimised by using a comparable range of gel and electrode buffers. Although no single buffer system revealed all variation present, the results shown are the best available for each gel system [viz. (a) cellulose acetate gel Tris–glycine, pH 8.4; (b) starch gel Tris–maleic acid, pH 7.4].

biological control strategies. Since such sibling taxa may differ in their ecology, behaviour, or physiology, control programmes designed to be effective against one species can appear to be totally ineffective against one or more others. This can result in partial or complete failure of the control measures as well as increase the actual pest problems by freeing the sibling taxa from competition, thus allowing their populations to grow (e.g. Makela and Richardson, 1978).

Similar problems can also arise due to intraspecific differentiation (e.g. Claridge, 1989). If, for example, a pest species is subdivided into 'biotypes' that differ in traits relevant to biological control, this could mean that different controlling agents and/or strategies are required for biological control. As a consequence, even the controlling agents themselves possibly are subject to biotype formation.

Sibling species usually do not appear out of thin air during an electrophoretic analysis, but rather there are *a priori* indications (from morphology, ecology or behaviour) for suspecting more than one species in the sample. *Stigmella basiguttella* Von Heinemann (Lepidoptera: Nepticulidae), consisting of two reproductively isolated species in various localities in the Netherlands, is an example of the first group:

only forewing coloration is slightly variable in *S. basiguttella* s.l. (E. van Nieukerken, personal communication). However, this variation pattern is clinal rather than different colour forms co-occurring in one population. Out of nine allozyme loci, the two cryptic species were fixed for different alleles at two loci (viz. leucine aminopeptidase and catalase), nearly fixed at the glucosephophate isomerase locus and with different major alleles (frequency >0.50) at the phosphoglucomutase locus (Cronau and Menken, 1990).

Another extreme example is *Capitella capitata* (F.) (Annelida: Polychaeta), once used as a bio-indicator of marine pollution. Since different samples of what was once thought to be one species reacted differently to the same kind of pollution, a genetic analysis was performed. The 'species' appeared to consist of six sibling species with different life characteristics which have virtually no allozymes in common between any two species (Grassle and Grassle, 1976).

On the other hand, variability in *Anastrepha fraterculus* (Wiedemann) (Diptera: Tephritidae) has long been interpreted as an indication that in nature a complex of cryptic species exists (Steck, 1991, and references therein). A population genetical survey over Central and South America indeed revealed strong genetic differentiation among samples (Steck, 1991). Further studies are needed, especially in economically important fruit-growing areas, because *A. fraterculus* s.l. is a highly polyphagous insect taxon, known to attack dozens of cultivated plants (Norrbom and Kim, 1988).

Differing host plant use might indicate that one is dealing with different biological species. In the *Yponomeuta padellus* complex (Lepidoptera: Yponomeutidae), for example, sibling species have specialized on plants from the Rosaceae, Celastraceae and Salicaceae (Menken *et al.*, 1992 and below). Since many species of phytophagous insects have been described solely on the basis of their host plant relationship, the opposite can also be true (Menken, 1992 and unpublished results). For instance, specimens of the monophagous nepticulid leafminer species *Stigmella aurella* (F.) (feeding on *Rubus* L.), *S. gei* (Wocke) (on *Geum* L.), *S. fragariella* (Von Heinemann) (on *Fragaria* L.), and *S. nitens* (Fologne) (on *Agrimonia* L.) all belong to one oligophagous species, *S. aurella.* A related species, *S. splendidissimella* (Herrich–Schaffer) occurs sympatrically in much (10–100 times) higher densities on *Rubus*, but has occasionally also been observed on *Geum* and *Agrimonia.* Despite a low Nei's genetic identity of 0.211 between *S. aurella* and *S. splendidissimella* some interspecific hybrids have been encountered in nature (Menken, 1992) and hybrids could also be produced in the laboratory (G. Bryan and S.B.J. Menken, unpublished results).

Sympatric taxa that are alternatively fixed at one locus (a so-called diagnostic locus) can be considered to be reproductively isolated species. A closer look at the morphology of the separated groups often reveals consistent morphological differences between the taxa or stepwise discriminant analysis on diverging morphological characters produce a key for the separation (Schutz *et al.*, 1989). Unless banding patterns are well interpretable and the inheritance is known, a minimum of two diagnostic loci should be used before establishing specific status, because the originally observed fixed difference might not be under simple monogenic control or banding patterns might have been misinterpreted.

Ideally, a diagnostic locus has the following characteristics: present in all members of both sexes of a taxon, absent from all other (related) taxa, monomorphic, developmentally constant, and not subject to the influence of environmental factors (Menken and Ulenberg, 1987).

Diagnostic loci produce error-free determination keys. Small probabilities of error in assigning an individual to the correct taxon occur at loci that show some amount of overlap in allele frequency (Miles, 1979; Berlocher, 1980; Oudman *et al.*, in press).

Species of the leafminer fly genus *Liriomyza* are pests on a variety of food plants like chrysanthemum, tomato, cucumber, lettuce, endives, etc. with a large overlap in host plant range. They can serve as a typical example of the use of diagnostic loci and biochemical keys in agricultural pests. *Liriomyza huidobrensis* (Blanchard), *L. sativae* Blanchard, and *L. trifolii* (Burgess) are on the EPPO and EC lists of quarantine species (Anonymous, 1992). The quarantine status of these three species necessitates correct identification of all developmental stages. Moreover, they do require different phytosanitary measures: e.g. cold storage for a few weeks kills all stages of *L. sativae* and *L. huidobrensis* but is not effective against all stages of *L. trifolii*. Morphological identification is restricted to adult males and based on differences in genitalia (Spencer, 1973). Larvae, pupae, and female adults can only be classified into two species groups: *L. bryoniae* (Kaltenbach), *L. strigata* (Meigen) and *L. huidobrensis* versus *L. trifolii* and *L. sativae*.

Over the years we have developed a protocol based on allozyme variability for individual characterization of all stages of the five *Liriomyza* species (Menken and Ulenberg, 1983, 1986; Oudman, 1992; Oudman *et al.*, in press). Characterization is possible by means of CA gel electrophoresis at three loci (viz. glucose-6-phosphate and isocitrate dehydrogenase and malic enzyme), but with this procedure it is impossible to separate *L. strigata* from *L. bryoniae* (*L. strigata* is the only species in this group feeding on the same host plants that is not an economically important pest). The additional electrophoresis of adenylate kinase allows for the discrimination between the last two species. Identification after allozyme analysis is facilitated by a biochemical key, using *L. bryonia* as a standard (Oudman *et al.*, in press). This identification procedure is now required in all EPPO countries. Obviously, knowledge of the level of intraspecific variability of a particular species is indispensable before reliable keys can be developed (Oudman *et al.*, 1993).

Enzymatic characterization of some endo- and ectoparasitoids of *Liriomyza* species has also been achieved (Oudman, 1992). In field samples of insects, one of the possible sources of error may arise from endoparasite proteins which can be erroneously interpreted as allelic variants of the host. This is because the general electrophoretic picture of a developing endoparasite is an increase in staining intensity of the parasite-specific bands with a concomitant reduction in intensity of those of the hosts (May *et al.*, 1977; Menken, 1982).

1.4 POPULATION STRUCTURE ANALYSIS

Genetic variation is widely assumed to be crucial as a prerequisite for evolutionary change as it provides the raw material for selection of better-adapted genotypes. The degree and distribution of genetic variation determine the evolutionary potential of a taxon. The degree is the result of effective population size, historical events like bottlenecks, mating system, and population structure (i.e. deviation from panmixis). The distribution patterns are the results of three interacting forces, viz. gene flow, genetic drift, and natural selection.

Gene flow among natural populations is of prime importance in evolution both below and above the species level; it can act as a creative as well as a constraining force (Slatkin, 1987). From one point of view, gene flow is an essential factor in maintaining species integrity (Mayr, 1970) and in spreading advantageous genes (such as pesticide resistance in agricultural pests) that could carry a population to a higher adaptive peak (Wright, 1932). From another, (near)-cessation of gene flow is implicit in most population genetic models of speciation (Mayr, 1963). Hence, studies of population structure and levels of gene flow are central not only in evolutionary biology, but also in agricultural entomology. Agricultural practice has benefited those phytophages that are good dispersers and insecticide resistance management requires, among other things, an understanding of the patterns and magnitude of gene flow (Roush and McKenzie, 1987; Daly and Trowell, 1996, this volume). Knowledge of population differentiation of both pest species and parasitoids is fundamental to successful integrated and biological pest management (Menken and Ulenberg, 1987; Pinto *et al.*, 1993).

1.4.1 Measurements of gene flow

Population structure and associated gene flow can be investigated using two approaches, one ecological (direct measurements through mark–

release–recapture methods) and one genetic (indirect measurements by analysing spatial patterns of gene frequency variation using theoretical models). Direct measurements pose the following interpretation problems.

1. Movement of individuals (dispersal) is not necessarily the same as the movement of genes. Only gene flow, the successful breeding of immigrants and the persistence of their genes in the new population, is what matters in evolution.
2. Investigations are necessarily restricted in space and time. Often long-distance, intermittent, and historical gene flow will go undetected during the period of observation.
3. Marking and handling, overcrowding at the site of release, and the use of non-native specimens can all contribute to abnormal behaviour of the study species (Coyne and Milstead, 1987).

Two indirect estimation methods of gene flow are commonly used; in both cases multilocus data from allozyme studies are employed. One applies F_{ST}, the among-population variance in allele frequency, to describe population structures following the theoretical relationship $F_{ST} = 1/(1+4N_em)$ (Wright, 1943; N_em is the average number of effective migrants exchanged between any two populations per generation). F_{ST} is best averaged over loci, as single locus values have a large variance due to genetic drift when the number of populations is small (Nei and Chakravarti, 1977).

Slatkin (1981, 1985) developed another estimator of gene flow based on the spatial frequency distribution of private alleles (i.e. alleles found in only one population). He found that the logarithm of N_em shows an approximately linear relationship to the logarithm of $\overline{p}(1)$, the average frequency of private alleles. According to computer simulations this method is largely unaffected by variation in mutation rate or natural selection (Slatkin, 1985). It is sensitive to changes in the sample size per population; a correction factor for differences in sample size has been suggested (Slatkin, 1985). N_em estimates fall inside the range of reliability only if adequate samples (>20) of private alleles are taken and if $0.1 < N_em < 10$, the range in which the relationship between $\ln N_em$ and $\ln \overline{p}(1)$ is linear. Such a large number of private alleles is achieved in few studies, especially not in cases where localities have been intensively sampled.

Indirect measurements depend on the reliability of the underlying models. The main assumption is that allele frequencies are directly related to the level of gene flow. Both models describe equilibrium conditions in an island model. However, simulation studies have shown that under a wide range of assumptions concerning selection regimes, mutation, and population structure both F_{ST} and $\overline{p}(1)$ can provide reasonably accurate N_em estimates (Slatkin and Barton, 1989). Furthermore, they assume gene flow to exceed the effects of natural selection and to be random with respect to the genotypes studied. These assumptions are usually met, as most allozymes are considered to be neutral or slightly deleterious (Kimura, 1983; Lewontin, 1985); however, an allozyme locus could be linked to and in linkage disequilibrium with a gene that influences dispersal. Finally, they contain largely integrated information on gene flow over historical time and therefore may be more meaningful to evolutionary studies than direct estimates.

Yet a general good agreement between indirect and direct measurements has been found (Daly, 1989), although F_{ST} values often yield somewhat higher N_em estimates than $\bar{p}(1)$ (Larson *et al.*, 1984; Caccone and Sbordoni, 1987; Singh and Rhomberg, 1987; Menken, 1992). For instance, North American populations of *Danaus plexippus* L. (Lepidoptera: Nymphalidae) migrate each year to south-central Mexico where they overwinter and interbreed before they return north (Urquhart and Urquhart, 1976). This breeding structure is reflected in low F_{ST} values (0.009; Eanes and Koehn, 1978). The same holds for *Spodoptera exempta* (Walker) (Lepidoptera: Noctuidae), the highly migratory African armyworm (F_{ST} = 0.006; Den Boer, 1978) and *Heliothis virescens* (F.) (Lepidoptera: Noctuidae; F_{ST} = 0.002; Korman *et al.*, 1993). On the contrary, *Euphydryas editha* Boisduval (Lepidoptera: Nymphalidae) depends on a host plant with a patchy spatial distribution (Harrison *et al.*, 1988). Consequently this species is organized into highly localized populations, typical for a species with a 'closed' population structure (F_{ST} = 0.118; McKechnie *et al.*, 1975). Such sedentary species combine a mosaic of transient and genetically differentiated local populations with low dispersal and colonization abilities (Levins, 1970).

1.4.2 Potential sources of bias

Genetic drift and gene flow affect all loci in the same way. A large variance in F_{ST} values among loci or in $\bar{p}(1)$ frequencies may be the result of natural selection acting on a locus or on a locus that is closely linked to and in linkage disequilibrium with that locus. Indeed, such patterns can be used as indirect evidence for selection. This might be the case for hexokinase in *Euphydryas editha* (McKechnie *et al.*, 1975) and esterase-3 in *Spodoptera frugiperda* (J.E. Smith) (Pashley *et al.*, 1985). The presence of clinal variation is strong evidence for the influence of selection. However, a general disagreement between direct and indirect estimates would not prove selection is operating but may merely hint at non-equilibrium conditions. The fact is that indirect methods assume populations to satisfy genetic and demographic equilibrium. This may not always be true (Slatkin, 1977, 1981). For instance, if, in the case of a recent invasion of a new area, all populations originated from a single ancestral population, $\bar{p}(1)$ frequencies will largely be determined by the time since radiation, even if actual gene flow is totally absent (Slatkin, 1981). It takes a long time for this effect to disappear (Slatkin, 1985). Bottlenecks may further complicate the pattern.

Another potential source of bias is the sampling scheme. Most schemes include populations that are patchily distributed resulting in an underestimate of gene flow levels. A more fine-scaled population sampling scheme results in higher values of N_em. An increase in the number of individuals sampled per population has two effects (Slatkin, 1985). It increases the number of (very low frequency) private alleles and thus the accuracy of the estimates of N_em. At the same time, former (mostly high frequency) private alleles will appear at low frequencies in other populations and thus disappear from the private allele pool. The combined effect is a decrease in the average frequency of private alleles and this means a higher estimate of gene flow.

Finally, estimates of N_em are averages, leaving room for local differences (Singh and Rhomberg, 1987). Thus, moderately large estimates of gene flow do not exclude some pairs of populations from being completely isolated. Alternatively, low estimates may be caused by one or a few isolated populations (Pashley *et al.*, 1985; Herrebout and Menken, 1990).

1.4.3 Practical examples

As stated before, the differentiation of an insect pest species into 'biotypes' that differ in traits relevant to control measures could require different controlling measures or strategies. A particular case of biotype is the formation of host races (i.e. populations that are partially reproductively isolated from other conspecific sympatric sister populations as a direct consequence of adaptation to a particular host; Diehl and Bush, 1984). Host races are considered to be an important step towards (sympatric) speciation (for a recent review on evidence for host race formation and sympatric speciation, see Bush, 1994). The idea that host races may appear from adaptation to a new host plant within the distribution area of the population and subsequently may diverge into distinct species dates back to the 1860s (Walsh, 1864). The most detailed and convincing example of the existence of host races comes from the very same group of insects for which this scenario originally has been proposed, viz. true fruit flies of the genus *Rhagoletis* L. (Diptera: Tephritidae). Feder and colleagues (e.g. Feder *et al.*, 1990, 1993) have collected an impressive amount of data on how host fidelity, host plant phenology, fly development, and climate interact in the differentiation of populations of *R. pomonella* (Walsh) on apple and hawthorn.

For more than 20 years, the palearctic genus *Yponomeuta* Latreille (Lepidoptera: Yponomeutidae) has been studied as a model system for the evolution of insect plant associations and speciation in phytophagous pest insects (for an overview, see Menken *et al.*, 1992). With the exception of the oligophagous *Y. padellus*, all taxa are monophagous on trees or shrubs, the great majority of which belong to the Celastraceae and Rosaceae. In Europe, *Y. cagnagellus* (Hübner) (on spindle tree, Celastraceae), *Y. evonymellus* (L.) (on bird cherry), *Y. padellus* (on hawthorn, blackthorn, mountain ash and other Rosaceae), *Y. rorellus* (Hübner) (on willow), and *Y. malinellus* Zeller (on apple) can cause the complete defoliation of their hosts. From Western Europe several species have been introduced into a number of places (North America and South East Asia: Herrebout and Menken, 1990; Sperling *et al.*, 1995), where they cause considerable economic losses as such outbreaks include apple and plum, or occur as cosmetic pests on various ornamentals (USDA, 1985).

Y. cagnagellus, *Y. padellus*, *Y. rorellus*, *Y. malinellus* and *Y. mahalebellus* Guenée (on *P. mahaleb* L.) form a complex of taxa, the taxonomic and evolutionary status of which has long been a matter of dispute (Menken *et al.*, 1992). Evaluations varied from five different species or five incipient species to one polyphagous and polytypic species. In the case of one polyphagous species, measures to control *Y. malinellus* in apple orchards needed to be extended to the windbreaks where the related *Yponomeuta* taxa occur. All five taxa, however, appeared to be genuine biological species with diagnostic enzyme loci (Menken, 1982; Arduino and Bullini, 1985), differences in the morphology of larvae, pupae, and imagines, and differing sex pheromone communication systems (Menken *et al.*, 1992).

Y. padellus, the only oligophagous moth in the genus, probably consists of host races. Thorpe (1930) and Pag (1959) experimentally observed in *Y. padellus* an oviposition preference for the host plant on which the ovipositing female developed, but given the percentages preference and variability of the data, their claims are not very convincing. Patterns of larval preference and performance appeared to be fairly consistent, showing a similar hierarchy as in their oviposition preference. Almost all *Y. padellus* prefer and perform best on *Prunus spinosa* L., whatever the host on which they developed was; next come *Crataegus* L. and *P. cerasifera* J.F. Ehrhart, which in turn are more suitable than *Prunus domestica* L., *Amelanchier lamarckii*, F.G. Schroeder and *Sorbus aucuparia* L. (Kooi, 1990; Menken *et al.*, 1992). Differences in oviposition behaviour among supposed host races have only been observed in localities where *Y. padellus* feeds on *P. cerasifera* or *S. aucuparia* (Kooi, 1990), the very populations that are genetically differentiated as indicated by allozyme analyses (Menken, 1981, 1989, and this chapter). Not only were allele frequencies distributed heterogeneously over the host samples, but there were also some host-specific alleles occurring at a lower frequency; this pattern was consistent over a number of years. The results could not be explained by sampling error or host-specific selection against certain alleles at the study loci or at loci closely linked to and in linkage disequilibrium with them and, therefore, reduced gene flow was the most likely explanation.

Recently, we started a detailed analysis of the population structure of *Y. padellus*; allozyme electrophoretic procedures were further optimized and the number of polymorphic loci was increased to more than 20 (Raijmann, 1992; Raijmann and Menken, 1992). Investigating as many loci as possible is desirable in order to minimize locus-specific effects (e.g. if a locus is under the action of natural selection). The necessity to sample a large number of loci automatically follows from the observation that the reliability of many summary statistics, such as F_{ST} and average heterozygosity, depends on the number of loci more than on the number of individuals (Nei, 1978; Chakraborty and Leimar, 1987).

(a) Genetic differences in one year at one locality

In many localities where *Y. padellus* can be found feeding on different food plants, statistically significant differences in allele frequencies can be encountered (Menken, 1981, 1989) (Table 1.1). For instance, in the De Lutte population in the eastern part of The Netherlands allele frequencies at 50% of the allozyme loci (i.e. 8 out of 16) are statistically significantly different between sympatric samples from *Crataegus* spp. and *P. cerasifera*; N_em is 1.30 according to the private allele method (corrected for sample size) and 2.68 (based on F_{ST}), which are both lower than average N_em values averaged over years and food plants (just over 8.00; Table 1.2).

(b) Genetic differences in time within and between populations

In order to find out if such sympatric differentiation is consistent in time we compared allele frequency distribution over some 12–13 generations (i.e. years in the univoltine *Y. padellus*) in two localities (Table 1.1; for part of the loci comparisons with the years 1975–1978 can be found in Menken, 1981). Although it is generally true that heterogeneity between host races was consistent over years, spatial fluctuations in allele frequencies were observed within a single population. According to expectation, the variability in populations on *Crataegus* tends to fluctuate less than that on *S. aucuparia* and *P. cerasifera*, which have smaller population sizes (the population in Zuidoost Beemster on *P. cerasifera* became so small that we had to stop further sampling). Some allele frequencies exhibit large differences between 1979 and 1990 such as *Pgi* allele 93 which steadily increased and became the most common allele on *S. aucuparia,* whereas the population on *Crataegus* did not change concomitantly (Table 1.1). This can be viewed as further evidence for the (partial) reproductive isolation of the host races and/or the greater gene exchange between geographic populations on *Crataegus* (leading to larger effective population sizes with less influence of random changes due to drift).

Host races in *Y. padellus* do not exhibit significant reductions in average heterozygosity or in the effective number of alleles relative to closely related species (Menken, 1989; Menken *et al.*, 1992). Thus, the origin of these races shows no sign of severe bottlenecking at their origin, a pattern similar to the one encountered in *Rhagoletis* (Feder *et al.*, 1990).

(c) *Relative levels of gene flow*

Absolute values of gene flow levels are difficult to interpret. Theoretically it has been shown that a minimum of one individual on average being exchanged between two populations per generation will prevent neutral alleles at a locus from differential fixation, independent of population size (Wright, 1932). On the other hand, selection causes divergence only if the selection coefficient is greater than the migration rate (Slatkin, 1977). Relative gene flow levels, however, can be used to evaluate the status of *Y. padellus* populations (Table 1.2).

The mean F_{ST} and corresponding N_em values averaged over two generations (1990 and 1991) and 10 loci and all food plants are 0.032 and 7.61, respectively (Raijmann, 1992). These generally high levels of gene flow are expected since *Y. padellus* is known to be a rather good disperser (Menken *et al.*, 1992). The values are in agreement with earlier reports (Menken, 1981, 1989) based upon five loci and averaged over five generations (1975–1979; 0.030 and 8.08, respectively). However, the intra-host F_{ST} values vary considerably: from 0.003 for *Crataegus* to 0.023 for *S. aucuparia* and 0.038 for *P. spinosa. P. spinosa* is the most preferred food plant for *Y. padellus* (Kooi, 1990), but is less widespread than *Crataegus* and this fact could partly explain the observed values: *P. spinosa* populations face larger distances between geographic populations, resulting in less gene exchange.

In conclusion, intra-host levels of gene flow are usually significantly higher than inter-host levels. When all available data are included the message remains essentially the same: intra-*Crataegus* F_{ST} < average F_{ST} and intra-*P. spinosa* > average F_{ST}. For private alleles again intra-*Crataegus* but also intra-*P. spinosa* $\overline{p}(1)$ values are smaller than average

Table 1.1 Allele frequencies at five polymorphic loci in two sympatric populations of *Yponomeuta padellus* sampled from *Sorbus aucuparia* and *Crataegus* in Malden, The Netherlands

		Sorbus aucuparia					*Crataegus*									
Year		*1979*	*1988*	*1990*	*1991*	*1992*	*1979*	*1988*	*1990*	*1991*	*1992.*					
Number of moths		*20*	*20*	*60*	*40*	*60*	*20*	*30*	*48*	*45*	*55*					
Locus	*Allele*											G-79[a]	G-88	G-90	G-91	G-92
Pgi	87	–	0.025	0.033	0.051	–	–	–	–	–	0.045	2.01	10.16*	19.69**	21.23***	18.47**
	93	0.125	0.425	0.433	0.487	0.567	0.250	0.167	0.208	0.233	0.232	(1)	(3)	(5)	(5)	(4)
	96	–	–	–	–	–	–	–	0.010	–	–					
	98	–	–	–	0.026	–	–	–	0.010	–	0.009					
	100	0.875	0.550	0.533	0.410	0.424	0.750	0.783	0.740	0.711	0.679					
	103.5	–	–	–	–	–	–	0.050	–	–	–					
	105	–	–	–	0.026	–	–	–	0.031	0.044	0.036					
	108	–	–	–	–	–	–	–	–	0.011	–					
		a[b]	bc	bc	b	c	ab	a	ab	b	b					
Me	98	0.575	0.450	0.345	0.188	0.325	0.100	0.317	0.273	0.389	0.500	21.28**	1.79	0.76	8.38**	7.32**
	100	0.425	0.550	0.655	0.813	0.675	0.900	0.683	0.727	0.611	0.500	(1)	(1)	(1)	(1)	(1)
		a	ab	b	c	b	a	b	b	bc	c					

Hbdh	91	–	–	0.017	0.050	0.033	–	–	0.045	0.068	0.009	6.50*	6.71	6.38*	2.50	8.83*
	96	–	–	–	–	–	–	0.033	–	–	–	(2)	(3)	(2)	(2)	(2)
	98	–	–	–	–	–	0.075	0.083	–	–	–					
	100	0.875	0.875	0.862	0.800	0.633	0.600	0.750	0.716	0.693	0.809					
	102	0.125	0.125	0.121	0.150	0.333	0.225	0.133	0.263	0.239	0.182					
		a	a	a	a	b	ad	a	bc	c	bd					
Mdh-2	90	–	–	–	0.050	0.025	–	–	0.021	–	–	–	1.64	2.60	5.45*	3.42
	100	1.000	1.000	1.000	0.950	0.975	1.000	0.970	0.979	1.000	1.000	–	(1)	(1)	(1)	(1)
	106	–	–	–	–	–	–	0.030	–	–	–					
		ab	ab	a	b	ab	a	a	a	a	a					
Fudh	94	–	–	0.075	0.125	0.142	–	–	0.063	0.105	–	1.86	2.66	12.05**	2.78	26.69***
	97	0.970	1.000	0.925	0.850	0.783	0.900	0.950	0.896	0.895	0.856	(1)	(1)	(2)	(2)	(3)
	100	0.030	–	–	0.025	0.075	0.100	0.050	0.042	–	0.106					
	105	–	–	–	–	–	–	–	–	–	0.038					
		a	a	b	c	c	ac	ac	ab	b	c					

[a]Comparison of allele frequencies between *Sorbus aucuparia* and *Crataegus* for each year of study
Probability levels for individual G-tests after Williams' correction are given as $^{*}P < 0.05$, $^{**}P < 0.01$ and $^{***}P < 0.001$. Degrees of freedom in parentheses.
[b]Comparison between years on the same food plant: data with the same letter were not significantly different as determined by G-test.

Table 1.2 F-statistic parameters in *Yponomeuta padellus*. **Upper table:** inter-host levels of gene flow as determined at three different sample sites and for several years. The N_em values calculated according to Slatkin's private allele method are corrected for differences in number of sampled individuals. # p. a. indicate the number of private alleles on which N_em is based. **Lower table:** First, the inter-host N_em and F_{ST} values calculated over two different time intervals are shown. Second, the intra-host F_{ST} values are given for three host plants and for two successive years. # loci indicates the number of loci on which F_{ST} values are based.

		F-statistics		*Private allele*	*# p. a.*
		F_{ST}	N_em	N_em *(corr.)*	
Crataegus – P. cerasifera					
De Lutte	1993	0.085	2.68	1.30	7
Crataegus – P. cerasifera					
Z.O.-beemster	1978	–	–	2.49	4
	1979	–	–	5.84	4
	1990	0.155	1.36	5.36	5
Crataegus – S. aucuparia					
Malden	1979	–	–	1.68	1
	1988	–	–	3.24	6
	1990	0.044	5.44	6.17	6
	1991	0.043	5.51	4.27	5
	1992	0.043	5.56	1.37	6

Average N_em and F_{ST} values

Period	*Host plants*	N_em	F_{ST}	*# loci*
Inter-host				
1975–1979		8.08	0.030	5
1990–1993		8.34	0.029	10
Intra-host				
1990–1991	intra-*Crataegus*	73.28	0.003	10
1990–1991	intra-*P. spinosa*	10.62	0.038	10
1990–1991	intra-*S. aucuparia*	6.33	0.023	10

values (Table 1.2). These values corroborate the general finding that gene flow estimates based on private alleles are often lower than those based on inbreeding coefficients (Caccone and Sbordoni, 1987; Singh and Rhomberg, 1987; Menken, 1989, 1992).

(d) Comparison with direct dispersal data

Direct estimates of male dispersal (through mark–release–recapture) do not seem to support our estimates inferred from population genetic data. In a detailed analysis, Brookes and Butlin (1994 and unpublished results) found that *Y. padellus* appeared to be a sedentary species, with in two consecutive years 77% and 91% of the males dispersing less than 20 metres, but occasionally more than 250 metres was observed (strongly leptokurtic distribution); females tend to migrate even less. *Y. padellus* shares this combination of low directly observed levels of migration with high levels of inferred gene flow with among other things *Drosophila pseudoobscura* Frolowa (Coyne *et al.*, 1982; Slatkin, 1994).

There are three obvious explanations for the disparity in amount of gene exchange observed in direct and indirect measurements (see also Slatkin, 1994).

1. The field experiment is restricted in space and time. It fails, by definition, to measure dispersal and associated gene flow out of the study population and historical gene flow. We know from personal observation that *Y. padellus* individuals sometimes move around a lot, often following or 'looking' for hedgerows of *Crataegus* or *P. spinosa* and other suitable habitat, and individual, isolated food plants are successfully found by dispersers. In this context it is interesting to note that in the very part of the study area where host density was the least the greatest dispersal was observed (Brookes and Butlin, 1994).
2. In the genetic studies, individuals from a number of nests in an area were sampled, i.e. from somewhat arbitrarily defined populations. This might have overlooked the fine-scale population structure by lumping different neighbourhoods. However, a thorough allozyme analysis using samples from areas the size of the estimated neighbourhood (according to the method of Wright, 1946; Brookes and Butlin, 1994) did not reveal much homogeneity. Pronounced genetic structure is only encountered at the level of the nest (i.e. sibs and possibly half-sibs). This genetic structure arises from kin association: a *Y. padellus* nest comprises a family, the mother of which might have mated more than once. A similar pattern of subdivision was found by Costa and Ross (1994) in their hierarchical population structure analysis of *Malacosoma americanum* (F.) (Lepidoptera: Lasiocampidae).
3. *Y. padellus* is not yet at a genetic equilibrium, for instance, because of a recent expansion of its geographic range. Since a good historical record for this species is lacking, not much can be said about this possibility.

We think that it is generally true that most *Y. padellus* individuals do not disperse much under normal conditions: adults eclose over a short period of time from communal webs and, after a maturation period of some days, mating takes place on the host plants very soon after a female starts calling (Hendrikse, 1986; De Jong, 1988). However, if long distance dispersal is not too rare and results in actual gene flow, then we expect the observed homogeneous distribution of genetic variation to occur. The *Y. padellus* example makes clear that direct observations of long-distance dispersal are required in order to correctly evaluate the indirect analyses of gene flow which are based on assumption-laden theories (see also Cockerham and Weir, 1993).

It should be stressed that the great majority of allozymes probably are being maintained by mutation/drift balance so that they should be considered as neutral markers of the process of population differentiation. If true, so much the better as the probability that different populations converge on the same allele frequencies due to drift will be exceedingly low. The rare case where under a number of assumptions a gene that is under selection pressure (e.g. a gene for insecticide resistance) can provide insight into gene flow and population substructuring is discussed by Daly and Trowell (1996, this volume).

Acknowledgements

We appreciate the help of Wil van Ginkel in the allozyme analyses. We also appreciate our discussions with Erik van Nieukerken and Sandrine Ulenberg.

REFERENCES

Anonymous (1992) Quarantine Pests for Europe. Data sheets on quarantine pests for the European Communities and for the European and Mediterranean Plant Protection Organization. Prepared by CABI and EPPO for the European Communities. Editorial Committee: I.M. Smith and D.G. McNamara (EPPO) and P.R. Scott and K.M. Harris, CABI International, UK, 1992.

Arduino, P. and Bullini, L. (1985) Reproductive isolation and genetic divergence between the small ermine moths *Yponomeuta padellus* and *Yponomeuta malinellus*. *Atti Academia Nazionale dei Lincei*, **18**, 33–61.

Berlocher, S.H. (1980) An electrophoretic key for distinguishing species of the genus *Rhagoletis* (Diptera: Tephritidae) as larvae, pupae, or adults. *Annals of the Entomological Society of America*, **73**, 131–7.

Brookes, M.I. and Butlin, R.K. (1994) Population structure in the small ermine moth *Yponomeuta padellus*: an estimate of male dispersal. *Ecological Entomology*, **19**, 97–107.

Bush, G.L. (1994) Sympatric speciation in animals: new wine in old bottles. *Trends in Ecology and Evolution*, **9**, 285–8.

Caccone, A. and Sbordoni, V. (1987) Molecular evolutionary divergence among North American cave crickets. I. Allozyme variation. *Evolution*, **41**, 1198–214.

Chakraborty, R. and Leimar, O. (1987) Genetic variation within a subdivided population, in *Population Genetics and Fisheries Management*, (eds N. Ryman and F. Utter), University of Washington Press, Seattle, pp. 90–120.

Choudhary, M., Coulthart, M.B. and Singh, R.S. (1992) A comprehensive study of genic variation in natural populations of *Drosophila melanogaster*. VI. Patterns and processes of genic divergence between *D. melanogaster* and its sibling species, *Drosophila simulans*. *Genetics*, **130**, 843–53.

Claridge, M.F. (1989) Electrophoresis in agricultural pest research – a technique of evolutionary biology, in *Electrophoretic Studies on Agricultural Pests*, (eds H.D. Loxdale and J. Den-Hollander), Clarendon Press, Oxford, pp. 1–6.

Cockerham, C.C. and Weir, B.S. (1993) Estimation of gene flow from F-statistics. *Evolution*, **47**, 855–63.

Cossu, G. and Righetti, P.G. (1987) Resolution of Gg and Ag foetal haemoglobin tetramers in immobulized pH gradients. *Journal of Chromatography*, **398**, 211–16.

Costa, J.T. III. and Ross, K.G. (1994) Hierarchical genetic structure and gene flow in macrogeographic populations of the eastern tent caterpillar (*Malacosoma americanum*). *Evolution*, **48**, 1158–67.

Coyne, J.A. and Milstead, B. (1987) Long-distance migration of *Drosophila*. 3. Dispersal of *D. melanogaster* alleles from a Maryland orchard. *American Naturalist*, **130**, 70–82.

Coyne, J.A., Boussy, I.A., Prout, T., Bryant, S.H., Jones, J.S. and Moore, J.A. (1982) Long distance migration of *Drosophila*. *American Naturalist*, **119**, 589–95.

Cronau, J.Ph. and Menken, S.B.J. (1990) Biochemical systematics of the leaf mining moth family Nepticulidae (Lepidoptera). II. Allozyme variability in the *Stigmella ruficapitella* group. *Netherlands Journal of Zoology*, **40**, 499–513.

Daly, J.C. (1989) The use of electrophoretic data in a study of gene flow in the pest species *Heliothis armigera* (Hübner) and *H. punctigera* Wallengren (Lepidoptera: Noctuidae), in *Electrophoretic Studies on Agricultural Pests*, (eds H.D. Loxdale and J. Den Hollander), Clarendon Press, Oxford, pp. 115–41.

Daly, J. and Trowell, S. (1996) Biochemical approaches to the study of ecological genetics: the role of selection and gene flow in the evolution of insecticide resistance, in *The Ecology of Agricultural Pests: Biochemical Approaches*, (eds W.O.C. Symondson and J.E. Liddell), Chapman & Hall, London, pp. 73–92.

De Jong, M.C.M. (1988) Evolutionary approaches to insect communication system. PhD thesis, University of Leiden, The Netherlands, 163 pp.

Den Boer, M. (1978) Isoenzymes and migration in the African armyworm *Spodoptera exempta* (Lepidoptera, Noctuidae). *Journal of Zoology, London*, **185**, 539–53.

Diehl, S.R. and Bush, G.L. (1984) An evolutionary and applied perspective of insect biotypes. *Annual Review of Entomology*, **29**, 471–504.

Eanes, W.F. and Koehn, R.K. (1978) An analysis of genetic structure in the monarch butterfly, *Danaus plexippus* L. *Evolution*, **32**, 784–97.

Feder, J.L., Chilcote, C.A. and Bush, G.L. (1990) Regional, local and microgeographic allele frequency variation between apple and hawthorn populations of *Rhagoletis pomonella* in western Michigan. *Evolution*, **44**, 595–608.

Feder, J.L., Hunt, T.A. and Bush, G.L. (1993) The effects of climate, host plant phenology and host fidelity on the genetics of apples and hawthorn infesting races of *Rhagoletis pomonella*. *Entomologia experimentalis et applicata*, **69**, 117–35.

Gotoh, T., Bruin, J., Sabelis, M.W. and Menken, S.B.J. (1993) Host race formation in *Tetranychus urticae*: genetic differentiation, host plant preference, and mate choice in a tomato and a cucumber strain. *Entomologia experimentalis et applicata*, **68**, 171–8.

Grassle, J.P. and Grassle, J.F. (1976) Sibling species in the marine pollution indicator *Capitella* (Polychaeta). *Science*, **192**, 567–9.

Harrison, S., Murphy, D.D. and Ehrlich, P.R. (1988) Distribution of the bay checkerspot butterfly, *Euphydryas editha bayensis*: evidence for a metapopulation model. *American Naturalist*, **132**, 360–82.

Hendrikse, A. (1986) Intra- and interspecific sex-pheromone communication in the genus *Yponomeuta. Physiological Entomology*, **11**, 159–69.

Herrebout, W.M. and Menken, S.B.J. (1990) Preliminary data on the origin of small ermine moths introduced into North America (Lepidoptera, Yponomeutidae). *Proceedings of Experimental and Applied Entomology*, **1**, 146–51.

Hung, A.C.F. and Huo, S. (1985) Malic enzyme, phosphoglucomutase and phosphoglucose isomerase isozymes in *Trichogramma* (Hym.: Trichogrammatidae). *Entomophaga*, **30**, 143–9.

Johnson, G.B. (1977) Assessing electrophoretic similarity: the problem of hidden heterogeneity. *Annual Review of Ecology and Systematics*, **8**, 309–28.

Kazmar, D.J. (1991) Isoelectric focusing procedures for the analysis of allozyme variation in minute arthropods. *Annals of the Entomological Society of America*, **84**, 332–9.

Kimura, M. (1983) *The Neutral Theory of Evolution*, Cambridge University Press, Cambridge.

King, J.L. and Ohta, T. (1975) Polyallelic mutational equilibria. *Genetics*, **79**, 681–91.

Kooi, R.E. (1990) Host-plant selection and larval food-acceptance by small ermine moths. PhD thesis, University of Leiden, 151 pp.

Korman, A.K., Mallet, J., Goodenough, J.L., Graves, J.B., Hayes, J.L., Hendricks, D.E., Luttrell, R., Pair, S.D. and Wall, M. (1993) Population structure in *Heliothis virescens* (Lepidoptera: Noctuidae): an estimate of gene flow. *Annals of the Entomological Society of America*, **86**, 182–8.

Larson, A., Wak, D.B. and Yanev, K.P. (1984) Measuring gene flow among populations having high levels of genetic fragmentation. *Genetics*, **106**, 293–308.

Lester, L.J. and Selander, R.K. (1979) Population genetics of haplodiploid insects. *Genetics*, **92**, 1329–45.

Levins, R. (1970) Extinction, in *Lectures on Mathematics in the Life Sciences*, (ed. M. Gerstenhaber), Vol. 2, pp. 77–107.

Lewontin, R.C. (1985) Population genetics. *Annual Review of Genetics*, **19**, 81–102.

Loxdale, H.D. and Den Hollander, J. (1989) *Electrophoretic Studies on Agricultural Pests*, Clarendon Press, Oxford.

Loxdale, H.D., Brookes, C.P. and de Barro, P.J. (1996) Application of novel molecular markers (DNA) in agricultural entomology, in *The Ecology of Agricultural Pests: Biochemical Approaches*, (eds W.O.C. Symondson and J.E. Liddell), Chapman & Hall, London, pp. 149–98.

Makela, M.E. and Richardson, R.H. (1978) Hidden, reproductively isolated populations: one of nature's countermeasures to genetic pest control, in *The Screwworm Problem: Evolution of Resistance to Biological Control*, (ed. R.H. Richardson), University of Texas Press, Austin, pp. 49–66.

May, B., Leonard, D.E. and Vadas, R.L. (1977) Electrophoretic variation and sex linkage in spruce budworm. *Journal of Heredity*, **68**, 355–9.

Mayr, E. (1963) *Animals Species and Evolution*, Harvard University Press, Cambridge, Massachusetts.

Mayr, E. (1970) *Populations, Species, and Evolution*, Harvard University Press, Cambridge, Massachusetts.

McKechnie, S.W., Ehrlich, P.R. and White, R.R. (1975) Population genetics of

Euphydryas butterflies. I. Genetic variation and the neutral hypothesis. *Genetics*, **81**, 571–94.

McLellan, T. (1984) Molecular charge and electrophoretic mobility in cetacean myoglobins of known sequence. *Biochemical Genetics*, **22**, 181–200.

Menken, S.B.J. (1981) Host races and sympatric speciation in small ermine moths. *Entomologia experimentalis et applicata*, **30**, 280–92.

Menken, S.B.J. (1982) Enzymatic characterization of nine endoparasite species of small ermine moths (Yponomeutidae). *Experientia*, **38**, 1461–2.

Menken, S.B.J. (1987) Is the extremely low heterozygosity level in *Yponomeuta rorellus* caused by bottlenecks? *Evolution*, **41**, 630–7.

Menken, S.B.J. (1989) Electrophoretic studies on geographic populations, host races, and sibling species in insect pests, in *Electrophoretic Studies on Agricultural Pests*, (eds H.D. Loxdale and J. Den Hollander), Clarendon Press, Oxford, pp. 179–200.

Menken, S.B.J. (1991) Identification of immature stages through iso- and allozyme analysis, in *Tortricid Pests, Their Biology, Natural Enemies and Control*, (eds L.P.S. van der Geest and H.H. Evenhuis), World Crop Pest Series, Elsevier, pp. 77–88.

Menken, S.B.J. (1992) Population structure dynamics and conservation of sexual and parthenogenetic Lepidoptera, in *Future of Butterflies in Europe: Strategies for Survival* (eds T. Pavlicek-van Beek, A.H. Ovaa and J. van der Made), Department of Nature Conservation, Wageningen Agricultural University, pp. 222–30.

Menken, S.B.J. and Ulenberg, S.A. (1983) Diagnosis of the agromyzids *Liriomyza bryoniae* and *L. trifolii* by means of starch gel electrophoresis. *Entomologia experimentalis et applicata*, **34**, 205–8.

Menken, S.B.J. and Ulenberg, S.A. (1986) Allozymic diagnosis of four economically important *Liriomyza* species (Diptera, Agromyzidae). *Annals of Applied Biology*, **109**, 41–7.

Menken, S.B.J. and Ulenberg, S.A. (1987) Biochemical characters in agricultural entomology. *Agricultural Zoology Reviews*, **2**, 305–60.

Menken, S.B.J., Herrebout, W.M. and Wiebes, J.T. (1992) Small ermine moths (*Yponomeuta*): their host relations and evolution. *Annual Review of Entomology*, **37**, 41–66.

Miles, S.J. (1979) A biochemical key to members of the *Anopheles gambiae* group of species (Diptera: Culicidae). *Journal of Medical Entomology*, **15**, 297–9.

Murphy, R.W., Sites, J.W. Jr, Buth, D.G., Haufler, C.H. (1990) Proteins I: Isozyme electrophoresis, in *Molecular Systematics*, (eds D.M. Hillis and C. Moritz), Sinauer Associates, Sunderland, Mass, pp. 45–126.

Nei, M. (1978) Estimation of average heterozygosity and genetic distance from a small number of individuals. *Genetics*, **89**, 583–90.

Nei, M. and Chakravarti, A. (1977) Drift variances of F_{ST} and G_{ST} statistics obtained from a finite number of isolated populations. *Theoretical Population Biology*, **11**, 307–25.

Norrbom, A.L. and Kim, K.C. (1988) A list of the reported host plants of the species of *Anastrepha* (Diptera: Tephritidae). USDA, APHIS-PPQ, Hyattsville, Md.

Oudman, L. (1992) Identification of economically important *Liriomyza* species (Diptera, Agromyzidae) and their parasitoids using enzyme electrophoresis. *Proceedings of Experimental and Applied Entomology*, **3**, 135–9.

Oudman, L., Aukema, B., Stigter, H. and Menken, S.B.J. (1993) Spatial and temporal variation in populations of *Liriomyza huidobrensis* (Diptera, Agromyzidae). *Proceedings of Experimental and Applied Entomology*, **4**, 169–74.

Oudman, L., Aukema, B., Menken, S.B.J. and Ulenberg, S.A. A procedure for identification of polyphagous *Liriomyza* species (Diptera, Agromyzidae) using enzyme electrophoresis. *EPPO Bulletin* (in press).

Packer, L. and Owen, R.E. (1992) Variable enzyme systems in the Hymenoptera. *Biochemical Systematics and Ecology*, **20**, 1–7.

Pag, H. (1959) *Hyponomeuta*-Arten als Schädlinge im Obstbau. *Zeitschrift für angewandte Zoologie*, **46**, 129–89.

Pashley, D.P., Johnson, S.J. and Sparks, A.N. (1985) Genetic population structure of migratory moths: the fall armyworm (Lepidoptera: Noctuidae). *Annals of the Entomological Society of America*, **78**, 756–62.

Pinto, J.D., Platner, G.R. and Sassaman, C.A. (1993) Electrophoretic study of two closely related species of North American *Trichogramma: T. pretiosum* and *T. deion* (Hymenoptera: Trichogrammatidae). *Annals of the Entomological Society of America*, **86**, 702–9.

Raijmann, L.E.L. (1992) Genetical population structure of small ermine moth *Yponomeuta padellus*. *Proceedings of Experimental and Applied Entomology*, **3**, 94–8.

Raijmann, L.E.L. and Menken, S.B.J. (1992) Population genetical evidence for host-race formation in *Yponomeuta padellus*, in *Proceedings, 8th International Symposium on Insect–Plant Relationships*, (eds S.B.J. Menken, J.H. Visser and P. Harrewijn), Kluwer Academic Publishers, Dordrecht, pp. 209–11.

Richardson, B.J., Baverstock, P.R. and Adams, M. (1986) *Allozyme Electrophoresis. A Handbook for Animal Systematics and Population Studies*, Academic Press, San Diego, CA.

Roush, R.T. and McKenzie, J.A. (1987) Ecological genetics of insecticide and acaricide resistance. *Annual Review of Entomology*, **32**, 361–80.

Schutz, S.J., Gaugler, R.R. and Vrijenhoek, R.C. (1989) Genetic and morphometric discrimination of coastal and inland *Tabanus lineola* (Diptera: Tabanidae). *Annals of the Entomological Society of America*, **82**, 220–4.

Shoemaker, D.D., Costa, J.T. III and Ross, K.G. (1992) Estimates of heterozygosity in two social insects using a large number of electrophoretic markers. *Heredity* **69**, 573–82.

Singh, R.S. and Rhomberg, L.R. (1987) A comprehensive study of genic variation in natural populations of *Drosophila melanogaster*. I. Estimates of gene flow from rare alleles. *Genetics*, **115**, 313–22.

Slatkin, M. (1977) Gene flow and genetic drift in a species subject to frequent local extinctions. *Theoretical Population Biology*, **12**, 253–62.

Slatkin, M. (1981) Estimating levels of gene flow in natural populations. *Genetics*, **99**, 323–35.

Slatkin, M. (1985) Rare alleles as indicators of gene flow. *Evolution*, **39**, 53–65.

Slatkin, M. (1987) Gene flow and the geographic structure of natural populations. *Science*, **236**, 787–92.

Slatkin, M. (1994) Gene flow and population structure, in *Ecological Genetics*, (ed. L.A. Real), Princeton University Press, Princeton, pp. 3–17.

Slatkin, M. and Barton, N. (1989) A comparison of three indirect methods for estimating average levels of gene flow. *Evolution*, **43**, 1349–68.

Spencer, K.A. (1973) *Agromyzidae (Diptera) of economic importance*, Series Entomologica 9, Junk, The Hague.

Sperling, F.A.H., Landry, J.-F. and Hickey, D.A. (1995) DNA-based identification of introduced ermine moth species in North America (Lepidoptera: Yponomeutidae: *Yponomeuta padellus* complex). *Annals of the Entomological Society of America*, **88**, 155–62.

Steck, G.J. (1991) Biochemical systematics and population genetic structure of

Anastrepha fraterculus and related species (Diptera: Tephritidae). *Annals of the Entomological Society of America*, **84**, 10–28.

Thorpe, W.H. (1930) Further observations on biological races in *Hyponomeuta padella* L. *Journal of the Linnean Society, London, Zoology*, **37**, 489–92.

Urquhart, F.A. and Urquhart, N.R. (1976) The overwintering site of the eastern populations of the monarch butterfly (*Danaus p. plexippus* L; Danaidae) in southern Mexico. *Journal of the Lepidopterists' Society*, **30**, 153–8.

USDA (1985) Apple ermine moth new to the United States. USDA Animal Plant Health Inspection Service, *Plant Prot. Quar. Plant Pest Updates*, October 1985, 1-2.

Walsh, B.D. (1864) On phytophagic varieties and phytophagic species. *Proceedings of the Entomological Society of Philadelphia*, **3**, 403–30.

Ward, P.S., Boussy, I.A. and Swincer, D.E. (1982) Electrophoretic detection of enzyme polymorphism and differentiation in three species of spider mites (*Tetranychus*) (Acari: Tetranychidae). *Annals of the Entomological Society of America*, **75**, 595–8.

Wright, S. (1932) The roles of mutation, inbreeding, crossbreeding and selection in evolution. *Proceedings 6th International Congress of Genetics*, **1**, 356–66.

Wright, S. (1943) Isolation by distance. *Genetics*, **28**, 114–38.

Wright, S. (1946) Isolation by distance under diverse systems of mating. *Genetics*, **31**, 39–59.

2

Combined use of biochemical, immunological and molecular assays for infection, species identification and resistance detection in field populations of *Anopheles* (Diptera: Culicidae)

J. Hemingway, G.J. Small, S.W. Lindsay and F.H. Collins

ABSTRACT

A number of biochemical and molecular techniques are now available for analysis of different factors in insect populations. The majority of these techniques are compatible with each other, and if used in combination can generate large amounts of data, covering different aspects of an individual insect's biology. For example, rDNA–PCR can be performed on an insect leg, leaving the remainder of the insect for the generation of other information.

In the field of medical entomology, techniques are available for the identification of insects in species complexes, detection of the insects' pesticide resitance status, identification of the source of the insects' bloodmeal and calculation of the infection rates of the insects with malaria parasites. Techniques using microsatellite DNA have also been developed for analysis of rates of gene flow between insect populations. These techniques are discussed, along with some of their advantages and disadvantages. One advantage of many of the new techniques is their independency of the species or stage of insect being analysed. Hence, techniques developed for medically

The Ecology of Agricultural Pests, Edited by W.O.C. Symondson and J.E. Liddell.
Published in 1996 by Chapman & Hall, London. ISBN 0 412 62190 8

important insects can be rapidly adapted for use with agricultural pests and vice versa. For example, biochemical and molecular techniques to detect insecticide resistance are being developed for use in both 'agricultural' and 'medical' insect pests.

A major rationale for devising such techniques is their use in the field, or in the laboratory using field collections of insects, to generate information on population biology. Many of these techniques are now being combined routinely. To demonstrate the type of information which can be generated from field insect collections, this paper describes a published study of *Anopheles gambiae* (Giles) in The Gambia which utilized many of these techniques. In this study, the relative frequencies of three members of the *An. gambiae* species complex were determined by rDNA–PCR. The study was undertaken in an area of The Gambia where village-scale use of insecticide-impregnated bednets was being implemented. In the pre- and post-bednet intervention year the susceptibility of the insect populations to a range of insecticides was determined using a series of biochemical assays.

In addition to determining the species of insect and its insecticide resistance status, it was possible to look at the infection rates in the same insects with the two major malaria parasites, *Plasmodium vivax* and *P. falciparum,* using an immunological assay. The origin of the insects' blood meal using a DNA–PCR-based assay can also be determined.

2.1 INTRODUCTION

The use of combinations of new techniques examining various aspects of insect population biology has revolutionised the amount of information that can be obtained rapidly from single insects. Earlier techniques generally needed the whole insect, or insects had to be used in a particular physiological state. For example, identification of insect species on the basis of polytene chromosomes required the use of females which were semi-gravid, so that the polytenes in the ovarian nurse cells could be used. The advent of DNA technology to differentiate between insects in species complexes means that identification can now be made both on males and females and on a fraction of an insect at any life stage. This then allows the DNA technique for species identification to be used in combination with other techniques looking at different aspects of the insects' biology. This chapter briefly reviews the types of techniques currently available to examine insect species identification, insecticide resistance, parasite infection and insect bloodmeal analysis, and discusses their relevance to field situations. As an example of the level of information that can be achieved from a field sample of insects, a recently published study of the *Anopheles gambiae* complex in The Gambia using many of these techniques is detailed (Hemingway *et al.*, 1995).

2.1 THE METHODS AVAILABLE

2.1.1 Species identification

Six different species of the *An. gambiae* complex have been identified on the basis of mating incompatibility (Davidson, 1966; Davidson and Hunt, 1973). Of these, two are the principal vectors of malaria and they are widely distributed throughout tropical Africa, being sympatric over much of their range. The different species within the complex cannot be distinguished morphologically. However, the different behaviours, vectorial capacities and different ecologies of the species make identification to the species level essential in vector surveillance and malaria vector ecology studies.

(a) Polytene chromosome analysis

The polytene chromosome banding analysis technique is limited to the use of the ovarian nurse cells of adult female insects or the salivary glands of fourth instar larvae. Differentiation of some members of the complex can be done on the basis of the banding pattern on the X-chromosome, while others require the skill to recognize specific banding patterns associated with fixed inversions on the autosomes (Robert *et al.*, 1989; Coluzzi and Sabatini, 1967). The method requires access to a good-quality microscope and is very labour intensive. One advantage of the polytene chromosome method is that ovaries can be collected from a range of samples and preserved at room temperature until analysis. However, an obvious disadvantage is that the insect needs to be in a particular physiological state before the polytenized cells can be removed; hence, the technique is not compatible with other assays which require the insects to be in a different state. Maintaining the insects until they reach the required physiological state can also be difficult in field conditions.

(b) Isozyme analysis

Starch gel electrophoresis has been used as a method of separation of the species within the *An. gambiae* complex. This biochemical key relies on the use of four different enzyme systems (esterase 1, glutamate–oxaloacetate transaminase, octanol dehydrogenase and superoxide dismutase) (Miles, 1979). The technique requires a single gel to be run, which is then sliced and the two gel halves treated separately to visualize the different electromorphs of the four diagnostic enzymes. It has the advantage over the polytene system that different ages of adult insect can be used. As the enzyme activity is concentrated in the thorax, it is also possible to remove the head and thorax and examine these for oocysts and sporozoites to determine the vectorial status of the insect. However, the alleles for identification of the two major species within the complex (*An. gambiae* (Giles) and *An. arabiensis* (Patton)) are not absolute and there is the possibility of a small number of insects being misidentified. This technique is expensive in both time and reagents and requires specialized electrophoretic equipment. It also requires live insects, or insects that have been deep-frozen since collection, so that enzyme activities are maintained. Although this technique requires a reasonable level of skill, the end results are easier to score than the polytene chromosome preparations, and the technique is being used satisfactorily in the field in Tanzania.

Isozyme banding patterns have also been suggested as a technique for measuring gene flow between species, a topic that is very important if we are to attempt to manipulate populations by introduction of 'favourable' genes. However, this is a relatively inefficient technique, which detects as little as 25% of the actual amino acid polymorphism of enzymes (Selander, 1976). In contrast, the use of microsatellite DNA polymorphism allows detection of greater variability, and this is likely to become the method of choice for future gene flow studies. Microsatellites have already been used to produce a high-density genome map of species such as *An. gambiae* as the first step in this process (Zheng *et al.*, 1993).

(c) DNA methods

DNA probes have now been developed and successfully used for the identification of specimens of a large number of important vectors, which occur as part of species complexes. These include *An. gambiae, Simulium damnosum* (Theobald), *An. dirus* (Peyton and Harrison) and *Phlebotomus papatasi* (Scopoli) (Gale and Crampton, 1987). Early methods required the use of radiolabelled probes, but these are now being replaced by non-radiolabelled methods, for example the enhanced chemiluminescent methods (ECL, Amersham).

rDNA–PCR

Analysis of the repeated RNA genes on the ribosomal (rDNA) of members of the *An. gambiae* complex revealed areas of species specific sequence in the intergenic spacers (IGS). The rDNA was chosen as the basis for this diagnostic method, as the rDNA is present in hundreds of tandem copies per cell nucleus, so a small amount of nuclear DNA can be used as a template in a polymerase chain reaction (PCR) assay. The genes are highly conserved but contain spacers with evolutionarily labile sequences, which might be expected to differ between species. Intraspecies variation in rDNA sequences is also potentially less of a complicating factor than for a single copy locus due to the molecular homogenization of tandem arrays in rDNA (Dover and Coen, 1981; Dover, 1982, 1986). The 5′ IGS sequences were not subject to significant intraspecies geographic variation but showed good interspecies differences; therefore, a PCR-based assay was developed using primers derived from this region of the rDNA. The method distinguishes the different species by amplification of rDNA fragments of different lengths. In the original method using three different primers a 1.3 kb band was produced if *An. gambiae* DNA was used as the template, while an 0.5 kb fragment resulted from *An. arabiensis* DNA (Paskewitz and Collins, 1990). Since the original method was published several primers have been made which improve on the reliability and differentiation of this technique. The five morphologically indistinguishable members of the *An. gambiae* complex can now be distinguished using a cocktail of five 20-base oligonucleotide primers (Scott *et al.*, 1993).

To use this technique in the Gambia study, individual abdomens were placed in separate Eppendorf tubes and homogenized in 30 μl 10% sodium dodecyl sulphate. The homogenate was placed on ice for 30 min and then microfuged at 6000 *g* for 10 min. The supernatants were transferred to fresh Eppendorf tubes and 1 ml phenol:chloroform/ isoamyl alcohol (1:1) was added to each supernatant. The upper aqueous layer was transferred to a fresh tube, 1.25 ml absolute ethanol added and the Eppendorfs stored at −20°C until required for PCR. A 1 μl portion of 1 mg/ml tRNA was added to each tube to facilitate pelleting of the DNA, which was done by microfuging at 10 000 *g* for 10 min. DNA pellets were washed with 70% ethanol, dried and taken up in 50 μl Tris–EDTA (pH 8.0). Samples (1 μl) of these were used as the template for PCR as described (Paskewitz and Collins, 1990; Hemingway *et al.*, 1995). However, to simplify the technique DNA can be extracted from a single leg or intact tissue as small as an egg or a segment of one leg can be used rather than DNA which has already been purified using this method (Scott *et al.*, 1993).

DNA dot blots

A simple squash blot system is available for the identification of different members of the *An. gambiae* complex. The technique has now been refined to avoid the need for complex hybridization solutions and the whole procedure can be carried out at a single temperature (25°C). The system uses alkaline phosphatase-labelled oligonucleotides. The kit for detection is being commercially produced and field tested in various parts of Africa (Hill *et al.*, 1992; Hill and Crampton, 1994). The advantage of this technique compared with the rDNA method is that a PCR step is not required, thus reducing the time and equipment needed. The disadvantage is that it requires the DNA from the majority of the insect to be squashed onto the filter, making it incompatible with other biochemical assay systems.

2.1.2 Insecticide resistance detection

A range of different techniques is available to determine the resistance status of an insect. These can be used in combination to determine rapidly the resistance status of any insect population to any pesticide.

(a) Bioassays

Batches of insects of a known age are exposed to insecticide-impregnated filter papers under defined conditions for 1 h. Mortalities are scored after a 24 h recovery period (WHO, 1970). This type of test has been in use for many years and is limited in the amount of information it gives. With a single insect, information can be gained on possible resistance to only one pesticide, and this may be inaccurate depending on the age and physiological state of the insect (Rowland and Hemingway, 1987; Lines and Nassor, 1991). If prior information is available on the insect species being tested, a discriminating dose of pesticide should be available. In theory this should discriminate between resistant and susceptible individuals. In reality, the sensitivity of the test is rarely this good and insects which express a low level of resistance will remain undetected.

(b) Biochemical assays

A series of basic biochemical assays are now available which allows a mechanistic approach to be taken to resistance detection (Brogdon and Barber, 1987, 1900; Brown and Brogdon, 1987; Hemingway *et al.*, 1987; Hemingway, 1989; Dary *et al.*, 1991; Peiris and Hemingway, 1993). The tests have been developed to work at the level of the individual insect, and will work on fractions of insects. All the tests are compatible with each other, so that a battery of tests can be run on a single insect. Hence, a single insect can be analysed for resistance to a range of different insecticides taking this type of mechanistic approach. As yet the tests do not entirely replace the bioassays, as tests are not available for all possible resistance mechanisms, although they are available for the majority. New developments in the molecular biology of insecticide resistance mechanisms may shortly lead to a full battery of tests becoming available using a mixture of biochemical and molecular biological approaches. For example, the nerve insensitivity-based 'kdr'-type mechanism, could lend itself to a DNA-based detection system, such as that already applied to the altered gamma-aminobutyric acid (GABA) receptor mechanism, if only a small number of possible nucleotide changes underlie the *kdr* phenotype (ffrench-Constant *et al.*, 1993a,b; Aronstein *et al.*, 1994).

The current tests are quick and easy to run. For example, to analyse an insect population for several resistance mechanisms, the insect is homogenized on ice in 200 μl of distilled water. A 50 μl portion of homogenate is reserved for the acetylcholinesterase (AChE) assay. The remaining solution is transferred to an Eppendorf tube, centrifuged at 12 000 *g* for 3 min and the supernatant used for the esterase, glutathione S-transferase and protein assays. AChE activity is measured by the microplate assay of ffrench-Constant and Bonning (1989). Two 25 μl aliquots of insect homogenate are placed in separate wells of a microtitre plate and 140 μl of Triton-phosphate buffer (pH 7.2, 0.1M, 1% Triton X-100) added to solubilize the enzyme. Reaction rates in the presence and absence of the insecticide propoxur are measured on a UVmax microtitre plate reader at 405 nm. The technique is readily adaptable to any insect, although the concentration of propoxur may need to be adjusted for each species, to that which gives approximately 80% inhibition of the susceptible insect AChE (Hemingway *et al.*, 1993).

Elevated esterase activity is measured as described by Peiris and Hemingway (1990) in 25 μl replicates of insect supernatant with the substrate β-naphthyl acetate. Alternatively the substrates α-naphthyl acetate or *p*-nitrophenyl acetate can be used (Karunaratne *et al.*, 1993). This assay can be undertaken in either a microtitre plate or filter paper-based assay format (Georghiou and Pasteur, 1989; Dary *et al.*, 1990).

Glutathione S-transferase activity is measured as described by Lindsay *et al.* (1993a) in 20 μl replicates of insect supernatant. The extinction coefficient of 9.6 $mM^{-1} cm^{-1}$ for the product of the chlorodinitrobenzene/glutathione reaction, with a calculated light path of 0.6 cm for the reaction mixture in the microtitre plate, is used to convert absorbance to μmoles.

Protein concentration is calculated for each sample by analysing two 10 μl replicates of supernatant with the Bio-Pad Protein estimation kit. Alternatively, a number of other commercially available protein estimation kits can be used, many of which are sold in a microtitre plate format. Protein values are determined from a bovine serum albumin standard curve.

(c) Molecular analysis of resistance

Two molecular assays are already available for field detection of resistance. One detects alterations in the GABA receptors, the target site for the cyclodienes; the other detects amplification of the esterases involved in organophosphate and pyrethroid resistance. Other assays, such as those for monoxygenases, *kdr* and glutathione S-transferases (GSTs) are being developed. The GABA receptor assay is PCR-based as described by ffrench-Constant *et al.* (1994). The assay relies on the fact that there are very few possible changes associated with this type of mechanism. This assumption seems to hold in a range of different insect species examined to date. The molecular assay for amplified esterases is a simple dot-blot type system, which detects an increased level of signal from resistant insects from DNA immobilized onto nitrocellulose filters (Field *et al.*, 1989). At its simplest level the DNA can be released from the insects by squashing whole organisms onto the filters, as described for *Drosophila* by Tchen *et al.*, (1985).

Assay systems for GSTs and monoxygenases are under development, but initial studies are concentrating on ascertaining the level of diversity of these enzymes within and between insects. Multiple forms of both enzyme systems are known to occur, and a range of these are over-expressed in resistant insects. A strategy of cloning a variety of these enzymes has been employed, and the expression levels of these and their ability to interact with insecticides is now being determined (Scott *et al.*, in press; Ranson *et al.*, in press).

2.1.3 Immunological assays for the detection of infection with malaria parasites

Detection of infection rates of mosquitoes formerly relied on microscopic analysis of prepared mosquito tissue. This type of analysis is both time consuming and requires a reasonable level of skill to prepare the material and recognize accurately the parasites in the preparations. The situation is also complicated as malaria is not a single disease involving only one parasite. In some parts of the world several species of malaria parasites are found sympatrically. The two most common species are *Plasmodium falciparum* and *P. vivax*. It is important to differentiate between these, as *P. falciparum* can be rapidly fatal, whereas fatalities from *P. vivax* are rare.

Monoclonal antibodies have been developed to the major malaria parasites and are now available for field monitoring of infection of mosquito populations using a simple ELISA-based assay. Details of the assay are given in Beier *et al.* (1988), Wirtz *et al.* (1987) and Burkot *et al.* (1984).

DNA-based PCR methods are now available for the detection and differentiation of *P. falciparum* and *P. vivax* in blood samples of patients (Barker *et al.*, 1994). This type of methodology is likely to be readily adaptable for use in insect studies, and may replace the current immunological assays in the medium term. DNA analysis is already being undertaken on mosquito bloodmeals. This type of analysis not only allows identification of the organism group from which blood came (e.g. human, cow or chicken), but also identifies the exact 'donor' of the blood. Hence, with recent bednet studies it was possible to determine whether mosquitoes had fed on the human sleeping under the treated net, or had entered the house after feeding on another human (Gokol *et al.*, 1993). The techniques used $(CA)_n$ dinucleotide repeats to profile mosquito bloodmeal DNA extracted from field specimens of insects up to 15 h after the bloodmeal has been taken. As the PCR reaction is used small quantities of DNA can be analysed. To preclude amplification of mosquito DNA, primer pairs for the variable numbers of dinucleotide repeat blocks $(CA)_n$ associated with the human genes encoding an insulin-like growth factor, somatostatin and apolipoprotein II were used to amplify selectively bands of a suitable size for analysis (Table 2.1).

2.2 PRACTICAL USES OF THE TECHNIQUES – AN EXAMPLE OF A FIELD STUDY IN THE GAMBIA

Pyrethroid-impregnated bednets are being used nationwide in The Gambia. The future success of this malaria control programme depends partly on the vectors remaining susceptible to those insecticides used for treating the nets. This study was carried out on the south bank of the River Gambia, during the first large-scale trial of nets in this country (Greenwood and Baker, 1993, Hemingway *et al.*, 1995). Thus, the area represents a sentinel site for insecticide resistance in local vectors. The study gives an example of how a system of early detection for resistance problems can be set up in a relatively complex situation where multiple vectors and non-vectors are present. It was undertaken in the pre- and post-intervention years of a large-scale field trial of permethrin-impregnated bednets in The Gambia (Greenwood and Baker, 1993). In this area there are three morphologically indistinguishable sibling species of the *An. gambiae* complex, which are not all equally efficient as vectors (Lindsay *et al.*, 1993b). It was necessary to determine the insecticide susceptibility status separately for each of the species within the *An. gambiae* complex due to their different vectorial status, to ensure that no species already had pyrethroid resistance as a result of prior DDT exposure. Samples of the *An. gambiae* complex were caught indoors using light traps in 12 villages used in the bednet study.

The possible cross-resistance between DDT and pyrethroids due to a *kdr*-type nerve insensitivity was first shown in houseflies and is well documented in this species (Sawicki, 1975; Farnham and Sawicki, 1976). *Kdr*-type resistance is one of the few examples of a completely recessive resistance mechanism. Hence its detection by bioassay at low frequen-

Table 2.1 Primers used to DNA fingerprint human bloodmeal DNA from mosquitoes, and the expected band sizes and allele numbers at each locus

Locus	*Primers**	*Repeat sequence*	*No. of alleles*
Insulin-like growth factor	GCFAGCCAGCTGGTGTTATT ACCACTCTGGGAGAAGGGTA	$CATA(CA)_{19}$	5
Somatostatin	GCTCAAATGTTTCTGCAACC CTTTGTAGCTCGTGATGTGA	$(AC)_{12}GCACAA(AC)_{13}$	6
Apolipoprotein	GGTCTGGAAGTACTGAGAAA GATTCACTGCTGTGGACCCA	$(AC)_{16}$	6

*The sequence of the CA strand of each pair of primers is given first.

cies is impractical. This type of resistance appears to be relatively common in *Aedes aegypti* (L), but there is no well-documented case in *Anopheles* despite reports of pyrethroid resistance in this genus (Rongsriyam and Busvine, 1975; Prassitisuk and Busvine, 1977; Herath *et al.*, 1988; Hemingway *et al.*, 1989).

Additionally, assays for possible organophosphate and carbamate resistance mechanisms were undertaken to determine whether these pesticides could still be utilized in the event of bednet failure. Biochemical assays were used to allow smaller insect groups to be tested. It would not have been possible to generate log–dosage probit lines from bioassays of individual species of the *An. gambiae* complex from the field-caught material, in combination with the PCR-based identification system, because the number of individuals which would have to be identified is prohibitively large.

Six pairs of villages were used for the mosquito collections. These were sited on the south bank of the River Gambia, approximately 100 km from the coast. Collections were made in October 1988, the pre-intervention year, and October 1989, the post-intervention year. Live-caught *An. gambia* s.l. were sorted from the other mosquitoes collected indoors in light traps. They were stunned, snap-frozen in liquid nitrogen and transported to London for further analysis. A sample of live adult *An. gambiae* complex collected in exit traps from three villages, Sitahuma, Pakali Ba and Male Kunda, were tested for susceptibility to permethrin and DDT using standard WHO test kits.

Individual unfed male and female adults were assigned to a species of the *An. gambiae* complex using rDNA probes in an rDNA–PCR diagnostic assay. The tip of the abdomen of each sample was removed and DNA extracted from it. Individual DNA samples were carefully coded so that this information could be linked retrospectively with the biochemical assays carried out with the same insect.

A total of 2036 *An. gambiae* complex adults were identified from the 12 villages using DNA probes, 1094 from the pre-intervention year (1988), and 942 from the post-intervention year (1989). The identity of a further 198 mosquitoes could not be determined due to lack of PCR products. *An. gambiae* s.s. (86.1%) was the most common member of the *An. gambiae* complex found in the study area in both post- and pre-intervention years, followed by *An. arabiensis* (10.4%). *An. melas*, which breeds in brackish water and is not a major vector of malaria, was found in significant numbers only in Sitahuma, a village on the western fringe of the study area, where the breeding sites are brackish during the early part of the rainy season (see Figure 2.1). The overall population of the three species was similar in both the pre- and post-intervention years, although minor variations were seen between individual villages in 1988 and 1989.

There were some survivors of the *An. gambiae* complex from three villages at the WHO discriminating dosage of permethrin (0.25% for 1 h), suggesting that resistance may be present. However, log–time probit mortality lines for mosquitoes from the three villages were not significantly different (Figure 2.2). The slopes of the lines and low χ^2 values suggest that there is little heterogeneity within the populations. Hence, there is little direct evidence to suggest that this population is developing resistance to permethrin, despite survival at the discriminating dosage. These survivors may be due to several factors, for example, variation in testing temperature (23–32°C in this study). Survival is unlikely to be due to early knock-down and subsequent recovery of susceptible insects within the 1 h exposure period, as test kits were laid horizontally during the exposure period, so contact with the treated surface occurred after knock-down.

A sample of mosquitoes collected from Sitahuma was tested on DDT-impregnated papers. There were no survivors at the WHO discriminating dosage (4% DDT for 1 h). This suggests that DDT resistance, if present, is neither at a high frequency nor at a high level. However, the lack of survivors at the WHO discriminating dosage of DDT in a mixed age field collection of *An. gambiae* complex cannot be taken as definitive proof of DDT-susceptibility in *An. melas* (Theobald) from this village. Lines and Nassor (1991) have shown that significant mortalities of >4-day-old DDT-resistant *An. gambiae* adults occur on the distinguishing dose. In 12–14-day-old DDT-resistant adults mortality with 4% DDT was 92%. The sample of *An. gambiae* complex adults tested in this study would have been a mixture of different ages and different species, only a small percentage of which would have been *An. melas*.

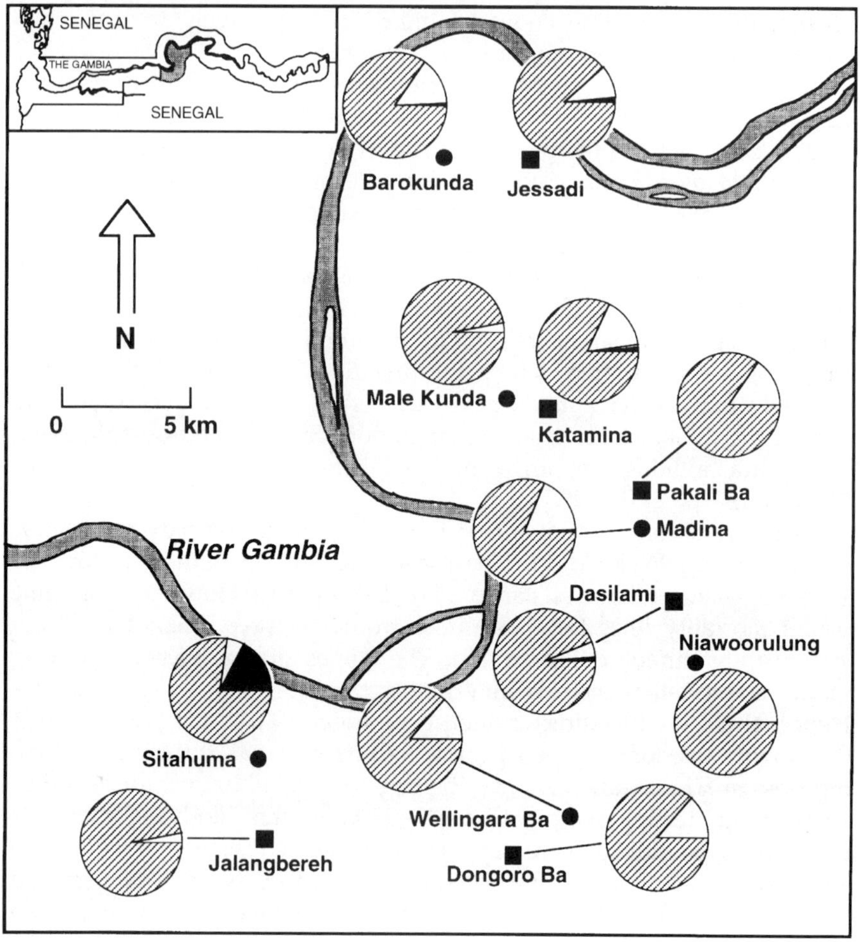

Figure 2.1 Location of the Gambian villages used in this study. The relative abundance of the different species in them, as determined by rDNA–PCR are shown by the inset pie diagrams. Hatched area, *Anopheles gambiae s.s.*; clear area, *Anopheles arabiensis*; solid area, *Anopheles melas*. (Reproduced with permission from the *Bulletin of Entomological Research* (1995), **85**, 229–34.)

Glutathione S-transferases are frequently involved in DDT resistance in *Anopheles* (Hemingway *et al.*, 1985; Herath *et al.*, 1988; Prapanthadara *et al.*, 1995). This is usually associated with an increase in total GST activity with substrates such as CDNB and DCNB in the resistant strain. Figure 2.3 gives the GST activity levels with CDNB for the three species of the *An. gambiae* complex. Data for different villages for each species in both years are pooled as there was no significant difference

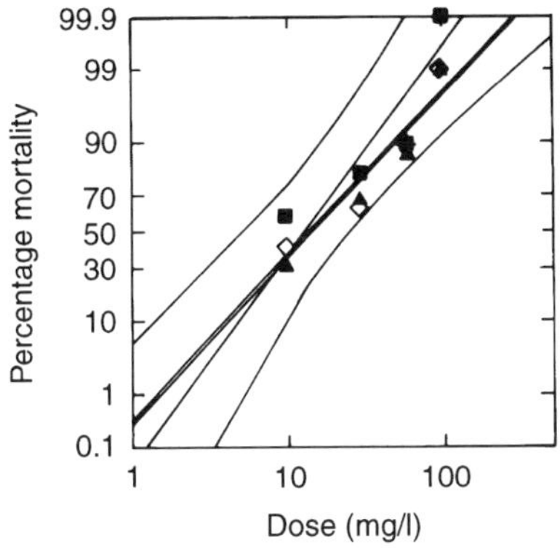

Figure 2.2 Log–time probit mortality lines for the *An-gambiae* complex tested with permethrin. (◇, *Anopheles melas;* ■, *Anopheles gambiae;* ▲, *Anopheles arabiensis*: 95% confidence interval for the lines is shown as a dotted line.)

in the activity between villages or years. The graphs for *An. gambiae* and *An. arabiensis* show the normal baseline activity level for GST generally seen in mosquitoes. The range of GST activity in *An. melas* is significantly different from that for the other two members of the *An. gambiae* complex. Our data on GST levels indicates that if DDT resistance is present, it is present only in *An. melas* and not in *An. arabiensis* or *An. gambiae*. DDT metabolism data indicated that the pooled remaining supernatant of *An. melas* individuals from Sitahuma with high GST activity were able to degrade DDT at a faster rate than pooled supernatant from low activity individuals (Lindsay *et al.*, 1993b). If DDT resistance is present in *An. melas* from Sitahuma, it has not yet been detected by bioassay. Alternatively, DDT resistance may not be present; if so, the baseline susceptibility range for *An. melas* is higher than that found in the majority of *Anopheles*, both for GST activity and DDT metabolic rates. With the present data it is not possible to differentiate between these two alternatives, as baseline data for *An. melas* from other regions are not available. The current evidence for DDT resistance in the *An. gambiae* complex in The Gambia suggests that this is not a complicating factor in the use of pyrethroid-impregnated nets.

The general esterase activity levels for the three species are not elevated compared with a normal susceptible insect. There was no significant difference in the ranges of activity with β-naphthyl acetate in any species, although the mean activity levels in *An. melas* were slightly lower than those in *An. gambiae* and *An. arabiensis*. There was no evidence of an altered AChE-based resistance mechanism in any of the three species tested. Biochemical tests for altered AChE and elevated esterases showed no evidence for either of these insecticide resistance

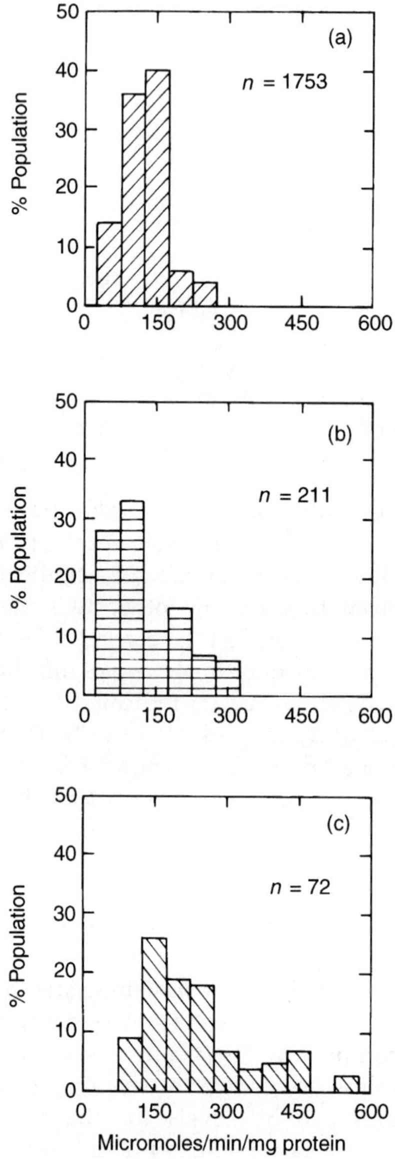

Figure 2.3 Glutathione S-transferase activity in three different species (A–C) of the *An. gambiae* complex. Data for the pre- and post-intervention years are pooled for each species as there was no significant difference in activity patterns between years.

mechanisms being present in *An. gambiae, An. arabiensis* or *An. melas,* which suggests that the mosquitoes of the *An. gambiae* complex in the study area are susceptible to organophosphates and carbamate insecticides. This reflects the minimal exposure to these classes of insecticides through both agricultural and public health pesticide use in The Gambia. Elevated esterases also confer pyrethroid resistance in some insects; however, this mechanism does not appear to be present in any member of the *An. gambiae* complex tested in this study.

The similarity of the data for the pre- and post-intervention years for all three species of the *An. gambiae* complex reflects both the basic susceptibility status of these mosquitoes and the current lack of any strong selection pressure for resistance in the field.

2.3 CONCLUSIONS

A number of biochemical and molecular assays are now available for looking at a range of problems in insect field biology. A great deal of research is under way in the molecular biology of various pest species. This means that new diagnostic assays for separation of sibling species, detection of insecticide resistance, analysis of insect feeding and other related topics, will become available in the short to medium term. The further development of DNA and RNA detection systems, allowing us to move away from radiolabels and into systems where visual detection is possible will facilitate the transfer of this type of technology to the field. Meanwhile, a battery of molecular and biochemical assays is already being used as part of the large-scale insect control programmes in some areas.

Acknowledgements

J.H. was supported by The Royal Society and GJS by the Wellcome Trust.

REFERENCES

Aronstein, K., Ode, P. and ffrench-Constant, R.H. (1994) Direct comparison of PCR-based monitoring for cyclodiene resistance in *Drosophila* populations with insecticide bioassay. *Pesticide Biochemistry and Physiology*, **48**, 229–33.

Barker, R.H., Banchongaksorn, T., Courval, J.M., Suwonkerd, W., Rimwungtragoon, K. and Wirth, D.F. (1994) *Plasmodium falciparum* and *P. vivax*: factors affecting sensitivity and specificity of PCR-based diagnosis of malaria. *Experimental Parasitology*, **79**, 41–9.

Beier, J.C., Perkins, P.V., Wirtz, R.A., Koros, J., Diggs, D., Gargan, T.P. and Koech, D.K. (1988) Bloodmeal identification by direct enzyme-linked immunosorbent-assay (ELISA), tested on *Anopheles* (Diptera, Culicidae) in Kenya. *Journal of Medical Entomology*, **25**, 9–16.

Brogdon, W.G. and Barber, A.M. (1987) Microplate assay of acetylcholinesterase inhibition kinetics in single-mosquito homogenates. *Pesticide Biochemistry and Physiology*, **29**, 252–9.

Brogdon, W.G. and Barber, A.M. (1990) Microplate assay of glutathione S-transferase activity for resistance detection in single mosquito triturates. *Comparative Biochemistry and Physiology*, **96**, 339–42.

Brown, T.M. and Brogdon, W.G. (1987) Improved detection of insecticide resistance through conventional and molecular techniques. *Annual Review of Entomology*, **32**, 145–62.

Burkot, T.R., Williams, J.L. and Schneider, L. (1984) Identification of *Plasmodium falciparum* infected mosquitoes by a double antibody enzyme-linked immunosorbant assay. *American Journal of Tropical Medicine and Hygiene*, **33**, 783–8.

Coluzzi, M. and Sabatini, A. (1967) Cytogenetic observations on species A and B of the *Anopheles gambiae* complex *Parasitologia*, **9**, 73–88.

Dary, O., Georghiou, G.P., Parsons, E. and Pasteur, N. (1990) Microplate adaptation of Gomori's assay for quantitative determination of general esterase activity in single insects. *Journal of Economic Entomology*, **83**, 2187–92.

Dary, O., Georghiou, G.P., Parsons, E. and Pasteur, N. (1991) Dot-blot test for identification of insecticide-resistant acetylcholinesterase in single insects. *Journal of Economic Entomology*, **84**, 28–33.

Davidson, G. (1966) *Anopheles gambiae*, a complex of species. *Bulletin of the World Health Organization*, **31**, 625–34.

Davidson, G. and Hunt, R.H. (1973) The crossing and chromosome characteristics of a new sixth species in the *Anopheles gambiae* complex. *Parasitologia*, **15**, 121–8.

Dover, G.A. (1982) Molecular drive: a cohesive mode of species evolution. *Nature*, **299**, 111–7.

Dover, G.A. (1986) Molecular drive in multigene families: how biological novelties arise, spread and are assimilated. *Current Trends in Genetics*, **8**, 159–65.

Dover, G. and Coen, E. (1981) Springcleaning ribosomal DNA: a model for multigene evolution? *Nature*, **290**, 731–2.

Farnham, A.W. and Sawicki, R.M. (1976) Development of resistance to pyrethroids in insects resistant to other insecticides. *Pesticide Science*, **7**, 278–82.

ffrench-Constant, R.H. and Bonning, B.C. (1989) Rapid microtitre plate test distinguishes insecticide resistant acetylcholinesterase genotypes in the mosquitoes. *Anopheles albimanus*, *An. nigerrimus* and *Culex pipiens*. *Medical and Veterinary Entomology*, **3**, 9–16.

ffrench-Constant, R.H., Rocheleau, T.A., Steichen, J.C. and Chalmers, A.E. (1993a) A point mutation in a *Drosophila*, GABA receptor confers insecticide resistance. *Nature*, **363**, 449–51.

ffrench-Constant, R.H., Steichen, J.C., Rocheleau, T.A., Aronstein, K. and Roush, R.T. (1993b) A single-amino acid substitution in a gamma-aminobutyric acid subtype A receptor locus is associated with cyclodiene resistance in *Drosophila* populations. *Proceedings of the National Academy of Sciences USA*, **90**, 1957–61.

ffrench-Constant, R.H., Steichen, J.C. and Shotkoski, F. (1994) Polymerase chain reaction diagnostic for cyclodiene insecticide resistance in the mosquito *Aedes aegypti*. *Medical and Veterinary Entomology*, **8**, 99–100.

Field, L.M., Devonshire, A.L., ffrench-Constant, R.H. and Forde, B.G. (1989) The combined use of immunoassay and a DNA diagnostic technique to identify insecticide-resistant genotypes in the peach-potato aphid, *Myzus persicae* (Sulz.). *Pesticide Biochemistry and Physiology*, **34**, 174–8.

Gale, K.R. and Crampton, J.M. (1987) A DNA probe to distinguish the species *Anopheles quadriannulatus* from other species of the *Anopheles gambiae* complex. *Transactions of the Royal Society of Tropical Medicine and Hygiene*, **81**, 842–6.

Georghiou, G.P. and Pasteur, N. (1989) Novel tests of organophosphate insecticide resistance in single mosquitoes: an overview of recent progress and outline of filter paper test. 57th Annual Conference of the California Mosquito and Vector Control Association, American Mosquito Control Association, pp. 174–8.

Gokool, S., Curtis, C.F. and Smith, D.F. (1993) Analysis of mosquito bloodmeals by DNA profiling. *Medical and Veterinary Entomology*, **7**, 208–15.

Greenwood, B.M. and Baker, J.R. (1993) A malaria control trial using insecticide-treated bednets and targeted chemoprophylaxis in a rural area of The Gambia, West Africa. *Transactions of the Royal Society of Tropical Medicine and Hygiene*, **87** (Suppl.) 2–60.

Hemingway, J. (1989) A note on simple biochemical methods for resistance detection and their field application in Sri Lanka. *Pesticide Science*, **27**, 281–5.

Hemingway, J., Malcolm, C.A., Kissoon, K.E., Boddington, R.G., Curtis, C.F. and Hill, N. (1985) The biochemistry of insecticide resistance in *Anopheles sacharovi*: comparative studies with a range of insecticide susceptible and resistant *Anopheles* and *Culex* species. *Pesticide Biochemistry and Physiology*, **24**, 68–76.

Hemingway, J., Jayawardena, K.G.I., Weerasinghe, I. and Herath, P.R.J. (1987) The use of biochemical tests to identify multiple resistance mechanisms in field-selection populations of *Anopheles subpictus* Grassi (Diptera: Culicidae). *Bulletin of Entomological Research*, **77**, 57–66.

Hemingway, J., Boddington, R.G., Harris, J. and Dunbar, S.J. (1989) Mechanisms of insecticide resistance in *Aedes aegypti* (L.)(Diptera: Culicidae) from Puerto Rico. *Bulletin of Entomological Research*, **79**, 123–30.

Hemingway, J., Small, G.J. and Monro, A.G. (1993) Possible mechanisms of organophosphorus and carbamate insecticide resistance in German cockroaches (Dictyoptera: Blattelidae) from different geographical areas. *Journal of Economic Entomology*, **86**, 1623–30.

Hemingway, J., Lindsay, S.W., Small, G.J., Jawara, M. and Collins, F.H. (1995) Insecticide susceptibility status in individual species of the *Anopheles gambiae* complex where pyrethroid impregnated bednets are used extensively for malaria control. *Bulletin of Entomological Research*, **85**, 229–34.

Herath, P.R.J., Jayawardena, K.G.I., Hemingway, J. and Harris, J. (1988) DDT resistance in *Anopheles culcifacies* Giles and *A. subpictus* Grassi (Diptera: Culicidae) from Sri Lanka: a field study on the mechanisms and changes in gene frequency after cessation of DDT spraying. *Bulletin of Entomological Research*, **78**, 717–23.

Hill, S.M. and Crampton, J.M. (1994) Synthetic DNA probes to identify members of the *Anopheles gambiae* complex and to distinguish the two major vectors of malaria within the complex, *An. gambiae* SS and *An. arabiensis*. *American Journal of Tropical Medicine and Hygiene*, **50**, 312–21.

Hill, S.M., Urwin, R. and Crampton, J.M. (1992) A simplified non-radioactive DNA probe control for the field identification of insect vector specimens. *Transactions of the Royal Society of Tropical Medicine and Hygiene*, **86**, 213–15.

Karunaratne, S.H.P.P., Jayawardena, K.G.I., Hemingway, J. and Ketterman, A.J. (1993) Characterization of a B-type esterase involved in insecticide resistance from the mosquito *Culex quinquefasciatus*. *Biochemical Journal*, **294**, 575–9.

Lines, J.D. and Nassor, N.S. (1991) DDT resistance in *Anopheles gambiae* declines with mosquito age. *Medical and Veterinary Entomology*, **5**, 261–5.

Lindsay, S.W., Alonso, P.L., Armstrong Schellenberg, J.R.M., Hemingway, J., Adimah, J., Shenton, F.C., Jawars, M. and Greenwood, B.M. (1993a) A malaria control trial using insecticide-impregnated bed nets and targeted chemoprophylaxis in a rural area of The Gambia, West Africa. 7. Impact of

permethrin impregnated bed nets on malaria vectors. *Transactions of the Royal Society of Tropical Medicine and Hygiene*, **87**, 45–52.
Lindsay, S.W., Alonso, P.L., Armstrong Schellenberg, J.R.M., Hemingway, J., Thomas, P.J., Shenton, F.C. and Greenwood, B.M. (1993b) A malaria control trial using insecticide-impregnated bed nets and targeted chemoprophylaxis in a rural area of The Gambia, West Africa. 3. Entomological characteristics of the study area. *Transactions of the Royal Society of Tropical Medicine and Hygiene*, **87**, 19–24.
Miles, S.J. (1979) A biochemical key to adult members of the *Anopheles gambiae* group of species (Diptera: Culicidae). *Journal of Medical Entomology*, **15**, 297–9.
Paskewitz, S.M. and Collins, F.H. (1990) Use of the polymerase chain reaction to identify mosquito species of the *Anopheles gambiae* complex. *Medical and Veterinary Entomology*, **4**, 367–73.
Peiris, H.T.R. and Hemingway, J. (1990) Mechanisms of insecticide resistance in a temephos selected *Culex quinquefasciatus* (Diptera: Culicidae) strain from Sri Lanka. *Bulletin of Entomological Research*, **80**, 453–7.
Peiris, H.T.R. and Hemingway, J. (1993) Characterization and inheritance of elevated esterases in organophosphorus and carbamate insecticide resistant *Culex quinquefasciatus* (Diptera: Culicidae) from Sri Lanka. *Bulletin of Entomological Research*, **83**, 127–32.
Prapanthadara, L., Hemingway, J. and Ketterman, A.J. (1995) DDT-resistance in *Anopheles gambiae* Giles from Zanzibar, Tanzania, based on increased DDT-dehydrochlorinase activity of glutathione S-transferase. *Bulletin of Entomological Research*, **85**, 267–74.
Prassitisuk, C. and Busvine, J.R. (1977) DDT-resistant mosquito strains with cross-resistance to pyrethroids. *Pesticide Science*, **8**, 527–33.
Ranson, H., Cornel, A.J., Fournier, D., Vaughan, A. and Hemingway, J. (in press) Cloning and isolation of a glutathione S-transferase class I gene from *Anopheles gambiae. Biochemistry Journal.*
Robert, V., Petrarca, V., Carnevale, P., Ovazza, L. and Coluzzi, M. (1989) Cytogenetic investigations on the *Anopheles gambiae* complex in Bobo-dioulasso region (Burkina faso). *Annales de Parasitologie Humaine et Comparee*, **64**, 290–311.
Rodriguez, M., Ortiz, E., Bisset, J.A., Hemingway, J. and Saledo, E. (1993) Changes in malathion and pyrethroid resistance after cypermethrin selection of *Culex quinquefasciatus* field populations of Cuba. *Medical and Veterinary Entomology*, **7**, 117–21.
Rongsriyam, Y. and Busvine, J.R. (1975) Cross-resistance in DDT-resistant strains of various mosquitoes (Diptera: Culicidae). *Bulletin of Entomological Research*, **65**, 459–71.
Rowland, M. and Hemingway, J. (1987) Changes in malathion resistance with age in *Anopheles stephensi* from Pakistan. *Pesticide Biochemistry and Physiology*, **28**, 239–47.
Sawicki, R.M. (1975) Some aspects of the genetics and biochemistry of resistance of houseflies to insecticides. Mimeographed document of the World Health Organization WHO/VBC/EC, **75.10**.
Scott, J.A., Brogdon, W.G. and Collins, F.H. (1993) Identification of single specimens of the *Anopheles gambiae* complex by the polymerase chain reaction. *American Journal of Tropical Medicine and Hygiene*, **49**, 520–9.
Scott, J.A., Collins, F.H. and Feyereisen, R. (1994). Diversity of cytochrome P450 genes in the mosquito *Anopheles albimanus. Biochemica et Biophysica Research Communications*, **205**, 1452–9.
Selander, R.K. (1976) Genic variation in natural populations, in *Molecular Evolution* (ed. F.J. Ayala), Sinauer, Sunderland, Mass, USA, pp. 21–45.

Tchen, P., Anoxolabehere, D., Nouaud, D. and Periquet, G. (1985) Hybridization on squashed flies: a method to detect gene sequences in individual *Drosophila. Analytical Biochemistry*, **150**, 415–20.

WHO (1970) Instructions for determining the susceptibility or resistance of adult mosquitos to organochlorine insecticides. *World Health Organization Technical Report Series*, **443**, 47–56.

Wirtz, R.A., Burkot, T.R., Graves, P.M. and Andrc, R. G. (1987) Field evaluation of enzyme-linked immunosorbant assays for *Plasmodium falciparum* and *Plasmodium vivax* sporozoites in mosquitoes (Diptera: Culicidae) from Papua New Guinea. *Journal of Medical Entomology*, **24**, 433–7.

Zheng, L., Collins, F.H., Kumar, V. and Kafatos, F.C. (1993) A detailed genetic map of the X-chromosome of the malaria vector *Anopheles gambiae. Science*, **261**, 605–8.

3

Studies of interactions between alfalfa weevil strains, *Wolbachia* endosymbionts and parasitoids

T.H. Hsiao

ABSTRACT

The alfalfa weevil, *Hypera postica* (Gyllenhal), a Palearctic pest of legumes, was accidentally introduced into the United States on three separate occasions. These introductions are now recognized as western, Egyptian and eastern weevil strains. The genetic relatedness between these weevil strains and two weevil populations from Europe and Asia was investigated by comparing sequences of their mtDNA (12S-, 16S-rRNA, COI, and URF1-CytoB gene coding regions) and nuclear rDNA (first internal transcribed spacer region) using polymerase chain reaction (PCR) and direct double-stranded DNA sequencing techniques. The aligned sequences of 1302 base pairs from the mtDNA regions were combined to reconstruct gene trees using neighbour-joining and parsimony methods. Gene trees generated from DNA data were in agreement with allozyme data. DNA sequence data provided diagnostic markers for distinguishing between weevil strains.

A PCR-based technique was used to detect *Wolbachia* symbionts among weevil populations. The origin of *Wolbachia* was traced back to Europe and Asia. The results confirmed that population mixing zones exist between weevil strains in western United States. A second *Wolbachia*, found in weevils from Xinjiang, China, did not cause cytoplasmic incompatibility. Both symbiotic and aposymbiotic individuals were found in this population.

The parasitoid, *Microctonus aethiopoides* Loan, an effective natural enemy against the eastern and Egyptian weevils, was unable to

The Ecology of Agricultural Pests, Edited by W.O.C. Symondson and J.E. Liddell.
Published in 1996 by Chapman & Hall, London. ISBN 0 412 62190 8

develop normally in the adult western alfalfa weevil due to *Wolbachia*. Normal development of *M. aethiopoides* can be restored by rearing with aposymbiotic western weevils. The interactions between alfalfa weevil strains, *Wolbachia*, and parasitoids illustrate the complexity of biotic relationships in an agricultural pest and will be useful for developing a more effective pest management strategy.

3.1 INTRODUCTION

Exotic agricultural insect pests are good models for studying intraspecific genetic variation among insects. These organisms are often introduced from diverse geographic regions of the world and have since adapted to their new habitats. There are often biological differences between geographic populations of exotic pests, and the terms 'biotype, strain, or race' have been used to distinguish among them (Hsiao, 1988). These biological variations among pest populations must be recognized to develop management strategies. Traditionally, morphological criteria are used to determine intraspecific variations of insects. Where distinct morphological variations are lacking, ecological, physiological, and behavioural criteria are used to distinguish pest populations. Biochemical differences such as protein polymorphism as revealed by gel electrophoresis of enzymes and immunological tests provide more precise information on population variations (Hsiao, 1989; Menken, 1989; Menken and Raijmann, 1996, this volume). DNA sequencing techniques now provide a direct means of assessing genetic variation among populations. The polymerase chain reaction (PCR) for DNA amplification makes it possible to analyse the sequences of selected nucleotide regions (Innis *et al.*, 1990; Mullis *et al.*, 1994). The rapid rate of evolution in mitochondrial DNA (mtDNA) means it is well suited for assessing variations at the population and species level (Simon, 1991).

The alfalfa weevil, *Hypera postica* (Gyllenhal), an exotic alfalfa pest of Palearctic origin, was introduced into the United States on three separate occasions (Salt Lake City, Utah, 1904; Yuma, Arizona, 1939; Maryland, 1951) (Hsiao, 1993). These three introductions are now known as western, Egyptian, and eastern strains and they have spread to all 48 contiguous states, Canada and Mexico. Mixing zones between weevil strains have been recorded in several western states since early 1970s (see Hsiao, 1993, Figure 15.3). Presently, weevil strains are distinguished by their geographic distribution and a rickettsia, *Wolbachia* sp., which is found only in the western strain (Hsiao and Hsiao, 1985a; Leu *et al.*, 1989). Cytogenetic and allozyme analyses show that all three strains belong to the same species and that the Egyptian and eastern weevils are most closely related (Hsiao and Hsiao, 1984; Hsiao and Stutz, 1985; Hsiao, 1989). There is no distinct genetic marker to discriminate between weevil strains, however. The symbiotic rickettsia is absent in the Egyptian and eastern strains (Hsiao and Hsiao, 1985b; Hsiao, 1993). The rickettsia is responsible for the unidirectional reproductive incompatibility between weevil strains and also affects the effectiveness of parasitoids of the alfalfa weevil. We investigated some of these relationships using molecular technologies involving PCR and direct double-standard DNA sequencing techniques. This discussion is an update of a recent review on alfalfa weevil strains (Hsiao, 1993).

3.2 MOLECULAR STUDIES OF ALFALFA WEEVIL STRAINS

3.2.1 Alfalfa weevil biotype problem

Based on morphological characteristics, distribution and host affinity, the alfalfa weevil has been recognized under several species names. In the Palearctic region, the species name *Phytonomus variabilis* Herbst is most commonly used, followed by *P. posticus* Gyllenhal, *P. transsylvanicus* Petri, *P. brunneipennis* Boheman, and *P. murina* F. (Csiki, 1934; Zaslavskiy, 1961). After its introduction into the Nearctic region, Titus (1911) adopted the name *Hypera postica* (Gyllenhal) for the introduction in the western United States. Subsequently, *H. brunneipennis* Boheman was used for the Egyptian alfalfa weevil found in Yuma, Arizona (Wehrle, 1940). The alfalfa weevil found in the eastern United States is called the eastern strain of *H. postica* to distinguish it from the western strain (Blickenstaff, 1965). Two species of alfalfa weevils (the alfalfa weevil and the Egyptian alfalfa weevil) are currently recognized in the North American literature, even though studies on strain hybridization, cytogenetics and allozyme analysis indicate that all North American weevils are strains of the same species (Hsiao and Hsiao, 1984, 1985a,b; Hsiao and Stutz, 1985). To confirm this contention, we examined DNA sequence variation among the three North American weevil strains and two Old World weevil populations (Budapest, Hungary and Xinjiang, China). The clover leaf weevil, *Hypera punctata* (F.), was used as an out-group to assess sequence divergence between *Hypera* species.

3.2.2 DNA sequence analysis

(a) Insect source

Adult weevils were field-collected, starved, and stored at –70°C until use. The sources for the five alfalfa weevil populations were: Yuma, Arizona for the Egyptian weevil (AZ); Logan, Utah for the western weevil (UT); Beltsville, Maryland for the eastern weevil (MD);

Budapest, Hungary for the European weevil (BU); Wulumulgi, Xinjiang, China for the Asian weevil (XC). The five weevil populations are subsequently referred to as strains. The clover leaf weevil, *Hypera punctata* (HP), was collected from Logan, Utah.

(b) DNA primers, PCR and sequencing protocol

Primers of three mtDNA regions and one nuclear DNA region were used for DNA amplification (Table 3.1). The mtDNA primers were named using the convention of Simon *et al.* (1994), which indicates the position and size of the DNA fragment amplified by *Drosophila yakuba* Burla (Clary and Wolstenholme, 1985). The 12S rRNA gene primers were designed by Simon *et al* (1991) and amplified a portion of the 12S region. The 16S rRNA gene primers were modified from those used by Xiong and Kocher (1991) for the black flies. The URF1-cytochrome B primers were designed by Pruess *et al.* (1992) and amplified a region covering a portion of URF1 and cytochrome B gene regions. The primers for the internal transcribed spacers (ITS) of the nuclear rDNA were designed by White *et al.* (1990). These primers were prepared by Operon Inc., CA., and were used without further purification.

DNA extraction, PCR amplification, template purification, and

Table 3.1 Primers used in this study

Primer name	*Primer sequence*	*Reference*
Mitochondrial DNA		
12S rRNA		Simon *et al.*, 1991
SR-N-14759	5′-GACAAAATTCGTGCCAGCAGT-3′	
SR-J-14233	5′-AAGAGCGACGGGCGATGTGT-3′	
16S rRNA		Xiong and Kocher, 1991
LR-N-13398	5′-CACCTGTTTATCAAAAACAT-3′	
LR-J-12883	5′-CACCGGTTTGAACTCAGATC-3′	
URF1-cytochrome B		Pruess *et al.*, 1992
N1-N-11841	5′-GGTACATTACCTCGGTTTCGTTATGAT-3′	
CB-J-11545	5′-ACATGAATTGGAGCTCGACCAGT-3′	
Nuclear DNA		
ITS rDNA		White *et al.*, 1990
ITS3	5-′GCATCGATGAAGAACGCAGC-3′	
ITS4	5′-TCCTCCGCTTATTAATATGC-3′	
Wolbachia 16S rRNA		*O'Neill et al.*, 1992
99F	5′-TTGTAGCCTGCTATGGTATAACT-3′	
994R	5′-GAATAGGTATGATTTTCATGT-3′	

manual sequencing followed protocols described previously (Hsiao, 1994a,b). The PCR-amplified mtDNA templates were sequenced manually. The ITS-amplified templates were sequenced using an ABI 373A autosequencer.

(c) Sequence data analysis

DNA sequences from weevil populations were aligned using ClustalV (Higgins *et al.*, 1992) and the GCG pileup programs (Wisconsin Sequence Analysis Package, 1992). Phylogenetic reconstruction of sequence data was conducted by the neighbour-joining method (MEGA program; Kumar *et al.*, 1993) and maximum parsimony method (PAUP v. 3.1.1; Swofford, 1991).

(d) Sequencing results

The sizes of DNA fragments amplified by each primer set were identical for all weevil strains and resembled those from chrysomelid beetles (Hsiao, 1994a). No sequence variation of the amplified fragments was detected among five individuals examined from the same strain. Hence, two to three individuals of each strain were used to obtain the DNA sequences of the amplified fragments. The four sequenced regions produced 1820 base pairs (bp) of aligned sequences. Of these, 42 bases or 2.3% were variable between the five weevil strains (Table 3.2). This compares with 11% base variation between the two *Hypera* species (alfalfa weevil versus clover leaf weevil). Among weevil strains, there were no base substitutions in the ITS2 region, but there was a six-base deletion, involving three AT repeats in the XC and BU strains. A single deletion was found in the 12S rRNA region of XC, BU and UT strains. The URF1-cytochrome B region comprises an amino acid coding (32 codons) and a tRNA non-coding region (194 bp). The amino acid sequence of the coding region was identical in all five weevil strains. The base variations shown in Table 3.2 included the entire URF1-cytochrome B region. Among the five weevil strains, 80% of the base substitutions were transition and 20% were transversion.

Of the five weevil strains, the DNA sequences of AZ and MD were most similar, showing only a single base substitution in the URF1-cytochrome B region (Table 3.2), which provides a diagnostic marker for these two strains. Compared with the AZ strain, the most base substitutions and gaps were in the UT strain (30), followed by BU (24) and the XC strains (15). The 2% base substitution rate between alfalfa weevil strains was similar to intraspecific variation in other beetle species (Hsiao, 1994b) and supports the hypothesis that the five weevil strains belong to the same species.

Table 3.2 Base variation of aligned DNA sequences among alfalfa weevil strains. Vertical numbers toward the top of the table refer to the nucleotide position within that sequenced region (. = identical; - = gap)

	12S (513bp)	16S (499bp)	Cytob (290bp)	ITS2 (518bp)
	1 1 2 3 3 3 3 4 4 4	2 2 2 2 2 3 3	1 1 1 1 1 2 2 2 2 2 2	3 3 3 3 3 3
	7 8 7 0 3 4 6 3 8 9	5 2 5 5 9 9 0 4	3 4 5 6 7 8 5 5 9 9 9 0 1 1 1 2 4 6	3 3 3 4 4 4
	8 0 1 4 8 5 0 8 2 0	3 3 7 9 5 6 7 7	0 8 4 3 5 7 4 5 5 7 9 3 0 8 9 2 0 1	7 8 9 0 1 2
AZ	GAGGCGTATG	CGGTCAGA	GTACGGTTACTTCCTCTC	ATATAT
MD	. . . .	. . . .	. . . . . . C.	. . . . . .
XC	. G. A. . – . . .	T. . . T. . G	. C. . . . . . G. . . . . . . T	––––––
BU	A. A. T A–GAA	.TAATTA.	AC. . . . . . . .T. . . . . . T	––––––
UT	A. A. T A–GAA	.TAATTA.	ACTAAA. C.TCATTCTCT	. . . . . .
HP	T. A. TA– . . .	TAT . TTT .	ACTTAA . A.TGATA.T.T	–. GCG.

GenBank accession No. (in order shown above):
AZ, U16960, U16966, U16972, U16978; **MD**, U16963, U16969, U16975, U16981;
XC, U16965, U16971, U16977, U16983; **BU**, U16961, U16967, U16973, U16979;
UT, U16964, U16970, U16976, U16982; **HP**, U16962, U16968, U16974, U16980.

(e) Genetic relationship between weevil strains

Aligned sequences of 1302 bp from the three mtDNA regions were combined to reconstruct gene trees. The ITS sequence data were not included in the analysis because they lacked sequence divergence in this species as compared with mitochondrial genes. Gene trees from both the neighbour-joining and PAUP programs were nearly identical in topology and branching patterns. Figure 3.1 represents a PAUP consensus tree generated by the branch and bound option with bootstrap values from 500 replicates. The high bootstrap values support the branching pattern of the gene tree. The Egyptian and eastern strains were closest allied (Table 3.2). The western strain was the most distinct from the other strains, followed by the Budapest strain. The Xinjiang strain was placed next to the Egyptian and eastern strains in the gene trees, which is consistent with the few nucleotide substitutions between these populations. There was extensive genetic divergence between alfalfa weevil and the clover leaf weevil (*H. punctata*). The gene tree generated by mtDNA sequence data was identical in topology to the tree produced by UPGMA (unweighted pair-group arithmetic average) method from allozyme data (Hsiao, 1993). The addition of the Budapest

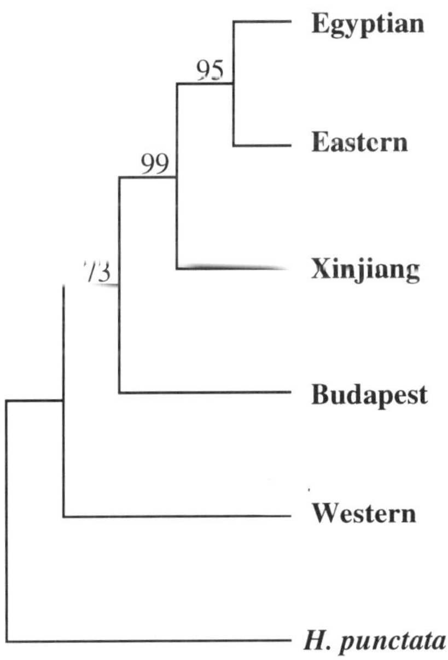

Figure 3.1 PAUP phylogenetic tree of alfalfa weevil strains generated from mtDNA sequence data using the branch and bound option, 50% consensus. Numbers are bootstrap values from 500 replications.

and Xinjiang strains in the gene tree further illustrated the extent of genetic divergence between the western strain and the Egyptian and eastern strains.

3.2.3 Implication of molecular data

The three alfalfa weevil strains in North America were introduced from the Old World during this century but the precise origin of these introductions is not known (Hsiao, 1993). The phylogenetic relationships generated from DNA data suggest a likely geographic origin of these weevils. The nearly identical DNA sequences of the eastern and Egyptian weevils imply that these two weevil strains probably came from the same region of southern Europe or Eurasia. The weevils from Xinjiang, the most easterly distribution of the alfalfa weevil in Asia, were genetically affiliated with the eastern and Egyptian weevils and probably came from Eurasia. The sequence similarity between the western and Budapest strains suggests a common geographic origin.

DNA sequence data are useful for the identification of alfalfa weevil strains (Table 3.2). The results of our study demonstrated that these molecular techniques accurately identify intraspecific variation among insects. A more practical approach involves combining PCR amplification of these mtDNA regions with restriction enzyme digestion to produce restriction fragment length polymorphism. The availability of sequence data on these mtDNA regions allows several choices of restriction enzymes to identify weevil strains. This technique has been used to identify Africanized and European honey bees (Crozier *et al.*, 1991).

3.3 WOLBACHIA ENDOSYMBIONTS AND CYTOPLASMIC INCOMPATIBILITY AMONG ALFALFA WEEVIL STRAINS

3.3.1 Studies of Wolbachia endosymbionts in insects

Reproductive incompatibility between North American weevil strains is caused by a rickettsia-like endosymbiont in the western weevil (Hsiao and Hsiao, 1985a,b). The endosymbiont, identified as *Wolbachia* sp. (*postica*), is found only in the western strain and is maternally inherited. While aposymbiotic (*Wolbachia*-free) eastern and Egyptian weevil are fully reproductively compatible, the crosses between females of eastern or Egyptian strains and males of western weevils are infertile. The reciprocal crosses produce viable progeny but have a female-biased sex ratio.

The alfalfa weevil was the first agricultural pest in which an endosymbiont was identified as a reproductive barrier. Other insect examples include mosquitoes, *Culex pipiens* Say and *Aedes scutellaris*

(Walker), and moths, *Ephestia cautella* (Walker) and *Cadra figulilella* (Gregson) (Hsiao, 1985a). *Wolbachia* and other rickettsia-like endosymbionts that induce cytoplasmic incompatibility have now been reported in over 20 insect species in five orders (Diptera, Lepidoptera, Coleoptera, Hymenoptera and Homoptera) and also in terrestrial isopods (Stevens and Wade, 1990; Rousset *et al.*, 1992a; Breeuwer and Werren, 1993; Wade and Stevens, 1994). Both unidirectional and bidirectional incompatibilities have been attributed to Wolbachiae. These endosymbionts have also been implicated in female-biased sex ratios, parthenogenesis and sterility. Based on 16S rRNA phylogenetic analysis, O'Neill *et al.* (1992) found that these endosymbionts belong to the α-subdivision of the proteobacteria. They considered the bacteria responsible for cytoplasmic incompatibility in different insect hosts (*C. pipiens, Tribolium confusum* Jacqueline duVal, *Hypera postica, Aedes albopictus* Skuse, *Drosophila simulans* (Sturtevant), *E. cautella*) to be closely related and provisionally classified them as members of the same species, *Wolbachia pipientis*. They also developed a PCR-based assay using species-specific oligonucleotide primers that distinguish the *Wolbachia* from other members of the α-Proteobacteria. The assay has been used for detection of endosymbiont infection in *Drosophila* species (O'Neill *et al.*, 1992; Rousset *et al.*, 1992b; Boyle *et al.*, 1993; Braig *et al.*, 1994) and *Nasonia* wasp (Breeuwer *et al.*, 1992). Horizontal transfers of *Wolbachia* have been conducted by microinjection of cytoplasm from symbiotic to aposymbiotic host strains, species, and even distantly related taxa (e.g. *Drosophila* and mosquitoes) (Boyle *et al.*, 1993; Williams *et al.*, 1993; Braig *et al.*, 1994; Rousset and de Stordeur, 1994). The endosymbionts can be eliminated from infected hosts by treatment with antibiotics (Stevens and Wade, 1988; Hsiao, 1993; Braig *et al.*, 1994).

3.3.2 Detection of Wolbachia endosymbionts among alfalfa weevil strains

The Giemsa staining technique was used initially to detect *Wolbachia* among alfalfa weevil strains. This assay involved the dissection of ovarian nurse cells, preparation and staining of microscopic slides, and visual inspection under a microscope (Hsiao and Hsiao, 1985a). The technique was not as reliable in males because of their low level of infection. Using this technique, our laboratory has surveyed the occurrence of *Wolbachia* among weevil populations in the United States and identified the mixing zones between eastern and western, and western and Egyptian strains (Hsiao, 1993). The mixing zone between eastern and western weevils now includes central Nebraska and Kansas to eastern Colorado. Two mixing zones between western and Egyptian weevils have been detected; one in southern Utah bordering Nevada and Arizona and the other in the four-corner area of Utah, Colorado, Arizona and New Mexico.

An immunoblotting assay using polyclonal antibody was developed to improve the reliability and sensitivity of *Wolbachia* detection (Leu *et al.*, 1989). This assay can detect infection in females, males and immature stages, but proteins must be extracted from live or frozen specimens, a time-consuming procedure.

We have adopted a PCR-based *Wolbachia* detection method for the alfalfa weevil. The technique used 16S rRNA primers designed by O'Neill *et al.* (1992) (Table 3.1). The primer 99F sequence is from the variable V1 region and corresponds to *Escherichia coli* position 76–99 forward, and the 994R primer is from the reverse complement of the variable V6 region and corresponds to *E. coli* position 1012–994 reverse. DNAs were extracted from individual weevils (Hsiao, 1994a) and standard PCR was performed using a temperature profile of 93°C (35 s), 52°C (35 s), and 72° (1 min) for 30 cycles. Agarose gel electrophoresis revealed a single PCR-amplified product with a fragment size slightly less than 1000 bp, which is identical to that reported by O'Neill *et al.* (1992) for *W. pipientis*. The PCR detection is extremely sensitive and can be used for all weevil stages. Weevils preserved in alcohol were also suitable for the PCR-based assay.

We compared the three *Wolbachia* detection methods among different weevil strains and geographic populations (Table 3.3). All detected *Wolbachia* among North American weevil populations. Thus, the western weevils were completely symbiotic, while the eastern and Egyptian weevils from different geographic areas were aposymbiotic. However, some variations existed in the mixing zones due to the presence of different levels of *Wolbachia* infection among individuals. The Giemsa staining showed more variable results between samples. At extremely low level of infection, the Giemsa method is not as sensitive as the PCR assay. The immunoblotting method is highly specific showing a 135-kDa protein band in the symbiotic western weevil (Leu *et al.*, 1989). However, the test requires a considerable amount of proteinaceous materials. The PCR method requires minute amount of tissues and is well suited for studying weevils in the mixing zones.

The Xinjiang weevils showed a positive *Wolbachia* infection with the Giemsa and PCR methods (Table 3.3). The immunoblotting test was negative for the 135 kDa protein band but showed a smaller band of about 100 kDa (Figure 3.2). The weevils from Wageningen, The Netherlands, showed both immunoreactive protein bands. These data suggest that more than one strain of *Wolbachia* exists in the alfalfa weevil. The PCR method detected 25% aposymbiotic individuals in another sample of the Xinjiang weevil, indicating that it is a mixed population (Table 3.3). In the mixing zones, about 85% of the weevils were symbiotic in

Table 3.3 Comparison of three techniques for detecting *Wolbachia* among alfalfa weevil populations

Strain	Source	Giemsa staining		Immunoblotting		PCR detection	
		No. examined (♀)	% positive	No. examined (♀ + ♂)	% positive	No. examined (♀ + ♂)	% positive
Western	Logan, UT	30	100	46	100*	20	100
Eastern	Niles, MI	30	0	35	0	–	–
Eastern	Beltsville, MD	30	0	–	–	20	0
Egyptian	Yuma, AZ	31	0	–	–	20	0
Egyptian	Blythe, CA	30	0	25	0	–	–
Egyptian	Las Vegas, NV	30	0	20	0	20	0
Eastern Europe	Budapest, Hungary	33	100	31	100‡	–	–
Western Europe	Wageningen, Netherland	31	100	27	100‡	–	–
Asia	Xinjiang, China	31	100	40	100†	20	75
Western mixing zone	St. Geoge, UT	47	87.2	47	83*	20	95
Eastern mixing zone	Julesburg, CO	71	22.8	10	20*	20	10

*Exhibiting an immunoreactive protein band of 135 kDa.
†Exhibiting an immunoreactive protein band of ~100 kDa rather than 135 kDa.
‡Exhibiting both ~100 and 135 kDa immunoreactive protein bands.

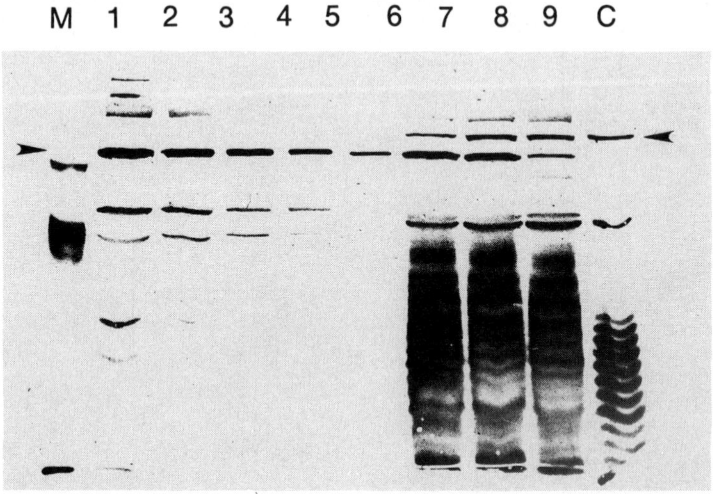

Figure 3.2 Immunoblotting detection of Wolbachiae from alfalfa weevils. Lane M represents the rainbow protein market. Lane C is from the partial purified Wolbachiae of western weevils used as a positive control; the right arrow indicates the 135-kDa immunoreactive protein band. Lanes 1–6 represent serial twofold dilution of a weevil homogenate from Xinjiang, China; the left arrow indicates an immunoreactive protein band of ~100 kDa. Note the absence of the 135-kDa band. Lanes 7–9 are weevil homogenates from Wageningen, The Netherlands showing both immunoreactive protein bands.

the St George, Utah, weevils, while only 10–20% were symbiotic in the Julesburg, Colorado, weevils (Table 3.3). The fact that both Giemsa and immunoblotting tests detected symbiotic weevils in the Wageningen population is consistent with the hypothesis that the western strain originated from western Europe (Hsiao, 1993).

3.3.3 Cross-compatibility among weevil strains

Wolbachia causes cytoplasmic incompatibility between North American weevil strains. The discovery of a second *Wolbachia* in the XC weevils prompted us to examine its cross-compatibility with other weevil strains. Cross-breeding tests were conducted between the XC weevil and the three North American weevil strains as described by Hsiao and Hsiao (1985b). The cross between XC females and western males was infertile, but the reciprocal cross produced viable progeny showing the typical unidirectional incompatibility (Table 3.4). Surprisingly, crosses between XC and the aposymbiotic eastern and Egyptian strains were

Table 3.4 Cross-compatibility between North American weevil strains and weevils from Xinjiang, China

Cross (♀ × ♂)	*Total no. of eggs*	*% Fertile*	*% Hatch*
Western A × Xinjiang	1128	55.8	53.3
Xinjiang × Western A	1249	6.2	2.7
Xinjiang × Western B	999	0	0
Egyptian × Xinjiang	1885	64.9	52.8
Xinjiang × Egyptian	1410	84.4	71.4
Eastern × Xinjiang	1851	57.8	51.7
Xinjiang × Eastern	565	92.4	86.4

Western A, St. George, UT: Western B, Logan, UT; Egyptian, Las Vegas, NV; Eastern, Beltsville, MD.

fully compatible, an indication that the *Wolbachia* in the XC weevils does not cause cytoplasmic incompatibility.

Wolbachia symbiotic and aposymbiotic insect populations have been reported in *Drosophila* (Turelli and Hoffman, 1991) and *Tribolium* (Stevens and Wade, 1990; Stevens and Wicklow, 1992). In *D. simulans*, as many as three different symbiotic Wolbachiae cause bidirectional cytoplasmic incompatibility between conspecific populations (O'Neill and Karr, 1990; Montchamp-Moreau *et al.*, 1991).

3.4 INTERACTIONS BETWEEN WOLBACHIA AND PARASITOIDS

3.4.1 Biological control of the alfalfa weevil

Because the alfalfa weevil is an exotic pest in North America, natural enemies have to be introduced for its biological control. Over the years, some 20 parasitoids have been released, but only a few have become established (Day, 1981; Bryan *et al.*, 1993). The earliest attempt was the introduction of several parasitoids into Utah from Europe during 1911–1914 to control the western weevil (Chamberlin, 1924). Among these, the hymenopterous parasitoid, *Bathyplectes curculionis* (Thomson), was the only larval parasitoid to become well established in western United States. It was subsequently introduced into the habitats of Egyptian and eastern weevil strains through releases and natural dispersal (Dysart and Day, 1976; Day, 1981; Bryan *et al.*, 1993). However, *B. curculionis* was ineffective against the Egyptian and eastern weevils because most of its eggs were encapsulated by these weevils, preventing the development of the parasitoid (Maund and Hsiao, 1991).

Between 1957 and 1979, 12 parasitoid species were imported and released in the eastern United States against the alfalfa weevil (Dysart and Day, 1976; Bryan *et al.*, 1993). Of these, six parasitoids (*B. annurus* (Thomson), *B. curculionis, B. stenostigma* (Thompson), *Microctonus aethiopoides* Loan, *M. colesi* Drea and *Tetrastichus incertus* (Ratzeburg)) became established (Day, 1981). These parasitoids have since been released in many western states where western and Egyptian weevil strains occur. The adult parasitoid, *M. aethiopoides*, has been established in the eastern United States since 1970 and is the most effective natural enemy against the eastern weevil (Day, 1981; Bryan *et al.*, 1993). It failed to establish in areas occupied by the western weevils in spite of repeated releases, apparently due to differences in biotypes of the alfalfa weevil. We have investigated the roles of *Wolbachia* in connection with parasitism by *M. aethiopoides* and *B. curculionis*.

3.4.2 Effects of Wolbachia on parasitism

Studies on *B. curculionis* evaluated egg encapsulation among alfalfa weevil strains in the laboratory and among field populations from weevil mixing zones (Maund and Hsiao, 1991). The laboratory study also determined whether the *Wolbachia* of the western weevil plays a role in the absence of egg encapsulation by *B. curculionis*. There were significant differences in egg encapsulation rates among weevils from different regions. There was no encapsulation of *B. curculionis* in the western strain, a significant level of encapsulation (14%) in the eastern strain, and nearly complete encapsulation in the Egyptian strain (96.1%) (Maund and Hsiao, 1991). An intermediate level of encapsulation existed in the weevil populations from zones in which western and Egyptian strains overlap in southern Utah, and between eastern and western strains in Colorado, but no egg encapsulation occurred in western weevils that were free of *Wolbachia* (laboratory-reared and antibiotic-treated). This provides conclusive evidence that encapsulation of *B. curculionis* eggs is due to biological differences among weevil strains and is not related to the presence of *Wolbachia*.

A similar study was conducted to evaluate the suitability of weevil strains as hosts for the adult parasite *M. aethiopoides*. Laboratory rearings indicated that both eastern and Egyptian weevils were suitable hosts of the parasite (Table 3.5), but the western weevil was unsuitable. The parasites reared from western weevils spun small and thin cocoons and many died during the prepupal and pupal stages. Those parasites (10.8%) that survived to adults were unable to reproduce. Cocoon lengths reliably indicated suitability of a host (Figure 3.3). Typically, parasite cocoons reared from eastern and Egyptian weevils were from 3–4 mm long and most developed into normal adults (Figure 3.3, shaded bar). In western weevils, most cocoons were less than 3 mm and did not yield viable adults. The failure of *M. aethiopoides* to develop

Table 3.5 Development of *Microctonus aethiopoides* (California stocks) in three alfalfa weevil strains. W+, *Wolbachia* symbiotic; W–, *Wolbachia* aposymbiotic

Host strain	*No. parasitized*	*Cocoon (%)*	*Adult (%)*	*Cocoon size (mm)* ± S.E.*
Western, W+	74	59.5	10.8	2.7 ± 0.04a
Western, W–	113	97.3	82.3	3.4 ± 0.02b
Eastern	80	88.8	56.3	3.4 ± 0.03b
Egyptian	139	87.8	69.1	3.5 ± 0.02b

*Duncan's multiple range test. Means followed by same letters are not significant at the 5% level.

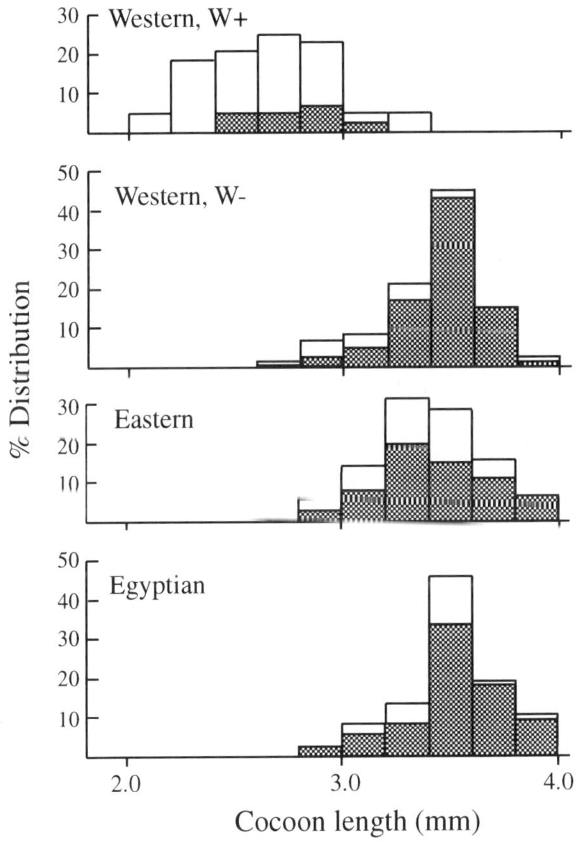

Figure 3.3 Frequency distribution of cocoon sizes of *Microctonus aethiopoides* reared from three alfalfa weevil strains. Shaded areas indicate the proportion of adults emerged from each cocoon-size group. W+, *Wolbachia* symbiotic; W–, *Wolbachia* aposymbiotic.

normally in the western weevil was related to the *Wolbachia*. When aposymbiotic western weevils (laboratory-reared and antibiotic-treated) were the hosts, the parasite developed normally (Table 3.5; Figure 3.3). Apparently, the *Wolbachia* somehow alters the physiological conditions of the weevil to make it unsuitable for the development of the parasite, thereby contributing to the weevil's defence against an insect parasitoid.

Another instance concerning the involvement of microorganisms in insect parasitism has been demonstrated. A virus is implicated as a suppressing agent of cellular immune defence in the tobacco budworm, *Heliothis virescens* (F.), thus facilitating the development of its parasitoid, *Campoletis sonorensis* (Cameron) (Vinson, 1993). This virus is transmitted from the parasite to its host during oviposition. In contrast, the *Wolbachia* increases the defence reaction of the western alfalfa weevil against its parasite.

3.5 CONCLUSION

The alfalfa weevil is a Palearctic pest of *Medicago* and related legumes. Its colonization in the Nearctic region during this century has greatly expanded its geographic distribution and elevated it into a cosmopolitan pest of alfalfas. Genetic and molecular studies of geographic populations from Asia, Europe and North America have revealed considerable genetic and ecological differentiation among populations. Based on their geographic origin, previous taxonomic treatments have placed these populations into several species, a view consistent with the discovery of a unidirectional reproductive incompatibility between eastern and western weevil strains (Blickenstaff, 1965). The characterization of the *Wolbachia* in the western weevil and cross-compatibility tests between weevil populations have clarified the taxonomy of this insect. The reproductive compatibility between weevil populations from China, Europe and North America indicates that they belong to the same species. The similarity of allozyme and mtDNA sequence data between these weevil populations further supports this contention, and should prompt a re-evaluation of morphological traits for identification of the alfalfa weevil complex.

The association between the alfalfa weevil and *Wolbachia* endosymbionts is an intriguing phenomenon both from basic and applied perspectives. *Wolbachia*-induced unidirectional incompatibility provides a reproductive barrier between weevil strains in North America. Since only the cross with symbiotic female weevils produces viable progeny and the *Wolbachia* is maternally inherited, the aposymbiotic eastern and Egyptian weevils are at a disadvantage and will be replaced gradually by the western weevil in areas where these strains are mixing. A similar phenomenon occurred with *Drosophila simulans* in California, where an aposymbiotic has been replaced by a symbiotic strain (Turelli and Hoffman, 1991). At present, there is no large-scale displacement by the western strain in the mixing zones between western and Egyptian strains in southern Utah, and between western and eastern strains in eastern Colorado (Hsiao, 1993 and T. Hsiao, unpublished data). The ineffective displacement may be related to ecological, physiological, and behavioural differences among weevil strains. For example, the western weevil did not adapt well to the hot arid environment, as shown by its slow spread in central California in 1930–1940, and was subsequently displaced by the northward spread of the Egyptian weevil in 1950–1960 (Michelbacher and Leighly, 1940; van den Bosch and Marble, 1971). Also, the eastern and Egyptian weevils are larger, and more fecund than the western weevil. Female weevils mate repeatedly during an oviposition period of several weeks in the field. Hence, an incompatible mating could be followed by a compatible intrastrain mating, thereby restoring fertility. Under these conditions, interstrain hybridization would have little effect on the coexistence of the two strains.

The geographic origin and distribution of the *Wolbachia* symbionts also warrant study. As in other insects, the symbiont is transmitted maternally. In North America, only the western weevil harbours the symbiont. In Europe and Asia, both symbiotic and aposymbiotic individuals were found within the same population, probably due to the presence of more than one *Wolbachia* strain, as occurs in weevils from Xinjiang, China, Wageningen, The Netherlands and Budapest, Hungary. Horizontal transmission of *Wolbachia* between insect species has been demonstrated experimentally in *Drosophila* and mosquitoes and is suggested as a mechanism of natural infection (Boyle *et al.*, 1993; Williams *et al.*, 1993; Braig *et al.*, 1994; Rousset and Stordeur, 1994). We have not detected horizontal transmission of *Wolbachia* in the alfalfa weevil in our survey.

The effect of *Wolbachia* on insect parasitoids illustrates another intriguing biotic interaction. In this instance, *Wolbachia* confers protection to the western weevil against the parasitoid, *M. aethiopoides*, and thus prevents the parasitoid's establishment in the western United States. In biological control practices, many attempts to colonize natural enemies have failed and few have been studied. The biotic interaction cited above demonstrates the need to match biotypes of the pest and its parasitoids. Our study of interactions between alfalfa weevil strains, *Wolbachia*, and parasitoids illustrates the complexity of these biotic relationships and will aid the development of a more effective pest management strategy.

Acknowledgements

I thank Catherine Hsiao for technical assistance in DNA sequencing, Professor Liu Fang-Zin of August 1st Agricultural College, Wulumugi,

Xinjiang, China for weevils, Dr Sy-Jye Christine Leu for immunoblotting tests, and Dr Scott L. O'Neill for *Wolbachia* primers. This research was supported by the Utah Agricultural Experiment Station, Utah State University, Logan, UT 84322-4810 and approved as journal paper no. 4701.

REFERENCES

Blickenstaff, C.C. (1965) Partial intersterility of eastern and western U.S. strains of the alfalfa weevil. *Annals of the Entomological Society of America*, **58**, 523–6.

Boyle, L., O'Neill, S.L., Robertson, H.M. and Karr, T.L. (1993) Interspecific and intraspecific horizontal transfer of *Wolbachia* in *Drosophila*. *Science*, **260**, 1796–9.

Braig, H.R., Guzman, H., Tesh, R.B. and O'Neill, S.L. (1994) Replacement of the natural Wolbachia symbiont of *Drosophila simulans* with a mosquito counterpart. *Nature*, **367**, 453–5.

Breeuwer, J.A. and Werren, J.H. (1993) Effect of genotype of cytoplasmic incompatibility between two species of *Nasonia*. *Heredity*, **70**, 428–36.

Breeuwer, J.A., Stouthamer, R., Barns, S.M., Pelletier, D.A., Weisburg, W.G. and Werren, J.H. (1992) Phylogeny of cytoplasmic incompatibility micro-organisms in the parasitoid wasp genus *Nasonia* (Hymenoptera: Pteromalidae) based on 16S ribosomal DNA sequences. *Insect Molecular Biology*, **1**, 25–36.

Bryan, M.D., Dysart, R.J. and Burger, T.L. (1993) Releases of introduced parasites of the alfalfa weevil in the United States, 1957–88. United States Department of Agriculture, APHIS, Miscellaneous Publication No. 1504, 203 pp.

Chamberlin, T.R. (1924) Introduction of parasites of the alfalfa weevil into the United States. United States Department of Agriculture, Circular 301, 9 pp.

Clary, D.O. and Wolstenholme, D.R. (1985) The mitochondrial DNA molecule of *Drosophila yakuba*: nucleotide sequence, gene organization, and genetic code. *Journal of Molecular Evolution*, **22**, 252–71.

Crozier, Y.C., Koulianos, S. and Crozier, R.H. (1991) An improved test for Africanized honeybee mitochondrial DNA. *Experientia*, **47**, 968–9.

Csiki, E. (1934) Coleoptera, Curculionidae, Subfamily Hyperinae, in *Coleopterorum Catalogus*, (ed. S. Schenkling), W. Junk, Berlin, pp. 1–66.

Day, W.H. (1981) Biological control of the alfalfa weevil in the northeastern United States, in *Biological Control in Crop Protection, Beltsville Symposia in Agricultural Research*, Vol. 5, (ed. G.C. Papavizas), Allanheld, Osmun & Co. Publishers, London, pp. 361–74.

Dysart, R.J. and Day, W.H. (1976) The release and recovery of introduced parasites of the alfalfa weevil in eastern North America. United States Department of Agriculture, Production Research Report 167, Washington, DC, 61 pp.

Higgins, D.G., Bleasby, A.J. and Fuchs, R. (1992) Clustal V: improved software for multiple sequence alignment. *Computer Application in Bioscience (CABIOS)*, **8**, 189–91.

Hsiao, T.H. (1988) Physiological and biochemical studies of insect biotypes, relevance to pest management. Symposium on Insect Biochemistry and Physiology, Institute of Zoology, Academia Sinica, Taipei, Taiwan, Symposium **13**, National Science Council, Taipei, Taiwan, pp. 1–9.

Hsiao, T.H. (1989) Estimation of genetic variability amongst Coleoptera, in *Electrophoretic Studies on Agricultural Pests*, (eds H.D. Loxdale and J. Den Hollander), Clarendon Press, Oxford, UK, pp. 143–80.

Hsiao, T.H. (1993) Geographic and genetic variation among alfalfa weevil strains, in *Evolution of Insect Pests: Patterns of Variation*, (eds K.C. Kim and B.A. McPheron), John Wiley & Sons, Inc., New York, pp. 311–27.

Hsiao, T.H. (1994a) Molecular techniques for studying systematics and phylogeny of Chrysomelidae, in *Novel Aspects of Biology of Chrysomelidae*, (eds P. Jolivet, P. Petitpierre and M.L. Cox), Kluwer Academic Publishers, Dordrecht, NL, pp. 237–48.

Hsiao, T.H. (1994b) Molecular phylogeny of chrysomelid beetles inferred from mitochondrial DNA sequence data (Coleoptera: Chrysomelidae), in *Proceedings of the 3rd International Symposium on the Chrysomelidae*, (ed. D. Furth), Backhuys Publishers, Leiden, NL, pp. 9–17.

Hsiao, C. and Hsiao, T.H. (1984) Cytogenetic studies of alfalfa weevil (*Hypera postica*) strains (Coleoptera: Curculionidae). *Canadian Journal of Genetics and Cytology*, **26**, 348–53.

Hsiao, C. and Hsiao, T.H. (1985a) Rickettsia as the cause of cytoplasmic incompatibility in the alfalfa weevil, *Hypera postica* (Gyllenhal). *Journal of Invertebrate Pathology*, **45**, 244–6.

Hsiao, T.H. and Hsiao, C. (1985b) Hybridization and cytoplasmic incompatibility among alfalfa weevil strains. *Entomologia Experimentalis et Applicata*, **37**, 155–9.

Hsiao, T.H. and Stutz, J.M. (1985) Discrimination of alfalfa weevil strains by allozyme analysis. *Entomologia Experimentalis et Applicata*, **37**, 113–21.

Innis, M.A., Gelfand, D.H., Sninsky, J.J. and White, T.J. (eds) (1990) *PCR Protocols: a Guide to Methods and Applications*, Academic Press, NY.

Kumar, S., Tamura, K. and Nei, M. (1993) *MEGA: Molecular Evolutionary Genetics Analysis*, version 1.0. Penn State University, University Park, PA.

Leu, C.S.-S., Li, J.K.-K. and Hsiao, T.H. (1989) Characterization of *Wolbachia postica*, the cause of reproductive incompatibility among alfalfa weevil strains. *Journal of Invertebrate Pathology*, **54**, 248–59.

Maund, C.M. and Hsiao, T.H. (1991) Differential encapsulation of two *Bathyplectes* parasitoids among alfalfa weevil strains, *Hypera postica* (Gyllenhal). *Canadian Entomologist*, **123**, 197–203.

Menken, S.B.J. (1989) Electrophoretic studies on geographic populations, host races, and sibling species of insect pests, in *Electrophoretic Studies on Agricultural Pests*, (eds H.D. Loxdale and J. Den Hollander), Clarendon Press, Oxford, UK, pp. 181–202.

Menken, S.B.J. and Raijmann, L.E.L. (1996) Biochemical systematics: principles and perspectives for pest management, in *The Ecology of Agricultural Pests: Biochemical Approaches*, (eds W.O.C. Symondson and J.E. Liddell), Chapman & Hall, London, pp. 7–29.

Michelbacher, A.E. and Leighly, J. (1940) The apparent climatic limitations of the alfalfa weevil in California. *Hilgardia*, **13**, 101–39.

Montchamp-Moreau, C., Ferveur, J.F. and Jacques, M. (1991) Geographic distribution and inheritance of three cytoplasmic incompatibility types in *Drosophila simulans*. *Genetics*, **129**, 399–407.

Mullis, K.B., Ferre, F. and Gibbs, R.A. (eds) (1994) *The Polymerase Chain Reaction*, Birkhauser, Boston.

O'Neill, S.L. and Karr, T.L. (1990) Bidirectional incompatibility between conspecific populations of *Drosophila simulans*. *Nature*, **348**, 178–80.

O'Neill, S.L., Giordano, R., Colbert, A.M.E., Karr, T.L. and Robertson, H.M. (1992) 16S rRNA phylogenetic analysis of the bacterial endosymbionts associated with cytoplasmic incompatibility in insects. *Proceedings of The National Academy of Sciences USA*, **89**, 2699–702.

Pruess, K.P., Zhu, X. and Powers, T.O. (1992) Mitochondrial transfer RNA genes in a black fly, *Simulium vittatum* (Diptera: Simuliidae), indicate long divergence from mosquito (Diptera: Culicidae) and fruit fly (Diptera: Drosophilidae). *Journal of Medical Entomology*, **29**, 644–51.

Rousset, F. and de Stordeur, E. (1994) Properties of *Drosophila simulans* strains experimentally infected by different clones of the bacterium Wolbachia. *Heredity*, **72**, 325–31.

Rousset, F., Bouchon, D., Pintureau, B., Juchault, P. and Solignac, M. (1992a) Wolbachia endosymbionts responsible for various alterations of sexuality in arthropods. *Proceedings of The Royal Society of London Series B. Biological Sciences*, **250**, 91–8.

Rousset, F., Vautrin, D. and Solignac, M. (1992b) Molecular identification of *Wolbachia*, the agent of cytoplasmic incompatibility in *Drosophila simulans*, and variability in relation with host mitochondrial types. *Proceedings of The Royal Society of London Series B. Biological Sciences*, **247**, 163–8.

Simon, C. (1991) Molecular systematics at the species boundary: exploiting conserved and variable regions of mitochondrial genome of animals via direct sequencing from amplified DNA, in *Molecular Techniques in Taxonomy*, (eds. G.M. Hewitt, A.W.B. Johnston and J.P.W. Young), NATO Advanced Studies Institutes, Springer-Verlag, H57, pp. 33–71.

Simon, C., Franke, A. and Martin, A. (1991) The polymerase chain reaction: DNA extraction and amplification, in *Molecular Techniques in Taxonomy*, (eds G.M. Hewitt, A.W.B. Johnston and J.P.W. Young), NATO Advanced Studies Institutes, Springer Verlag, H57, pp. 329–55.

Simon, C., Frati, F., Flook, P., Beckenbach, A., Crespi, B. and Liu, H. (1994) Assessing the phylogenetic usefulness of specific mitochondrial genes and a compilation of conserved PCR primers. *Annals of Entomological Society of America*, **87**, 1–57.

Stevens, L. and Wade, M.J. (1988) Effect of antibiotics on the productivity of genetic strains of *Tribolium confusum* and *Tribolium castaneum* (Coleoptera: Tenebrionidae). *Environmental Entomology*, **17**, 115–19.

Stevens, L. and Wade, M.J. (1990) Cytoplasmically inherited reproductive incompatibility in *Tribolium* flour beetles: the rate of spread and effect on population size. *Genetics*, **124**, 367–72.

Stevens, L. and Wicklow, D.T. (1992) Multispecies interactions affect cytoplasmic incompatibility in *Tribolium* flour beetles. *American Naturalist*, **140**, 642–53.

Swofford, D.L. (1991) *PAUP, phylogenetic analysis using parsimony*, Version 3.0s. for Apple Macintosh. Illinois Natural History Survey, Champaign, IL.

Titus, E.G. (1911) The genera *Hypera* and *Phytonomus* (Coleoptera, family Curculionidae) in America, north of Mexico. *Annals of the Entomological Society of America*, **4**, 383–494.

Turelli, M. and Hoffman, A.A. (1991) Rapid spread of an inherited incompatibility factor in California *Drosophila. Nature*, **353**, 440–2.

Van den Bosch, R. and Marble, V.L. (1971) The Egyptian alfalfa weevil, the threat to California alfalfa. *California Agriculture*, **25**, 3–4.

Vinson, S.B. (1993). The suppression of the insect immune system by parasitic Hymenoptera, in *Insect Immunity*, (ed. J.P.N. Pathak), Kluwer Academic Publishers, Dordrecht, The Netherlands.

Wade, M.J. and Stevens, L. (1994) The effect of population subdivision on the rate of spread of parasite-mediated cytoplasmic incompatibility. *Journal of Theoretical Biology*, **167**, 81–7.

Wehrle, L.P. (1940) The discovery of an alfalfa weevil (*Hypera brunneipennis* Boheman) in Arizona. *Journal of Economic Entomology*, **33**, 119–21.

White, T.J., Bruns, T., Lee, S. and Taylor, J. (1990) Amplification and direct sequencing of fugal ribosomal RNA genes for phylogenetics, in *PCR Protocols: a Guide to Methods and Applications*, (eds M.A. Innis, D.H. Gelfand, J.J. Sninsky and T.J. White), Academic Press, San Diego, pp. 315–22.

Williams, E.H., Fields, S. and Saul, G.B. II, (1993) Transfer of incompatibility factors between stocks of *Nasonia (=Mormoniella) vitripennis*. *Journal of Invertebrate Pathology*, **61**, 206–10.

Xiong, B. and Kocher, T.D. (1991) Comparison of mitochondrial DNA sequences of seven morphospecies of black flies (Diptera: Simuliidae). *Genome*, **34**, 306–11.

Zaslavskiy, V.A. (1961) Review of the species of the genus *Phytonomus* Schonh. (Coleoptera: Curculionidae) in the fauna of the USSR. *Entomological Review*, **40**, 342–8.

4

Biochemical approaches to the study of ecological genetics: the role of selection and gene flow in the evolution of insecticide resistance

J.C. Daly and S. Trowell

ABSTRACT

The selection and spread of insecticide resistance is an evolutionary phenomenon. Resistance becomes a problem for pest control when an allele conferring resistance increases in frequency within populations and then spreads more widely. Ecological genetics studies the interaction between population structure and the factors – mutation, selection, genetic drift and gene flow – that produce these changes. Traditionally, resistance studies identified only two phenotypes, susceptible or resistant, and made simplifying assumptions about the inheritance of resistance. Such an approach has impeded our ability to test alternative hypotheses about the evolution of resistance because the resistance phenotype can involve more than one major gene; the genetic basis of resistance can change over time; and the expression of resistance can vary throughout the life-cycle. More detailed evolutionary studies of resistance require suitable tools. Biochemical techniques, such as electrophoresis, antibody, protein and DNA-based technologies, provide the means to identify unambiguously gene products, or genes and their allelic variants. The techniques can address a wide range of issues such as: separation of sibling species, detection and monitoring of resistant individuals, and the distribution of resistance genes in populations.

The Ecology of Agricultural Pests, Edited by W.O.C. Symondson and J.E. Liddell.
Published in 1996 by Chapman & Hall, London. ISBN 0 412 62190 8

Biochemical markers can also be used to make inferences about population subdivision and gene flow. Each biochemical approach, however, can have its own limitations and these must be understood if the appropriate tool is to be deployed. These points can be illustrated from studies of pesticide resistance in *Helicoverpa armigera*.

4.1 INTRODUCTION

Insecticide resistance is a serious and widespread problem in insect pests. Resistance can reduce the effectiveness of pest control and lead farmers to use alternative pesticides that may be more expensive, or have longer residual life or broad-spectrum activity. Unfortunately, some of the world's major agricultural, veterinary and medical insect pests have developed resistance to a wide range of chemical pesticides. For example, the diamond back moth, *Plutella xylostella* L., is resistant to all chemical insecticides that have been used for its control (Sun *et al.*, 1985), and recently has developed resistance to the bioinsecticide, *Bacillus thuringiensis* Berliner (Tabashnik, 1994).

Ecology is the study of factors that influence the distribution and abundance of organisms. Ecological studies benefit our understanding of resistance in many ways. They can identify key factors in insect life-histories that influence evolution of resistance and the conditions under which resistance poses problems for pest control (Daly, 1994). These factors include dispersal rates, host preferences, overwintering strategies, age structure of populations and so on. Ecological considerations are implicit in the management of resistance because strategies are designed to manipulate mortality factors or to enhance the effects of immigration (May and Dobson, 1986; Roush and Croft, 1986). In the future, ecological studies will be needed to identify critical insect–plant interactions, such as the feeding patterns of the insect or interplant movement, that will determine if genetically engineered plants can be deployed without selecting for resistance (Daly, 1994).

The aim of this review is to assess the contributions of different biochemical methods to the study of the ecological genetics of insecticide resistance using Australian heliothine pests as an example. Biochemical methods can be used in a myriad of ways to study resistance so we will limit our review to three specific applications: separation of sibling species, measurement of population subdivision and gene flow, and detection and monitoring of resistant individuals. We will show that some of the newer techniques allow greater precision in identifying genes. However, this precision is not always justified in terms of cost or information obtained so that the more conventional biological and biochemical methods (such as such bioassays and isozyme electrophoresis) still have a place.

4.1.1 Insecticide Resistance in Heliothine Pests

The caterpillar stages of *Helicoverpa armigera* (Hübner) (Lepidoptera: Noctuidae) and *H. punctigera* (Wallengren) are pests of broad acre crops in the tropics and sub-tropics of Australia. *H. armigera* is also a pest in Asia and Africa. In North America, two other heliothine species, *Heliothis virescens* (F.) and *Helicoverpa zea* (Boddie), occupy similar ecological niches. In high-value crops such as cotton and grain legumes, pest outbreaks are controlled with frequent use of broad-spectrum insecticides. Resistance has evolved to a number of these insecticide groups in *H. armigera* and *Heliothis virescens* but not in *H. punctigera* and only to a limited extent in *H. zea* (McCaffery *et al.*, 1989; Plapp *et al.*, 1990; Forrester *et al.*, 1993). The evolutionary response of the resistant versus susceptible species is thought to depend on different aspects of their life-history strategies (Daly, 1993; Forrester *et al.*, 1993).

Management of resistance has been undertaken successfully with both Australian *H. armigera* (Forrester *et al.*, 1993) and American *Heliothis virescens* (Plapp *et al.*, 1990). In Australia, the rotational use of insecticide groups with unrelated modes of action has been encouraged in the management programs for *H. armigera*. This 'strategy' is supported by a monitoring programme by research into the underlying evolutionary, ecological and biochemical processes associated with resistance (Fitt, 1989; Gunning and Easton, 1989; Gunning *et al.*, 1991; Forrester and Fitt, 1992; Daly, 1993; Forrester *et al.*, 1993).

4.2 SPECIES IDENTIFICATION – ISOENZYMES VERSUS ANTIBODIES

The first major impediment to an understanding of resistance in *H. armigera* is a practical one. *H. armigera* is difficult to distinguish morphologically from its sibling *H. punctigera*. Zalucki and coworkers (1986) summarized the available information on the taxonomy and distribution of the two pest species: they are sympatric both in space and time, although *H. punctigera* is more common in spring and *H. armigera* is more prevalent in summer and autumn; they differ in their host range but they share some major hosts such as cotton and sunflowers; in laboratory cultures the two species can be readily distinguished morphologically in middle to late instar larval, pupal and adult life-stages. In the field, however, identity of eggs, larvae and pupae is either impossible or unreliable using morphological characteristics (Daly and Gregg, 1985; Zalucki *et al.*, 1986). Adults can be distinguished from their wing markings, although even these differences are hard to see if individuals have been damaged in light- or pheromone traps (Twine and Kay, 1973). Dissection of genitalia is then needed to distinguish the two species.

The similarity of the two species morphologically not only limits ecological research, particularly in life-table studies of early life-stages, it also makes it difficult to monitor for resistance quickly and inexpensively. At present, eggs are collected from the field and then reared for 10 days until their species can be determined. The *H. armigera* individuals are then screened for resistance using a standard topical dose bioassay (Gunning *et al.*, 1984). This process is inefficient until midsummer when *H. armigera* becomes abundant – during spring many *H. punctigera* are reared unnecessarily in order that sufficient *H. armigera* are screened to give a reliable estimate of resistance (Forrester *et al.*, 1993). A second problem is a more practical one for the farmer. Pest infestations of *H. punctigera* can be controlled using insecticides that are less expensive and less damaging to the environment than those chosen for *H. armigera* control. However, in the absence of reliable information on species identity, he or she can only make educated guesses about the species present, based on the time of year. Yet even within a single week, species composition of egg lays can vary from 10–100% *H. armigera* in sites separated by no more than 4 km (Daly and Fitt, 1993).

4.2.1 Isoenzymes

A variety of biochemical methods could be used to separate the species. In this case, enzyme electrophoresis and an immunosquash technique using monoclonal antibodies have been used. Daly and Gregg (1985) used cellulose acetate gel electrophoresis to identify the species at all life-stages. The two species differ at seven of the 28 isoenzymes and three of these were chosen as diagnostic (isocitrate dehydrogenase-1, phosphogluconate dehydrogenase and glucose-6-phosphate dehydrogenase) for eggs. Electrophoretic techniques for species identification are simple, quick, inexpensive (Loxdale, 1994) and, in the case of *Helicoverpa* species, it only took a few weeks to develop the application. Fitt *et al.* (1988) evaluated the technique for commercial application. They concluded that it was feasible to screen up to 1000 eggs per day, at an estimated total cost of less than 10 pence per egg (G. Fitt and Daly, unpublished results). The technique proved to be a valuable research tool because it enabled Fitt and coworkers to calibrate the more traditional approach using pheromone and light-trap catches. They demonstrated that the relative abundance of the two species in pheromone traps was not a reliable indicator of the relative abundance of the two species in eggs laid in adjacent fields (Fitt *et al.*, 1988; Daly and Fitt, 1993). Subsequently, pheromone traps were abandoned as a monitoring

Table 4.1 Comparison of the three valid methods for discriminating eggs or small larvae of *H. punctigera* and *H. armigera*. Comments relate to a sample size of 100

	Method		
	*Immunosquash**	*Electrophoresis*†	*Rearing*†
Times to complete	30 min	60 min	10 days
Equipment required	Pocket-sized base plate	Power pack and gel rig	Containers, diet preparation, etc.
Facilities required	Water source	Laboratory	Laboratory
Skill level required	Intelligent lay person	Laboratory assistant	Laboratory assistant
Material costs	A85¢/egg	A20¢/egg	A$1–$2

*S. Trowell, unpublished results.
†J. Daly, unpublished results.

tool for species composition. Daly and Gregg (1985) used the technique to demonstrate that morphological criteria used to distinguish 3rd–6th instar larvae collected from field populations were unreliable. In some samples collected by field entomologists up to 70% of individuals classified as *H. punctigera* were shown to be *H. armigera*.

4.2.2 Immunosquash technique

While isoenzyme electrophoresis was useful to the researcher, it was not readily adapted to use by the farmer or their consultants because of the need to undertake the identifications in a laboratory. In response to a perceived need of the farming community, Trowell and colleagues developed an immunosquash test for field use by lay people (Trowell *et al.*, 1992). This method requires eggs or small larvae to be squashed onto a nitrocellulose membrane. Larger larvae can be tested by applying a small drop of haemolymph to the membrane. An enzyme-linked monoclonal antibody detects individuals of the species *H. armigera*, but gives no signal with *H. punctigera*. This method was made available commercially within Australia as the LepTon™ Test Kit (patent pending) (Abbott Laboratories).

Compared with the other valid methods for discriminating the two species, the immunosquash test is straightforward and rapid. The species are identified uniquely by a single variable (antibody binding) and this makes interpretation simple. The test requires minimal equipment and can be performed next to a vehicle in the field. It can also be used as a research tool in similar ways to electrophoresis. For example, Greenstone and coworkers (1991) used another monoclonal antibody in an ELISA test to examine the phylogeny of the heliothines.

The relative advantages of the two biochemical methods and of the rearing method are listed in Table 4.1. The principal disadvantage of the immunosquash is that its development required a major commitment of time and resources. It took 18 months to obtain an antibody that could reliably distinguish both eggs and larvae. A further 18 months of development time was required to make the performance of the test reliable under field conditions. The most serious problems proved to be interference from endogenous larval enzymes (Trowell *et al.*, 1994). Peroxidases present in the leaves of cotton plants grown under field conditions also interfered with the signal (S.C. Trowell, unpublished results). As in this example, availability of an immunosquash test for commercial application may require the backing of a commercial partner. Therefore, commercial tests of this type may only be applicable where there is an important economic decision to be made. An additional disadvantage, relative to the rearing method, is that when used on eggs or small larvae the test is destructive so individuals cannot be assayed for resistance using standard bioassay tests (but see Hemingway *et al.*, 1996, this volume, for an alternative approach). However, no matter what technique is used, the time and inconvenience of sample collection will always be unavoidable.

In technical terms, the kit has proven extremely reliable. In control tests using eggs of known provenance, the rate of false positive/negative was very low (<1%). In field trials the agreement between the test and the results from rearing of parallel samples was high (N.W. Forrester, personal communication). Assays of larvae produced rates of false positive/negative rates of up to 5% and <2%, respectively (Trowell *et al.*, 1994). While it is too early to decide how the test kit will fare commercially, it has now been available to farmers for one summer growing season. In its first season of commercial releases, the test was used in the management of 30–50% of the cotton-growing area (N.W.Forrester, personal communication).

While the kit was designed to identify the species of the eggs, the information is being used to infer the presence of resistant *H. armigera* individuals in the crop (Forrester, 1994). Previous results suggest that resistance frequencies are low in spring, medium in summer and high in autumn (Forrester *et al.*, 1993). This information, coupled with the projected survival of eggs to hatchlings in different seasons (Dillon, 1993) can be used to predict the density of resistant *H. armigera* larvae 3 days after an egg-lay. Until it is possible to monitor resistance more directly (see below) the species kit provides farmers with the capacity to make informed choices about the best insecticide to use that minimizes the risk of economic losses due to resistance. This in turn reduces selection pressure for resistance (Daly, 1993).

4.3 POPULATION AND GENE FLOW

The distribution of genetic variation between populations results from an interaction of selection, mutation, gene flow and random genetic effects. If variation is assumed to be a neutral, or a nearly neutral character then the geographic pattern of variation can be used as a measure of population sub-structure (Slatkin, 1985). Differences in gene frequency between populations are seen to be a result of processes that contribute to effective population size and the amount of genetic interchange between populations. Nevertheless, studies of genetic structure can only establish hypotheses about population structure and these hypotheses need independent corroboration from ecological studies (Slatkin, 1985; Daly, 1989). This is because a given population's structure may result from a variety of interactions between effective population size and gene flow.

Additional difficulties occur with agricultural pests (Loxdale, 1994). In particular, much of the theory of population genetics that provides the basis for interpretation assumes that populations are at equilibrium. This may not be true of many pest species of annual crops which must recolonize the host plant annually (Southwood, 1971). This, coupled with the widespread use of insecticides, may result in large fluctuations in population size every few generations.

4.3.1 Isoenzyme electrophoresis

Until recently, little direct information was available about the dispersal of the Australian heliothines except that *H. punctigera* was considered to be highly migratory compared with the more sedentary *H. armigera* (Farrow and Daly, 1987). Biochemical methods were useful in testing hypotheses about the spread and dilution of resistance alleles among populations. Isoenzyme electrophoresis of four to seven loci failed to detect any geographic variation in either *H. armigera* or *H. punctigera* populations (Daly and Gregg, 1985; Daly, 1989). Indeed, the distribution of isoenzyme variants was consistent with high levels of gene flow in both species (Daly and Gregg, 1985). This certainly could explain why resistance to pyrethroids had spread so rapidly among *H. armigera* populations but could not account for any role gene flow played in retarding the evolution of resistance in *H. punctigera*.

Recent ecological studies of dispersal, however, have confirmed that major differences in dispersal strategy do occur between *H. armigera* and *H. punctigera*. Both species are capable of long-distance migrations of 1000–2000 km (Farrow and Daly, 1987) and spring invasions occur of both species into cropping areas (Fitt and Daly, 1990; Gregg, 1993). However, *H. armigera* is a facultative migrant and a substantial proportion of the population overwinters as resident populations in cropping areas. Fitt and Daly (1990) observed large numbers of *H. armigera* but not *H. punctigera* to overwinter as pupae in the Namoi Valley cotton growing area in eastern Australia. *H. punctigera* on the other hand is an obligate disperser. Each winter large populations breed in the interior of Australia on ephemeral plants (Zalucki *et al.*, 1994). In spring, billions of moths migrate eastwards and westwards to re-invade the cropping areas (Gregg *et al.*, 1993 and in press). This different migratory strategy may well be sufficient to account for the occurrence of resistance in *H. armigera* but not in *H. punctigera* (Daly and Gregg, 1985). Even within a season, it was observed that, on emergence, *H. punctigera* tended to take-off and disperse compared with *H. armigera*, which remained locally if nearby crops were in flower (G.P. Fitt, personal communication).

4.3.2 mtDNA

McKechnie and coworkers (1993) have recently re-examined population structure using mtDNA. This form of DNA has the potential to detect population substructure even when isozymes cannot; its role in revealing population structure has been reviewed extensively (Avise *et al.*, 1987; Moritz *et al.*, 1987; Moritz, 1994). mtDNA is maternally inherited and is particularly useful for tracing maternal genealogy, for identifying the source population of a recent introduction or range expansion (Harrison, 1989). Typically, a greater proportion of the variation in mtDNA is distributed among, rather than within, populations (Moritz, 1994). McKechnie and coworkers used the polymerase chain reaction (PCR) to amplify from individual moths a highly variable region of mtDNA, 700 bp long. The PCR products from some moths were further characterized for variation by restriction fragment length polymorphism (RFLP) analysis using six different restriction enzymes, McKechnie *et al.* (1993) observed one haplotype common in *H. punctigera* populations. Eleven other variants were also detected but there was no indication of any geographic pattern. Variation within *H. armigera* was greater – 33 haplotypes were found among 113 moths. Most of the haplotypes were found widely distributed in eastern Australia but there was some evidence of geographic variation with latitude (Spackman and McKechnie, in press).

Thus, analysis of variation in mtDNA has revealed a minor difference in population structure between *H. armigera* and *H. punctigera* that was not detected using isozyme electrophoresis. However, neither technique could detect or quantify the different dispersal strategies of the two pest species because with neutral genetic variation the consequence of both strategies is similar – genetic homogeneity. Detection of the consequences of gene flow in which the habitat is recolonized each year, from that in which perhaps only 50% of the population are immigrants, would appear to be beyond the sensitivity of either biochemical method. It is ironic that the most sensitive measure of population structure in *H. armigera* is the geographic variation in the frequency of resistance genes (Gunning and Easton, 1989; Forrester *et al.*, 1993) that are under strong selection. Ironic, because we would consider that their lack of neutrality would exclude them as indicators of population subdivision. However, it appears that gene flow is sufficiently common in both species that only major selection with insecticides will maintain gene frequency differences between populations.

4.4 DETECTING RESISTANCE

4.4.1 The techniques

Traditionally, resistance to insecticides has been detected and monitored in field populations using bioassays of whole insects (ffrench-Constant and Roush, 1990) but biochemical and DNA methods are now being applied (Brown and Brogdon, 1987; Devonshire, 1990a,b; ffrench-Constant and Roush, 1990; Hemingway *et al.*, 1996, this volume).

Bioassays include all techniques in which live insects are exposed to a known dose of insecticide. Their response is monitored, usually whether the insect is alive or dead, although other criteria such as time to recovery, can also be used. Bioassays can be used to characterize the response of different genotypes/phenotypes to a range of insecticide concentrations (a dose–response bioassay, see ffrench-Constant and Roush, 1990) or may measure the frequency of resistance using a discriminating dose test (Roush and Miller, 1986). They have also been used to reveal the genetic basis of resistance. There are many problems associated with the use of these traditional techniques (Brown and Brogdon, 1987).

More recently, a range of biochemical methods, such as enzyme electrophoresis, enzyme assays and immunoassays have been used to monitor metabolic resistance (Brown and Brogdon, 1987; Field *et al.*, 1989). Esterases have been most amenable to this method, particularly where resistance involves an over-expression or amplification of the gene (Hemingway *et al.*, 1996, this volume). Esterase-based resistance can be detected also on polyacrylamide gels using standard esterase staining procedures (Hughes and Raftos, 1985) or by direct squashes of individuals with colorimetric test (Pasteur and Georghiou, 1981). In *H.*

armigera, R.V. Gunning and coworkers (personal communication) have observed variation for resistance to the synthetic pyrethroids that appears to be associated with elevated esterases. They have detected this variation using a variety of methods: electrophoresis, a squash colorimetric test, and through measurements of enzyme activity.

DNA techniques extend our capability to detect and monitor target-site mechanisms (Brown and Brogdon, 1987; Devonshire *et al.*, 1990a,b; Aronstein *et al.*, 1994) in which the difference between susceptible and resistant individuals may involve one, or a few, base pair changes in a single gene. Techniques include PCR that amplifies DNA variants detected by RFLPs or randomly amplified polymorphic DNA (RAPDs). In *Drosophila melanogaster* Meigen, ffrench-Constant *et al.* (1993) observed that the point mutation conferring cyclodiene resistance results in the loss of a *Hae* II restriction endonuclease (REN) site in exon 7 of the *Rdl* gene. A second polymorphism, the gain of an *Eco*RI site 770 bp from the resistance mutation, was closely but not absolutely, linked to resistance. Aronstein *et al.* (1994) used PCR methods to amplify exon 7 from field-collected individuals and then REN methods to distinguish between phenotypes. More recently, they have refined their methods further and applied the PASA (PCR amplification of specific alleles; Sarkar *et al.*, 1990) technique in which PCR primers are made that will amplify one of the alleles, but not the other due to a mismatch with the region of DNA containing the resistance mutation. Thus, they eliminated the need for postamplification processing of samples (Zetterquist and Olerup, 1992) and directly detected the base pair change that confers resistance.

4.4.2 Biossays versus DNA Methods

The choice of technique to monitor resistance will be influenced by many factors, including the nature of the resistance mechanism (see earlier), the cost and convenience of the method, and the accuracy and precision required (Devonshire, 1990a,b). We have compared the main advantages and disadvantages of bioassays versus DNA methods as follows:

1. Phenotypes. Bioassays may only detect differences between phenotypes, resistant versus susceptible, unless resistance factors are very large (Roush and Daly, 1990). For example, the discriminating dose test used to detect metabolic resistance to the pyrethroids in *H. armigera* kills 99% of susceptible, a variable proportion of heterozygous, but few homozygous resistant individuals (Daly and Fisk, 1992). While information on the frequency of phenotypes is useful, particularly for pest management, it is necessary to be able to

classify unambiguously individuals into genotype if we are to measure characteristics such as relative fitness and gene flow (Daly, 1993). In contrast, most DNA methods will identify, unambiguously, the different genotypes, with the exception of RAPDs (see Loxdale *et al.*, 1996, this volume). This characteristic is independent of the relative tolerance of the genotypes to the insecticide.

2. Mechanisms. Bioassay results are generally independent of the mechanism of resistance, because the method of pesticide application, usually to the outside of the insect or through ingestion, will ensure that different phenotypes can be detected no matter what the mechanism/s (penetration, metabolism, or target site insensitivity). This is a major advantage of bioassays in cases where the mechanism is not known, or in the case where bioassays are used to detect resistance in the early stages of evolution, or if there is sequential evolution of resistance, as certainly has been the case with pyrethroid resistance in *H. armigera* (Daly, 1993) and other species such as *Culex pipiens* L. (Raymond *et al.*, 1987; Roush and Daly, 1990). Biossays can be modified to increase their discrimination, through for example, the addition of synergists. Forrester *et al.* (1993) monitored pyrethroid resistance in *H. armigera* from 1983 to 1993 using a discriminating dose bioassay with larvae. From 1987 onwards they used the synergist, piperonyl butoxide, in combination with the pyrethroid fenvalerate, to determine the frequency of the component of resistance attributable to the mixed function oxidases. Alternatively, the bioassay method may be modified to detect other mechanisms, e.g. nerve insensitivity was detected in *Heliothis virescens* using the time to knockdown, rather than death (F. W. Plapp, personal communication). However, in general, information on different mechanisms require multiple samples of the population because only one insecticide can be assayed per insect (Brown and Brogdon, 1987).

DNA methods are generally specific for a single mechanism of resistance because the DNA variation that is used either maps close to, or is within, a single resistance locus. Multiple markers would be required if different loci were associated with resistance. However, the technique of denaturing gradient gel electrophoresis (DeMarini and Fuscoe, 1991) offers the possibility of monitoring a number of alleles, or even mutations in a number of genes, simultaneously.

Not all mechanisms of resistance are equally amenable to the use of PCR methods. For example, if resistance involves an amplification rather than a base pair change, the difference between phenotypes is the amount of a protein, such as an esterase, rather than a qualitative change in the gene (Raymond *et al.*, 1987; Field *et al.*, 1988). In such cases, antibody-based techniques can be used (Field *et al.*, 1989).

In some dipteran species, such as *Drosophila* spp., *Lucilia cuprina*

(Wiedemann) and *Musca domestica* L., in which detailed chromosome maps are available through traditional cytogenetic studies, it is possible to map resistance genes to chromosomes, and the contribution of each chromosome to resistance can be measured. So in these species, the number and contribution of major and minor genes can be determined accurately (Sawicki *et al.*, 1984; McKenzie and Game, 1987). The traditional approaches to mapping of genes are not possible in most pest species because little is known about their genetic structure, or because – as is the case in noctuid moths – there are a large number of small chromosomes undifferentiated by standard staining techniques. A chromosomal map of *Heliothis virescens* has recently been developed using a combination of isoenzymes, and the molecular variation of RFLPs and RAPDs (D. Heckel, personal communication). This map will aid future resistance studies in heliothine moths (see later).

3. Lack of generality. Bioassays are developed for standardized conditions, such as age of insect, or temperature. This greatly limits the utility of the test in the field. For example, resistance in *H. armigera* is monitored in 4th instar larvae 28–40 mg in body mass, yet the field collections are taken of eggs to avoid the risk of prior selection with insecticides (Forrester *et al.*, 1993). This technique is less satisfactory if large larvae, pupae or adults are collected because it requires testing of the F_1 generation. Daly and Fitt (1990) applied an adult test developed for monitoring resistance in *Heliothis virescens*. While this test was satisfactory in laboratory studies, Daly (1992) showed that the expression of resistance declined with the age of the moth. Attempts to validate the vial test for field studies failed because there was a poor relationship between resistance frequencies obtained from larval monitoring results and frequencies in the adults (Daly and Fitt, 1993). In other situations, it may be difficult to collect live individuals. For example, in inland Australia, *H. armigera* are uncommon and high mortality of field-collected larvae can occur before resistance tests are done (G.P. Fitt and P. Gregg, personal communication). Pheromone traps can trap adult males, but flights of adults are unpredictable events.

DNA markers are particularly useful because they can indicate the resistance status of an individual independently of its age, life-stage or condition. These also offer the advantage of determining the genetic make-up non-destructively if only a leg or some haemolymph can be obtained. Further, DNA can be prepared from a small amount of tissue (Dowling *et al.*, 1990) and amplified using PCR. DNA technologies of this kind could be of great advantage in pest populations in which the easiest life-stage to collect may not be the best one in which to test for resistance using standard biochemical tests. Thus, with the low-density populations of *H. armigera* from inland regions, it could be possible to leave pheromone traps in place for long periods and then monitor resistance in dead samples using PCR methods. Indeed, a combination of molecular technologies could provide the only means by which we can assess the resistance status of these inland populations.

4. Development time. Bioassays are usually rapid to develop – a few days to a few weeks may be all that is required. Standard techniques are available for many pest species. DNA-based techniques currently require considerably longer to develop, particularly if the resistance mechanism is not known.
5. Species barrier. DNA-based methods may allow the researcher to apply information gained in one species to other, unrelated species. Precise genetic analyses can be done in dipteran species such as *Drosophila* spp. and *L. cuprina*. Molecular genetic techniques, such as gene cloning, coupled with DNA technologies such as PCR, enable us to apply knowledge obtained readily with these dipteran species to agricultural pests. Genes associated with target-site insensitivity to pyrethroids and DDT (*para*), to cyclodienes (*Rdl*) and organophosphates (*ace*) can be cloned from genetically amenable species, notably *Drosophila*, and then used to design degenerate PCR primers or as probes for the same loci in other species (e.g. Doyle and Knipple, 1991; Taylor *et al.*, 1993; Thompson *et al.*, 1993; Williamson *et al.*, 1993). This approach will yield insights into the molecular basis of insecticide resistance in many more species than could be contemplated if each species had to be studied independently.

Some DNA markers will be species-specific so will need to be derived independently for new pest species. For example, in the case of cyclodiene resistance, in which the same amino acid change seems to be associated with resistance in a range of insects (Thompson *et al.*, 1993), it is clear that the polymorphic restriction site that is used to tag resistance in *D. melanogaster* is only present in very closely related species. However, if resistance has a common genetic origin, as is reported for cyclodiene resistance, the gene of interest may be analysed quickly in a new species using PCR or heterologous hybridization, as described above. PCR methods coupled with RFLP/REN or mismatch methods may then yield a genetic marker that is linked to resistance.

Recently, Trowell *et al.*, (1994) have investigated the molecular basis of cyclodiene resistance in *H. armigera* using degenerate PCR primers for conserved regions of the gene. They found that resistance was not associated with the base change in the location observed by Thompson *et al.* (1993) in other species. Both susceptible and resistant *H. armigera* had the amino acid, glutamine, compared with the change from an alanine to serine in resistant *D. melanogaster*. There is also genetic evidence indicating the *Rdl* gene is not involved in cyclodiene resistance in *H. armigera*. J. Daly and J. Fisk (submitted) found that cyclodiene resistance in *H. armigera* is sex-linked while D. Heckel and S. Trowell (unpublished results) mapped the *H. armigera 'Rdl'* homologue to autosomal linkage group 8 in *Heliothis virescens*.

6. Cost. Bioassay techniques are generally simple and therefore inexpensive. DNA techniques are more expensive, both in terms of the personnel costs and the laboratory consumables and equipment. While detailed information may be required for research studies of resistance, a crude detection system may suffice for the detection of resistant insects in a pest management programme. Thus, the choice of resistance monitoring or detection system will depend on the requirements of the data collection. DNA techniques are fashionable but their cost may not justify the extra expense over the simpler bioassay methods.

4.4.3 Monitoring cyclodiene resistance

While methods that detect variation at the level of the DNA are very powerful it is unlikely that they could replace completely traditional bioassays. In general, their very precision means that DNA methods cannot detect mechanisms as they arise, without doing a whole range of tests for all possible mechanisms. On the other hand, McKenzie's laboratory has been successful in duplicating field resistance mechanisms (Smyth *et al.*, 1992; personal communication Prof J.A. McKenzie) using EMS mutagenesis and selection in the laboratory, so that predictive work may in future be possible. However, DNA methods may be superior to bioassays for monitoring resistance, at least in some cases. Aronstein *et al.* (1994) state that PCR could be more accurate than bioassay in detecting dieldrin resistance in field populations of *Drosophila*. They compared estimates based on traditional bioassays, in which flies are exposed to vials treated with insecticide, to estimates obtained using PCR/REN methods. They observed a very large spread of estimates with bioassays from 0–12% with bioassays compared with estimates of 0.5–1.5% from PCR; bioassays also failed to detect resistance in about one-third of samples, compared with a 100% detection rate using PCR. Roush and Daly (1996), however, calculated that the larger variance for bioassays was expected from sampling theory: Aronstein *et al.* (1994) had used smaller sample sizes for bioassays compared with those for PCR. Furthermore, Roush and Daly (1996) report that when sample sizes of less than 100 individuals are used, and the frequency of resistance is of the order of 1%, then it is expected that about 37% of samples will not contain any resistant individuals. Thus, these authors argued that bioassays were just as efficacious for monitoring dieldrin resistance in *D. melanogaster* as the PCR method and were considerably more cost-effective. In the future however, quantitative allele-specific PCR techniques (e.g. Sabbatini *et al.*, 1993) applied to groups of individuals, which overcome some of the practical limitations of current PCR/REN methods, may be advantageously adapted to the monitoring of insecticide resistance.

So, while molecular techniques have an important role in the detection and monitoring of resistance as well as in research they will not replace completely the more imprecise bioassay techniques. Bioassay tests are quick to devise and are relatively independent of the mechanism of resistance. However, they cannot detect changes in mechanisms that may be occurring and the techniques may restrict the availability of material for testing, and may fail to distinguish heterozygotes from homozygotes. Molecular techniques on the other hand are time-consuming to devise, particularly those that involve antibodies. Their value lies in their ability to tease apart the evolutionary story of resistance, and to identify the nature of the genetic change, which is only possible if individuals can be classified unambiguously with respect to their genotype. Development of molecular techniques to monitor resistance, though costly, will be justified for some major insect pests of agriculture, live stock and human health.

CONCLUSIONS

Biochemical methods have much to contribute to a study of insecticide resistance. The newer DNA technologies have revealed the underlying genetic changes involved in resistance. They have enabled rapid and simple means to identify species and in the future will facilitate resistance monitoring. The utility of the methods will increase rapidly over the next decade. At the same time their cost and time to develop will decrease. However, traditional biochemical techniques are still important and will be preferred when the mechanism of resistance is not known, and because of their simplicity, speed of application to a new problem and their low cost. Future studies of resistance will need to evaluate carefully the relative benefits of the different techniques to maximize information at the minimum cost of time and money.

Acknowledgements

We thank Anna Grabowski, Eva Zinkovsky, Jenny Fisk and Karen Paschalidis for technical assistance. The Systematics Association is thanked for its support for J.C.D. to attend the workshop in Cardiff. A number of referees made useful comments. Our research is funded by the Australian Cotton Research and Development Corporation.

REFERENCES

Aronstein, K., Ode, P. and ffrench-Constant, R.H. (1994) Direct comparison of PCR-based monitoring for cyclodiene resistance in *Drosophila* populations with insecticide bioassay. *Pesticide Biochemistry and Physiology*, **48**, 229–33.

Avise, J.C., Arnold, J., Ball, R.M., Bermingham, E., Lamb, T., Neigel, J.E., Reeb, C.A. and Saunders, N.C. (1987) Intraspecific phylogeography: the mitochondrial DNA bridge between population genetics and systematics. *Annual Review of Ecology and Systematics*, **18**, 489–522.

Brown, T.M. and Brogdon, W.G. (1987) Improved detection of insecticide resistance through conventional and molecular techniques. *Annual Review of Entomology*, **32**, 145–62.

Daly, J.C. (1989) The use of electrophoretic data in a study of gene flow in the pest species *Heliothis armigera* (Hübner) and *H. punctigera* Wallengren (Lepidoptera: Noctuidae), in *Electrophoretic Studies on Agricultural Pests*, Systematics Special Volume No. 39, (eds H.D. Loxdale and J. Den Hollander), Clarendon Press, Oxford, pp. 115–41.

Daly, J.C. (1992) Decline in survival with age of resistant adult *Helicoverpa armigera* (Hübner) (Lepidoptera: Noctuidae) at a discriminating dose of the pyrethroid, fenvalerate. *Journal of Economic Entomology*, **85**, 1583–6.

Daly, J.C. (1993) Ecology and genetics of insecticide resistance in *Helicoverpa armigera*: interactions between selection and gene flow. *Genetica*, **90**, 217–26.

Daly, J.C. (1994) Ecology and resistance management for Bt transgenic plants. *Biocontrol Science and Technology*, **4**, 563–71.

Daly, J.C. and Fisk, J.H. (1992) Inheritance of metabolic resistance to the synthetic pyrethroids in Australian *Helicoverpa armigera* (Hübner) (Lepidoptera: Noctuidae). *Bulletin of Entomological Research*, **82**, 5–12.

Daly, J.C. and Fitt, G.P. (1990) Monitoring for pyrethroid resistance in relation to body weight in adult *Helicoverpa armigera* (Hübner) (Lepidoptera: Noctuidae). *Journal of Economic Entomology*, **83**, 705–9.

Daly, J.C. and Fitt, G.P. (1993) The need to validate monitoring techniques for insect populations, in *Pest Control and Sustainable Agriculture*, (eds S.A. Corey, D.J. Dall and W.M. Milne), CSIRO, Melbourne, pp. 414–16.

Daly, J.C. and Gregg, P. (1985) Genetic variation in *Heliothis* in Australia: species identification and gene flow in the two pest species *H. armigera* (Hübner) and *H. punctigera* Wallengren (Lepidoptera: Noctuidae). *Bulletin of Entomological Research*, **75, 169–84.**

DeMarini, D. M. and Fuscoe, J. C. (1991) Workshop overview: new molecular techniques in genome analysis. *Environmental and Molecular Mutagenesis*, **18**, 222–3.

Devonshire, A.L. (1990a) Biochemical and DNA probing techniques for monitoring insecticide resistance genes, in *Pesticides and Alternatives*, (ed. J.E. Casida), Elsevier Science Publishers, pp. 421–31.

Devonshire, A.L. (1990b) Biochemical and molecular genetic analyses of insect populations resistant to insecticides, in *Proceedings of Brighton Crop Protection, Crops and Diseases*, British Crop Protection Council Publications, Thornton Heath, UK. pp. 889–96.

Dillon, G. (1993) Predicting natural heliothis egg mortality. *Australian Cotton Grower*, **14**, 59–60.

Dowling, T.E., Moritz, C. and Palmer, J.D. (1990) Nucleic Acids II. Restriction Site analysis, in *Molecular Systematics* (eds D.M. Hillis and C. Moritz), Sunderland, Mass., pp. 250–317.

Doyle, K.E. and Knipple, D.C. (1991) PCR-based phylogenetic walking: isolation of *para*-homologous sodium channel gene sequences from 7 insect species and an arachnid. *Insect Biochemistry*, **21**, 689–96.

Farrow, R.A. and Daly, J.C. (1987) Long-range movements as an adaptive strategy in the genus *Heliothis* (Hübner) (Lepidoptera: Noctuidae): a review of its occurrence and detection in four pest species. *Australian Journal of Zoology*, **35**, 1–24.

Ffrench-Constant, R.H. and R.T. Roush (1990) Resistance detection and documentation: the relative roles of pesticidal and biochemical assays, in *Pesticide Resistance in Arthropods*, (eds R.T. Roush and B. Tabashnik), Chapman & Hall, New York, pp. 4–38.

Ffrench-Constant, R.H., Steichen, J.C., Rocheleau, T.A., Aronstein, K. and Roush, R.T. (1993) A single-amino acid substitution in a γ-aminobutyric acid subtype A receptor locus is associated with cyclodiene insecticide resistance in *Drosophila* populations. *Proceedings of the National Academy of Sciences USA*, **90**, 1957–61.

Field, L.M., Devonshire, A.L., and Forde, B.G. (1988) Molecular evidence that insecticide resistance in peach-potato aphids (*Myzus persicae*, Sulz.) results from an amplification of an esterase gene. *Biochemical Journal*, **251**, 309–12.

Field L.M., Devonshire, A.L., ffrench-Constant, R.H., and Forde, B.G. (1989) The combined use of immunoassay and a DNA diagnostic technique to identify insecticide-resistant genotypes in the peach-potato aphid *Myzus persicae* (Sulz.). *Pesticide Biochemistry and Physiology*, **34**, 174–8.

Fitt, G.P. (1989) The ecology of *Heliothis* species in relation to agro-ecosystems. *Annual Review of Entomology*, **34**, 17–52.

Fitt, G.P. and Daly, J.C. (1990) Abundance of overwintering pupae and the spring generation of *Helicoverpa* spp. (Lepidoptera: Noctuidae) in northern New South Wales, Australia: implications for pest management. *Journal of Economic Entomology*, **83**, 1827–36.

Fitt, G.P., Daly, J.C., Wilson, A.G.L. and Terras, M. (1988) Identification of heliothis eggs using electrophoresis. Australian Cotton Grower, **8**, 54–6.

Forrester, N. (1994) Guidelines for using the Lepton Test kit. *Australian Cotton Grower*, **15**, 18–21.

Forrester, N.W. and Fitt, G.P. (1992) Management of insecticide resistance in *Heliothis armigera* – ecological and chemical countermeasures, in *Resistance '91: Achievements and Developments in Combating Pesticide Resistance*, (eds I. Denholm, A.L. Devonshire and D.W. Holloman), Elsevier Applied Science, London, pp. 88-99.

Forrester, N.W., Cahill, M., Bird, L.J. and Layland, J.K. (1993) Management of pyrethroid and endosulfan resistance in *Helicoverpa armigera* (Lepidoptera: Noctuidae) in Australia. *Bulletin of Entomological Research, Supplement Series No 1*. pp. 132. CAB International, Wallingford, Oxon.

Greenstone, M.H., Stuart, M.K., and Haunerland, N.H. (1991) Using monoclonal antibodies for phylogenetic analysis: an example from the Heliothinae (Lepidoptera: Noctuidae). *Annals of The Entomological Society of America*, **84**, 457–64.

Gregg, P.C. (1993) Pollen as a marker for *Helicoverpa armigera* (Hübner) and *H. punctigera* Wallengren (Lepidoptera: Noctuidae) emigrating from western Queensland. *Australian Journal of Ecology*, **18**, 151–72.

Gregg, P.C., Fitt, G.P., Zalucki, M.P., Murray, D.A.H. and McDonald, G. (1993)

Winter breeding and spring migration of *Helicoverpa* spp. in inland Australia, 1989–1991, in *Pest Control and Sustainable Agriculture*, (eds S.A. Corey, D.J. Dall and W.M. Milne), CSIRO, Canberra, Melbourne, pp. 460–3.

Gregg, P.C., Fitt, G.P., Zalucki, M.P. and Murray, D.A.H. (in press). Insect migration in an arid continent. II. *Helicoverpa* spp. in eastern Australia, in *Insect Migration: Tracking Resources in Space and Time*, (eds V.A. Drake and A.G. Gatehouse), Cambridge University Press.

Gunning, R.V. and Easton, C.S. (1989). Pyrethroid resistance in *Helicoverpa armigera* (Hübner) collected from unsprayed maize crops in New South Wales 1983–1997. *Journal of the Australian Entomological Society*, **28**, 57–61.

Gunning, R.V., Easton, C.S., Greenup, L.R., and Edge, V.R. (1984) Pyrethroid resistance in *Heliothis armigera* (Hübner) (Lepidoptera: Noctuidae) in Australia. *Journal of Economic Entomology*, **77**, 1283–7.

Gunning, R.V., Easton, C.S., Balfe, M.E. and Ferris, I.G. (1991) Pyrethroid resistance mechanisms in Australian *Helicoverpa armigera. Pesticide Science*, **33**, 473–90.

Harrison, R.G. (1989) Animal mitochrondial DNA as a genetic marker in population and evolutionary biology. *Trends in Ecology and Evolution*, **4**, 6–12.

Hemingway, J., Small, G.J., Lindsay, S.W. and Collins, F.H. (1996) Combined use of biochemical, immunological and molecular assays for infection, species identification and resistance detection in field populations of *Anopheles* (Diptera: Culicidae), in *The Ecology of Agricultural Pests: Biochemical Approaches*, (eds W.O.C. Symondson and J.E. Liddell), Chapman & Hall, London, pp. 31–49.

Hughes, P.B. and Raftos, D.A. (1985) Genetics of an esterase associated with resistance to organophosphorous insecticides in the sheep blowfly, *Lucilia cuprina* (Wiedemann) (Diptera: Calliphoridae). *Bulletin of Entomological Research*, **75**, 535–44.

Loxdale, H.D. (1994) Isozyme and protein profiles of insects of agricultural and horticultural importance, in *The Identification and Characterization of Pest Organisms*, (ed. D.L. Hawksworth), CAB International, Wallingford, pp. 337–75.

Loxdale, H.D., Brookes, C.P. and de Barro, P.J. (1996) Application of novel molecular markers (DNA) in agricultural entomology, in *The Ecology of Agricultural Pests: Biochemical Approaches*, (eds W.O.C. Symondson and J.E. Liddell), Chapman & Hall, London, pp. 149–198.

May, R.M. and Dobson, A.P. (1986). Population dynamics and the rate of evolution of pesticide resistance, in *Pesticide Resistance: Strategies and Tactics for Management*, National Research Council, Committee on Strategies for the Management of Pesticide Resistant Pest Populations, National Academy Press, Washington, DC, pp. 170–93.

McCaffery, A.R., King, A.B.S., Walker, A.J. and El-Nayir, H. (1989) Resistance to synthetic pyrethroids in the bollworm, *Heliothis armigera* from Andhra Pradesh, India. *Pesticide Science*, **27**, 65–76.

McKechnie, S.W., Spackman, M.E., Naughton, M.E., Kovacs, I.V., Ghosn, M. and Hoffman, A.A. (1993) Assessing budworm population structure in Australia using AT-rich region of mitochondrial DNA, in *Proceedings of the Beltwide Cotton Conference*, pp. 838–40.

McKenzie, J.A. and Game, A.Y. (1987) Diazinon resistance in *Lucilia cuprina*; mapping of a fitness modifer. *Heredity*, **59**, 371–81.

Moritz, C., Dowling, T.E. and Brown, W.M. (1987) Evolution of animal mitochondrial DNA: relevance for population biology and systematics. *Annual Reviews of Ecology and Systematics*, **18**, 269–92.

Moritz, C. (1994) Applications of mitochrondrial DNA analysis in conservation: a critical review. *Molecular Ecology*, **3**, 401–41.

Pasteur, N. and Georghiou, G.P. (1981) Filter paper test for rapid determination of phenotypes with high esterase activity in organophosphate resistant mosquitoes. *Mosquito News*, **41**, 181–3.

Plapp, F.W., Jr, Campanhola, C., Bagwell, R.D. and McCutcheon, B.F. (1990) Management of pyrethroid-resistant tobacco budworms on cotton in the United States, in *Pesticide Resistance in Arthropods*, (eds R.T. Roush and B. Tabashnik), Chapman & Hall, New York, pp. 237–60.

Raymond, M., Pasteur, N. and Georghiou, G.P. (1987) Inheritance of chlorpyrifos resistance in *Culex pipiens* L. (Diptera: Culicidae) and estimation of the number of genes involved. *Heredity*, **58**, 351–6.

Roush, R.T. and Croft, B.A. (1986) Experimental population genetics and ecological studies of pesticide resistance in insects and mites, in *Pesticide Resistance: Strategies and Tactics for Management*, National Research Council, Committee on Strategies for the Management of Pesticide Resistant Pest Populations, National Academy Press, Washington, DC, pp. 57–70.

Roush, R.T and Daly, J.C. (1990) The role of population genetics in resistance research and management, in *Pesticide Resistance in Arthropods*, (eds R.T. Roush and B. Tabashnik), Chapman & Hall, New York, pp. 97–153.

Roush, R.T. and Daly, J.C. (1996) Considerations for comparing bioassays and biochemically-based resistance monitoring methods: lessons from a study of cyclodiene resistance. *Pesticide Biochemistry and Physiology*, in press.

Roush, R.T. and Miller, G.L. (1986) Considerations for design of insecticide resistance monitoring programs. *Journal of Economic Entomology*, **79**, 293–8.

Sabbatini, A. R., Werner, P. A., Guha, C., Paddock, G. V. and Galbraith, R. M. (1993) The vitamin D-binding protein gene: quantitation of amplified nucleic acids by ELISA. *Biotechniques*, **15**, 706–13.

Sarkar, G., Cassady, J., Bottema, C. D. and Sommer, S. S. (1990) Characterization of polymerase chain reaction amplification of specific alleles. *Analytical Biochemistry*, **186**, 64–8.

Sawicki, R.M., Devonshire, A.L., Farnham, A.W., O'Dell, K.E., Moores, G.D. and Denholm, I. (1984) Factors affecting resistance to insecticides in house-flies, *Musca domestica* L. (Diptera: Muscidae). II Close linkage on autosome 2 between an esterase and resistance to trichlorophon and pyrethroids. *Bulletin of Entomological Research*, **74**, 197–206.

Slatkin, M. (1985) Gene flow in natural populations. *Annual Reviews of Ecology and Systematics*, **16**, 393–430.

Smyth, K.-A., Parker, A. G., Yen, J. L. and McKenzie, J. A. (1992) Selection of dieldrin-resistant strains of *Lucilia cuprina* (Diptera: Calliphoridae) after ethyl methanesulfonate mutagenesis of a susceptible strain. *Journal of Economic Entomology*, **85**, 352–8.

Spackman, M.E. and McKechnie, S.W. (in press) Assessing population structure in the cotton bollworm, *Helicoverpa armigera* (Lepidoptera: Noctuidae), using variation in the A+T-rich region of mitochondrial DNA, in *Proceedings World Cotton Conference 1. Challenging the future* (eds E.A. Constable and N.W. Forrester), CSIRO Publications, Melbourne.

Southwood, T.R.E. (1971) The role and measurement of migration in the population system of an insect pest. *Tropical Science*, **13**, 275–8.

Sun, C.N., Wu., T.K., Chen, J.S. and Lee, WT. (1985) Insecticide resistance in diamondback moth, in *Proceedings of 1st International Workshop on Diamondback Moth Management*, (ed. N.S. Talekar), AVRDC Publication No. 86-248 (The Asian Vegetable Research and Development Centre), Shanua, Taiwan, pp. 359–71.

Tabashnik, B.E. (1994) Evolution of resistance to *Bacillus thuringiensis*. *Annual Review of Entomology*, **39**, 47–79.

Taylor, M. F. J., Heckel, D. G., Brown, T. M., Kreitman, M. E. and Black, B. (1993) Linkage of pyrethroid insecticide resistance to a sodium channel focus in the tobacco budworm. *Insect Biochemistry and Molecular Biology*, **23**, 763–75.

Thompson, M., Steichen, J.C. and ffrench-Constant, R.H. (1993) Conservation of cyclodiene insecticide resistance associated mutations in insects. *Insect Molecular Biology*, **2**, 149–54.

Trowell, S., Forrester, N., Daly, J., Garsia, K., Bird, L. and Lang, G. (1992) Developmental trials of the Heliothis ID kit. Sixth Australian Cotton Conference, Broadbeach, Queensland, Australian Cotton Growers Research Association, pp. 209–16.

Trowell, S., Garsia, K., Skerritt, J., Hill, A., Forrester, N. and Bird, L. (1994) The LepTon™ test kit (Pat. Pending) – the heliothis ID project comes to fruition. The Seventh Australian Cotton Research Conference, Broadbeach, Queensland, Australian Cotton Growers Research Association, pp. 17–20.

Trowell, S. C., Lang, J. C. and Garsia, K. A. (1993) A Heliothis identification kit, in *Pest Control and Sustainable Agriculture*, (eds S.A. Corey, D. J. Dall and W. A. Milne), CSIRO Publications, Melbourne, pp. 176–9.

Twine, P. and Kay, I.R. (1973) *Heliothis* species and DDT resistance. *Queensland Agricultural Journal*, **99**, 520–2.

Williamson, M. S., Denholm, I., Bell, C. A. and Devonshire, A. L. (1993) Knockdown resistance (*kdr*) to DDT and pyrethroid insecticides maps to a sodium channel gene locus in the housefly (*Musca domestica*). *Molecular and General Genetics*, **240**, 17–22.

Zalucki, M.P., Daglish, G., Firempong, S. and Twine, P. (1986) The biology and ecology of *Heliothis armigera* (Hübner) and *H. punctigera* Wallengren (Lepidoptera: Noctuidae) in Australia: what do we know? *Australian Journal of Zoology*, **34**, 779–814.

Zalucki,M.P., Murray, D.A.H., Gregg, P.C., Fitt, G.P., Twine, P.H. and Jones, C. (1994) Ecology of *Helicoverpa armigera* (Hübner) and *H. punctigera* (Wallengren) in the inland of Australia: larval sampling and host plant relationships during winter/spring. *Australian Journal of Zoology*, **42**, 329–46.

Zetterquist, H. and Olerup, O. (1992) Identification of the HLA-DRB1*04, -DRB1*07, and -DRB1*09 alleles by PCR amplification with sequence-specific primers (PCR-SSP) in 2 hours. *Human Immunology*, **34**, 64–74.

5

A molecular and ecological investigation of the large arionid slugs of North-West Europe: the potential for new pests

L.R. Noble and C.S. Jones

ABSTRACT

Some large arionid slugs are now considered to be important agricultural pests in many areas of Continental Europe, where 'slug plagues' have become a common phenomenon. However, identification of arionid taxa is problematic, complicated by common colour varieties and morphology, confounding attempts to map their distribution and differentiate pests and non-pests. We have used a PCR-based molecular approach combined with anatomical investigations, first as an aid to identification and subsequently as a tool for investigating the ecology and reproductive strategies and phylogeny of this group. Our evidence suggests that the large arionids are divisible into four distinct taxonomic groups, of which three appear capable of generating pests, distinguishable from their congeners on the basis of reproductive strategy and life-history data.

Using this integrated approach we have produced a concordance between colour varieties and taxa, allowing use of varietal distribution records from the early 19th century to the present day. Combining these data with our own pan-European survey has made it possible to identify the rapid recent advance of some taxa throughout the UK. Their life-history and colonization profiles closely parallel those of more strictly Continental species shortly before they began to produce major outbreaks and became serious agricultural pests, suggesting that these taxa deserve further attention

The Ecology of Agricultural Pests, Edited by W.O.C. Symondson and J.E. Liddell.
Published in 1996 by Chapman & Hall, London. ISBN 0 412 62190 8

from agriculturalists. Future studies may be assisted by the taxon-specific molecular markers we have identified.

5.1 INTRODUCTION

It is widely thought that few species of slug have been adversely affected by human activities, while many have actually had their range greatly extended by such disturbances, often travelling many miles with arable produce (Chichester and Getz, 1968; Smith, 1989). However, despite some range expansion the overall effect of cultivation is to reduce species diversity, and promote the proliferation of those few taxa adapted to these conditions. In this way human activity can be seen to lead to the whole-scale perturbation of the malacofauna of an area, increasing gene flow between populations via occasional dispersal of individuals in a patchwork of sites, some of which may become nuclei for the introduction and establishment of exotic species, together with their attendant parasites and pathogens. The distribution of several relatively recent introductions, such as that of *Deroceras caruanae* (Pollonera) from ports in north-west England (Kerney, 1976), and the extreme genetic variability of *D. reticulatum* (Müller) (Foltz *et al.*, 1984; Fleming, 1989) suggest that cultivation does indeed favour long-range dispersal and enhances gene flow (De Wilde *et al.*, 1983). These are characteristics more generally associated with highly mobile insect pests, rather than slugs.

Although slugs are considered to be serious pests of arable crops in most temperate agricultural systems, attention has focused on the study and control of many of the smaller species of *Deroceras, Milax* and *Arion*. Their proliferation has coincided with increasing slug damage to commercial arable crops over the past 30–40 years (South, 1992) and has shifted attention away from the larger species of arionid. The large arions were once considered uncommon pests, except perhaps locally on allotments or in market gardens, which represent a mosaic of habitats and offer more shelter (Atkinson *et al.*, 1979). However, in recent years (1970 onwards) the often neglected threat posed by some of these species has been forcibly impressed upon agriculturalists and horticulturalists in North-Western Europe, where 'slug plagues' have had widespread and often devastating effects.

In most outbreaks the species responsible has been referred to as the 'Spanish slug' or the 'German slug', reflecting the distribution of outbreaks and the investigators' belief in the slugs' origins, which to date remain unsubstantiated. Reischütz (1984a,b) describes how *Arion lusitanicus* Mabille, 1868 has become a serious pest species in alpine meadows in Austria and Switzerland where the endemic *Arion ater rufus* (L.) are now thought to be under threat of replacement. *A. lusitani-*

cus has heavily damaged pastures and fields in the Vorarlberg and Tirol, where it spread rapidly from gardens to farms in 1982–1983, acquiring the popular name of 'Kapuzinerschnecke' or 'Spanische Wegschnecke'. The slug plagues of Venice (Cesari, 1978) and the recent outbreaks of large black slugs in Mediterranean holiday resorts (publicized in the popular press in spring/summer 1992), may both be attributable to *A. lusitanicus*, which has reached particularly high densities on farmland in the foot hills of the French Pyrenees, south-eastern France and Switzerland (L.R. Noble, personal observations).

The considerable speed and distances over which horticultural and vegetable produce are transported and the dispersal ability of juvenile slugs (Chichester and Getz, 1968) suggest that the threat posed by large arionid taxa deserves further investigation. However, the series of events which lead up to these outbreaks and the factors which trigger them have been ignored. This may be because, in spite of their pan-European distribution (Noble, in press), conspicuous colouring and size, and their long history of study (Lister, 1678) the systematics and taxonomy of the large arionids remains surprisingly unreliable (but see Noble, 1992 and in press). Hence, although most identification records suggest that *A. lusitanicus* is the culprit there is good evidence (Noble, 1992) to suppose that this 'species', in common with many others in the Arionidae, may actually comprise several closely related yet discrete taxa – some pest and some innocuous – with distinctly different ecologies and distributions. The ecology and anatomy of at least one taxon (*A. flagellus*; Collinge, 1893) are considered sufficiently distinct by some authorities (e.g. Davies, 1987) to warrant its removal to a different subgenus. The extent to which relationships within the *A. lusitanicus* aggregate may be further complicated by hybridization is unknown.

Hence, dubious identifications will hamper pest recognition and, more importantly, the study of circumstances leading to 'slug plagues'. To resolve this unsatisfactory situation we have initiated a pan-European survey of arionid taxa (Noble, in press), augmented by a simple unambiguous molecular methodology for identification and assessing relationships and population structure. Once the identification problem has been resolved the ecologies of pest and non-pest taxa, in particular their reproductive strategies, may be compared to indicate those factors responsible for a species attaining pest status. Only by reliably identifying the component taxa and understanding something of their ecology will the threat posed by the large arions be fully appreciated.

Although anatomical characters are valuable taxonomic markers their use often requires experience and an intimate knowledge of a specific group, especially where hybridization is a possibility. Molecular information from samples validated by specialists using proven anatomical characters is one way of standardizing identification by non-specialists using techniques which are becoming more widely available. If suitable molecular markers are chosen this approach has the advantage of being quick, unequivocal, and capable of detecting relatively minor amounts of introgression from an anatomically similar sibling species. Population analyses using sequence data derived from such markers promises the possibility of identifying the origin and subsequent spread of pest taxa.

5.2 THE LARGE ARIONIDS

Hesse (1926) regarded the arionids as the most difficult genus of slugs to identify. (Throughout this chapter we shall use the words 'arion' and 'arions' to refer exclusively to taxa of the subgenera *Lochea* and *Mesarion* of the genus *Arion*, while the term 'arionid' may include taxa of the genus *Geomalacus*.) As colour is extremely variable and changeable with age, immature slugs have often been described as new species, leading to many superfluous names and uncertain species. Earlier workers variously divided the genus *Arion* according to physical properties of the vestigial shell (Moquin-Tandon, 1855), on internal characters (Simroth, 1885) or by external characters such as colour (Linneaus, 1758). However, the various subgenera proposed by Hesse (1926) have not been satisfactorily delimited (for a full discussion see Backeljau and De Bruyn, 1990) and nowadays, largely for convenience, European fieldworkers (e.g. Cameron *et al.*, 1983) arbitrarily divide the arions according to adult size. Adult large arions, corresponding to the subgenus *Lochea* (Moquin-Tandon, 1855) of the subfamily Arioninae, are typically greater than 5 cm long when fully extended.

Slugs, particularly the arionids, are inherently difficult to identify to the specific level. The multiplicity of common colour forms with a plethora of varietal names, frequent hybridization and a lack of taxonomically useful hard parts, compound identification difficulties in a group consisting of aggregates of morphologically similar species. The high incidence of facultative self-fertilization (Williamson, 1959; McCracken and Selander, 1980) complicates matters further in this already difficult group, since isolated peripheral populations may produce phenotypically similar, yet genetically distinct forms. If these persist they may eventually diverge sufficiently from their progenitor to be recognized as separate taxa, confounding systematicists further by hybridizing with other closely related forms. These problems have led to specific names being used in accordance with local or national tradition, without critical evaluation of diagnostic characters, leaving the nominal species poorly understood. Without a reliable systematic approach the identification, and hence the distribution, of large arionid taxa must remain uncertain.

Most workers agree that the large arionids are divisible into three major groups. Two, the *Arion ater* aggregate and the *A. lusitanicus* aggregate, are of the subgenus *Lochea*. The third includes taxa of the genus *Geomalacus*, which Quick (1960) states exhibit the most primitive characters, a solid internal shell, anterior reproductive orifice, narrow foot fringe and restricted Lusitanian distribution, traits which indicate *Geomalacus* taxa may lie closer to the ancestral line of the family Arionidae than other north-west European representatives. However, work by specialists is showing that many species of both genera must now be considered species aggregates (Evans, 1986; Noble, 1992; Castillejo and Rodriguez, 1993) making classification to the subgenus controversial (Davies, 1987).

5.3 IDENTIFICATION

5.3.1 Anatomical characters

Many authors have attempted to classify the large arions on the basis of colour and/or genital anatomy, and although the value of these various characters and approaches is discussed here a detailed account of how they may be used to effect an identification is inappropriate in a review chapter; the interested reader is referred to Noble (1992) and Noble (in press).

British workers usually consider the large arions to consist of two aggregates, the *Arion ater* agg. and the *A. lusitanicus* agg., based on the position of the ligula, a holdfast/stimulatory device in the distal genitalia. The *A. lusitanicus* aggregate is distinguished from the *A. ater* aggregate by a prominent proximal oviducal bulge marking the position of the ligula. Continental authors often include these aggregates under *A. empiricorum* Férussac, 1819. The *A. ater* agg. is a synthesis of *A. ater* (L.) and *A. rufus* (L.), and many earlier workers assigned material to these divisions on the basis of colour. In the tenth edition of *Systema Naturae* (1758, p. 652) Linnaeus distinguished *Limax ater* and *Limax rufus* (now *Arion ater* (L.) and *Arion rufus* (L.)) by external characters only. In his ideal case the former was black and the latter orange, but the variety of colour forms often made classification difficult; Taylor (1907) describes 14 colour varieties assignable to *A. ater* (s.l.). The first attempts to solve this problem by anatomical studies were made by Nunneley (1837), and in the same year by M.C. Verloren (in litt., Quick, 1947), who figured the genitalia and copulation of Dutch *Limax rufus*. However, when Pollonera (1889) distinguished the two forms for the first time on anatomical grounds it became apparent that both taxa share the black colour variety. This was confirmed by Gerhardt (1940), who correlated anatomical differences with differences in time spent in coitus. However, the *A. lusitanicus* agg. shares many colour varieties with the *A. ater* agg., confounding identification on external characters alone.

Quick (1952, 1960) attempted to remedy this situation by producing a clear description of each species of large arionid then recognized, giving diagnostic characters based on the relative proportions of various parts of the distal genitalia. Cain and Williamson (1958), Roach (1971) and Evans (1977, 1986) examined the distal genitalia and colour varieties in more detail, suggesting that measurement of several superficial features of the distal genitalia, together with scores of juvenile and adult coloration were often required for differentiation of taxa. These studies demonstrated that identification of the *A. ater* agg. taxa was further complicated, at least in Britain (Burnet, 1972; Evans, 1977, 1986), by frequent hybridization between *A. ater* and *A. rufus*. These findings supported the proposition of Cain and Williamson (1958) that *ater* and *rufus* should be given sub-specific status, hence the trinomials *A. ater ater* and *A. ater rufus* were adopted. However, without a reliable species-specific character or multivariate ordination procedures, unambiguous classification of individual taxa and recognition of some *ater-rufus* hybrids remained problematic.

An observation by Cain (1980) partially resolved this issue, suggesting that dissection of the distal genitalia to reveal the shape and form of the ligula was useful in differentiating species of the *A. lusitanicus* aggregate. Earlier workers had considered this species-specific device too variable and plastic to be a useful taxonomic tool (e.g. Scharff, 1891; Altena, 1954; Chevallier, 1969). In fact Chevallier (1981), in a continuation of his work on the *A. empiricorum* complex (Chevallier, 1969, 1972), preferred to use superficial and subjective criteria based on the distal genitalia, colour variety and radula teeth, to establish four subspecies and two forms of *A. rufus*. His attempted revision did not clarify the taxonomy of the large arionids, confusing *ater-rufus* hybrids and *A. lusitanicus* species. With the application of multivariate ordination procedures to the large arions Noble (1987) and L.R. Noble and N.J. Evans (in preparation) were able to assess objectively the reliability of a variety of anatomical and biochemical characters by their ability, singly and in combination, to produce discrete groupings on an *a priori* basis. These operational taxonomic units (OTUs) generally corresponded with available topotypic material and the accepted published descriptions of the various taxa (Quick, 1960). Classification was simplified when, following on from Cain's (1980) observations, it was found that the shape and form of the ligula was the only single anatomical feature which showed discrete variation corresponding to the OTUs defined by multivariate analyses. The ligula is now recognized as a species-specific character (Noble, 1992).

The *A. lusitanicus* agg. consists of an unknown number of taxa of undecided taxonomic status (Castillejo and Rodriguez, 1993), several of which have been found recently in large numbers on marginal agricultural land in southern France (Noble, 1992). The nominal taxon, *A. lusitanicus* Mabille 1868, was illustrated and described by Pollonera (1889, 1890) and Simroth (1891) who suggested it could be identified by its ligula 'of *A. empiricorum*' and oviduct 'of *A. subfuscus*'. Davies (1987) adds that it can be distinguished from other large arionids by the shape and form of its spermatophore, which corresponds with ligula morphology (using the nomenclature of Noble, 1992). On the shape and form of the ligula there appear to be several, as yet, undescribed taxa which probably have a very restricted distribution in the Eastern Pyrenees (Noble, 1992). Although the ligula of one of these (*A. lusitanicus* form A; Noble, 1992) was figured by Altena (1955) and described by him as *A. lusitanicus* Mabille, 1868 its ligula is very unlike that of what Davies (1987) considers to be the nominal taxon. The ligula of another taxon (*A. lusitanicus* form B; Noble, 1992) is unique and has not been described or figured elsewhere. The precise distribution and pest potential of many '*lusitanicus*' taxa is therefore unknown.

Early British records of *A. flagellus* Collinge, 1893 were identified as either *A. subfuscus* (Draparnaud, 1805) or *A. lusitanicus* (Mabille ss.) because of common colour varieties and the similarity of the distal genitalia. However, Cain (1980) differentiated them on details of the ligula, suggesting that *flagellus*' ligula was more similar to that of *A. subfuscus* than to *A. lusitanicus* (Mabille), a supposition supported by external features, eggs and functional anatomy of the genitalia (Davies, 1987).

Most useful pulmonate classification schemes are based on features of the distal genitalia, and this approach has been useful in differentiating the large arions. Identification to the specific level is difficult, requiring detailed knowledge of the slug's internal anatomy, but attempts to classify specimens on their colour have been generally unreliable, as many colour varieties are common to taxa of both aggregates, and no generally agreed concordance of taxa and colour variety has been produced (but see Noble, in press, summarized in Table 5.1).

5.3.2 The value of colour varieties in identification

Although the various taxa can be defined on the anatomy of the distal genitalia and a single specimen identified by the shape and form of its ligula, unambiguous external species-specific characters would be useful identification aids for non-specialists. A concordance between the many colour varieties identified by early 18th and 19th century naturalists and taxa delimited using a proven method of classification would indicate those varieties which are taxon-specific. Furthermore, as the majority of accounts of large arions from this era are based on records of colour varieties, these data could be interpreted to reveal a reliable chronicle of the distribution and colour of large arions in Britain at this time.

A concordance of colour varieties (as described and figured in Taylor, 1907) and taxa defined on ligula morphology was produced (Noble, in press). Depending upon the level of differentiation required (either between aggregates or species) many varieties can be useful taxonomic markers (see Table 5.1 for concordance and Table 5.2 for descriptions) (Noble, in press); much of the earlier confusion resulted from an unreliable taxonomy and parochial surveys. Clearly some taxa can be unambiguously identified on colour alone. For instance var. *aterrima* is diagnostic of *A. a. ater*, while taxa of the *A. ater* agg. may be distinguished from those of the *A. lusitanicus* agg. by the presence of vars. *alba, plumbea* and *rubra. A. flagellus* is perhaps the most clearly recognized species, being the only large arionid to have a greenish hue.

5.3.3 Soluble enzyme systems

Early protein electrophoresis of the *A. ater* agg. (Burnet, 1972) necessarily concentrated on a limited number of enzyme systems from only a few populations in Derbyshire and Yorkshire. However, this study, of what were later identified as *ater* and *ater* × *rufus* hybrids (Evans, 1977), clearly showed that these two taxa differed little at the enzyme level; *ater* was fixed for alleles found in *rufus*/hybrids at all enzyme loci investigated. Foltz *et al.* (1982) extended the survey of the large arions in Britain; running nine enzyme systems which gave four polymorphic (indophenol oxidase [EC 1.15.1.1] – Ipo; phosphoglucomutase [EC 2.7.5.1] – Pgm; acetylesterases [EC 3.1.1.1] – Est-1 and Est-2) and nine monomorphic loci (α-glycerophosphate dehydrogenase [EC 1.2.1.12] – a-Gpd; malate dehydrogenase [EC 1.1.1.37] – Mdh-1 and Mdh-2; glutamic-oxaloacetic transaminase [EC 2.6.1.1] – Got; acetylesterase – Est-3; two unspecified peptidase loci – Pep-1 and Pep-2; leucine amino peptidase [EC 3.4.13] – Lap; mannose phosphate isomerase [EC 5.3.1.8] – Mpi) in *A. ater* agg. taxa from 33 sites, and six systems giving two polymorphic (Est-2 and Lap) and six monomorphic loci (a-Gpd, Ipo, Pgm, Est-1, Est-3 and Pep) in three neighbouring introduced populations of *A. lusitanicus*. However, the value of this work was limited because they did not differentiate between *ater* and *ater* × *rufus* hybrids, nor did they collect widely; most of their samples were from south-west Ireland, with a limited number from South Wales and southern England. Consequently, it was impossible to suggest which, if any, enzyme systems differentiated *ater* and *rufus*, and as the *lusitanicus* populations sampled were from introduced garden populations, which

Table 5.1 Concordance of colour varieties recognized by Taylor (1907) with taxa classified on ligula morphology as described in Noble (1992). (Reproduced with permission from Noble (in press) *Malacologia*.)

Colour variety	Taxon						
	A.a.ater	A.a.rufus	ater × rufus hybrids	A. lusitanicus (Mabille)	A. lusitanicus (variant A)	A. lusitanicus (variant B)	A. flagellus
aterrima[a]	✓	✗	✗	✗	?	?	✗
albolateralis[ah]	✓	✗	✓	✗	?	?	✗
castanea[rh]	✗	✓	✓	✗	?	?	✗
marginella[rh]	✗	✓	✓	✗	?	?	✗
bicolor[h]	✗	✗	✓	✗	?	?	✗
alba[Aa]	✓	✓	✓	✗	?	?	✗
plumbea[Aa]	✓	✗	✓	✗	?	?	✗
rubra(sub var.)[Aa]	✓	✗	✓	✗	?	?	✗
fasciata[f]	✗	✗	✗	✗	?	?	✓
reticulata[f]	✗	✗	✗	✗	?	?	✓
hiberna[(f)]	✗	✗	✗	✗	?	?	✓
atra[A]	✓	✗	✓	✓	?	?	✓†
rufa[A]	✗	✓	✓	✓	✓	✓	✓
succinea[A]	✗	✓	✓	✗	?	?	✓

✓ – colour variety typical of this taxon; ✗ – colour variety not typical of this taxon; ? – little known taxon; † – appears very dark green on close inspection.

a – diagnostic of *A.a.ater*; ah – diagnostic of *ater* and hybrids; rh – diagnostic of *rufus* and hybrids; h – diagnostic of hybrids; Aa – distinguishes *A.ater* agg. from *A. lusitanicus* agg.; f – diagnostic of *A. flagellus*; (f) – probably diagnostic of *A. flagellus* but common also to *A. subfuscus*; A – common to both *A. ater* and *A. lusitanicus* aggregates.

Table 5.2 Colour varieties ascribed to the large arions (modified after Taylor, 1907)

Colour variety	Foot fringe	Foot-sole	Dorsum	Shield	Sides	Sub-varieties	Other remarks
aterrima	black	black	black	black	black	none	Probably the *Arion hispanicus* of Simroth.
albolateralis	orange	pale	black	black	white	*media* – pale brown on sides of body and shield	Restricted to N-W Britain.
castanea	lighter/paler brown/orange than rest of body.	pale	brown	brown	brown	*semi-nigra* – black body dark brown shield. *olivacea* – greenish body.	*olivacea* – probably *A. flagellus*. Restricted to W coast of Eire Scharff (1891).
marginella	yellow/vermilion red	pale	black	black	black	*griseo-marginata* – footfringe greyish.	*marginella* is a variety indicative of hybrid swarms in Britain.
bicolor	yellow/orange	pale	brown	brown	yellow/orange	*scharffi* – black dorsum.	Restricted to W in Britain. Easily confused with *A. subfuscus*.
alba	white/yellowish lineolation clear.	pale	white	white	white	*simplex* – uniform white. *marginata/albida* – yellow fringe. *elegans* – orange head + fringe. *oculata* – black tentacles. *cinereo-nebulosa* – brown side spots, yellow fringe.	*alba* and its varieties are almost certainly the result of a mutation in the biochemical pathway for melanin production. Rare, except at periphery of species' range where inbreeding may promote appearance of this deleterious recessive gene.

plumbea	dull yellow	pale	lead grey	pale grey	pale grey	*cinerea/cinerascens – dull* brown foot fringe.	Problematic variety as common to all taxa of the *A. ater* agg.
fasciata	various	pale	various	dark bands	dark bands	*subdeleta* – greenish or yellowish-grey body.	Most records are *A. flagellus*, and the subvar. cannot be anything else
reticulata	pale/orange	pale	yellowish and reticulated	yellowish	yellowish	*subreticulatus* – less clearly reticulate.	Restricted to West Britain and Eire. Probably *A. flagellus*.
hiberna	claret coloured	pale	purple	purple	pale purple	none	Rare, *A. flagellus* (Davies, 1987)?
atra	black	pale median area	black	black	black	none	The *A. empericorum* of many authors.
rufa	bright red	pale	orange/red	orange/red	orange/red	*rufula* – pale form. *coccinea, lamarkii+ rubra* – bright red. *draparnaudi* – red, yellow fringe. *nigrescens* – red/brown with red fringe.	Taylor (1907) notes that this var. is seldom found above 1,000m. The black varieties dominant above this altitude. *Rubra* is rare and characteristic of *ater*.
succinea	orange/red	pale	yellowish	yellowish	yellowish	*melanocephala* – head black. *aurantia* – body orange. *livida* – livid yellow body, greenish grey dorsum. *pallescens* – pale yellow body.	Coloration common in juv. hybrids and *rufus*, although may occasionally persist in mature individuals.

undoubtedly had atypical gene frequencies, these results could not be used to identify species-specific alleles.

Noble (1987) electrophoresed 500 population samples (a minimum of 20 individuals per population) of *A. a. ater* and *A. a. rufus* for 13 soluble enzyme systems, representing 20 presumptive loci (Figure 5.1). *A. a. ater* was fixed at all loci for the most common allele in Continental *rufus* populations, with the exception of Got; *ater* is fixed for an allele which appears to be uncommon in *rufus* (Figure 5.1), although more extensive population surveys may reveal this not to be so. Similarly the unique Pgm allele apparently characteristic of Norwegian *ater* may eventually be found in *rufus*. However, overall *ater* displays a striking genomic depauperacy and close genetic affinity to the morphologically similar *rufus*, suggesting that *ater* may have been derived from *rufus* in the relatively recent past in a series of saltational founder events in peripheral populations, perhaps driven by selection for self-compatibility following catastrophic population crashes (see later). Consequently the similarity of the genomes (Figure 5.1) means that there are no *ater*-specific alleles; although heterozygosity at enzyme loci may exclude an unknown specimen(s) from *ater*.

Like *ater* and *rufus, lusitanicus* and *flagellus* are not easily differentiated by enzyme analysis. Although *lusitanicus* has a greater variety of alleles at most loci only two species-specific alleles, Est-2^{a} (*lusitanicus*) and Lap^{s-} (*flagellus*), are sufficiently common to be useful in differentiating these taxa from each other (Figure 5.1). Protein electrophoresis may be an appropriate tool for identification when specimens are too small for examination of anatomical markers. However, routine enzyme analysis is laborious and its opponents convincingly argue that as a highly specific class of genes soluble enzymes do not accurately represent the total genetic variation, much of which goes undetected (enzyme electrophoresis is estimated to detect about one-third of the total genetic variation; Hartl, 1980).

However, soluble enzymes act as co-dominant Mendelian markers and are useful in breeding system studies, where it is necessary to distinguish selfed and outcrossed progeny. However, as non-destructive sampling of a prospective parent is impossible and the number of parent-specific polymorphisms is strictly limited, many mating pairs possess insufficient genetic differences to differentiate selfed from crossed progeny. Hence, the value of this approach is strictly limited in a complex of closely related, often highly inbred taxa.

5.3.4 A molecular approach

Extensive inbreeding and bottlenecking in isolated populations and the potential for selfing have probably been responsible for the formation of the closely related taxa characteristic of species aggregates of the large arions (Noble, in press). Hence, it is not surprising that conventional genetic markers, such as enzymes, reveal limited variation within and between these taxa and their populations. Clearly a more sensitive technique is required, one that can identify markers in any part of the genome, reliably to reveal diagnostic genetic differences between the large arions. Ideally the technique should permit the derivation of taxon-specific markers which can be developed to give unambiguous results, preferably via a quick and simple methodology which may be used under field conditions.

Molecular analysis offers several advantages over conventional genetic markers such as soluble enzyme systems for population analyses and species identification. An investigation need not necessarily be limited to one class of highly specific markers: DNA from a variety of coding or non-coding regions may be utilized to give a truly random distribution of markers throughout a genome. The presence of multilocus systems means that more independently segregating markers can be scored in a survey, thereby improving the possibility of identifying taxon-specific markers. A minimum of technical sophistication is then required to use probes derived from these unique fragments for unambiguous identification of taxa in the field. The advent of the polymerase chain reaction (PCR) has further refined approaches to identification, permitting amplification of species-specific markers from minute tissue biopsies. However, conventional PCR requires sequence information so that oligonucleotide primer pairs can be designed to amplify specific regions, thereby limiting this approach to those organisms considered sufficiently important to benefit from the investment of time and effort required for cloning and sequencing.

There is little or no useful molecular data which can be used to provide a convenient entry point into the genome of arionid slugs for systematic purposes without subsequently sequencing the PCR products; as is the case for well characterized gene regions such as ribosomal RNA genes and mtDNA cytochrome oxidase subunit 1. A technically simpler and less involved approach employs low-stringency PCR using short primers of arbitrary sequence, a technique known as *r*andom *a*mplified *p*olymorphic *D*NA (RAPD) (Williams *et al.*, 1990), including its many synonyms, arbitrarily primed PCR (AP–PCR) (Welsh and McClelland, 1990) and DNA amplification fingerprinting (DAF) (Caetano-Anolles *et al.*, 1991; Caetano-Anolles, 1993) (for recent reviews, see Williams *et al.*, 1993; Smith and Williams, 1994). Unlike conventional PCR-based analyses, the RAPD assay uses single, arbitrary 5–20 base pair (bp) oligonucleotide primers to amplify regions between small inverted repeats that are scattered throughout the genome. Polymorphisms in the fragments so generated (ca. 0.2–ca. 2 kbp length) are

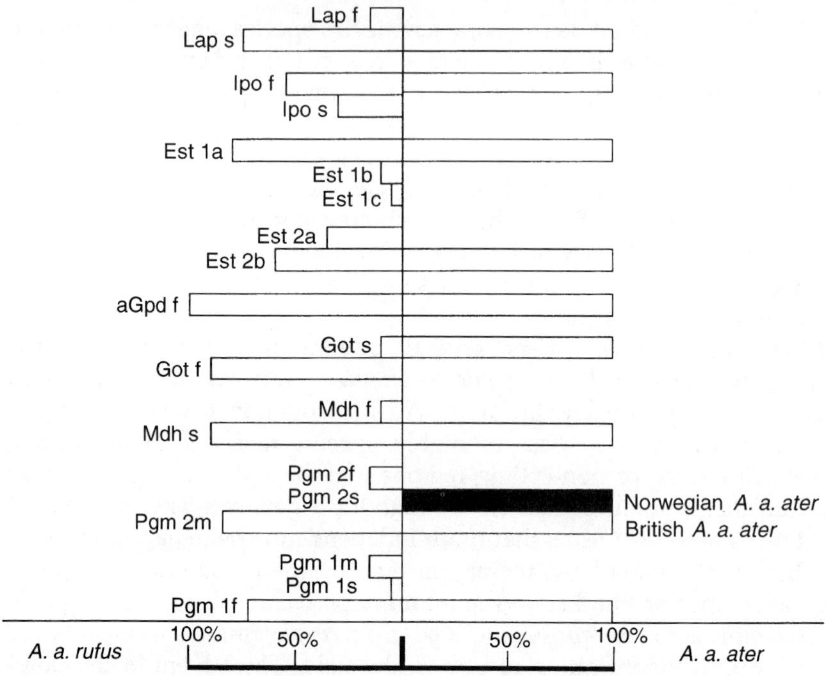
Lap f
Lap s
Ipo f
Ipo s
Est 1a
Est 1b
Est 1c
Est 2a
Est 2b
aGpd f
Got s
Got f
Mdh f
Mdh s
Pgm 2f
Pgm 2s
Norwegian A. a. ater
Pgm 2m
British A. a. ater
Pgm 1m
Pgm 1s
Pgm 1f
A. a. rufus
100%
50%
50%
100%
A. a. ater

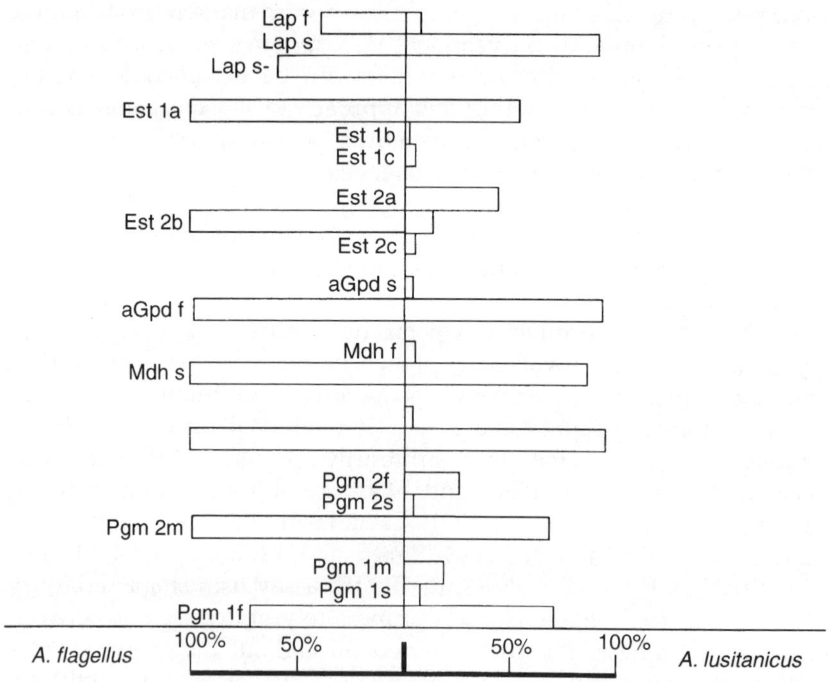
Lap f
Lap s
Lap s-
Est 1a
Est 1b
Est 1c
Est 2a
Est 2b
Est 2c
aGpd s
aGpd f
Mdh f
Mdh s
Pgm 2f
Pgm 2s
Pgm 2m
Pgm 1m
Pgm 1s
Pgm 1f
A. flagellus
100%
50%
50%
100%
A. lusitanicus

frequent, being caused by single base pair changes in the priming sites, and serve as dominant Mendelian markers (Williams *et al.*, 1990; Hadrys *et al.*, 1992; Hedrick, 1992). The multiple products generated can be separated electrophoretically and are generally visualized on agarose or polyacrylamide gels, stained with ethidium bromide or with silver. Amplification of organellar DNA is apparently negligible (Hadrys *et al.*, 1992; Heun and Helentjaris, 1993) possibly because of its relatively small size or due to a shortage of the inverted repeats necessary for the production of discrete fragments.

RAPD polymorphisms have been shown to be useful in cultivar (Welsh and McClelland, 1990), pathotype (Goodwin and Annis, 1991) and clone (Okamura *et al.*, 1993) identification. Hence, the RAPD assay provides great potential for population-wide surveys of highly inbred taxa because large numbers of closely related individuals can be characterized rapidly. In addition, no prior molecular knowledge of the study organism is required due to the arbitrary design of the primers, and the small quantity of template DNA required allows work on tiny individuals or biopsies (Welsh and McClelland, 1990), permitting the development of a non-destructive approach.

RAPDs have produced useful diagnostic profiles which may be of value for discriminating between closely related species. Ten putative taxa of large arionids, identified using traditional anatomical markers or ligula morphology, were screened with a total of 20 decamer primers (L.R. Noble *et al.*, in prep). Nine primers revealed unambiguous and informative banding patterns, and generated consistent useful polymorphisms for distinguishing large arionid species. Consistent with Hadrys *et al.* (1992), who suggested that certain RAPD–PCR products could be exploited to generate markers that are characteristic of different taxonomic levels, we have found RAPD markers that appear to differentiate clearly anatomically distinct species of different genera, such as *A. lusitanicus* taxa from *G. maculosus* (shown in Figure 5.2), while others distinguish the anatomically similar taxa within an aggregate, shown for *A. lusitanicus* Mabille, 1868 and *A. flagellus* in Figure 5.3. Additionally, we have found RAPD fragments that are common to all the species within an aggregate (Figure 5.3).

Figure 5.1 Soluble enzyme systems assayed for *A. a. ater*, *A. a. rufus*, *A. lusitanicus* and *A. flagellus*. The horizontal bars represent the frequency of each allele at a locus averaged over all populations sampled. The different alleles of each enzyme system are denoted by the following suffixes: in order of anodal mobility, **f** (fast); **s** (slow); **s**$^-$ indicates a slightly more cathodal allele than an **s**. For esterase systems only, the notations **a**, **b**, **c** apply, again in order of anodal mobility.

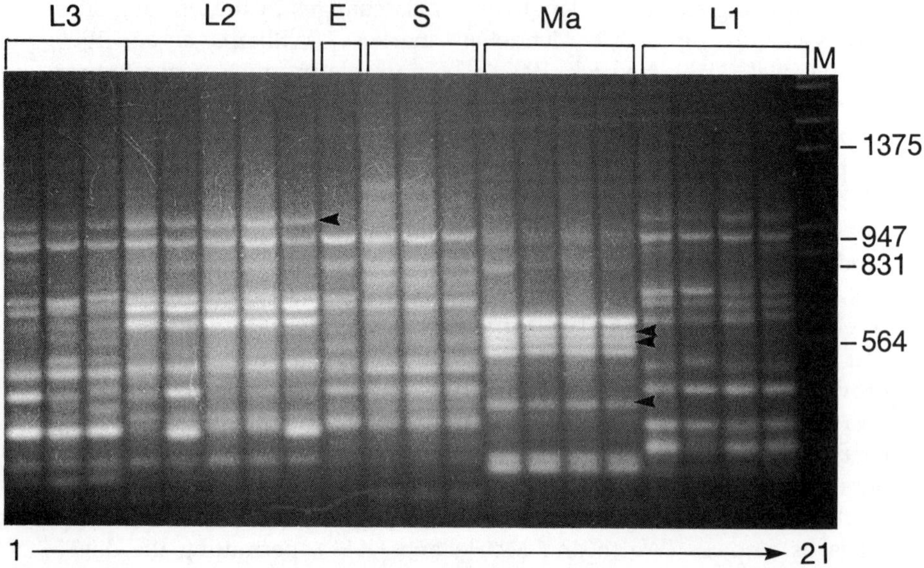

Figure 5.2 Taxon-diagnostic RAPD markers for different taxa of the large arionids, amplified with RAPD primer 18 (5'-AATCGGGCTG-3'). The species are: *Arion lusitanicus* (L1–*A. lusitanicus* form B, Mount Louis, France; L2–*A. lusitanicus* form A, Sapin Fleuri, S. France; L3–*A. lusitanicus* Mabille, 1868, Olette, S. France); *Arion sulcatus* Morelet, 1845 (S–Picos de Europe, Valley of Rio Deva, Spain); *Arion empiricorum* Férussac, 1819 (E–Heulgoat Forest, Brittany), and *Geomalacus maculosus* Allman, 1843 (Ma–Ballinskelligs, Eire). Arrows indicate a taxon-specific fragment of *Arion lusitanicus* aggregate (1000 bp fragment found in all populations, see L1, L2 and L3) and several species-specific fragments from *Geomalacus maculosus* (600 bp, 560 bp and 350 bp). M, molecular weight marker, lambda DNA restricted with *Eco*RI/*Hind*III.

Although comparison of RAPD profiles is useful, standardizing PCR parameters between laboratories and interpretation of the resulting banding patterns may be problematic, due particularly to the variability of band intensities between runs. Scoring ambiguities may be resolved by identifying and excising 'unique species-specific' fragments which are then radiolabelled for use as species probes on full Southern blots of the RAPD gels.

Previous studies have made use of RAPD analysis to reconstruct phylogenies (Puterka *et al.*, 1993; Tibayrenc *et al.*, 1993; Wilkie *et al.*, 1993) and in many cases the resulting pattern agrees with that derived by more conventional means (Kazan *et al.*, 1993). Although these studies relied upon phenetic techniques to interpret the RAPD data, a cladistic approach has recently been evaluated (Megnegneau *et al.*, 1993; Ralph *et al.*, 1993; Schierwater *et al.*, 1994), although many workers remain cautious about including RAPD fragments in such analyses (see Clark and Lanigan, 1993). The use of RAPDs in phylogenetics is limited to relatively closely related species, and systematic characters are defined as RAPD bands of a particular molecular weight and character states as the presence or absence of that fragment (Bowditch *et al.*, 1993). It is legitimate to use RAPD markers in a study of the large arionids. Here we have groups of taxa which, on any anatomical, biochemical genetic and distributional criteria, are clearly closely related. Although *Geomalacus* species are clearly distinct from other arions they are included in the family Arionidae and so can be considered in this analysis.

However, RAPDs are only considered useful for estimating nucleotide divergence for closely related taxa if a number of criteria are satisfied (Clark and Lanigan, 1993). In particular, the demonstration that co-migrating bands of the same size in different species are homologous. The inference of homology is strong when the total sequence divergence between taxa is low and many RAPD fragments are shared. When higher taxonomic levels are examined only a few bands will be shared and their homology is in doubt. Therefore, when scoring these character states it is essential to check the homology of a shared fragment. Apparently 'shared' fragments between taxa can be investigated by Southern analysis of RAPD gels using the isolated 'shared' bands as radiolabelled probes.

In this study we have used cluster analysis of genetic distances derived from similarity coefficients (Nei and Li, 1979) based on the presence–absence of RAPD fragments, after verifying ambiguous common fragments by probing (data not shown, see L.R. Noble *et al.*, in prep). Phylogenetic analysis of the distance matrix using the FITCH, KITCH and neighbour-joining options in the phylogenetic inference package, PHYLIP (Felsenstein, 1993) produced the same groupings. These divisions correspond to those delimited by morphometric and biochemical analysis (Noble, 1987), ligula morphology (Noble 1992) and distributional (Noble, in press) data, suggesting that each OTU recognized on the basis of RAPD markers is a good biological species or subspecies. The inference is that this molecular analysis recognizes the same taxonomic discontinuities as the classical systematic approaches.

Hence, the approximate phylogenetic distances between these groups can be regarded as preliminary indicators of their relationships, and hypotheses can be framed regarding the evolution and ecological characteristics of particular taxa. The phylogenetic analysis produced an unrooted tree with four discrete branches (Figure 5.4). Each branch corresponded to a major subdivision of the large arions, as defined by anatomy or ecology, with the twigs representing individual taxa. As anticipated, the two representatives of *Geomalacus maculosus* Allman,

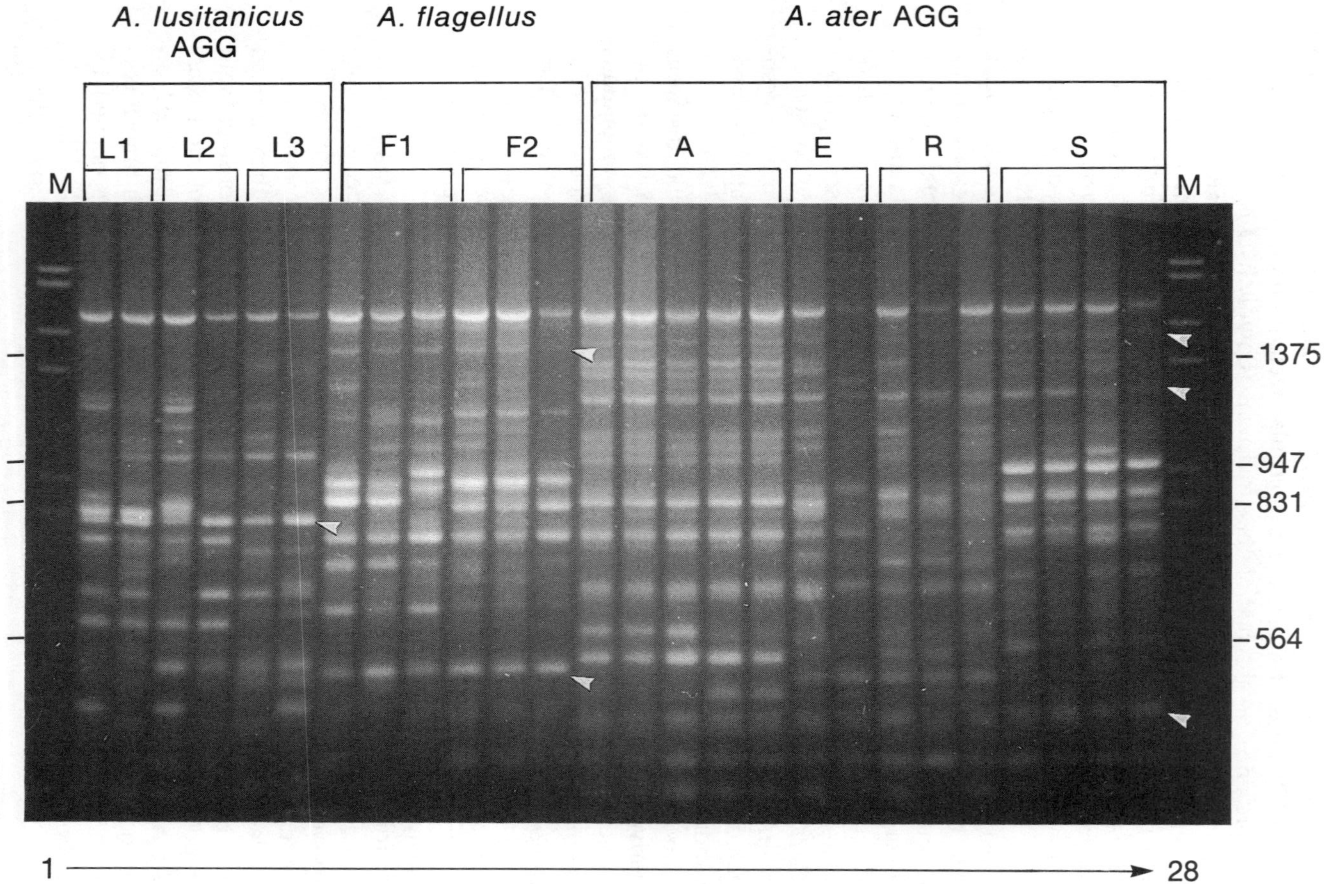
A. lusitanicus AGG
A. flagellus
A. ater AGG
L1
L2
L3
F1
F2
A
E
R
S
M
M
1375
947
831
564
1
28

Figure 5.3 Taxon- and species-aggregate-specific RAPD markers for the large arionids, amplified with RAPD primer 9 (5′-AGCAGCGTGG-3′). *Arion lusitanicus* aggregate (lanes 2–3, *A. lusitanicus* form B, L1 – Mount Louis, France; lanes 4–5, *A. lusitanicus* form A, L2 – Sapin Fleuri, S. France; lanes 6–7, *A. lusitanicus* Mabille, 1868 (non-pest taxon), L3, Olette, France), *Arion flagellus* (lanes 8–10, F1 – Thurstaston Common, Wirral, UK; lanes 11–13, F2, Wirral Country Park, Wirral, UK), *Arion ater* aggregate including the sub-species *Arion ater ater* (lanes 14–18, A, Snowshill, Cotswolds, UK), *Arion empiricorum* Férussac, 1819 (lanes 19–20, E – Heulgoat Forest, Brittany), *Arion ater rufus* (lanes 21–23, R – Brackenwood, Wirral, UK) and *Arion sulcatus* Morelet, 1845 (lanes 24–27, S – Picos de Europe, Valley of Rio Deva, Spain). Arrows indicate species-specific markers differentiating taxa of the *Arion lusitanicus* agg. and *Arion flagellus*, and aggregate-specific markers common to taxa of the *Arion ater* aggregate. M, molecular weight marker, lambda DNA restricted with *Eco*RI/*Hind*III.

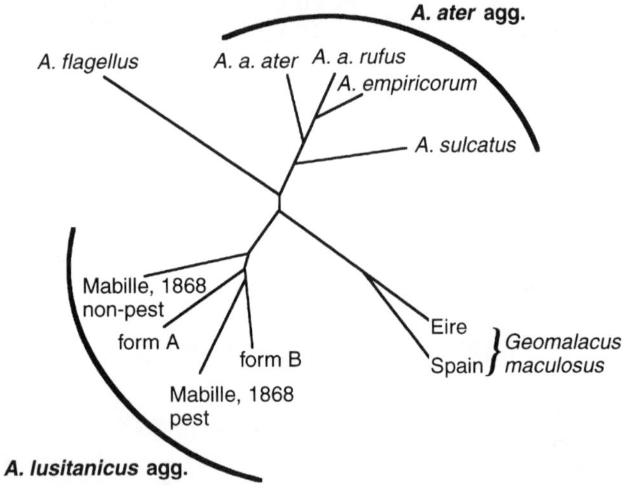

Figure 5.4 Distance dendrogram generated from ethidium bromide-stained/agarose RAPD data using nine primers. All major fragments apparently shared between taxa were checked for homology by probing before their inclusion in the binary data set. This tree has no outgroup and was arbitrarily rooted using the first species listed in the input data (*A. a. ater*). Four branches are apparent. Taxa included in the *A. ater* and *A. lusitanicus* aggregates are indicated by the thick curves. Within the *A. lusitanicus* agg. the Olette, S. France samples of *A. lusitanicus* Mabille, 1868 are labelled 'Mabille, 1868 non-pest' and those from Igls, Austria are labelled 'Mabille, 1868 pest' to emphasize their agricultural status (see section 5.7). Ligula morphology of 'Mabille, 1868 pest' samples was most similar to 'Mabille 1868, non-pest', yet cluster analysis shows them to be genetically distinct, the former having affinities with the relictual *A. lusitanicus* form B. The *A. empiricorum* samples are actually classed as *A. a. rufus* on ligula morphology, which corresponds with the clustering shown here. *A. sulcatus* Morelet, 1845 is possibly a relictual taxon which, on ligula morphology, is related to *A. a. rufus*. *A. flagellus* forms a third branch and *G. maculosus* Allman, 1843 forms a fourth, represented by two populations, Santiago de Compostela, Spain (labelled 'Spain'), and Ballinskelligs Bay, Eire (labelled 'Eire').

1843 (from populations in Eire and Spain), thought to be closest to the ancestral lineage of the large arions, formed a distinct group (Figure 5.4) clearly separable from the *A. lusitanicus* and *A. ater* aggregates each of which lie on a discrete branch – suggesting that the infrageneric systematics of the sub-genus *Lochea* may need some revision (Backeljau and De Bruyn, 1990). Another representative of the *Lochea*, *A. flagellus* – anatomically similar to the *A. lusitanicus* taxa – did not group with other representatives of the *A. lusitanicus* aggregate, appearing to cluster more closely with the *A. ater* agg., yet remaining distinct. This situation warrants further research, especially in view of Davies' (1987) suggestion that this species may belong to a different sub-genus (*Mesarion* and not *Lochea*), and is therefore perhaps most closely allied to *A. subfuscus* (the type species of the *Mesarion*). If this were substantiated then it would have important implications for the taxonomic status of not only *A. flagellus*, but also the *A. ater* and *A. lusitanicus* aggregates. The sub-genus *Lochea* would then become a paraphyletic group, urging a review of the taxonomic position of the other large arions. However, this situation must remain unresolved until *A. subfuscus* can be added to the analysis. For the moment it is apparent that *flagellus* remains close to, but distinct from *A. empiricorum*, the type species of *Lochea*.

Both *A. empiricorum* and *A. sulcatus* Morelet, 1845 (supplied by Ms S. Davies, verified as from original sites and fitting the authors' descriptions) have a ligula morphology, body coloration and geographic distribution broadly within that of *A. a. rufus* as suggested by Noble (1992). Consequently, it is not surprising that the RAPD data suggest that *A. empiricorum*, as defined on topotypic material, is the closest OTU to *A. a. rufus* in the analysis, and Quick's (1960) suggestion that it is a synonym of *rufus* seems justified. However, the position of the little known taxon *A. sulcatus* Morelet, 1845 found in Iberia and along the Western Pyrenees and anatomically similar to *rufus*, of which Quick (1960) considered it a synonym, is more equivocal. The morphology of the ligula of *sulcatus* identifies it as *rufus*, but there are minor yet consistent differences which differentiate it from northern populations of *A. a. rufus* (Noble, 1992). Molecular data show that these specimens belong to a distinctive member of the *A. ater* aggregate (Figure 5.4), which must be investigated further if its taxonomic position is to be substantiated.

The *A. lusitanicus* agg. is similarly differentiated on ligula morphology and RAPD analysis. The nominal taxon. *A. lusitanicus* Mabille, 1868 is clearly separated from each of two relictual taxa, *A. lusitanicus* form A and form B (Noble, 1992). Specimens of *A. lusitanicus* taken from an area where it is a recognized pest (labelled Mabille 'pest' in Figure 5.4) produced a fourth group, more closely related to *A. lusitanicus* form B than to the nominal taxon. However, although the ligula morphology of these specimens (from Igls, Austria) was closest to that of *A. lusitanicus* Mabille, 1868 the ligula was markedly asymmetrical; a distinctive feature of form B.

Clearly then, at the level of individual taxa each twig within a major branch corresponds to an OTU distinguished by morphometric analysis of the distal genitalia and ligula morphology. This suggests that a multilocus approach using RAPD markers resolves the same genetic divisions as those highlighted by apparent discontinuities in anatomical systems probably under quantitative genetic control. Once it is established that RAPD analysis corresponds with an accepted anatomical classification scheme the search for taxon-specific markers may begin.

5.4 DISTRIBUTION AND ECOLOGY

The frequent occurrence of shell granules in Pleistocene and post-glacial deposits (Evans, 1972) suggests that the arionids may have had a long history in North-West Europe (but see Bräm, 1956), although fossil evidence suggests that they may have benefited rather less from the advent of agriculture than did *Deroceras reticulatum*, the shells of which steadily increase in number in post-glacial deposits. Nevertheless, some arions are associated with humans, where they find shelter, colonizing gardens and waste ground like *A. lusitanicus* and *A. flagellus*, or an agency of dispersal, *ater* × *rufus* hybrids have been introduced into America, Australia, New Zealand and South Africa by human activities. Some of these introductions have become established, often as pests, benefiting from the genetic diversity afforded by continued immigration. The efficiency with which some arionids are dispersed by human agencies is illustrated by the occurrence of a single record of *A. lusitanicus* in the Falkland Isles (Von Proschwitz, 1988). Conversely, *Geomalacus* species have a very restricted distribution; the most widespread, *G. maculosus* Allman 1843, is found only in south-west Eire and in the north-west of the Iberian peninsula (Platts and Speight, 1988).

As consequences of their dispersal capabilities, hybridization and recent range expansions are significantly modifying the distributions of the constituents of this taxonomically complex group, which is already bedevilled by identification problems. Such perturbations may prove important in the genesis of pests from closely related species aggregates, suggesting that the distribution and spread of many of the large arionids should be closely monitored.

The distributions of the large arions in North-West Europe have been revised using a taxonomy based on robust, relatively unambiguous taxon-specific morphological characters (Noble, 1987 and in press) (Figure 5.5). The discrete grouping of this complex group of taxa by anatomical methods corresponds well with that suggested by other lines of evidence, such as molecular analyses. A pan-European survey using this systematic approach (Noble, in press) allowed the first accurate investigation of the distribution and ecology of the large arionids. (It is our intention here to indicate in what ways this approach has altered perception of the European distribution of large arionids and to discuss the implications of these findings for pest

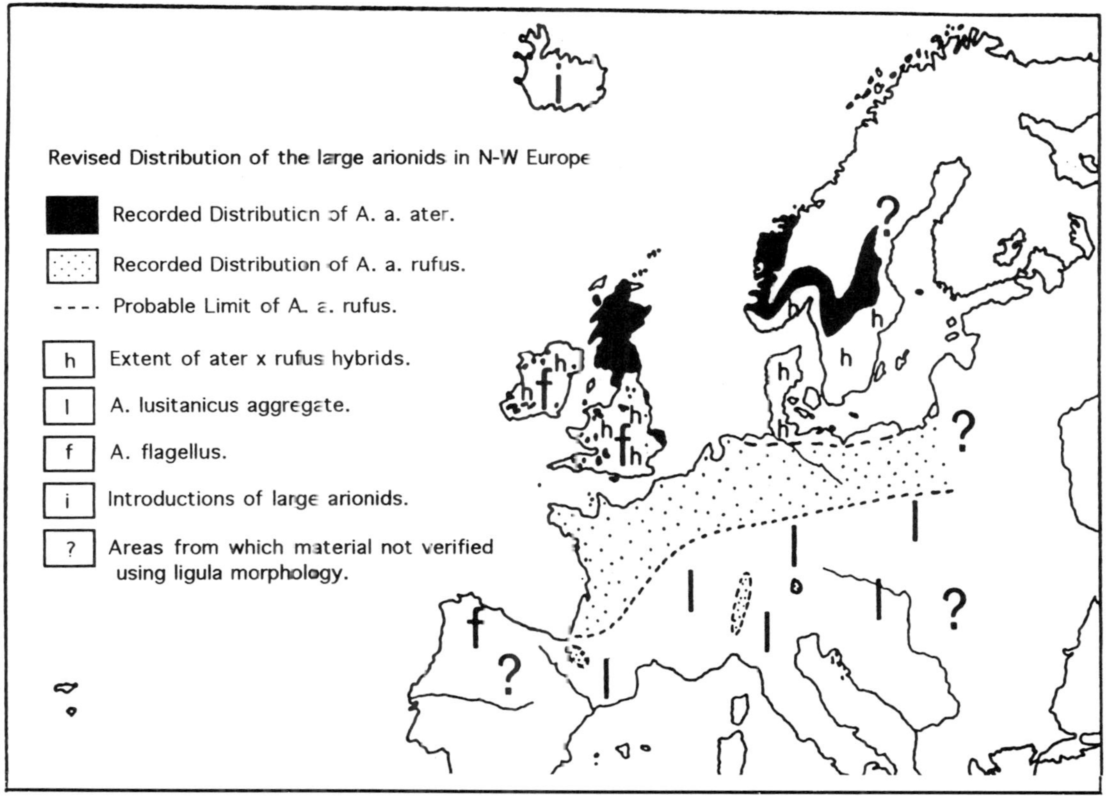

Figure 5.5 Map showing the revised distribution of the *Arion ater* and *Arion lusitanicus* aggregates in North-West Europe, based on a new systematic approach using the morphology of the ligula (Redrawn from Noble (in press), *Malacologia*.)

genesis; consequently, only the generalized distributions of taxa are discussed. Although it is implicit to our argument that human activities can and do often allow widespread dispersal it is impossible to consider all such cases.)

Much of the earlier confusion surrounding the distribution of these taxa was attributable to the frequent and widespread hybridization of *A. a. ater* and *A. a. rufus*, and the failure to recognize their hybrids. Consequently, pure *ater* was found to have a very restricted distribution, confined to upland and northern Britain and fjordlands in Norway. It is absent from continental Europe, where reports of *ater* proved to be little more than inaccurate descriptions of *A. a. rufus* or *A. lusitanicus* spp. Similarly *A. a. rufus* has a more restricted range than that suggested by earlier workers. It is virtually absent from Britain, surviving in its pure form in proximity to horticultural/agricultural activities in *ater*-free areas, mainly in the lowlands of England and Wales.

A. ater ater × *A. a. rufus* hybrids are distributed over a much greater part of Britain than was previously supposed. This discrepancy may be partially explained by the difficulty that earlier workers had in identifying hybrids, but may also be attributable to continuing multiple introductions of *rufus* and the rapid spread of the subsequent hybrids. Ellis (1969) suggested that *A. a. rufus'* continental distribution may have been south of *ater*'s and that it may have been a fairly recent introduction into the British Isles, a contention supported by Burnet's (1972) isozyme work which linked British and Dutch populations of *A. ater*. Most hybrid populations in Britain may have originated from local introductions of *rufus*, producing hybrid swarms which grade rapidly into pure *ater* on heath or moorland. This suggested that these taxa were each adapted to different habitats, and that if a continental hybrid zone existed it would be fairly narrow – reflecting the lowered fitness of the hybrids. However, in Denmark and north Germany the zone was found to be much wider and the intergradation of the taxa smoother than models incorporating hybrid inferiority would predict (Hewitt, 1989), *rufus* enzyme alleles being found in animals of intermediate anatomy in the extreme north of Denmark (Noble, 1987, in press) (Figure 5.5). These observations indicate either a zone of primary contact, and/or little fitness differential between the two taxa in the often disturbed continental lowland habitats.

It would appear that *ater* has survived on the periphery of *rufus* only in areas where it is protected from introgression; elsewhere, *rufus* has displaced it by extensive hybridization and range expansion, probably begun towards the end of the last glaciation. However, the greatest range expansion has come about through human agencies, with highly successful hybrid swarms producing slug plagues on the Pacific and Atlantic coasts of North America (Chichester and Getz, 1968; L.R.

Noble and A.J. Cain, personal observation). Here, as in Europe, human activity has served to accelerate gene flow between populations, encouraging the proliferation of hybrids.

Human activities may have similarly modified the distribution of *A. flagellus*, which is now perceived as largely synanthropic in lowland habitats of Britain. The continental affinities of *flagellus* remained obscure until Castillejo (1992) recorded it as common in Galicia in the north-west of the Iberian peninsula. Discounting recent synanthropic records of *flagellus* from mainland Britain, and considering only the observation of Scharff (1891) and his contemporaries (Taylor, 1907) on the distribution of the *flagellus*-specific green–yellow colour varieties, such as var. *reticulata*, suggests that *flagellus* was established in Eire in the late 19th century, a time when such colour varieties were absent from both disturbed and undisturbed habitats of England and Wales (Theobald, 1904; Taylor, 1907). Not until 1904 (Collinge, 1904) was *flagellus* (described as *subfuscus* at the time, Davies, 1987) first reported on mainland Britain. More recently (Evans, 1977; Davies, 1987 and references therein; Noble, in press) there has been a spate of new British records for this species. These data suggest that *flagellus* may have originally been restricted to Lusitanian habitats, from which it has spread. This contention is supported by Conroy's (1980) and Noble's (personal observations) findings that *flagellus* is often restricted to local sites with the warmest microclimate; a finding consonant with its continued absence from the continental climate of the east coast of Britain (Noble, in press) (Figure 5.5). This recent range expansion seems to have been at the expense of *ater* in many areas. Long-term study of several heathland habitats in lowland England (L.R. Noble, in preparation) has shown that *A. flagellus* spreads from the warmest microclimates to rapidly dominate areas once the preserve of *ater*.

A similar situation involving *lusitanicus* and *rufus* pertains in Austria, where the endemic *rufus* are now under threat of replacement by what is thought to be the nominative member of the *A. lusitanicus* aggregate, which has become a serious pest. Reischütz (1984a,b) describes damage to the pastures and fields of the Vorarlberg and Tirol by the 'Kapuzinerschnecke', which seems to have suddenly increased in numbers after spreading from gardens to farms in 1982–1983. *A. lusitanicus* species has also been implicated in the slug plagues of Venice (Cesari, 1978) and more recent outbreaks of large black slugs in Cyprus holiday resorts. With the exception of a few very localized garden populations, *A. lusitanicus* species are absent from mainland Britain, while the nominative taxon (*A. lusitanicus* Mabille, 1868) predominates south of the 20°C July isotherm in areas receiving less than 15 cm of July rainfall, in the drier continental climate of mid- and southern North-West Europe (Noble, in press) (Figure 5.5).

Although widespread and common in continental Europe this species appears to be replaced in certain areas of the eastern Pyrenees by a number of local undescribed forms, probably relict taxa once more widespread during the cold dry climate of the Pleistocene. Although their anatomy and RAPD profiles differentiate them, the taxonomic status of these relictual *A. lusitanicus* remains unknown, awaiting further studies on the extent to which they and the nominative taxon may introgress.

5.5 REPRODUCTIVE ECOLOGY

The population genetics and evolution of breeding systems in animals has received comparatively little attention, compared to those of plants (Allard, 1975; Solbrig, 1976). Only recently (Hedrick and Whittam, 1989; Browne and Hoopes, 1990; Hastings, 1991) has it been realized that uniparental reproduction, amixis or self-fertilization, in some animal systems may be part of an evolutionary strategy, rather than an auxiliary mode of reproduction employed when there is no opportunity for outcrossing. The frequent incidence of facultative self-fertilization in many higher animal groups is persuasive evidence that such systems obtain the population genetic advantages of both modes of reproduction. In addition, recent modelling (Lynch *et al.*, 1991) suggests that mutation, crossing over, and gene conversion during mitosis (discussed in Hastings, 1991) may generate significant levels of genetic variation, producing substantial evolutionary change in even clonal or obligatory selfing populations. Thus, it may emerge that invertebrate life-cycles involving obligate sex and/or outcrossing may be evolutionary oddities, whereas life-cycles that incorporate occasional sex may be the norm (Hastings, 1992).

Most pulmonate molluscs are either simultaneous or sequential hermaphrodites whose breeding systems are poorly understood. This is especially true for the many terrestrial taxa which require internal fertilization (Runham and Hunter, 1970; South, 1992). The conspicuous 'mating dances' of pulmonate snails and slugs led early researchers to consider cross-fertilization the norm, a suspicion reinforced by population genetic theory which suggests that selfing should be avoided, since outcrossing generates adaptive genetic variation and heterozygosity which masks recessive deleterious mutations. Although uniparental reproduction has been recorded in several species of slug *Limax flavus* L. and *L. pseudoflavus* (Evans, 1983), *Deroceras agreste* (Linné, 1758) (Maury and Reygrobellet, 1963) it is particularly prevalent in the arionids; *A. ater* (s.l.) (Wotton, 1893; Williamson, 1959; Evans, 1977; McCracken and Selander, 1980; Foltz *et al.*, 1982), and *A. lusitanicus* Mabille, 1868 (Noble, 1987). However, many investigations of pulmonate breeding systems have been of limited value, either because the method of assessment is indirect and hence inadequate (see below), or because different workers used incomparable methods of determining self-compatibility, invalidating comparison of their results. Generally an ecologically meaningful comparison of self-compatibility between taxa can only be made within the same study.

Early workers (such as Wotton, 1893) relied upon direct observation of a few isolated individuals, whose parentage and breeding history were well known. Although these often anecdotal studies of breeding system may seem invaluable evidence in determining if a species is at least capable of self-fertilization, doubts often persist as to the species involved. The absence of genetic markers or an understanding of Mendelian inheritance prevented any attempts to estimate a mated individual's propensity for selfing. The first genetic investigation of the breeding system of a large arionid (Williamson, 1959) was forced to rely upon the very few visible markers then available. *Arion ater ater* homozygous for a rare recessive body colour marker (white) were allowed to mate with individuals homozygous for the common dominant body coloration (black). However, for a definitive estimate of selfing in the parental cross the dominant black allele required that the F_1 produce an F_2 by selfing. Although the segregation of these markers gave an estimate of selfing and outcrossing the effort required to breed animals over a 2-year period prohibited its routine use, and its accuracy was degraded by the impossibility of rearing all F_1 individuals through to an F_2.

The co-dominant markers afforded by protein electrophoresis allowed a more pragmatic approach to the problem of breeding systems. Electrophoresis of soluble enzyme markers in natural populations appeared a convenient and relatively simple method of obtaining these data and the literature of the late 1970s and early 1980s is replete with such studies. Burnet (1972) and later Evans (1977) and Foltz *et al.* (1982) used soluble enzyme markers to determine the breeding systems of a variety of large arionid species. Evans (1977) studied inheritance of markers in laboratory-bred individuals, but Burnet (1972) and Foltz *et al.* (1982) sampled natural populations, attempting to infer breeding systems by estimating gene frequencies in a population and testing for significant departures of heterozygotes from Hardy–Weinberg expectation. However, a variety of phenomena affect the frequency of heterozygotes in natural populations, making it difficult to determine the preferred mode of reproduction without corroboration from laboratory breeding studies. Sampling bias, particularly where specimens are collected from linear habitats (such as a field edge or roadside), or cluster around suitable microhabitats or food resources (Runham and Hunter, 1970; South, 1992), means that demes with substantially different gene frequencies may be sampled as a single population. This produces a deficit of heterozygotes according to Hardy–Weinberg expectations, the so-called 'Wahlund Effect' (Hartl, 1980). Other explanations for sporadic heterozygote deficits include negative heterosis (which seems unlikely, but cannot be entirely discounted) and null alleles (particularly common in esterase systems) (Oxford, 1973). Hence, in real terms studies based solely on interpretation of population genetic enzyme data provide very little evidence as to the probable, or even potential, breeding system of a species. The relatively few genes which can be scored by this method may produce gross underestimates of genetic variation, making comparisons between sibling or similarly closely related taxa the only reliable studies.

The effects of continued self-fertilization on enzyme polymorphisms may be masked or mimicked by a variety of other population genetic events, even in obligatory or preferentially outcrossing taxa. For instance, after many generations of self-fertilization, a single generation of random mating can restore genotype frequencies to Hardy–Weinberg equilibria, thus masking the effects of self-compatibility, whereas polymorphisms may be lost through population bottlenecking. The effects of the latter are particularly well illustrated by the extreme loss of enzyme polymorphisms in California elephant seals (Bonnell and Selander, 1974) and in cheetahs (O'Brian *et al.*, 1986). These examples emphasize that caution is required when attempting to determine a species' breeding system in the absence of polymorphisms. Similarly, even when polymorphisms are present it is equally difficult to state categorically that self-fertilization alone is responsible for departures from Hardy–Weinberg equilibria. Hence, although it may be tempting to extrapolate a population's breeding system from enzyme surveys such data, considered in isolation, do not reliably distinguish between facultative selfers and obligate outcrossers. Before a taxon's breeding system can be determined electrophoretic surveys must be complemented by experimental investigation of the breeding system, by examining the self-compatibility of isolates and mating pairs derived recently from mated wild-caught individuals.

However, as population bottlenecking favours self-compatibility and prolonged episodes of reduced heterozygosity, populations of gastropods with little genetic variation might be expected to show the greatest propensity for self-fertilization. Therefore, a large-scale enzyme survey (Noble, 1987) of large arionids from a variety of habitats throughout North-West Europe was undertaken to indicate typical levels of genetic variability for each taxon, suggesting which might be capable of self-fertilization. This assumption was tested in the laboratory using enzyme or visible markers to determine levels of self-compatibility in ostensibly outcrossed clutches from mated pairs, derived from wild-caught individuals. The capacity of each taxon for self-fertilization was further tested by estimating the number of fertile eggs produced by isolated virgin slugs. Although not absolute such estimates are pertinent to comparative studies of the breeding systems of closely related taxa, allowing them to be ranked in order of their comparative self-compatibility.

No populations of *ater, rufus, lusitanicus* Mabille, 1868 or *flagellus* showed consistent evidence of multilocus heterozygote deficits according to Hardy–Weinberg expectations. Some esterase systems occasionally showed a deficiency of heterozygotes, but departures from Hardy–Weinberg equilibria at single loci are difficult to attribute to inbreeding effects; unless the locus is in disequilibria with genes for selfing (Maynard Smith, 1978). A sample of 200 populations of *ater* suggested all were fixed for the same allele at 20 enzyme loci; Norwegian and British *ater* were only differentiated by PGM-2 alleles of slightly different mobilities (Figure 5.1). Laboratory studies of isolated virgin slugs (Noble, 1987) showed a continuum in self-compatibility from *ater* (90% of isolates selfed) through *rufus* and *lusitanicus* (about 60% of isolates selfed in both species) to *flagellus* which appeared to be an obligate outcrosser; in each case self-compatibility increased in samples collected from higher altitudes. In all three self-compatible taxa protein electrophoresis suggested at least a proportion of the first clutch progeny were the product of self-fertilization, even when outcrossing had occurred, a result consonant with Williamson's (1959) findings for captive breeding of *ater*.

There is a clear increase in self-compatibility with altitude/latitude, one taxon gradually replacing another (Figure 5.5). This ecological replacement suggests that there are corresponding differences in life-history strategy which adapt each to the particular environment it inhabits. Life-history traits co-varied with self-compatibility across all four taxa, suggesting that outcrossing was associated with the highest fecundity and greatest reproductive investment, which necessarily led to production of the smallest eggs and the greatest number of clutches (Noble, 1987).

5.6 THE EVOLUTIONARY HISTORY AND ECOLOGICAL GENETICS OF THE LARGE ARIONS – ASSESSING THE POTENTIAL FOR PEST GENESIS

It would appear then that facultatively selfing taxa, in both the arionids and the distantly related deroceratids, are always peripherally located to their closely allied outcrossing taxa, which occupy lower altitudes and latitudes (Noble, in press). This concordance may reflect the history of the differentiation of selfing taxa, all of which occupy habitats on land only recently free of ice or tundra, suggesting that they arose towards the end of the glaciation of North-West Europe. Rapid variations in temperature and sudden climatic deteriorations may have driven repeated re-colonizations of marginal habitats at this time (Hewitt, 1989). Peripheral populations of *rufus* may have been largely destroyed by these changes, selecting for self-compatibility, as an adaptation to small population sizes and a low rate of increase, and adaptation to local extremes of climate. Populations of *A. sulcatus* in Iberia and the Pyrenees may effectively represent relict populations of *A. a. rufus* which are isolated from what are now the main populations further to the north – a southern equivalent of *ater*.

The histories of the *A. lusitanicus* aggregate and *A. flagellus* are less clear. Molecular and anatomical data suggest that they are quite distinct, and no clearly identifiable sibling taxa to *flagellus* are known. As yet, we lack comprehensive anatomical, distributional and genetic data on the *A. lusitanicus* agg. and it is unwise to speculate on the phylogeny of its taxa; although the distribution and self-compatibility of *A. lusitanicus* Mabille, 1868 suggest that it may have played a similar role to *rufus*'.

In the absence of useful fossil material the extent to which speciation in the large arionids is attributable to Pleistocene climatic fluctuations (Solem, 1978) must remain debatable. But clearly when self-compatible taxa occur on the periphery of anatomically and genetically similar but more variable taxa there are good grounds for believing that the latter has given rise to the former (Lewis, 1962, 1966; Gottlieb, 1973, 1974, 1976).

Theoretically, selfing should be the favoured strategy in a self-compatible hermaphrodite if there is no concomitant reduction in fitness (Maynard Smith, 1978), thereby avoiding both the physical and genetic costs of mating. However, some large arions are facultative selfers, suggesting that there may be short-term advantages to outcrossing, or there would frequently exist genetic variation capable of suppressing it (Williams, 1975; Maynard Smith, 1978). Ecological models which seek to explain the maintenance of sex (such as Van Valen's [1973] 'Red Queen' and Bell's [1982, 1985] 'Tangled Bank') suggest that the unpredictable and novel variation produced by continued outcrossing is essential to counteract new environmental challenges. Such challenges are perceived as originating with biotic components of an organism's environment, such as more rapidly evolving parasites or pathogens ('Red Queen'), or sib-competition for limited resources ('Tangled Bank'); the physical environment, unless it varies in a random and implausible way, does not in itself favour sex. The distributions of sexual and asexual (and self-compatible) taxa of both plants (Levin, 1975) and animals (Glesener and Tilman, 1978) argue that an environment deteriorates as other species in the ecosystem evolve; asexual and self-compatible forms are more common at higher altitudes and latitudes than their sexual and outcrossing counterparts. Thus, self-compatible populations might be expected to inhabit biotically simpler habitats, comparatively free of parasites and pathogens.

How far this scenario can explain the distribution and pest status of the large arions is only just becoming evident. Until a reliable and unambiguous method of identifying *ater*, *rufus* and their hybrids, *lusitanicus* (s.l.) and *flagellus* was determined (Noble, 1992) their pan-European distribution was uncertain, and their capacity for self-fertilization unknown (Noble, in press). However, it is now apparent that *ater*, the most self-compatible large arion, has a much more restricted distribution than was previously suspected (Figure 5.5), being confined to the undisturbed, biotically simpler habitats of the mountains of Norway and Britain. At lower altitudes and latitudes, where habitats are biotically more complex, it hybridizes with its progenitor *rufus*, over much of its former range. This introgression has produced a series of localized hybrid swarms, which over much of Britain are associated with agricultural/horticultural activity. On the Continent *lusitanicus* and *rufus* partition habitats on the basis of summer rainfall, *lusitanicus* occupying the drier eastern localities, most notably in the Pyrenees. Both species are equally self-compatible, and there seems little difference in their potential exposure to parasites and pathogens, suggesting *lusitanicus*' higher fecundity (Noble, 1987) may contribute to its spread at *rufus*' expense in marginal *rufus* habitats, and help to explain the well-documented *lusitanicus* slug plagues.

A. flagellus is the only large arion which is not self-compatible. Like *lusitanicus*, it is highly fecund, capable of producing many small eggs, and there is evidence of recent (since 1800) range expansion, principally in the British Isles (Noble, in press). It shows a Lusitanian distribution, restricted to north-west Spain/Portugal, not found north of the Pyrenees, but abundant in South-West Eire and becoming increasingly common on the British mainland (Noble, in press). Its increase in numbers has coincided with a decline in *ater* specifically where the two species occur together, suggesting that *flagellus*' presence may in some way affect *ater*. Certainly *flagellus* is more fecund than *ater*, but as both are generalists and omnivorous it is unlikely that one has displaced the other by competition for resources (Conroy, 1980).

These two taxa lie at opposite ends of a continuum of self-compatibility in the large arions (Noble, 1987), with all the concomitant differences in breeding system and ecology that implies. Hence, a very clear difference between them might be in their tolerance to the attacks of parasites and pathogens. Although there are few investigations of the parasitology and pathology of these taxa there is good circumstantial evidence to suggest that *flagellus* is an important intermediate host of at least one nematode of veterinary importance (Jones *et al.*, 1980; Simpson and Neal, 1982), acquired through the coprophagous habit common to all arionids. Investigation of several sites on Wirral (L.R. Noble, personal observation) suggests that dog faeces infected with a variety of nematodes is often consumed by representatives of all the locally common arionids. If parasites are infecting the local populations then there are good theoretical reasons for expecting their effect to be comparatively more marked on the predominantly self-compatible taxa. This is supported by our observations which suggest that *ater* populations infected with tissue nematodes suffer a higher mortality than *flagellus* when brought into the laboratory. However, further work, including infectivity studies, is clearly necessary before *ater*'s decline, and *flagellus*' success, can be attributed to parasite-mediated interactions.

5.7 WHAT MAKES A PEST?

The aim of this chapter is to illustrate the advantages that molecular characterization has for the unambiguous and rapid identification of a taxonomically complex group, containing both innocuous and pest taxa. Once these taxa have been identified those characteristics which predispose certain of them to becoming agricultural pests can be investigated, and control strategies formulated which target only pest taxa. As a taxonomically complex group indiscriminant control measures (particularly biological ones) might exacerbate the problem if, as evolutionary theory predicts (sensu Van Valen's 'Red Queen'), they have a more marked effect on inbred taxa which cannot evolve to meet the challenge as quickly as their outbred counterparts.

The taxonomic uncertainty surrounding the large arions hints at the complexity of the underlying population genetics which molecular techniques are only now revealing. Models of metapopulation structure (which have their origin with Wright, 1931) suggest that a population divided into spatially separated subunits, but still capable of occasional gene exchange through migration, is the best equipped for rapid evolution. The subunits share a common gene pool, with favoured mutations spreading throughout the entire population. Inbreeding within each subunit would permit the establishment of variants by chance, and encourage local adaptation – an important consideration in sessile species. Facultative self-fertilization may provide an opportunity to achieve the same population structure, albeit with subunits isolated by selfing, and gene exchange occurring by outcrossing rather than migration. At a higher taxonomic level the many distinguishable yet closely related taxa, particularly those of the *lusitanicus* agg., which exist in the Pyrenees may effectively form a similar metapopulation structure. How far this situation may have contributed to the recent slug plagues of continental Europe remains debatable. Nevertheless, it is perhaps not altogether without significance that it is *A. lusitanicus* Mabille, 1868, one member of a group of closely allied taxa, which has become an important pest species on the Continent. The fact that populations of *A. lusitanicus* Mabille, 1868 from Igls (Austria), where it is a well-known pest, cluster more closely to one of the relictual Pyrenean populations (*A. lusitanicus*, form B; Noble, 1992) than to non-pest *A. lusitanicus* Mabille, 1868 on molecular, but not anatomical, criteria is not altogether surprising if there has been some gene flow between the two taxa (Figure 5.4). The prolific *ater* × *rufus* hybrid swarms in the British lowlands and on the north-west coast of North America may similarly benefit from occasional gene flow. The genesis of pest species by this route in species aggregates of the large arions is an area we are actively working in, one where RAPD-derived markers will have great potential for further applied research.

There is clearly a case for further work on the taxonomy and distribution of the large arionids of the Iberian peninsula and the Pyrenees. Such studies may lead to an understanding of *A. flagellus'* origins, its relationship to other relictual taxa in this area (thought to exist on morphological grounds [Castillejo, 1992]) and its potential pest status. Only by an integrated approach using morphological and molecular markers (ligula morphology and RAPD-derived markers) will the potential for gene flow between them and their taxonomic status be recognized. With molecular characterizations of these taxa it may be possible to generate DNA probes which can be used for identification and to assess population genetic structure and direction and magnitude of gene flow in pest taxa. Similar studies using enzyme markers have provided data on population structure which have aided control programmes of insect pests, indicating how genes for insecticide resistance may spread. Such an approach to monitoring populations of slug pests might be timely with the advent of biological and more widespread chemical control measures in this area.

Acknowledgements

We would like to thank Jenny Vernon and Lisa Trueman for technical assistance, Arthur Cain and Nick Evans for their continuing interest and encouragement, S. Borfield, H. Burla, R.H. Cowie, Steve Jones, Siân Jones, Hans and Trudi Meyer, Dave Noble and Torstein Sølhoy for their companionship and help in the field, and Dave Bilton, Stella Davies, Beth Okamura and Liz Platts for specimens. The authors are grateful to Thierry Backeljau and Mary Seddon for their comments and recommendations which greatly improved the manuscript. C.S. Jones is supported by the Wellcome Trust Programme in Taxonomy and Systematics.

REFERENCES

Allard, R.W. (1975) The mating system and microevolution. *Genetics*, **79**, 115–26.

Allman, G.J. (1843) On a new genus of terrestrial gastropod. *Athenaeum* (Sept.), **851**, col. 3.

Altena, C.O. van Regteren. (1954) Faunistische aantekeningen, 3. Hoe moet onze grote *Arion* heten? *Basteria*, **18**, 19–21.

Altena, C.O. van Regteren, (1955) Notes sur les Limaces. 3. Sur la présence en France d'*Arion lusitanicus* Mabille. *Journal de Conchyliologie*, **95**, 89–99.

Atkinson, H.J., Gibson, N.H.E. and Evan, H. (1979) A study of common crop pests in allotment gardens around Leeds, Yorkshire, England, UK. *Plant Pathology*, **28**, 169–77.

Backeljau, R. and De Bruyn, L. (1990) On the infrageneric systematics of the genus *Arion* Férussac, 1819 (Mollusca, Pulmonata). *Bulletin Institute Royal Science Naturelle Belgique, Biologie*, **60**, 35–68.

Bell, G. (1982) *The Masterpiece of Nature: The Evolution and Genetics of Sexuality*, Croom Helm, London, Canberra.

Bell, G. (1985) Two theories of sex and variation. *Experientia*, **41**, 1235–45.

Bonnell, M.L. and Selander, R.K. (1974) Elephant seals: genetic variation and near extinction. *Science*, **184**, 908–9.

Bowditch, B.M., Albright, D.G., Williams, J.G.K. and Braun, M.J. (1993) Use of Randomly Amplified Polymorphic DNA markers in comparative genome studies. *Methods in Enzymology*, **224**, 294–309.

Bräm, H. (1956) Was sind *Arion kinkelini* Wenz und *A. hochheimiensis* Wenz? *Ecologae geologicae Helvetiae*, **49**, 593–8.

Browne, R.A. and Hoopes, C.W. (1990) Genotype diversity and selection in asexual brine shrimp (*Artemia*). *Evolution*, **44**, 1035–51.

Burnet, B. (1972) Enzyme protein polymorphisms in the slug *Arion ater*. *Genetical Research*, **20**, 161–73.

Caetano-Anolles, G. (1993) Amplifying DNA with arbitrary oligonucleotide primers. *PCR Methods and Applications*, **3**, 85–94.

Caetano-Anolles, G., Bassam, B.J. and Gresshoff, P.M. (1991) DNA amplification DNA fingerprinting using very short arbitrary oligonucleotide primers. *Biotechnology*, **9**, 553–7.

Cain, A.J. (1980) Addendum to Conroy, B.A. (1980) Co-existence of two closely related species of *Arion* in natural habitats. *Journal of Conchology*, **30**, 189–200.

Cain, A.J. and Williamson, M.H. (1958) Variation and specific limits in the *Arion ater* aggregate. *Proceedings of the Malacological Society of London*, **33**, 72–86.

Cameron, R.A.D., Jackson, N. and Eversham, B. (1983) A field key to the slugs of the British Isles. *Field Studies*, **5**, 807–24.

Castillejo, J. (1992) The anatomy of *Arion flagellus* Collinge, 1893, present on the Iberian peninsula (Gastropoda: Arionidae: Terrestria Nuda). *Veliger*, **35**, 146–56.

Castillejo, J. and Rodriguez, T. (1993) Las especies del genero *Arion* Férussac, 1819 en Portugal (Gastropoda, Pulmonata: Arionidae). *Graellsia*, **49**, 17–37.

Cesari, P. (1978) Nota preliminare sulla diffusione in Italia e l'esplosione demographica nel Veneto di *Arion lusitanicus* Mabille (Mollusca: Pulmonata). *Lavori Societa Veneziana de Scienze Naturali*, **3**, 3–7.

Chevallier, H. (1969) Taxonomie et biologie des grands *Arion* de France (Pulmonata: Arionidae). *Malacologia*, **9**, 73–8.

Chevallier, H. (1972) Arionidae (Mollusca, Pulmonata) des Alpes et du Jura français. *Haliotis*, **2**, 7–23.

Chevallier, H. (1981) Taxonomie des "Limaces Rouges" (genre *Arion*, sous-genre *Arion* s.s., *Mollusca, Pulmonata*). *Haliotis*, **11**, 87–99.

Chichester, L.F. and Getz, L.L. (1968) Terrestrial slugs (ecology, biology, vector studies, laboratory culture). *Biologist*, **50**, 148–66.

Clark, A.G. and Lanigan, C.M.S. (1993) Prospects for estimating nucleotide divergence with RAPDs. *Molecular Biology and Evolution*, **10**, 1096–111.

Collinge, W.E. (1893) On the occurrence of *Arion lusitanicus*, Mab., in the British Isles, and descriptions of four new varieties. *Annals and Magazine of Natural History*, **12**, 414–15.

Collinge, W.E. (1904) Note on two varieties of *Arion subfuscus* Drap. *Journal of Malacology*, **11**, 98.

Conroy, B.A. (1980) Co-existence of two closely related species of *Arion* in natural habitats. *Journal of Conchology*, **30**, 189–200.

Davies, S.M. (1987) *Arion flagellus* Collinge and *A. lusitanicus* Mabille in the British Isles: a morphological, biological and taxonomical investigation. *Journal of Conchology*, **32**, 339–54.

De Wilde, J.J., Van Gorthem, J.L. and Marquet, R. (1983) Distribution and dispersal of *Boettgerilla pallens* Simroth, 1912 in Belgium. *Studiedocumenten van het K.B.I.N.*, No. **12**.

Draparnaud, J.P.R. (1805) *Histoire naturelle des mollusques terrestres et fluviatiles de la France*, (Atlas), Paris.

Ellis, A.E. (1969) *British Snails*, Clarendon Press, Oxford.

Evans, J.G. (1972) *Land Snails in Archaeology*, Seminar Press, London, New York.

Evans, N.J. (1977) *Studies on the variation and taxonomy in two species aggregates of terrestrial slugs; Limax flavus (L.) aggregate and Arion ater (L.) aggregate*, Unpublished PhD Thesis, University of Liverpool.

Evans, N.J. (1983) Notes on self-fertilization and variation in body colour in *Limax flavus* L. and *L. pseudoflavus* Evans. *Irish Naturalist's Journal*, **21**, 37–40.

Evans, N.J. (1986) An investigation of the status of the terrestrial slugs *Arion ater ater* (L.) and *Arion ater rufus* (L.) (Mollusca, Gastropoda, Pulmonata) in Britain. *Zoological Scripta*, **15**, 313–22.

Felsenstein, J. (1993) *PHYLIP – plylogenetic inference package*, version 3.5, Seattle, Washington, Department of Genetics, University of Washington.

Férussac, J.B.L. d'Audeband de. (1819) *Histoire naturelle générale et particulière des mollusques terrestres et fluviatiles*, Volume 2, Fol, Paris.

Fleming, C.C. (1989) The population structure of *Deroceras reticulatum* (Mollusca: Pulmonata: Limacidae), in *Electrophoretic Studies on Agricultural Pests*, Systematics Association, Special Volume No. 39 (eds H.D. Loxdale and J. Den Hollander), Clarendon Press, Oxford, pp. 393–413.

Foltz, D.W., Ochman, H., Jones, J.S., Evangelisti, S.M. and Selander, R.K. (1982) Genetic population structure and breeding systems in arionid slugs (Mollusca: Pulmonata). *Biological Journal of the Linnean Society*, **17**, 225–41.

Foltz, D.W., Ochman, H., and Selander, R.K. (1984) Genetic diversity and breeding systems in terrestrial slugs of the families Limacidae and Arionidae. *Malacologia*, **25**, 593–605.

Gerhardt, U. (1940) Neue biologische Nacktschneckenstudien. *Zeitschrift fur Morph. und Oekologie*, **36**, 557–80.

Glesener, R.R. and Tilman, D. (1978) Sexuality and the components of environmental uncertainty: clues from geographic parthenogenesis in terrestrial animals. *American Naturalist*, **112**, 659–73.

Goodwin, P.H., and Annis, S.L. (1991) Rapid identification of genetic variation and pathotype of *Leptosphaeria maculans* by random amplified polymorphic DNA assay. *Applied Environmental Microbiology*, **57**, 2482–6.

Gottlieb, L.D. (1973) Enzyme differentiation and phylogeny in *Clarkia franciscana, C. rubicunda* and *C. amoena. Evolution*, **27**, 205–14.

Gottlieb, L.D. (1974) Genetic confirmation of the origin of *Clarkia lingulata. Evolution*, **28**, 244–50.

Gottlieb, L.D. (1976) Biochemical consequences of speciation in plants, in *Molecular Evolution*, (ed. F.J. Ayala), Sinauer Assocs, Mass, pp. 123–40.

Hadrys, H., Balick, M. and Schierwater, B. (1992) Applications of random amplified polymorphic DNA (RAPD) in molecular ecology. *Molecular Ecology*, **1**, 55–63.

Hartl, D.L. (1980) *Principles of Population Genetics*, Sinauer Associates, Sunderland.

Hastings, I.M. (1991) Germline selection: population genetic aspects of the sexual/asexual life cycle. *Genetics*, **129**, 1167–76.

Hastings, I.M. (1992) Why is sex so frequent? *Trends in Ecology and Evolution*, **7**, 278–9.

Hedrick, P. (1992) Shooting the RAPDS. *Nature*, **355**, 679–80.

Hedrick, P.W. and Whittam, T.S. (1989) Sex in diploids. *Nature*, **342**, 231.

Hesse, P. (1926) Die Nacktschnecken der palaearktischen Region. *Abhandlungen des Archiv fur Molluskenkunde*, **2**, 1–151.

Heun, M. and Helentjaris, T. (1993) Inheritance of RAPDs in F_1 hybrids of corn. *Theoretical and Applied Genetics*, **85**, 961–8.

Hewitt, G.M. (1989) The subdivision of species by hybrid zones, in *Speciation and its Consequences*, (eds D. Otte and J.A. Endler), Sinauer, Mass., pp. 85–110.

Jones, G.W., Neal, C. and Turner, G.R.A. (1980) *Angiostrongylus vascorum* infection in dogs in Cornwall. *The Veterinary Record*, **106**, 3.

Kazan, K., Manners, J.M. and Cameron, D.F. (1993) Genetic variation in agronomically important species of *Stylosanthes* determined using random amplified polymorphic DNA markers. *Theoretical and Applied Genetics*, **85**, 882–8.

Kerney, M.P. (1976) *Atlas of the Non-Marine Mollusca of the British Isles*, Institute of Terrestrial Ecology, Cambridge, England.

Levin, D.A. (1975) Pest pressure and recombination systems in plants. *American Naturalist*, **109**, 437–51.

Lewis, H. (1962) Catastrophic selection as a factor in speciation. *Evolution*, **16**, 257–71.

Lewis, H. (1966) Speciation in flowering plants. *Science*, **152**, 167–71.

Linnaeus, C. (1758) *Systema Naturae per Regna tria Naturae, secundum classes, ordines, genera, species, cum characteribus, differentiis, synonymis, locis. Editio decima, reformata.* Holmiae: Salvius. [Tomus 1, 1758, *Introitus, Imperium Naturae*, and *Regnum Animale*; 1956, facsimile edition London; British Museum (Natural History)], pp. 652.

Lister, M. (1678) *Historiae Animale Anglia*, pp. 131, pl. 2.

Lynch, M., Gabriel, W. and Wood, A.M. (1991) Adaptive and demographic responses of plankton populations to environmental change. *Limnology and Oceanography*, **36**, 1301–12.

Mabille, M.J. (1868) Des limaciens européens. *Revue et Magasin du Zoologie Pure et Appliquée*, **20**, 129–46.

Maury, M.F. and Reygrobellet, D. (1963) Sur les distinctions spécifique chez les mollusques limacidés du genre *Deroceras. Séances Academie Science*, **288**, 2902–3.

Maynard Smith, J. (1978) *The Evolution of Sex*, Cambridge University Press.

McCracken, G.F. and Selander, R.K. (1980) Self-fertilization and monogenic strains in natural populations of terrestrial slugs. *Proceedings of the National Academy of Sciences USA*, **77**, 684–8.

Megnegneau, B., Debets, F. and Hoekstra, R.E. (1993). Genetic variability and relatedness in the complex group of black *Aspergilli* based on random amplification of polymorphic DNA. *Current Genetics*, **23**, 323–9.

Moquin-Tandon, A. (1855) *Histoire naturelle des Mollusques terrestres et fluviatiles de France*, (Atlas), Paris.

Morelet, P.M.A. (1845) *Description des mollusques terrestres et fluviatiles du Portugal*, J.B. Baillière, Paris, 177 pp.

Nei, M. and Li, W.H. (1979) Mathematical model for studying genetic variation in terms of restriction endonucleases. *Proceedings of the National Academy of Sciences USA*, **76**, 5269–73.

Noble, L.R. (1987) *The ecological genetics of reproduction in North-West European arionids*, Unpublished PhD Thesis, University of London, UK.

Noble, L.R. (1992) Differentiation of large arionid slugs (Mollusca, Pulmonata) using ligula morphology. *Zoologica Scripta*, **21**, 255–63.

Noble, L.R. (in press) The distribution of large arionids in North-West Europe: a new systematic approach. *Malacologia*.

Nunneley, T. (1837) A description of the internal structure of various Limaces found in the neighbourhood of Leeds. *Transactions of the Leeds Philosophical and Literary Society*, **1837**, 41–79 (pl. 1–7).

O'Brian, S.J., Wildt, D.E. and Bush, M. (1986) The cheetah in genetic peril. *Scientific American*, **254**, 68–76.

Okamura, B., Jones, C.S. and Noble, L.R. (1993) Randomly amplified polymorphic DNA analysis of clonal population structure and geographic variation in a freshwater bryozoan. *Proceedings of the Royal Society of London, B*, **253**, 147–54.

Oxford, G.S. (1973) The biochemical properties of esterases in *Cepaea* (Mollusca: Helicidae). *Comparative Biochemistry and Physiology*, **45**, 529–38.

Platts, E.A. and Speight, M.C.D. (1988) The taxonomy and distribution of the Kerry slug *Geomalacus maculosus* Allman 1843 (Mollusca, Arionidae) with a discussion of its status as a threatened species. *Irish Naturalist's Journal*, **22**, 417–30.

Pollonera, C. (1889) Nuove contribuzioni allo studio degli *Arion* europei. *Atti Reale Accademia della Scienze di Torino*, **24**, 623–40, pl. 9.

Pollonera, C. (1890) Recensement des Arionidae de la Region Paléarctique. *Bollettino dei Musei di Zoologia ed Anatomia comparata della Universitá di Torino*, **87**, 1–40.

Puterka, G.J., Black, W.C. IV, Steiner, W.M. and Burton, R.L. (1993) Genetic variation and phylogenetic relationships among worldwide collections of the Russian wheat aphid, *Diuraphis noxia* (Mordvilko), inferred from allozyme and RAPD-PCR markers. *Heredity*, **70**, 604–18.

Quick, H.E. (1947) *Arion ater* (L.) and *A. rufus* (L.) in Britain and their specific differences. *Journal of Conchology*, **22**, 249–61.

Quick, H.E. (1952) Rediscovery of *Arion lusitanicus* Mabille in Britain. *Proceedings of the Malacological Society of London*, **29**, 93–101.

Quick, H.E. (1960) British Slugs (Pulmonata, Testacellidae, Arionidae, Limacidae). *Bulletin of the British Museum (Natural History) Zoology*, **6**, 135–47.

Ralph, D., McClelland, M., Welsh, J., Baranton, G. and Perolat, P. (1993) *Leptosira* species categorized by arbitrary primed polymerase chain reaction (PCR) and mapped restriction polymorphisms in PCR-amplified rRNA genes. *Journal of Bacteriology*, **175**, 973–81.

Reischütz, P.L. (1984a) Zum massenhaften Auftreten von *Arion lusitanicus* Mabille in den Jahren 1982 und 1983. *Mitteilungen aus dem Zoologischen Gesellschaft, Braunau*, **4**, 253–4.

Reischütz, P.L. (1984b) Zur Schadwirking der "Kapuzinerschnecke" *Arion lusitanicus*. Mabille im Alpengebiet. *Heldia*, **1**, 39.
Roach, D.K. (1971) The puzzle of the big arions. Slug Research Conference, U.C. Cardiff, September 21–23, 1971, 19–20.
Runham, N. and Hunter, P. (1970) *Terrestrial Slugs*, Hutchinson, London.
Scharff, R.F. (1891) The slugs of Ireland. *Scientific Transactions of the Royal Dublin Society*, **4**, 513–62.
Schierwater, B., Ender, A., Schwenk, K., Spaak, P. and Streit, B. (1994) The evolutionary ecology of *Daphnia*, in *Molecular Ecology and Evolution: Approaches and Applications*, (eds B. Schierwater, B. Streit, G.P. Wagner and R. Desalle), Birkhäuser, Basel, Switzerland, pp. 495–508.
Simpson, U.R. and Neal, C. (1982) *Angiostrongylus vascorum* infection in dogs and slugs. *The Veterinary Record*, **111**, 303–4.
Simroth, H. (1885) Versuch einer Naturgeschichte der deutschen Nacktschnecken und ihrer europäischen Verwandten. *Zeitschrift fuer wissenschaftliche Zoologie Leipzig*, **42**, 203–366, pl. 7–11.
Simroth, H. (1891) Die Nacktschenecken der portugiesisch-azorischen Fauna. *Nova Acta Leopoldina*, **56**, 201–424.
Smith, B.J. (1989) Travelling snails. *Journal of Medical and Applied Malacology*, **1**, 195–204.
Smith, J.S.C. and Williams, J.G.K. (1994). Arbitrary primer mediated fingerprinting in plants: case studies in plant breeding, taxonomy and phylogeny, in *Molecular Ecology and Evolution: Approaches and Applications*, (eds B. Schierwater, B. Streit, G.P. Wagner and R. Desalle), Birkhäuser, Basel, Switzerland, pp. 5–15.
Solbrig, O.T. (1976) Chromosomal cytology and evolution in the family Compositae, in *The Biology and Chemistry of the Compositae*, (eds V.H. Heywood and B.L. Turner), Academic Press, New York, pp. 267–81.
Solem, A. (1978) Classification of the land Mollusca, in *Pulmonates, vol. 2A, Systematics, Evolution and Ecology*, (eds V. Fretter and J.F. Peake), Academic Press, New York, pp. 123–40.
South, A. (1992) *Terrestrial Slugs, Biology, Ecology and Control*, Chapman & Hall, London.
Taylor, J.W. (1907) *Monograph of the land and freshwater Mollusca of the British Isles: Testacellidae, Limacidae, and Arionidae*, Taylor Bros., Leeds.
Theobald, F.V. (1904) *Second Report on Economic Zoology*, British Museum (Natural History), London.
Tibayrenc, M., Neubauer, K., Barnabe, C., Guerrini, F., Skarecky, D. and Ayala, F.J. (1993) Genetic characterization of six parasitic protozoa: parity between random-primer polymorphic DNA and multilocus enzyme electrophoresis. *Proceedings of the National Academy of Sciences USA*, **90**, 1335–9.
Van Valen, L. (1973) A new evolutionary law. *Evolutionary Theory*, **1**, 1–30.
Von Proschwitz, T. (1988) *Arion lusitanicus*(?) on the Falkland Islands. *Journal of Conchology*, **33**, 49.
Welsh, J. and McClelland, M. (1990) Fingerprinting genomes using PCR with arbitrary primers. *Nucleic Acids Research*, **18**, 7213–18.
Wilkie, S.E., Isaac, P.G. and Slater, R.J. (1993) Random amplified polymorphic DNA (RAPD) markers for genetic analysis in *Allium*. *Theoretical and Applied Genetics*, **86**, 497–504.
Williams, G.C. (1975) *Sex and Evolution*, Princeton University Press.
Williams, J.G.K., Kubelik, A.R., Livak, K.J., Rafalski, J.A. and Tingey, S.V. (1990) DNA polymorphisms amplified by arbitrary primers are useful as genetic markers. *Nucleic Acids Research*, **18**, 6531–5.

Williams, J.G.K., Hanafey, M.K., Rafalski, J.A. and Tingey, S.V. (1993) Genetic analysis using random amplified polymorphic DNA markers. *Methods in Enzymology*, **218**, 704–40.

Williamson, M.H. (1959) Studies on the colour and genetics of the black slug. *Proceedings of the Royal Society of Edinburgh*, **27**, 87–93.

Wotton, F.W. (1893) The life history of *Arion ater* and its power of self-fertilization. *Journal of Conchology*, **7**, 158–67.

Wright, S. (1931) Evolution in Mendelian populations. *Genetics*, **16**, 97–159.

6

Systematics of brown planthopper and related species using nuclear and mitochondrial DNA

P.L. Jones, P. Gacesa and R.K. Butlin

ABSTRACT

The brown planthopper, *Nilaparvata lugens*, is a serious pest of rice in tropical and temperate Asia, and northern Australia. This study aimed to derive a phylogeny of *N. lugens* and other species in the genus, and to obtain molecular markers for the different species. Four populations of *N. lugens* collected from rice, and three populations of a morphologically indistinguishable species, collected from a wild grass, *Leersia*, were analysed. Single populations of *N. bakeri*, a sibling species of *N. lugens, N. muiri* and a West African species, *N. maeander*, were also included in the study.

Molecular phylogenies were inferred following amplification and restriction site analysis of part of the 16S RNA gene and a region between the cytochrome oxidase subunits I and II of the mitochondrial DNA. The mitochondrial DNA showed limited variability between populations but the data implied that *N. bakeri* and *N. maeander* are closely related, as are *N. muiri* and *N. lugens*. Amplification and restriction site analysis of 57 sites in the nuclear 18S rRNA gene revealed one variable site in the DNA from *N. maeander*. DNA sequences of one of the nuclear ribosomal DNA internal transcribed spacers were used to infer a phylogeny which agreed with the predicted species relationships. The nuclear DNA phylogeny strongly supported a division between the *N. lugens* populations in Asia and Australia, regardless of host plant associations.

A partial genomic library was constructed from total *N. lugens* DNA, and nine low-copy DNA clones selected and partially

The Ecology of Agricultural Pests, Edited by W.O.C. Symondson and J.E. Liddell.
Published in 1996 by Chapman & Hall, London. ISBN 0 412 62190 8

sequenced. Seven of these loci were amplified from *N. lugens* DNA and four from *N. bakeri* DNA. Restriction site analysis of the loci revealed restriction sites characteristic for *N. lugens* populations from the alternative host plants. A phylogeny inferred from the data supported a division between the *N. lugens* populations from Asia and Australia. The combined analysis of all the molecular data supported the inference that the Asian and Australian *N. lugens* populations are separate species.

6.1 INTRODUCTION

Molecular data are being used to an increasing extent to reconstruct phylogenies and in many cases offer advantages over the use of morphological data. Molecular characters are effectively limited only by the amount of DNA in an organism, and contain heritable variation which can be used to describe relationships at all levels of divergence. Molecular variation is qualitatively the same both between and within species and suitable molecular characters are available to determine the evolutionary history of individuals, populations, species or higher taxonomic units (Hillis and Moritz, 1990; Crozier, 1991; Miyamoto and Cracraft, 1991). As well as inferring phylogeny, molecular data can be used to measure gene flow across populations and deduce patterns of colonization and migration of species or populations (Avise, 1989; Moritz and Hillis, 1990). Molecular data are ideal in a situation where species have very similar morphology or for the study of intra-specific variation. An appreciation of the relationships between species provided by phylogenetic analysis provides the means by which patterns in geographical distributions can be understood, for example, in the case of closely related allopatric species (DeSalle and Grimaldi, 1991). The report that follows describes a study on ten populations of planthoppers from the genus *Nilaparvata* using different types of molecular information, with the aim of reconstructing the phylogeny and obtaining molecular markers for the populations.

6.2 *NILAPARVATA* SPECIES IN THE STUDY

There are 19 families of planthoppers comprised of more than 9000 described species; the largest and most studied family, the Delphacidae, includes more than 2000 species (O'Brien and Wilson, 1985; Wilson *et al.*, 1994). A recent review has examined the ecology, behaviour and pest management of planthoppers, with emphasis on the Delphacidae (Denno and Perfect, 1994). Most delphacids are small insects, 2–4 mm in length, and feed on the phloem of members of the grass family. The family as a whole is important economically as it includes serious pests of sugarcane, rice, wheat, sorghum and other cereals (O'Brien and Wilson, 1985; Denno and Roderick, 1990).

The delphacids included in this study are some of the 16 named species from the genus *Nilaparvata*. The best-known species is the rice brown planthopper, *Nilaparvata lugens* (Stål) which is monophagous on rice, *Oryza sativa* L. (Mochida and Okada, 1979; Wilson and Claridge, 1991). *N. lugens* is a serious pest in tropical and temperate regions of Asia and Australia where losses of rice are estimated to be hundreds of millions of dollars per year (Dyck and Thomas, 1979; Kenmore, 1987; Benrey and Lamp, 1994; Gallagher *et al.*, 1994). High densities of insects feeding on the phloem and ovipositing on the rice plants lead to a condition known as hopperburn, which can kill the plant (Dyck and Thomas, 1979; Benrey and Lamp, 1994). More recently, a West African species, *N. maeander* (Fennah), has been reported as a pest on rice seed beds in Nigeria (Alam *et al.*, 1983). A further problem is the transmission of plant viruses. *N. lugens* and two other species, *N. bakeri* (Muir) and *N. muiri* (China) are vectors of the rice grassy stunt virus, and *N. lugens* and *N. bakeri* are vectors of the rice ragged stunt viruses (Wilson and Claridge, 1991).

Populations of *N. lugens, N. bakeri, N. muiri* and *N. maeander* were included in the analysis presented here (Table 6.1) and are distinguishable by the structure of the male genitalia. However, this method requires special techniques and often the females cannot be identified

Table 6.1 The species and populations of *Nilaparvata* used in the study

Species	*Site of collection*	*Host plant*	*Date of collection*
N. lugens	Puri, Orissa, India	*Oryza sativa*	June 1990
N. lugens	Puri, Orissa, India	*Leersia*	June 1990
N. lugens	Sukamandi, West Java	*Oryza sativa*	June 1990
N. lugens	Sukamandi, West Java	*Leersia*	April 1990
N. lugens	Chamariggu, Japan	*Oryza sativa*	1983
N. lugens	Near Ingham, Queensland, Australia	*Oryza sativa*	February 1988
N. lugens	Ross River, near James Cook University,Townsville, Queensland, Australia	*Leersia*	February 1988
N. bakeri	Puri, Orissa, India	*Leersia*	1990
N. muiri	Kyushu, Japan*	*Leersia*	April 1984
N. maeander	Ibadan, Nigeria	*Oryza*	September 1991

*A laboratory population.

without the associated male insects (Wilson and Claridge, 1991). *N. bakeri* (Muir) is a sibling species of *N. lugens* which is found in sympatry with *N. lugens* throughout much of tropical and temperate Asia. *N. muiri* has a distribution sympatric with *N. lugens* in China, Japan, Taiwan and Korea. *N. maeander* is morphologically distinct from the other species (Mochida and Okada, 1979; Wilson and Claridge, 1991) and as the only non-Asian species can be used as an outgroup for phylogenetic analysis.

Populations of planthoppers indistinguishable morphologically from *N. lugens* occur on *Leersia hexandra* (Schwartz), a common wild grass closely related to rice which is often found in the irrigation channels and reservoirs next to paddy fields in Asia and Australia. The two populations from rice and *Leersia* are thought to be separate species but the *Leersia* population has not yet been named and for convenience these populations are referred to as rice and *Leersia* populations of *N. lugens*. (Claridge *et al.*, 1985a., 1988; Wilson and Claridge, 1991). Three *Leersia* and four rice populations of *N. lugens* were examined in the current study.

6.3 REPRODUCTIVE ISOLATION OF *NILAPARVATA* POPULATIONS

The species *N. lugens, N. bakeri, N. muiri* and *N. maeander* can be distinguished on the basis of their calling behaviour, i.e. the pattern of signals exchanged by males and females on the same plant for mate location. The male call is particularly useful in discriminating between the species (Claridge and Morgan, 1987). The acoustic vibrations occur in pulses and there is good evidence that the pulse repetition frequency (prf) is an important mate recognition signal, at least for *N. lugens*. The prf of *N. lugens* rice populations differs from that of sympatric *Leersia* populations (Claridge *et al.*, 1985a, 1988). Moreover, the prf of Asian populations from both host plants differs from that of allopatric Australian populations from the corresponding host plants (Claridge *et al.* 1984, 1985b, 1988). Mate choice experiments have been used to demonstrate that the variation in prf contributes to reproductive isolation between the different populations. There is a significant preference for homogametic matings for the populations from the two host plants. Similarly, individuals from Asia and Australia prefer to mate with other individuals from the same broad geographic range (Claridge *et al.*, 1984, 1985a, 1988). Fertile hybrids between each of the populations can be obtained in the laboratory (Claridge *et al.*, 1985a,b; Heinrichs and Medrano, 1984) which have prfs intermediate to the parental populations, but extensive sampling of populations in the field has found no evidence of hybridization between *N. lugens* from rice and *Leersia*.

Furthermore, the populations from rice and *Leersia* survive better and prefer to oviposit on the host plant from which they were collected and Australian populations of *N. lugens* collected from *Oryza* do not survive on the cultivated *Oryza* races commonly grown in Asia. Thus, the sympatric *N. lugens* populations from rice and *Leersia* appear to be sibling species (Claridge *et al.*, 1985a, 1988). Although it is difficult to define allopatric populations as distinct species, it has been suggested that the Australian rice and *Leersia* populations may be sibling species, separate from the Asian populations (Diehl and Bush, 1984).

There is geographic variation in the acoustic signals of *N. bakeri* populations from India, Sri Lanka, Indonesia and the Philippines, but tests similar to those used for *N. lugens* populations suggest there is no reproductive isolation between the *N. bakeri* populations. Therefore, in this case the differences in male calls do not indicate separate species (Claridge and Morgan, 1993). The complex of rice- and *Leersia*-feeding planthoppers in Asia and Australia provide useful biological systems for testing alternative models of sympatric and allopatric speciation (Claridge and de Vrijer, 1994) and theories of the evolution of species recognition systems as species isolating mechanisms (Butlin, 1993).

6.4 TYPES OF MOLECULAR DATA

Analysis of the chromosome numbers of *N. lugens* reveals 14 pairs of autosomes and the sex determination system is XY in the male and XX in the female. The chromosomes are small and difficult to differentiate (den Hollander, 1982; Noda and Tatewaki, 1990) as is often the case in the Homoptera, thus precluding their use for routine species identification. One molecular study of Asian *N. lugens* populations from rice and *Leersia* has used allozymes and found values of the coefficient of genetic identity typical for populations of the same species of *Drosophila* (Saxena *et al.*, 1991). Another study of *N. lugens* populations from a number of geographical areas found frequency differences in electrophoretic variants between some populations, with the greatest differentiation between the Asian and Australian populations (den Hollander, 1989). Allozymes have been widely used and are often the method of choice for investigations of population structure and gene flow. However, there is greater variation in the DNA than allozymes, and with small organisms there is the possibility of examining a larger number of loci at the level of the genotype rather than the phenotype.

The character-based analysis of DNA can be conducted in a number of ways. Techniques such as single-strand conformation polymorphism (SSCP) analysis and denaturing gradient gel electrophoresis (DGGE) are probably more useful when applied to population genetics than phylogenetics (Burke *et al.*, 1991). Single-locus or multilocus DNA fingerprinting has been widely used in parentage analysis and for identifying close relatives. The usually high level of variation at the loci analysed precludes the use of this method for phylogeny reconstruction, with the possible exception of some closely related organisms (Brookfield, 1991). Random amplified polymorphic DNA (RAPD) has been used for constructing genetic maps and as molecular markers (e.g. Williams *et al.*, 1990). However, as it is difficult to assign homology to amplified fragments – particularly in interspecific comparisons – they are not ideal for reconstructing phylogenies.

Analysis of the actual DNA sequence can be done using restriction enzymes to sample variation in a sequence, or the sequence itself can be determined. The fragments produced in a restriction digest or the restriction sites which the enzyme recognizes can be compared, but the analysis of restriction sites is preferable because the methods of inferring phylogenies are more advanced and the problem of convergent fragment lengths is avoided (Dowling *et al.*, 1990; Crozier, 1991). Although the analysis of restriction sites only samples the variation at a locus, in general it appears to give a good estimate of the average level of sequence variation across a region of DNA (Aquadro, 1991). DNA sequencing provides the most detailed information that can be applied to molecular systematic questions, but restriction site analysis allows more samples to be examined for considerably less effort and expense (Dowling *et al.*, 1990).

The polymerase chain reaction (PCR) is a relatively new technique (Saiki *et al.*, 1988) which has revolutionized the application of molecular data to systematics. The presence of conserved regions in DNA sequences such as the mitochondrial DNA and nuclear ribosomal DNA makes it possible to amplify fragments from organisms for which there is no specific sequence information available (Sogin, 1990; White *et al.*, 1990). For example, DNA sequences were obtained from a 25 million-year-old termite preserved in amber using PCR (DeSalle *et al.*, 1992). Many microorganisms cannot be cultivated using standard techniques, but their DNA can be investigated using PCR. Thus, amplification of ribosomal DNA has been widely used to investigate species diversity in bacteria (e.g. Britschgi and Giovannoni, 1991) and unculturable endosymbiotes (e.g. Amann *et al.*, 1991; Campbell *et al.*, 1992).

The ability to study more than one locus in an individual makes molecular data more useful for population studies and the use of PCR-based methods makes this possible even in small organisms. A number of regions of the DNA can be investigated in individuals and the ability to directly sequence the resulting amplified fragments is particularly powerful (Arnheim *et al.*, 1990; Dowling *et al.*, 1990). Furthermore, the amplified products are free of methylation sites which can be a problem with the conventional approach of Southern blotting for assaying restriction sites (Dowling *et al.*, 1990). Amplification of mitochondrial DNA, nuclear ribosomal DNA and random low-copy nuclear DNA loci with subsequent restriction digests or direct sequencing of the products was used to analyse the *Nilaparvata* populations.

6.4.1 Mitochondrial DNA

Mitochondrial DNA (mtDNA) is a maternally inherited, non-recombining molecule which in many organisms (but not all) evolves more rapidly than the coding nuclear DNA. Consequently, mtDNA has been widely used in the reconstruction of phylogenies of populations and closely related species, in studies on population structure and gene flow and for investigating geographic variation (Avise *et al.*, 1987; Moritz *et al.*, 1987). Two regions of the mtDNA were studied; a section of the 16S ribosomal DNA (rDNA) gene, and the region between the cytochrome oxidase subunits I and II genes, including parts of the genes (CytO) (Jones, 1994). The estimated length of the 16S rDNA–PCR product from *Nilaparvata*, 738 bp, is identical in size to the product found in honeybees. The length of the CytO amplification product was 670 bp in all the populations of *Nilaparvata* species, whereas in honeybees (*Apis mellifera*, L.) the equivalent region occurs in discrete lengths from 848 to 1658 bp, and the frequencies of these different lengths have been used to distinguish the subspecies of *A. mellifera* (Hall and Smith, 1991). The 16S rDNA and CytO amplification products were digested with four to eight restriction enzymes and restriction sites were found which distinguished between *N. lugens*, *N. bakeri*, *N. muiri* and *N. maeander*. There was no variation in the restriction sites of the *N. lugens* populations. In contrast, geographic variation was found in the restriction sites in the CytO region and the 16S rDNA of *A. mellifera* (Hall and Smith, 1991). Geographic population structure has been found in many species, for example, *Drosophila mercatorum* (DeSalle *et al.*, 1987), horseshoe crabs (*Limulus*) and some freshwater fish. Conversely, in some species such as American eels (*Anguilla rostrata* Lesueur) and red-winged blackbirds (*Agelaius phoeniceus* (L.)), there is no apparent geographic correlation with mtDNA genotypes (Avise *et al.*, 1987). The lack of restriction site variation between the populations of *N. lugens* could indicate recent high levels of gene flow between the populations or possibly a population bottleneck (Harrison, 1989). Alternatively, these regions of the mtDNA may be too conserved to distinguish the populations.

6.4.2 Nuclear ribosomal DNA

The universal presence of rRNA genes in living organisms, and the variety of evolutionary rates found in the coding and non-coding regions of eukaryote rDNA make the study of this part of the nuclear genome applicable to most areas of molecular systematics (Hillis and Dixon, 1991). The 18S rRNA gene and the rDNA internal transcribed spacers (ITS) (Figure 6.1) were examined for species of *Nilaparvata* (Jones, 1964).

Ancient evolutionary events such as the divergence of prokaryotes and eukaryotes have been studied using the 18S rRNA gene (Pace *et al.*, 1986; Cedergren *et al.*, 1988) as well as divisions between more closely related taxa. Consequently, it was not unexpected when restriction site analysis of the amplified 18S gene of the *Nilaparvata* species revealed very little variation. Approximately 12% of the gene was covered in the study and only one variant site was found in the amplified product from *N. maeander*. This provided supporting evidence for the selection of *N. maeander* as the outgroup to the other species.

The ITS are more variable in sequence than the 18S gene and provided greater resolution of the relationships between the *Nilaparvata* species. Both ITS including the 5.8S gene were amplified and the product length alone is sufficient to differentiate between species (Table 6.2).

Restriction digests revealed some variation between the *N. lugens* populations and therefore, ITS 1 was chosen for more detailed examination in order to provide information that would more clearly delineate the populations. The amplified ITS 1 products of two individuals from each population were sequenced directly using thermocycle sequencing. There was no variation in length within populations except for the *N. bakeri* ITS 1 sequences (Table 6.2). Variations in the form of nucleotide substitutions and insertions or deletions separated *N. lugens, N. bakeri, N. muiri* and *N. maeander*. Similar variations were found in a comparison of ITS 1 and ITS 2 in seven *Drosophila* species (Schlötterer *et al.*,

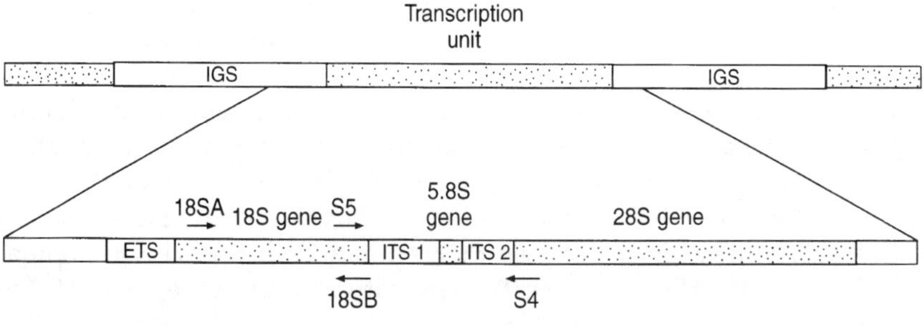

Figure 6.1 A schematic diagram of the ribosomal DNA repeat unit indicating the sites of primers used to amplify regions of the ribosomal DNA. Abbreviations: IGS, intergenic spacer; ETS, external transcribed spacer; ITS, internal transcribed spacer; 18SA and 18SB, 18S ribosomal DNA primers; S4 and S5, ITS primers.

Table 6.2 The length of the amplified products and ITS 1 sequences in *Nilaparvata* species

Species	*Length of amplified product (bp)**	*Length of ITS1 (bp)*
N. lugens	1510	537
N. bakeri	1550	564, 567†
N. muiri	1960	1026‡
N. maeander	2150	765

*Including the 5.8S gene and both ITS.
†The *N. bakeri* sequences differed by a 3-bp insertion/ deletion.
‡The total length of the *N. muiri* sequence assuming the 5′ end is the same as for *N. lugens* and *N. bakeri*.

1994). The ITS 1 of *N. lugens* populations differed by a small number of nucleotide substitutions, six at the most, and the populations from the same broad geographical area, regardless of host plant association, were the most alike.

6.4.3 Low-copy nuclear DNA

With the possible exception of DNA–DNA hybridization studies, the low-copy DNA fraction of the nuclear genome has been much less utilized in molecular systematics than analysis of nuclear rDNA and mtDNA (Dowling *et al.*, 1990; Crozier, 1991). Most studies on low-copy loci in insects have been on *Drosophila* species (Aquadro, 1991), for example, the alcohol dehydrogenase gene has been used to provide information on variation within and between species of *D. melanogaster* (e.g. Aquadro *et al.*, 1986) and the *planitibia* subgroup of Hawaiian *Drosophila* (Bishop and Hunt, 1988).

There was no sequence information available for *Nilaparvata* or related species and therefore low-copy nuclear DNA clones from a partial genomic DNA library prepared from *N. lugens* DNA were selected at random. Nine clones were partially sequenced and primers designed to amplify the loci from individuals of the *Nilaparvata* populations. Successfully amplified products were analysed by restriction enzyme digests (Jones, 1994). No products were obtained from two of the loci, but amplification of the remaining seven was achieved for at least the *N. lugens* populations. Length variation in the amplified products was observed at the DNL3 and DNL8 loci (Table 6.3). The restriction fragment profiles of all the loci amplified from *N. muiri* and *N. maeander* were very different from those from *N. bakeri* and *N. lugens*. Some restriction sites at four loci (DNL3, DNL6, DNL8, DNL9)

Table 6.3 The low-copy nuclear DNA loci analysed

	*Length of amplified products (bp)**			
Locus	N. lugens	N. bakeri	N. muiri	N. maeander
DNL2	885	855	–	–
DNL3	3440	–	3670	–
DNL4	2075	2075	–	–
DNL6	2260	–	–	–
DNL7	1625	1625	–	1625
DNL8	2490	2490	2490	2060
DNL9	2460	–	–	–

*– indicates that no product was amplified.

varied within and between populations of *N. lugens*, and many of these polymorphisms distinguished the Australian populations from some or all of the Asian populations. There appeared to be differing rates of evolution at some of the various loci examined. The 13 restriction sites at the DNL4 locus were invariant for the *N. lugens* populations, whereas five restriction sites of 11 at DNL8, and 13 of 28 at DNL9 were polymorphic. Many of the polymorphisms discovered could be used as molecular markers to examine the *N. lugens* population structure in greater detail.

6.5 PHYLOGENY OF *NILAPARVATA* POPULATIONS AND ITS IMPLICATIONS

Phylogenies were inferred separately from the restriction site or sequence data obtained from the mtDNA, low-copy nuclear DNA loci and nuclear rDNA. The mtDNA phylogeny implied that *N. lugens* and *N. muiri* are more closely related to each other than to *N. bakeri* or *N. maeander*, whereas a closer relationship between *N. bakeri* and *N. lugens* was expected from their morphology (section 6.2). However, as the phylogeny was based on only a few restriction sites this conclusion must be interpreted with caution. The expected relationships between *N. lugens, N. bakeri, N. muiri and N. maeander* are supported by the nuclear rDNA phylogeny (Figure 6.2). *N. bakeri* is most closely related to *N. lugens* followed by *N. muiri*, and *N. maeander* is the most distant relative. The Asian and Australian populations of *N. lugens* are more distantly related than the *Leersia* and rice *N. lugens* populations from the same geographical area. The low-copy nuclear DNA data for *N. muiri* and *N. maeander* were not included in the analysis because of the

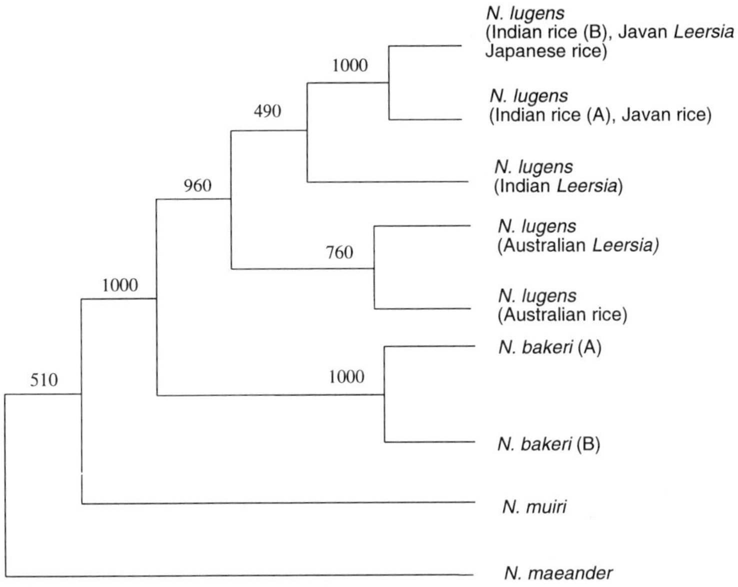

Figure 6.2 A phylogeny inferred from the ribosomal ITS sequences by maximum parsimony. One other most parsimonious tree was found which differed only in the order of the Asian populations of *N. lugens* in the phylogeny. Bootstrap values calculated for 1000 replicates are given on the branches. The PHYLIP package (version 3.5c) (Felsenstein, 1993) was used to derive the phylogeny (DNAPENNY) and bootstrap values (SEQBOOT, CONSENSE).

inability to amplify most of the loci. The low-copy nuclear DNA phylogeny corroborated the relationships between the *N. lugens* populations inferred from the rDNA ITS sequences. An analysis of the data combined from mtDNA, nuclear rDNA and low-copy nuclear DNA for *N. lugens* and *N. bakeri* gave support for the separation of both Australian *N. lugens* populations from the Asian populations of *N. lugens*. In other words, the *N. lugens* populations from the two host plants are more closely related, if they are from the same broad geographical area, than they are to populations from the same host and from a different geographical area.

At present all the populations of *N. lugens* are classified as the same species. There is good evidence from the experiments on reproductive isolation and mate choice that the populations from rice and *Leersia* are separate sibling species (section 6.3) (Claridge *et al.*, 1985a, 1988). The molecular data from the mtDNA, low-copy nuclear DNA and rDNA suggest the populations of *N. lugens* are very closely related. However, the sympatric populations are more closely related than the allopatric populations, from either host plant. Thus, it would be inconsistent with the evidence from the molecular data, to divide the *N. lugens* populations by host plant specificities; the division implied from the molecular phylogenies is between the populations from Asia and Australia. Therefore, the evidence suggests that the Australian and Asian *N. lugens* populations may be separate species. This has been suggested from the information on the acoustic mating signals of these allopatric populations (Diehl and Bush, 1984).

The occurrence of host- or habitat-associated sibling species in many phytophagous insects has been used as evidence in support of sympatric speciation (Tauber and Tauber, 1989). A recent study on an amphipod species, *Eogammarus confervicolous*, has examined shifts in habitat preference using a large number of genomic DNA restriction fragment length polymorphisms (Stanhope *et al.*, 1993). The results strongly supported two independent shifts in habitat preference in different localities. Furthermore, the authors argued that because of the potential for dispersal over large distances, the shifts in habitat preference of this species have occurred under sympatric conditions. Similarly, *N. lugens* can disperse over large distances (Perfect and Cook, 1994) and the rice and *Leersia* host plants are often in close proximity. Thus, the shifts in host plant specificities could be seen as having occurred in sympatry. The evolutionary history inferred from the molecular phylogenies implies there has been more than one host shift between rice and *Leersia*, with separate events in Asia and Australia.

REFERENCES

Alam, M.S., Gibbons, J.W., Akibo-Betts, D., Wasano, K., Zan Kuano, K. and Efron, Y. (1983) Hopperburn in rice caused by brown planthopper (*N. maeander*). *International Institute of Tropical Agriculture Research Briefs*, **4**, 7–8.

Amann, R., Springer, N., Ludwig, W., Gortz, H. and Schleifer, K. (1991) Identification *in situ* and phylogeny of uncultured bacterial endosymbionts. *Nature*, **351**, 161–4.

Aquadro, C.F. (1991) Molecular population genetics of *Drosophila*, in *Molecular Approaches to Pure and Applied Entomology*, (eds J. Oakeshott and M. Whitten), Springer–Verlag, New York, pp 222–66.

Aquadro, C.F., Desse, S.F., Bland, M.M., Langley, C.H. and Laurie-Ahlberg, C.C. (1986) Molecular population genetics of the alcohol dehydrogenase gene region of *Drosophila melanogaster*. *Genetics*, **114**, 1165–90.

Arnheim, N., White, T. and Rainey, W.E. (1990) Application of PCR: organismal and population biology. *BioScience*, **40**, 174–82.

Avise, J.C. (1989) Gene trees and organismal histories: a phylogenetic approach to population biology. *Evolution*, **43**, 1192–208.

Avise, J.C., Arnold, J., Ball, R.M., Eldredge, B., Lamb, T., Neigel, J.E., Reeb, C.A. and Saunders, N.C. (1987) Intraspecific phylogeography: the mitochondrial

DNA bridge between population genetics and systematics. *Annual Review of Ecology and Systematics*, **18**, 489–522.

Benrey, B. and Lamp, W.O. (1994) Biological control in the management of planthopper populations, in *Planthoppers: their ecology and management*, (eds R.F. Denno and J. Perfect), Chapman & Hall, New York, pp. 519–50.

Bishop, J.G. and Hunt, J.A. (1988) DNA divergence in and around the alcohol dehydrogenase locus in five closely related species of Hawaiian *Drosophila*. *Molecular Biology and Evolution*, **5**, 415–31.

Britschgi, T.B. and Giovannoni, S.J. (1991) Phylogenetic analysis of a natural marine bacterioplankton population by rRNA gene cloning and sequencing. *Applied Environmental Microbiology*, **57**, 1707–13.

Brookfield, J.F.Y. (1991) The statistical interpretation of hypervariable DNAs, in *Molecular Techniques in Taxonomy*, (eds G.M. Hewitt, A.W.B. Johnston and J.P.W. Young), Springer Verlag, Berlin, pp. 159-69.

Burke, T., Rainey, W.E. and White, T.J. (1991) Molecular variation and ecological problems, in *Molecular Techniques in Taxonomy*, (eds G.M. Hewitt, A.W.B. Johnston and J.P.W. Young), Springer Verlag, Berlin, pp. 229–54.

Butlin, R.K. (1993) The variability of mating signals and preferences in the brown planthopper, *Nilaparvata lugens* (Homoptera: Delphacidae). *Journal of Insect Behaviour*, **6**, 125–39.

Campbell, B.C., Bragg, T.S. and Turner, C.E. (1992) Phylogeny of symbiotic bacteria of four weevil species (Coleoptera: Curculionidae) based on analysis of 16S ribosomal DNA. *Insect Biochemistry and Molecular Biology*, **22**, 415–21.

Cedergren, R., Gray, M.W., Abel, Y. and Sankoff, D. (1988) The evolutionary relationships among known life forms. *Journal of Molecular Evolution*, **28**, 98–112.

Claridge, M.F. and de Vrijer, P.W.F. (1994) Reproductive behaviour: the role of acoustic signals in species recognition and speciation, in *Planthoppers: their Ecology and Management*, (eds R.F. Denno and J. Perfect), Chapman & Hall, New York, pp. 216–33.

Claridge, M.F. and Morgan, J.C. (1987) The brown planthopper, *Nilaparvata lugens* and related species: a biotaxonomic approach, in *Proceedings of the 2nd Intenational Workshop on Leafhoppers and Planthoppers of Economic Importance*, (eds M.R. Wilson and L.R. Nault), CAB Institute of Entomology, Oxford, pp. 19 32.

Claridge, M.F. and Morgan, J.C. (1993) Geographical variation in acoustic signals of the planthopper, *Nilaparvata bakeri* (Muir), in Asia: species recognition and sexual selection. *Biological Journal of the Linnean Society*, **48**, 267–81.

Claridge, M.F., Den Hollander, J. and Morgan, J.C. (1984) Specificity of acoustic signals and mate choice in the brown planthopper, *Nilaparvata lugens*. *Entomologia experimentalis et applicata*, **35**, 221–6.

Claridge, M.F., Den Hollander, J. and Morgan, J.C. (1985a) The status of weed-associated populations of the brown planthopper, *Nilaparvata lugens* (Stål) – host race or biological species? *Zoological Journal of the Linnean Society*, **84**, 77–90.

Claridge, M.F., Den Hollander, J. and Morgan, J.C. (1985b) Variation in courtship signals and hybridisation between geographically definable populations of the rice brown planthopper, *Nilaparvata lugens* (Stål). *Biological Journal of the Linnean Society*, **24**, 35–49.

Claridge, M.F., Den Hollander, J. and Morgan, J.C. (1988) Variation in host plant relations and courtship signals of weed-associated populations of the brown planthopper, *Nilaparvata lugens* (Stål), from Australia and Asia: a test of the recognition species concept. *Biological Journal of the Linnean Society*, **35**, 79–93.

Crozier, R.H. (1991) Molecular methods for insect phylogenetics, in *Molecular Approaches to Pure and Applied Entomology*, (eds J. Oakeshott and M. Whitten), Springer-Verlag, New York, pp. 164-221.

Den Hollander, J. (1982) The chromosomes of *Nilaparvata lugens* Stål and some other Auchenorrhyncha. *Cytologia*, **47**, 227–36.

Den Hollander, J. (1989) Electrophoretic studies on planthoppers and leafhoppers of agricultural importance, in *Electrophoretic Studies on Agricultural Pests*, (eds H.D. Loxdale and J. den Hollander), Clarendon Press, Oxford, pp. 297–315.

Denno, R.F. and Perfect, T.J. (eds) (1994) *Planthoppers: their Ecology and Management*, Chapman & Hall, New York.

Denno, R.F. and Roderick, G.K. (1990) Population biology of planthoppers. *Annual Review of Entomology*, **35**, 489–520.

DeSalle, R. and Grimaldi, D.A. (1991) Morphological and molecular systematics of the Drosophilidae. *Annual Review of Ecology and Systematics*, **22**, 447-75.

DeSalle, R., Templeton, A., Mori, I., Pletscher, S. and Johnston, J.S. (1987) Temporal and spatial heterogeneity of mitochondrial DNA polymorphisms in natural populations of *Drosophila mercatorum. Genetics*, **116**, 215–23.

DeSalle, R., Gatesy, J., Wheeler, W. and Grimaldi, D. (1992) DNA sequences from a fossil termite in oligo-miocene amber and their phylogenetic implications. *Science*, **257**, 1933–6.

Diehl, S.R. and Bush, G.L. (1984) An evolutionary and applied perspective of insect biotypes. *Annual Review of Entomology*, **29**, 471–504.

Dowling, T.E., Moritz, C. and Palmer, J.D. (1990) Nucleic acids II: restriction site analysis, in *Molecular Systematics*, (eds D.M. Hillis and C. Moritz), Sinauer, Sunderland, MA, USA, pp. 250–317.

Dyck, V.A. and Thomas, B. (1979) The brown planthopper problem, in *Brown Planthopper: Threat to Rice Production in Asia*, International Rice Research Institute, Los Banos, Philippines, pp. 3–17.

Felsenstein, J. (1993) *PHYLIP (Phylogeny in inference package)*, version 3.5c. Distributed by the author, Department of Genetics, University of Washington, Seattle.

Gallagher, K.D., Kenmore, P.E. and Sogawa, K. (1994) Judicial use of insecticides deters planthopper outbreaks and extends the life of resistant varieties in Southeast Asian rice, in *Planthoppers: their Ecology and Management*, (eds R.F. Denno and J. Perfect), Chapman & Hall, New York pp. 599-614.

Hall, H.G. and Smith, D.R. (1991) Distinguishing African and European honeybee matrilines using amplified mitochondrial DNA. *Proceedings of the National Academy of Science, USA*, **88**, 4548–52.

Harrison, R.G. (1989) Animal mitochondrial DNA as a genetic marker in population and evolutionary biology. *Trends in Ecology and Evolution*, **4**, 6–11.

Heinrichs, E.A. and Medrano, F.G. (1984) *Leersia hexandra*, a weed host of the brown planthopper *Nilaparvata lugens* (Stål). *Crop Protection*, **3**, 77–85.

Hillis, D.M. and Dixon, M.T. (1991) Ribosomal DNA: molecular evolution and phylogenetic inference. *Quarterly Review of Biology*, **66**, 411–53.

Hillis, D.M. and Moritz, C. (1990) An overview of applications of molecular systematics, in *Molecular Systematics*, (eds D.M. Hillis and C. Moritz), Sinauer, Sunderland, MA, USA), pp. 502–15.

Jones, P.L. (1994) Molecular studies on *Nilaparvata*. PhD Thesis, University of Wales.

Kenmore, P.E. (1987) Crop loss assessment in a practical integrated pest control program for tropical Asian rice, in *Crop Loss Assessment and Pest Management*, (ed. P.S. Teng), American Phytopathological Society, St. Paul, MN, USA pp. 225–41.

Miyamoto, M.M. and Cracraft, J. (eds) (1991) *Phylogenetic Analysis of DNA Sequences*, Oxford University Press, Oxford.

Mochida, O. and Okada, T. (1979) Taxonomy and biology of *Nilaparvata lugens*, in *Brown Planthopper. A Threat to Rice Production in Asia*, International Rice Research Institute, Los Banos, Philippines, pp. 21–43.

Moritz, C. and Hillis, D.M. (1990) Molecular systematics: context and controversies, in *Molecular Systematics*, (eds D.M. Hillis and C. Moritz), Sinauer, Sunderland, MA, USA, pp. 1–10.

Moritz, C., Dowling, T.E. and Brown, W.M. (1987) Evolution of animal mitochondrial DNA: relevance for population biology and systematics. *Annual Review of Ecology and Systematics*, **18**, 269–92.

Noda, H. and Tatewaki, R. (1990) Reexamination of chromosomes of three species of rice planthoppers (Homoptera: Delphacidae). *Applied Entomology and Zoology*, **25**, 538–40.

O'Brien, L.B. and Wilson, S.W. (1985) Planthopper systematics and external morphology, in *Leafhoppers and Planthoppers*, (eds L.R. Nault and J.G. Rodriguez), Wiley and Sons, London, pp. 61–102.

Pace, N.R., Olsen, G.J., and Woese, C.R. (1986) Ribosomal RNA phylogeny and the primary lines of evolutionary descent. *Cell*, **45** 325–6.

Perfect, T.J. and Cook, A.G. (1994) Rice planthopper population dynamics: a comparison between temperate and tropical regions, in *Planthoppers: their Ecology and Management*, (eds R.F. Denno and J. Perfect), Chapman & Hall, New York, pp. 282–301.

Saiki, R.K., Gelfand, D.H., Stoffel, S., Scharf, S.J., Higuichi, R., Horn, G.T., Mullis, K.B. and Erlich, H.A. (1988) Primer-directed enzymatic amplification of DNA with a thermostable DNA polymerase. *Science*, **239**, 487–91.

Saxena, R.C., Demayo, C.G. and Barrion, A.A. (1991) Allozyme variation among biotypes of the brown planthopper *Nilaparvata lugens* in the Philippines. *Biochemical Genetics*, **29**, 115–23.

Schlötterer, C., Hauser, M-T., von Haeseler, A. and Tautz, D. (1994) Comparative evolutionary analysis of rDNA ITS regions in *Drosophila*. *Molecular Biology and Evolution*, **11**, 513–22.

Sogin, M.L. (1990) Amplification of ribosomal RNA genes for molecular evolution studies, in *PCR Protocols: A Guide to Methods and Applications*, (eds M.A. Innis, G.H. Gelfand, J.J. Sninsky and T.J. White), Academic Press, New York, pp. 307–14.

Stanhope, M.J., Hartwick, B. and Baillie, D. (1993) Molecular phylogeographic evidence for multiple shifts in habitat preference in the diversification of an amphipod species. *Molecular Ecology*, **2**, 99-112.

Tauber, C.A. and Tauber, M.J. (1989) Sympatric speciation in insects: perception and perspective, in *Speciation and its Consequences*, (eds. D. Otte and J.A. Endler), Sinauer, Sunderland, MA, USA. pp. 307–44.

White, T.J., Bruns, T., Lee, S. and Tayler, J. (1990) Amplification and direct sequencing of fungal ribosomal RNA genes for phylogenetics, in *PCR Protocols: A Guide to Methods and Applications*, (eds M.A. Innis, G.H. Gelfand, J.J. Sninsky and T.J. White), Academic Press, New York, pp. 315–22.

Williams, J.G.K., Kubelik, A.R., Livak, K.J., Rafalski, J.A. and Tingey, S.V. (1990) DNA polymorphisms amplified by arbitrary primers are useful as genetic markers. *Nucleic Acids Research*, **18**, 6531–5.

Wilson, M.R. and Claridge, M.F. (1991) *Handbook for the Identification of Leafhoppers and Planthoppers of Rice,* C.A.B. International, Wallingford, Oxon, UK.

Wilson, S.W., Mitter, C., Denno, R.F. and Wilson, M.R. (1994) Evolutionary patterns of host plant use by delphacid planthoppers and their relatives, in *Planthoppers: their Ecology and Management,* (eds R.F. Denno and J. Perfect), Chapman & Hall, New York, pp. 7–45.

7

Application of novel molecular markers (DNA) in agricultural entomology

H.D. Loxdale, C.P. Brookes and P.J. De Barro

ABSTRACT

Isozymes have been extensively employed for the past 30 years to investigate a wide variety of aspects of pure and applied entomology, including phylogeny and taxonomy, population genetics, insecticide resistance, development and infection by hymenopterous parasitoids and entomopathogenic fungi. Recently, entomologists have also exploited the potential of DNA techniques for such studies. These have proved particularly useful to investigate mating relationships in the Hymenoptera and clonal structure in aphids, both groups having generally lower levels of enzyme polymorphism than many other insects, and in insect phylogeny. Since DNA techniques reveal greater amounts of genetic variability compared with isozymes, they allow examination of phylogenetic relationships and population structure and dynamics hitherto impossible with enzyme markers. Some DNA techniques, unlike isozymes, also work with dried or ancient samples. Examination of DNA variability from dried insect specimens has potential for study of long-term temporal as well as spatial genetic changes.

This paper briefly describes the main DNA techniques now in use and their application, particularly in relation to agricultural entomology.

The Ecology of Agricultural Pests, Edited by W.O.C. Symondson and J.E. Liddell.
Published in 1996 by Chapman & Hall, London. ISBN 0 412 62190 8

7.1 INTRODUCTION

Isozymes and allozymes (the phenotypic products of alleles segregating at various enzyme/isozyme loci), are well established as a means of investigating evolution and population genetics (Richardson *et al.*, 1986; Avise, 1994). For phylogenetic and taxonomic studies, allozyme polymorphisms may be summated over a number of enzyme loci and compared to provide estimates of genetic relatedness and of time since divergence (Nei, 1915); such comparisons often involve the construction of dendrograms (Avise, 1994). For population genetic studies, allozymes often give clear homozygous and heterozygous banding patterns at various loci; therefore allele and genotype frequencies may be assessed and, from the latter, deviations from Hardy–Weinberg expectations calculated (Hartl, 1980). From such data, *F*-statistics or inbreeding coefficients may be derived, which can be used to indicate levels of interpopulation gene flow (Workman and Niswander, 1970). Allele frequencies *per se* provide statistical measures of population similarity (Workman and Niswander, 1970), and qualitative and quantitative estimates of interpopulation gene flow and by inference, migration (Slatkin, 1985, 1987).

In applied entomology, isozyme/allozyme markers have proved useful in identifying levels of insecticide resistance within populations of pest insects (Devonshire, 1989; Hemingway *et al.*, 1996, this volume; Daly and Trowell, 1996, this volume); assessing hymenopterous parasitism within field populations of aphids (Powell and Walton, 1989); identifying the gut contents of polyphagous insect predators (Murray *et al.*, 1989; Solomon, 1996, this volume); detecting infection of aphids by various entomopathogenic fungi (Wilding *et al.*, 1993); and for studies of host adaptation (Loxdale, 1994; Atanassova *et al.*, submitted) and insect development (e.g. Colgan, 1986; Machado and Contel, 1991) (see also Loxdale and den Hollander, 1989).

If isozymes/allozymes are so versatile, why have so many pure and applied entomologists turned to DNA technologies in recent years? Certainly, DNA technologies are currently in vogue, but this alone is not the reason for their rapid adoption. Rather, these new techniques provide unique data unobtainable with isozymes (e.g. nucleotide sequence data), or provide further evidence to support or oppose conclusions derived from protein data. In addition, they provide a means to gather data on genetic, evolutionary and population parameters in species which show little or no enzyme polymorphism such as members of the Hymenoptera and Aphididae.

Here, the various DNA techniques now available are briefly described, with emphasis mainly on their application in entomological studies, especially those concerning insects of agricultural and horticultural importance.

7.2 rDNA GENES IN PHYLOGENETIC STUDIES AND AS TAXONOMIC MARKERS

Isozymes have proved valuable for examination of phylogenetic and taxonomic relationships in insects (Tomiuk and Wöhrmann, 1983; Eggers-Schumacher, 1987; Guldemond and Eggers-Schumacher, 1989; Futuyma and McCafferty, 1990; Pinto *et al.*, 1992; Spicer, 1992; Berlocher *et al.*, 1993). However, to estimate divergence between taxa requires assumptions as to the homology of bands, while arguments relating to molecular clocks and differential rates of protein evolution have also to be considered (Avise, 1994); all of these could lead to inaccuracies in phylogenetic assessments of genetic relatedness between insect species, families, etc. With DNA, the fact that different regions are evolving at different rates can be a distinct advantage in phylogenetic studies 'by permitting choice of the appropriate DNA regions geared to the time-scale requirements of a particular phylogenetic problem' (Avise, 1994).

The sequencing of the different regions of the ribosomal RNA gene clusters (rDNA) in the nuclear or mitochondrial genome allows comparison of these sequences from different insects, in turn, allowing phylogenetic relationships to be inferred (see Hamby *et al.*, 1988 and Fang *et al.*, 1993 for technical details concerning sequencing and section 7.8 for a brief description of, and references to, the polymerase chain reaction (PCR) which is involved in sequencing protocols).

In insect cells, as with all eukaryotes, the nuclear genome contains a cluster of several hundred tandemly-repeated copies of the rDNA transcription unit. Each unit contains 18S ('S' denotes sedimentation velocity in Svedbergs), 5.8S and 28S rRNA genes as well as two internal transcribed spacers (ITS-1 and ITS-2) which separate the 18S, 5.8S and 28S genes, an external transcribed spacer (ETS) upstream of the 18S gene, and a non-transcribed spacer (NTS or intergenic spacer, IGS) which in turn separates adjacent copies of the rRNA repeat units (Figure 7.1). In addition, the nuclear genome of eukaryotes contains a completely separate cluster of transcriptional units which code for 5S rRNA. The mitochondria have a large subunit rRNA gene (16S) as well as a smaller subunit gene coding for the 12S rRNA.

There are five main advantages of using rDNA for phylogenetic studies. First, the different regions of the repeated rDNA (18S, 5.8S and 28S rRNA genes) evolve at different rates and, collectively, more slowly than the transcribed and non-transcribed spacer regions, the last often showing significant interspecific length and sequence variation in insects (Black *et al.*, 1989). Second, rDNA evolves in a concerted fashion, meaning that the rDNA of an individual is generally representative of the species/taxa to which it belongs. Thus, only small numbers of individuals need to be examined for any comparative study. Third, because the rDNA sequence is moderately to highly repetitive with numerous copies present, hybridization analysis is assured (Hillis and Dixon, 1991). Fourth, the highly conserved nature of many rDNA gene sequences allows 'universal' primers to be produced for many species which, in turn, can be employed as primers for PCR or as probes in restriction fragment analysis (Hillis and Davis, 1986; Hillis and Dixon, 1991; Pashley *et al.*, 1993). Last, when assessing phylogenetic distance between widely divergent taxa, assumptions concerning the validity of gene homology, as required when using isozymes (Avise, 1994), are generally much less problematic because rDNA sequences may be directly compared and contrasted. However, caution is required if the nucleotide sequences under examination are extensively modified with insertions or deletions (Pashley *et al.*, 1993).

While rDNA has been widely investigated in phylogenetic studies of numerous pro- and eukaryotes (Hillis and Dixon, 1991), its examination in studies of insect phylogeny has so far been limited. Since the 18S gene is among the slowest evolving sequences found in living organisms, it has proved informative for examining ancient evolutionary events, i.e. divergence in the Palaeozoic and Mesozoic eras (Hillis and Dixon, 1991). It has recently been used, for example, to examine the ancient evolutionary divergence and phylogenetic relationships in holometabolous insects (i.e. those with a complete metamorphosis), which are considered to be physiological and morphological assemblages of monophyletic origin (Hennig, 1981). Carmean *et al.* (1992) sequenced approximately 1080 nucleotides of the 5′ end of the 18S rRNA gene of various diverse orders (Hymenoptera, Mecoptera, Neuroptera, Siphonaptera) and genera and families within the orders Hymenoptera and Neuroptera. Sequences were aligned and compared with published sequences for Coleoptera and Diptera and also with those of an aphid, shrimp and spider. These last three, which undergo incomplete metamorphosis, were used as outgroupings to 'root' the dendrograms (Figure 7.2a). They found the Diptera 18S rDNA had evolved rapidly, having in excess of three times as many substitutions among aligned sequences and two to three times more unalignable nucleotides than other Holometabola investigated. Such rapid, atypical evolution is at variance with the existence of an insect-wide molecular clock (Carmean *et al.*, 1992). In a similar study, 1330 nucleotide bases of the 18S gene from 19 diverse species were aligned (Pashley *et al.*, 1993). These represented nine holometabolous orders, and three outgroup orders (Odonata, Phasmida and Homoptera) were used to root the dendrograms (Figure 7.2b).

The dendrograms from the two studies have certain similarities (Figures 7.2a,b). For instance, scorpionflies (Mecoptera) and fleas (Siphonaptera) were closely related, while the Neuroptera and Hymenoptera

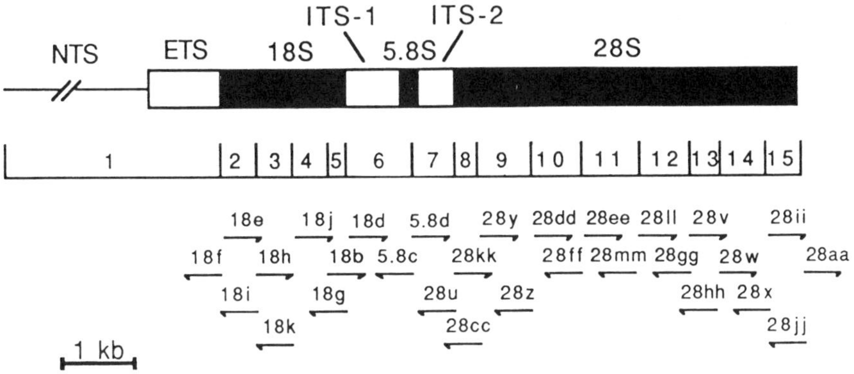

Figure 7.1 The ribosomal DNA array of a eukaryote (from Hillis and Dixon, 1991) (see text for explanation of abbreviations used). The entire array can be amplified (sections 1–15) using PCR and the primers indicated by arrows.

showed some degree of phylogenetic affinity. None of the various dendrograms, produced in these studies using parsimony and genetic distance analyses, especially favoured any of the morphologically based phylogenetic trees cited by Carmean *et al.* (1992) and Pashley *et al.* (1993). Both studies placed Coleoptera as basal to the other orders, but could not clearly resolve the taxonomic position of the Hymenoptera. This agrees with Boudreaux's (1979) hypotheses, that Coleoptera are the oldest of the extant holometabolous orders (based on characters at the bases of the legs), while characters of the male genitalia unite all the remaining holometabolous orders except the Hymenoptera (Carmean *et al.*, 1992).

Thus, the molecular data broadly support a monophyletic origin of holometabolous insect evolution. However, the dendrograms produced did not exactly correspond to any of the phylogenies previously proposed based on morphological characters (see Carmean *et al.*, 1992 for details). Probably, nucleotide substitution rates of the 18S gene are too low to resolve most of the interordinal relationships studied.

Because the 23–28S nuclear rDNA gene is larger and shows more variation in evolutionary rates within its different domains than the 18S subunit, it can be used to examine more recent events (for example, in the Cenozoic era). However, the 28S regions examined must be chosen with care if the taxa have diverged recently (Gutell and Fox, 1988; Hillis and Dixon, 1991). Mitochondrial (mt) rRNA genes have also evolved faster than the nuclear RNA genes and can also be used to compare taxa that may have diverged in the Cenozoic era, within the past 20 million years (Hillis and Dixon, 1991).

To date, few studies have exploited 28S gene sequences to investigate insect phylogeny. Two such studies concern Diptera (Vossbrinck and Friedman, 1989; Pelandakis and Solignac, 1993), while more recently, Schmitz and Moritz (1994) have investigated two variable domains (D1 and D2) near the 5′ end of the gene encoding the 28S large subunit of the hornet, *Vespa crabro* L. (Hymenoptera, Vespinae) and have compared its sequences with those of other Hymenoptera and with *Drosophila*.

Very recent evolutionary events may be examined by the sequencing of 16S mitochondrial rDNA (Hillis and Dixon, 1991). With insects, Fang *et al.* (1993) have used this approach to examine the phylogeny of New World *Deltocephalus*-like leafhoppers. Previously, after examination of the male genitalia, Oman (1949) had split the genera into four main groups, possibly of monophyletic origin. Fang *et al.* (1993) examined 562 nucleotides at the 3′ end of mt 16S rDNA for 21 species representing 19 New World genera. Two oligonucleotide PCR primers (16S-A and 16S-B; 20 and 22 bp in length, respectively) were used which aligned with 16S rDNA sequences of *Drosophila yakuba* Burla, *Locusta migratoria* L. and species of the genus *Magicicada* Davis. Following PCR amplification and sequence comparison, it was revealed that 18 out of 19 leafhopper genera belonged to a monophyletic group, a conclusion largely in agreement with Oman's (1949) earlier findings.

In addition to sequencing rDNA to examine phylogenetic relationships, these genes may also be exploited to provide marker probes to investigate taxonomic differences between closely-related insect species. Gargiulo *et al.* (1988) studied small hymenopterous aphid parasitoids of the genus *Aphidius* Nees (*A. ervi* Haliday, *A. microlophii* Pennacchio and Tremblay, *A. urticae* Haliday and *A. smithi* Sharma and Subba Rao) using rDNA probes. These probes were derived from *Drosophila* and consisted of the complete uninterrupted genetic ribosomal unit (cDm Y22) or its components: 28Sα (probe 1 which includes the ITS region); 28Sβ2 (probe 2, a region adjacent to the NTS region which is thus likely to show maximum inter- and intraspecific heterogeneity) and the 18S region (probe 3). Single and double restriction digests of pooled *A. ervi* DNA (*Bgl*II, *Pst*I and *Kpn*I) were used to produce a restriction map which was then compared with the corresponding map for *Drosophila melanogaster* Meigen. Certain restriction enzyme sites in the two species were identical and an intron within the 28S gene was also found to be conserved in both species.

A. ervi and *A. microlophii* from different geographical localities were examined using Southern blots (Sambrook *et al.*, 1989) and the NTS probe (2). *A. ervi* showed restriction fragment length polymorphisms (RFLPs), whereas the *A. microlophii* samples were monomorphic. The variation seen in *A. ervi* probably represents heterogeneity within and

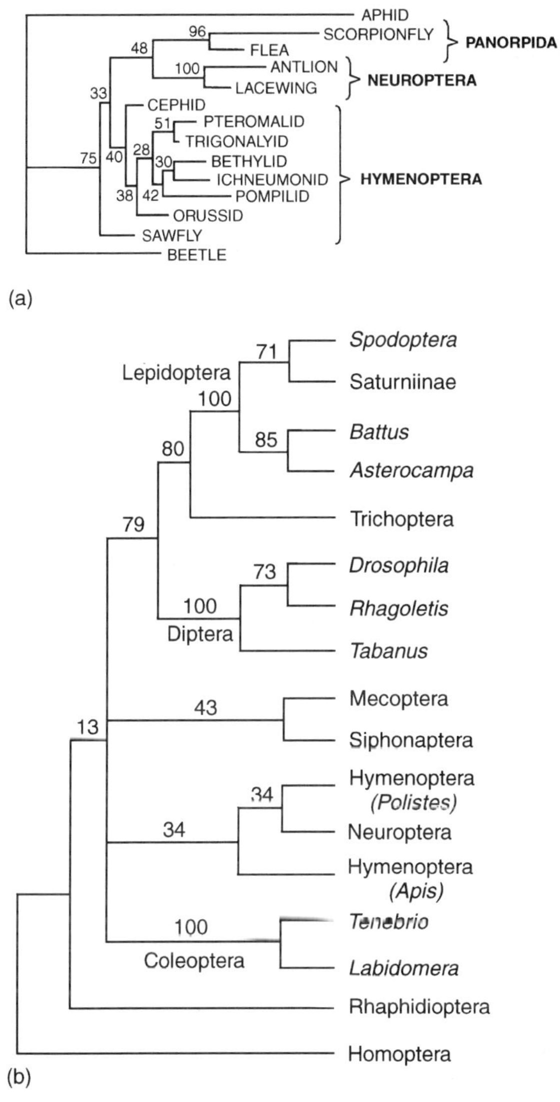

Figure 7.2 Phylogenetic trees constructed from 18S ribosomal RNA data showing relatedness of Holometabolous insect orders. (a) Bootstrap consensus tree with data sequences aligned with published sequences for Coleoptera (beetle) and with tree rooted using an outgrouping arthropod, i.e. aphid, *Acyrthosiphon pisum* (Heterometabola). (From Carmean *et al.*, 1992); (b) Bootstrap consensus tree rooted with the Spittlebug, *Prosapia bicincta*. (From Pashley *et al.*, 1993.)

between rDNA clusters. Further examination of the genetic variation between the four *Aphidius* species was done using the ITS region probe (1) and the NTS probe (2). This enabled reliable taxonomic discrimination of all the species, except *A. ervi* from *A. microlophii*. These last two species, which show host preference differences (Pennacchio *et al.*, 1994), may be differentiated electrophoretically by an apparently fixed allozyme difference for the enzyme phosphoglucomutase, PGM (Atanassova *et al.*, submitted).

An rDNA probe described by Collins *et al.* (1988) was, in contrast, able to separate unequivocally two species of mosquitoes in Kenya – *Anopheles gambiae* Giles and *A. arabiensis* Patton (vectors of malaria) (Figure 7.3), regardless of geographical location. Previously, chromosomal examination (Coluzzi and Sabatini, 1967; Coluzzi, 1968) or diagnostic octanol dehydrogenase (*Odh*) allozymes (Miles 1978, 1979) were used to distinguish these species. However, with the electrophoretic method and unlike the case of *Aphidius ervi/microlophii* cited above, both species shared certain alleles at the diagnostic locus, e.g. *Odh-95* and *Odh-100*, albeit at very different frequencies, *A. gambiae* having a low frequency of *Odh-95* and a high frequency of *Odh-100*, *A. arabiensis* the reverse. Hence differentiation using these particular markers was not absolute (Collins *et al.*, 1988). The rDNA technique has several advantages over chromosome analysis and allozymes: (i) it does not require appropriately preserved half gravid females and early 4th instars (chromosomes) or freshly killed or frozen material (allozymes), but can be used on dried specimens; (ii) it is extremely sensitive due to the highly repetitive nature of the rDNA target sites; and (iii), it can be used for all life-stages larger than the 1st instar larva, or a portion of an adult specimen (Collins *et al.*, 1988).

More recently, Scott *et al.* (1993) have employed an rDNA–PCR method to distinguish between the five most widespread members of the *A. gambiae* complex in Africa (*A. gambiae* s.s., *A. arabiensis, A. quadriannulatus* Theobald, *A. melas* Theobald and *A. merus* Donitz), an approach similar in some respects to that used by Fang *et al.*, (1993) in their study of leafhopper phylogeny. With the mosquito study, restriction fragments containing the 3′ end of the 28S coding region and the 5′ end of the NTS region were subcloned into plasmids and subsequently sequenced, the 3′ end of the 28S coding region being aligned with the rDNA 28S sequences for *D. melanogaster* and *Anopheles albimanus* (Wiedemann). From these sequences, 20-bp oligonucleotide primers were selected which, following PCR amplification, gave different DNA length products (ranging from 153–466 bp) for each species when separated electrophoretically and hence allowed species discrimination. As with the aforementioned ribosomal DNA probe technique, the method is very sensitive: here, even a single mosquito egg or a fragment of leg may be used to provide DNA for amplification and subsequent species determinations (Scott *et al.*, 1993).

7.3 USE OF rDNA FOR EXAMINING POPULATION STRUCTURE AND DYNAMICS

Ribosomal DNA probes have been used to investigate intergenic spacer heterogeneity within and between natural populations of insects and hence population structure and dynamics. Several studies (Lupoli *et al.*, 1990; Shufran *et al.*, 1991; Shufran and Wilde, 1994; Birch *et al.*, 1994) have used the approach to examine aphid populations, while Black *et al.* (1989) and Kambhampati and Rai (1991a) have studied populations of the mosquito, *Aedes albopictus* (Skuse).

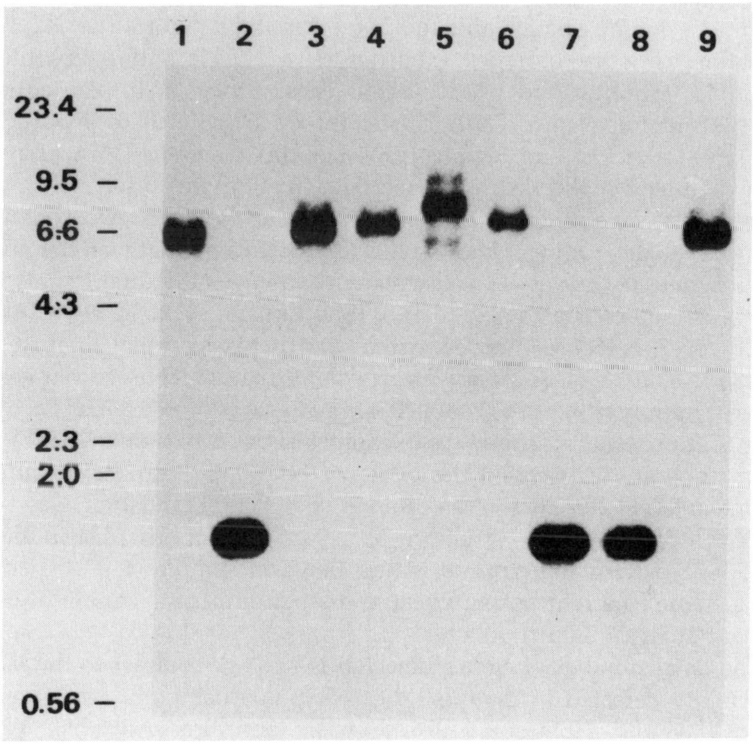

Figure 7.3 Agarose gel showing hybridization of a radiolabelled rDNA gene probe (pAg112A) to *Eco*RI digests in field-collected (Kenyan) specimens or morphologically identical mosquito species of the genus *Anopheles*, viz. *A. gambiae* and *A. arabiensis*. Individuals 2, 7 and 8 are scored for *A. gambiae* (1.4 kb *Eco*RI fragment), the remaining six are *A. arabiensis* with a cluster of bands centred in the 6–8-kb region. (From Collins *et al.*, 1988.)

Lupoli *et al.* (1990) used an approximately 8.5-kb cloned fragment (from *A. albopictus*), probably representing much of the rDNA gene cluster, including the rRNA coding genes (18S, 5.8S, 28Sα, and 28Sβ) to probe Southern blots of corn leaf aphid, *Rhopalosiphum maidis* (Fitch) DNA digested with various restriction enzymes. This revealed interclonal variability using several of the enzymes, but no intraclonal variation was demonstrated over 20 generations (restriction site polymorphisms and/or size polymorphism within restriction sites, among copies of the NTS regions of the rDNA multicopy gene complex). The advantage of this probe is that polymorphism may be detected in animals which are entirely parthenogenetic and which generally show little or no allozyme variation compared with other insects (Tomiuk and Wöhrmann, 1983; Loxdale *et al.*, 1985).

In similar studies of North American populations of the greenbug, *Schizaphis graminum* (Rondani) (Aphididae), two rDNA probes were used to examine spatial and temporal variation (Shufran *et al.*, 1991), variation between biotypes (Shufran *et al.*, 1992) and overwintering dynamics (Shufran and Wilde, 1994). These probes hybridize either to the NTS region of the rDNA cluster (probe pGBS1) or to a portion of the 18S and 28S coding regions (probe pGBE). Genomic DNA of aphids was digested by the restriction enzyme *Xho*I, giving three rDNA fragments, one containing the complete NTS, and the other two containing portions of the coding region (Black, 1993a). The length of the coding portion detected by pGBE was invariant among individual *S. graminum* and was thus used as a size marker (Shufran and Wilde, 1994).

However, NTS sizes varied within and between clonal lineages, but within a clone, spacer sizes were constant over six consecutive generations (Shufran *et al.*, 1991), which argues against recombination occurring within clonal lineages (i.e. endomeiosis; see Blackman, 1987). The population study revealed the existence of a large number of different *S. graminum* clones in Kansas, but no spatial differentiation was noted in terms of the genetic structure of clonal populations (distribution of NTS size pattern fingerprints) either between sample sites or between aphids from the two hosts (wheat and sorghum) using canonical discriminant analysis (Figure 7.4a,b).

The large number of clones detected is in stark contrast to the lack of variability detected in many studies of aphids using allozyme markers (e.g. May and Holbrook, 1978; Loxdale and Brookes, 1990; Wynne *et al.*, 1994). In a subsequent study of overwintering populations of *S. graminum* from Kansas, Oklahoma and Texas, Shufran and Wilde (1994) using the NTS probe, confirmed the existence of large numbers of clones, although here some spatial heterogeneity was observed and was assumed to be the result of founder effects.

Birch *et al.* (1994) used the rDNA probe pBG 35, which hybridizes to the 18S, 5.8S and 28S plus ITS and NTS spacer regions (Fenton *et al.*, 1994) and the endonuclease *Bgl*II, to examine biotypes of the large raspberry aphid, *Amphorophora idaei* (Börner). Using standard clones of biotypes susceptible and resistant to the A_1 host-resistance mechanism, they found consistent RFLP differences between the clones. However, in field populations, significant variation was frequently found which appeared unrelated to biotype classification, unlike the situation in *S. graminum* where characteristic biotype-specific IGS patterns are known (Black, 1993a). The authors attribute the within-population genetic variability to migration of parthenogenetic alatae in summer and males in the autumn.

In *Aedes albopictus*, NTS length variability was examined over three consecutive years and in eight populations from Texas (Kambhampati and Rai, 1991a). Variability was large, both within and between populations. However, the frequencies of the different length variants fluctuated in an apparently random, non-directional fashion with time. Statistically, nearly 80% of the total variance in frequencies of NTS variants was attributable to the within-population component, suggesting local differentiation of populations (Table 7.1). The results are very similar to allozyme studies performed on the species from 17 subpopulations in the US over 3 years (Kambhampati *et al.*, 1990). These also showed significant, directional changes in allele frequencies, with most variation occurring at the within-population level. Both these data sets imply that the populations of this mosquito are largely reproductively-isolated and that genetic drift is responsible for the pattern of within population variability observed, which fits with the fact that the breeding habitats of this insect are patchily distributed (Black *et al.*, 1988) and that adults have a limited ability to disperse (Kambhampati and Rai, 1991a).

7.4 OTHER DNA PROBES IN STUDIES OF TAXONOMY, POPULATION STRUCTURE AND DYNAMICS AND MATING SYSTEMS: RFLP (DNA) FINGERPRINTING

The development of the DNA fingerprint by A.J. Jeffreys and co-workers (1985a,b) has led to the rapid expansion of the use of molecular techniques in ecology and the Jeffreys system will be used to illustrate the basic model. Eukaryotic DNA is composed of firstly, single copy DNA for genes coding for enzymes and structural proteins; secondly, multiple copy DNA coding for a few genes where the products are required in large quantities, e.g. histones, ribosomal and transfer RNAs; and thirdly, stretches of non-coding sequences interspersed with repetitive sequences of varying length. In addition, eukaryotic genes often contain non-coding sequences (introns) interspersed between the coding

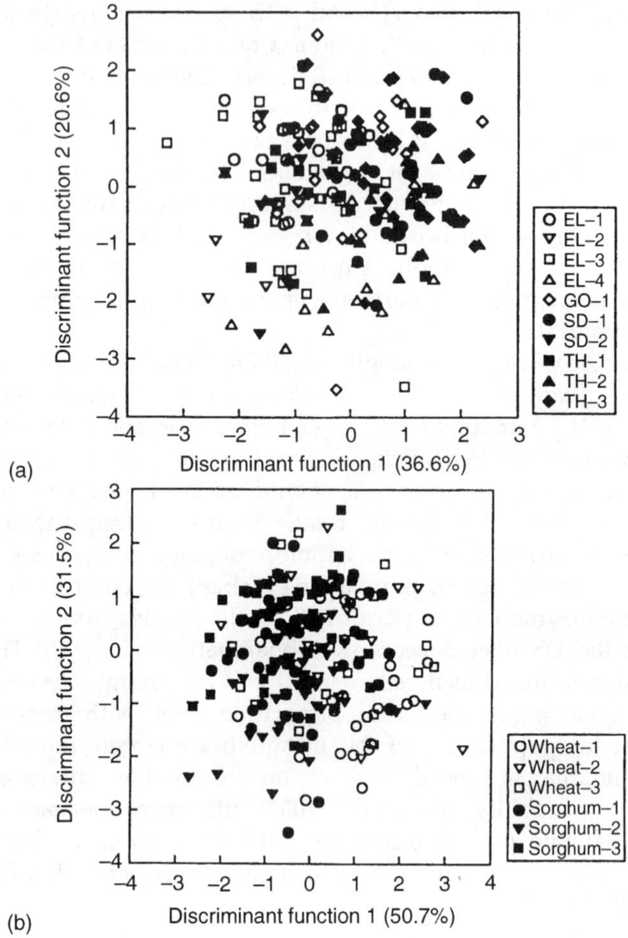

Figure 7.4 Canonical discriminant plots (first and second functions) for *Schizaphis graminum* (Aphididae) attacking cereal crops in Kansas, USA in 1989/90. (a) 1989 populations from 10 sorghum fields in Ellis, Gove, Sheridan and Thomas Counties. (b) 1990 populations from six wheat and sorghum fields in Riley County. See text for explanation of these plots. (From Shufran *et al.*, 1991.)

sequences (exons). In contrast, prokaryotic DNA is composed of 99% single copy DNA sequences (Kirby, 1992). In DNA fingerprinting it is the sequences of repeated DNA (tandem repeats) found in non-coding DNA which are of interest. These sequences of repeated DNA are assumed to be neutral, i.e. have no function. However, recent studies (Dorer and Henikoff, 1994; Stephan and Cho, 1994; Karpen, 1994) suggest that in some cases, long series of tandem repeats may silence

Table 7.1 Partitioning of variance in the NTS (non-transcribed spacer) region of the rDNA cistron of Texan *Aedes albopictus* determined using a specific radiolabelled probe into within-population, within-cities, and among-cities components for four sampling dates*. (From Kambhampati and Rai, 1991a.)

Variance component	*Fall 87*	*Summer 88*	*Fall 88*	*Summer 89*
Within-populations	70.8	82.9	70.4	86.1
Among locations within cities	2.7	3.5	0.02	3.1
Among cities	26.5	13.6	29.6	10.8
Total	100.0	100.0	100.0	100.0

*The values are given as percent contribution to the total. The partitioning is based on the Shannon diversity index. For calculations in the table, the raw, rather than the jack-knifed, values were used. (See text of original paper for details.)

genes by converting euchromatin to heterochromatin. These series may therefore have a selective influence should the silenced gene have a significant effect on fitness.

DNA fingerprinting using the Jeffreys system makes use of numerous hypervariable DNA sequences scattered across different loci throughout the eukaryotic genome (minisatellites). These different loci are often referred to as alleles, but this usage should not be confused with the traditional definition, i.e. genes which occupy a common locus on a particular chromosome. Hypervariable regions are pieces of DNA which do not code for a gene product and they comprise various numbers of identical repeated sequences (VNTR, variable number of tandem repeats). The Jeffreys system uses minisatellite repeats of 10–35 bp in length and contains a core sequence (10–15 bp) of highly conserved DNA. An example of such a core sequence is ACAGGGTTGGGGG. There are several core sequences which vary in base composition, but all share common base sequences (underlined) (Dover, 1989). Although the technique was first published by Jeffreys and co-workers in the mid-1980s, these highly polymorphic loci were known 5 years before (Wyman and White, 1980). Jeffreys *et al.* (1985a) obtained DNA fingerprints by using the minisatellite probes 33.15 and 33.6 which were derived from hypervariable regions within human minisatellite DNA.

When hypervariable DNA is restricted using endonucleases, a series of different length fragments of DNA results (RFLPs). Fragment length is dependent upon the number of repeats within the block as well as the choice of restriction enzyme. Fragment number also depends upon the restriction enzyme used: for instance, a frequent cutter yields more and smaller fragments than an infrequent cutter.

Because of the variation in size and number, RFLPs can be used as genetic markers. The cause of the variation in fragment length at specific loci is thought to be a result of one of three mechanisms:

1. The loss of a restriction site through base-pair mutation increases fragment length and reduces fragment number, whereas the addition of a site reduces fragment length, but increases fragment number.
2. Fixed (stable) restriction sites that flank tandem repeat sequences where the number of repeats within the sequence varies, probably due to unequal cross-over during meiosis. Unequal cross-over results in one strand of DNA losing a repeat to the complementary strand. This is thought to be the main mechanism by which RFLPs are formed and the one most useful in creating the heterozygosity necessary for a genetic marker to be of value in fingerprinting.
3. Insertion or deletion of nucleotides between fixed restriction sites.

Type (1) RFLPs derived from the presence or absence of a restriction enzyme recognition site have lower heterozygosity relative to those formed by VNTRs between fixed restriction enzyme recognition sites. According to the Hardy–Weinberg law, heterozygosity due to an unstable enzyme site between two fixed sites has a maximum value of 50% as the site can be either present or absent and so the probability for either allele is $p = q = 0.5$. Since these RFLPs have such low heterozygosity, their usefulness as markers for DNA fingerprinting is low. In contrast, hypervariable sites (type 2) have large numbers of different alleles and hence high heterozygosity, making them far more useful for fingerprinting. Genetic markers which are most useful for fingerprinting are highly polymorphic and easily detected and allow a sound statistical analysis of population and genotypic frequencies.

Many genetic markers are detected using single-stranded DNA or RNA probes which contain the complementary code for a specific sequence of genome bases. Two groups of fingerprint probes are currently in use. One detects tandem repeat sequences at a specific locus (single locus DNA fingerprinting) under high-stringency conditions and the other detects tandem repeats at numerous loci (multilocus DNA fingerprinting) under low-stringency conditions. In either case, each locus consists of many possible alleles, their frequency being dependent upon the specific population under study. The choice of system is dependent upon the questions being asked. The probes are either derived from genomic repeated DNA (from the organism under study or a different species), or are synthetic DNA oligonucleotides (short sequences) which are complementary to the repeated genomic DNA. Complementary DNA (cDNA) is not normally used as this is derived from mature RNA which does not contain repeated sequences. The restricted DNA is placed in an agarose gel, separated electrophoretically (predominantly on the basis of fragment size), and Southern blotted onto a nitrocellulose or nylon membrane (Sambrook *et al.*, 1989). Fragments of DNA containing the blocks of repeated DNA are then visualized using either a DNA or RNA probe (a sequence of nucleotides homologous to the core sequence within each repeat: in the original Jeffreys [1985a,b] system, an RNA probe was used) labelled with either a radioactive isotope, e.g. ^{32}P-ATP or a chemiluminescent label, e.g. Digoxigenin (DIG) (Boehringer-Mannheim).

Following hybridization of the labelled probe to the DNA on the membrane, and autoradiography to reveal the position of detected bands, which represent different length fragments (Post and Crampton, 1988; Sambrook *et al.*, 1989), a DNA fingerprint is produced (Jeffreys *et al.*, 1985a,b). The Jeffreys system produces multilocus fingerprints since the probe binds to the core sequence of repeats present in blocks located at many different positions (loci) in the DNA. The process is identical for single locus probes (see later).

Because it is simpler, multilocus DNA fingerprinting has been more widely used to study a range of entomological questions, although the best-known multilocus probes, the 33.15 and 33.6 probes discovered by A.J. Jeffreys, have been infrequently used to fingerprint insect genomes. Carvalho *et al.* (1991) used an RNA derivative of Jeffreys' 33.15 minisatellite core sequence as a probe to demonstrate that individuals of the grain aphid, *Sitobion avenae* (F.), inseparable using allozyme analysis, were in fact genetically distinct clones. Other authors who have also used the Jeffreys probe include Mueller *et al.* (1994) to examine kinship behaviour in the Halictid bee, *Augochlorella striata* (Provancher), Saccheri and Bruford (1993) in a mating study of the tropical Satyrid butterfly, *Bicyclus anynana* Butler, and Zeh *et al.* (1993) to investigate the ability of these probes to distinguish unrelated individuals of the Chrysomelid beetle, *Macrohaltica jamaicensis* (F.).

Endogenous sequences can be cloned from the target species and used to detect hypervariable sequences. They have been used by Haymer *et al.* (1992) to show genetic variation between different strains of the Mediterranean fruit fly, *Ceratitis capitata* (Weidemann) and Yang-jiang *et al.* (1992) used endogenous probes from the fall armyworm, *Spodoptera frugiperda* (J.E. Smith) to confirm the existence of the two forms of the moth previously demonstrated using allozymes. Species-specific endogenous probes have also been used to separate individuals of the *Anopheles dirus* Peyton and Harrison complex (Panyim *et al.*, 1988), while Post and Crampton (1988) and McLain *et al.* (1986) have used similar approaches to distinguish members of the medically important *Simulium damnosum* Theobald s.l. and *Aedes scutellarus* Edwards species complexes, respectively.

Smaller repeats (2–6 bp) arrayed in various length blocks and scattered throughout the genome are collectively known as oligonucleotide repeats. They are distinct from the Jeffreys minisatellite and NTS (=IGS) repeats and are more widely distributed in the genome than these larger repeat sequences. They are in the same class as the short microsatellite repeats such as $(CA)_{13}$, and $(AAT)_{10}$ described by Hughes and Queller (1993) and Queller *et al.* (1993). However, the method of isolation and detection of the latter is distinct from that of multilocus oligonucleotide repeats and so will be discussed later (section 7.5). These short repeats are also an important source of genomic variation and so complementary probes can be used to obtain DNA fingerprints. Like the Jeffreys system already discussed, oligonucleotide probes can be used to reveal inter- and intraspecific genetic variability (Weising *et al.*, 1991a).

Oligonucleotide fingerprinting has been used to obtain species, variety, accession, strain or individual specific fingerprints from a wide variety of plants, fungi, protozoa, insects, fish, amphibians, reptiles, birds and mammals (Weising *et al.*, 1991a; Epplen, 1992). One such family of repeats are the simple quadruple repeats such as GATA and GACA (Epplen, 1992), although others have also been detected (Schafer *et al.*, 1988; Weising *et al.*, 1991a,b). The number of repeats in each series is also hypervariable, making them useful as probes for DNA fingerprinting. The synthetic oligonucleotide $(GATA)_4$ [GATAGATAGATA-GATA] has been used as a probe in a process identical to the Jeffreys' and NTS probes. The resultant fingerprints are specific to an individual (Weising *et al.*, 1991a) or clone line.

Epplen (1992) states several advantages to fingerprinting using oligonucleotides including constant probe quality (as they are chemically synthesized and involve a shorter, simpler fingerprinting process) and absolute reproducibility (as even a single mismatch can prevent hybridization under conditions of high stringency).

In entomological studies, De Barro *et al.* (1994a) used the oligonucleotide, $(GATA)_4$ to demonstrate clonal fidelity in the grain aphid, *Sitobion avenae* (Figure 7.5a,b). In subsequent papers, De Barro *et al.* (1994b, 1995a) used $(GATA)_4$ to investigate geographic and micro-geographic genetic differentiation in *S. avenae* and the rose-grain aphid, *Metopolophium dirhodum* (Walker) as well as temporal variation in the clonal structure of *S. avenae* on wheat ears. $(GATA)_4$ is to date the most widely employed of these short motifs in the study of insect genomes and has also been used to investigate sperm precedence in the bush-cricket, *Poecilimon veluchianus* (Ramme) (Archmann *et al.*, 1992) and to demonstrate the absence of meiotic recombination in thelytocous parthenogenesis in the honey bee *Apis mellifera capensis* Escholtz (Moritz and Haberl, 1994). In addition, Traut *et al.* (1992) recently used a probe containing this repeat to look at patterns of inheritance and mutation in the moth, *Ephestia kuehniella* Zeller. Another probe $(TTTA)_7$, has also used by Mueller *et al.* (1994) to study kinship behaviour in bees.

The most widely used of the multilocus probes are those derived from the bacteriophage vector M13. It has been used to study strain variation in the Mediterranean fruit fly (Medfly), *Ceratitis capitata* (Haymer *et al.*, 1992); in population, kinship, mating and genetic relatedness studies of the social bee *Lasioglossum marginatum* Brullé (Blanchetot and Packer, 1992); the solitary (leafcutter) bee, *Megachile rotundata* (F.) (Blanchetot, 1992); subspecies of the honeybee, *Apis mellifera* L. (Blanchetot, 1991); and in the spread and level of hybridization (Figure 7.6) of the African 'killer' subspecies *A. m. scutellata* (Lepeletier) in South and Central America (Muralidharan and Hall, 1990; Hall, 1992a,b); M13 has proved especially useful for studying genetic variation in the Hymenoptera which usually show little or no allozyme variation (Metcalf *et al.*, 1975; Packer and Owen, 1990, 1992). Genetic variation in the tsetse fly complex, *Glossina* spp. (Blanchetot and Gooding, 1993), nepotism in the cicada killer wasp, *Sphecius speciosus* (Drury) (Pfennig and Reeve, 1993) and variation in offspring of the silkworm *Bombyx mori* L. as a result of ameiotic and meiotic parthenogenic reproduction (Tretjk *et al.*, 1992) have also been investigated using M13. Lastly, population-specific fingerprints have recently been demonstrated for Panama and French Guiana populations of the pseudoscorpion, *Cordylochernes scorpioides* (Hermann) (Zeh *et al.*, 1992) using this approach.

The analysis of multilocus fingerprinting data is source-independent and relies largely on comparison of the number of shared bands between two individuals using the mean similarity index, S (Wetton *et al.*, 1987; Lynch, 1991). The S index is calculated as: $S = 2N_{AB}/[N_A+N_B]$, where N_A and N_B are the number of bands in individuals A and B respectively and N_{AB} is the number of bands shared between A and B (Wetton *et al.*, 1987; Lynch, 1988, 1990, 1991). In clonal and partially sexual organisms, the index S is a function of the time since the individuals shared a common ancestor (Brookfield, 1992) and, despite methodological and theoretical reservations (Lynch, 1988), it is generally positively correlated with relatedness. In mating studies, most comparisons are simply based on the detection of bands from the putative parents in their offspring.

A less detailed comparison can be made using the average distance of band migration (ADBM) or individual mean fragment length (IMFL) of individuals (Zeh *et al.*, 1992 use this approach with the pseudoscorpion, *Cordylochernes scorpioides*). Compared with band-sharing indices, the ADBM is a very crude way to investigate fingerprints. Indeed, since different fingerprints can potentially have the same ADBM, discrimination between populations requires supportive data from other approaches. However, it allows an individual fingerprint to be summarized by a single variable.

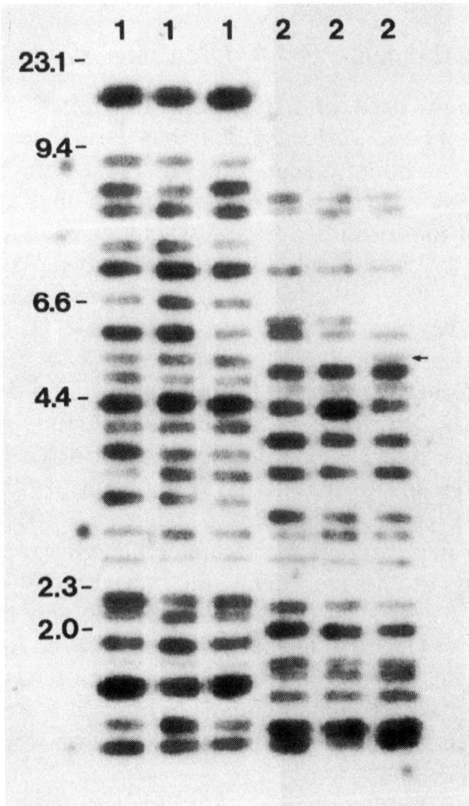
1 1 1 2 2 2
23.1
9.4
6.6
4.4
2.3
2.0

(a)

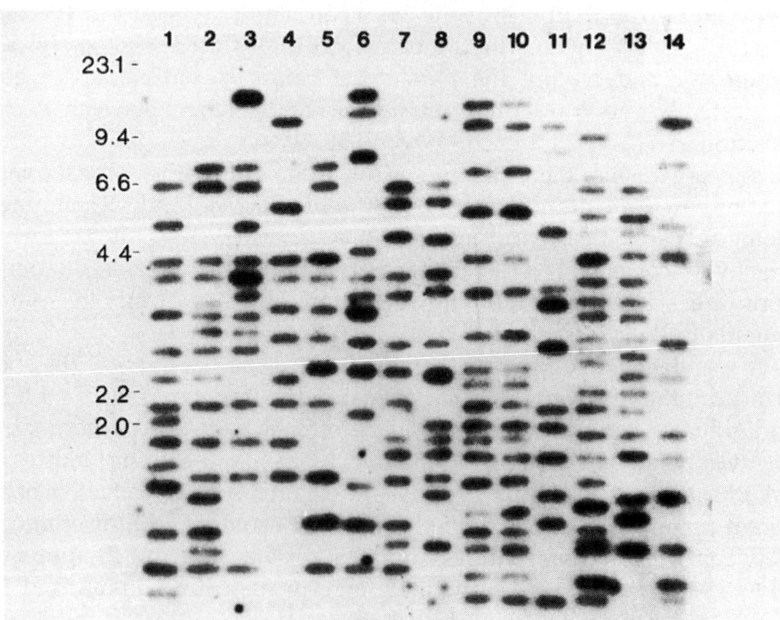
1 2 3 4 5 6 7 8 9 10 11 12 13 14
23.1
9.4
6.6
4.4
2.2
2.0

(b)

Multilocus fingerprinting has been widely used in pedigree (kinship) analysis and paternity studies. However, the rapid fixation of mutations relative to allozymes and rRNA limits its use in phylogenetic studies. In addition, the large amount of variability often demonstrated in outbreeding populations results in an unacceptably large number of individuals being required to demonstrate a significant difference. For this reason, population studies using multilocus techniques are uncommon. However, the reduced variability in inbred populations or reproductively isolated species, i.e. clonal species, allows the application of this technology (Burke *et al.*, 1991).

Probes for single loci can be used to rapidly screen populations for the presence of specific genes, for example, the amplified esterase-4 (E4) gene of the peach potato aphid, *Myzus persicae* (Sulzer), which confers insecticide resistance (Devonshire and Field, 1991; Field and Devonshire, 1992). Here, DNA from individuals can be blotted directly onto a membrane and then probed with a sequence homologous to the target gene. In the case of *M. persicae* the target gene can be detected as a dark dot on X-ray film, the intensity of which is directly related to the number of E4 genes present in the different resistant strains of the aphid (Figure 7.7) (see later).

To date, there has been only a limited application of single locus probes to insect genomes because detection and isolation is an involved process although recent advances (Armour *et al.*, 1990; Hughes and Queller, 1993) have made it easier. Burke *et al.* (1991), in their review, indicate that these locus-specific probes will be of use in studying evolution of repeat sequences, paternity, genetic relatedness and population structure, as well as in genome mapping. The principal advantage of locus-specific probes is that they can provide information on whether an individual is homozygous or heterozygous (Jeffreys *et al.*, 1991).

In a series of papers, Field *et al.* (1988, 1989, 1993, 1994) have demonstrated the use of probes which detect amplification of the E4 gene and its allele FE4 in *M. persicae*. This has allowed them rapidly to assess the presence of resistant strains, including highly resistant strains as well as the apparently susceptible revertant strains which contain unexpressed

Figure 7.5 (left) $(GATA)_4$ fingerprints of individual grain aphids, *Sitobion avenae*. (a) Showing clonal stability within two clones representing generations 1, 6 and 12. Clone 1 shows continuous stability compared with clone 2, which displays a single band shift in the twelfth generation. (b) Fingerprints from 14 putative clones collected from six populations at several localities across southern England. Lanes 1 and 4, 2 and 3, 5 and 6, 7 and 8, 9–11, 12–14 are from the same population. Kb values are shown vertically. (From De Barro *et al.*, 1994a.)

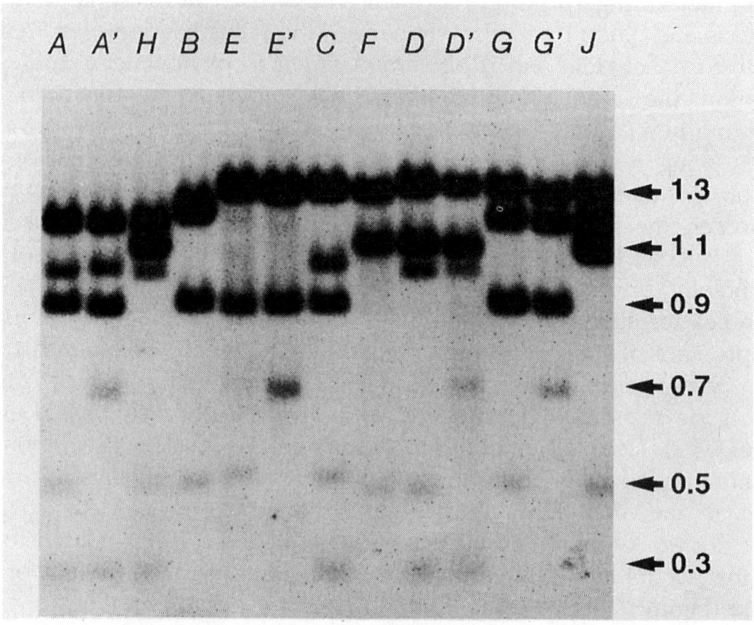

Figure 7.6 Use of a DNA probe to reveal RFLP polymorphisms in honey bees, *Apis mellifera*, collected from various sources world-wide (South Africa, Honduras, Mexico, USA, Hawaii, and western and eastern Europe). Probe 12R1C1, which hybridizes to DNA fragments cut using the restriction enzyme *Hha*I, revealed 14 RFLP allele patterns (haplotypes). These include 'African specific' alleles (A′,C′,D,D′, E,E′,F,G,G′,H,J), prevalent in samples from South Africa, some of which are also found in samples from Honduras (A′,C′,D′,E,G,G′,J; Africanized bees), but absent in Mexican, US, Hawaiian and European samples, while alleles A and C were found in most of the samples examined (allele A was absent from samples collected from Eastern Europe). Allele B was found in only one colony in southern Mexico, while allele C′ is not included in the figure. Fragment sizes indicated in kilobases; letter designation of alleles at top of each lane. Several allele pairs possess the same pattern except that one member carries a 0.5-kb fragment while the other (given a prime notation) carries a 0.7-kb fragment. The profiles shown represent patterns from haploid drones; hence, examination of patterns from drones from several colonies and from diploid females allows elucidation of homozygous and heterozygous patterns for alleles segregating at single loci. (From Hall, 1992a.)

amplified genes, a phenomenon which involves changes (Figure 7.7) in DNA methylation (Field *et al.*, 1989). The ability to detect revertants is important as they can be readily reselected by insecticide applications (ffrench-Constant *et al.*, 1988). Since revertants cannot be detected using enzyme electrophoresis, the use of the resistance probe is essential for

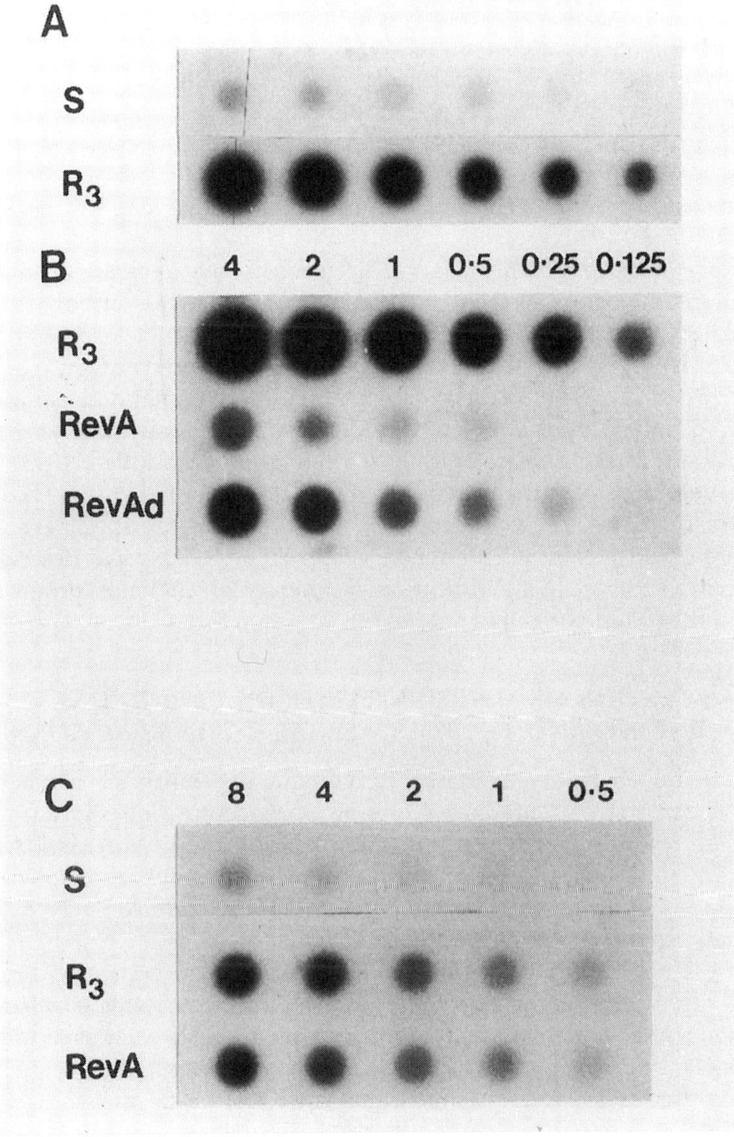

Figure 7.7 Dot-blot detection of level of amplification of E4 (esterase-4) gene in susceptible and insecticide-resistant variants of *Myzus persicae* (Aphididae), including revertants, using a radiolabelled cDNA probe (pMp24). (A,B) Hybridization to poly(A^+) RNA and (C) to DNA from susceptible (S), extremely resistant (R_3), revertant (RevA) and reselected revertant (RevAd) aphids. (From Field *et al*, 1989.)

an accurate estimate to be made of the frequency of resistant *M. persicae* strains within a population (Devonshire, 1989; Devonshire and Field, 1991). In a similar study of amplified genes in mosquitoes, Mouches *et al.* (1990) have detected the esterase B gene in the genus *Culex* L. using the B1 probe from Californian *Culex quinquefasciatus* Say. This has been used to study the spread of the amplified B2 gene through Africa, Asia and North America (Raymond *et al.*, 1991). Single copy DNA sequences from moths belonging to the genus *Helicoverpa* Hardwick (formerly *Heliothus*) have also been used to construct a linkage map for this genus (Zraket *et al.*, 1990).

Other areas in which nuclear DNA probes have potential in agricultural entomology is in the identification of gut contents of polyphagous predatory insects and in the detection of parasitism of insects, such as aphids by hymenopterous endoparasitoids. Both predator gut contents (Murray *et al.*, 1989; Solomon *et al.*, 1996, this volume) and parasitoids (Powell and Walton, 1989) can be detected using enzyme electrophoresis, the former also serologically (Symondson and Liddell, 1993, 1996, this volume), but these require fresh or deep-frozen material. By contrast, species-specific probes may potentially utilize specimens dried or preserved in alcohol. Recently, Clapp *et al.* (1993) have described a genomic subtractive hybridization technique which holds promise for gut content and parasitoid studies.

7.5 DETECTION OF MICROSATELLITE DNA USING DNA SEQUENCING AND THE POLYMERASE CHAIN REACTION (PCR)

As mentioned earlier, scattered throughout the eukaryote genome are numerous, highly variable regions of DNA comprising short (<6 bp) tandem repeat oligonucleotide sequences, for example $(CA)_{13}$, $(GATA)_4$, or $(AAT)_{10}$. Here we look briefly at a new method which simplifies the detection of variable length sequences (here termed microsatellite repeats) and also allows specific loci to be targeted.

In essence, polymorphic loci are detected by first extracting genomic DNA from the target organism. The DNA is cut with a restriction enzyme and separated by electrophoresis to isolate short fragments (300–600 bp, some of which will contain microsatellite sequences), which are ligated into plasmids and used to transform *Escherichia coli* and thus construct a genomic library. The colonies are then screened with a radiolabelled oligonucleotide repeat sequence, e.g. $(AAT)_{10}$ and positive colonies are sequenced, those containing ten or more repeats and showing microsatellite polymorphism being selected. From these, primers can be designed complementary to regions flanking the microsatellite repeats. These primers are then used in a PCR reaction with DNA from test individuals to amplify specific microsatellite loci. The end result upon electrophoresis will be either a single or double band (Queller *et al.*, 1993; Hughes and Queller, 1993).

Microsatellite markers have several advantages, including abundance, numerous alleles at specific loci and high heterozygosity. In the latter two respects, they are like allozymes and 'appear to be good co-dominant markers' (Queller *et al.*, 1993). In addition, they produce stable patterns which are easy to compare across gels by using a ladder or by counting up or down the bands which differ by 'uniform, known and resolvable distances' from a reference band(s) (Queller *et al.*, 1993). Ideally, banding patterns of individuals show, as with allozymes, either one allele or two equally dark ones (Figure 7.8). Another significant advantage is that these polymorphisms may be readily detected in dried material (see Hughes and Queller, 1993; Queller *et al.*, 1993 for further details and diagrammatic explanation of the technique for producing microsatellite fingerprints).

So far, the technique has not been used widely in entomology, but it is likely to become a very useful tool in the study of insects, especially with members of the Hymenoptera and Aphididae, both of which generally show a paucity of enzyme variation. One study (Hughes and Queller, 1993), used the AAT repeat to identify genetic variation in the social wasp, *Polistes annularis* L., a species displaying little enzyme polymorphism. Trinucleotide microsatellite loci (Figure 7.8) are abundant in this wasp and highly polymorphic. For example, of six loci tested, mean observed heterozygosity was 0.62 compared with 0.035 for 33 enzyme loci. In another recent study of Hymenoptera, Evans (1993) has used primers flanking CA/GT repeat loci in ants of the genus *Myrmica* Latreille to study parentage within colonies.

7.6 HYBRIDIZATION TECHNIQUES

In many ways, 'classical' hybridization techniques are a rather outdated approach, and although they have been used with great effect to elucidate certain aspects of avian and hominoid phylogeny (Sibley and Ahlquist, 1983, 1984; Sibley *et al.*, 1987), very few entomological studies have used them. Hunt *et al.* (1981) and Hunt and Carson (1983) have used hybridization to examine the evolutionary relationships of various species of Hawaiian *Drosophila*, while more recently, Broughton and Kistner (1991) have used it to discriminate between four species of North American termites of the genus *Zootermopsis* Emerson.

The technique is described in detail by Sibley and Ahlquist (1986) and Broughton and Kistner (1991). It relies on the shearing by sonication of DNA from a given species into fragments of about 500 bp. The sheared DNA is then separated into single bands by brief heating and, as it cools, re-annealing occurs with the most abundant sequences re-asso-

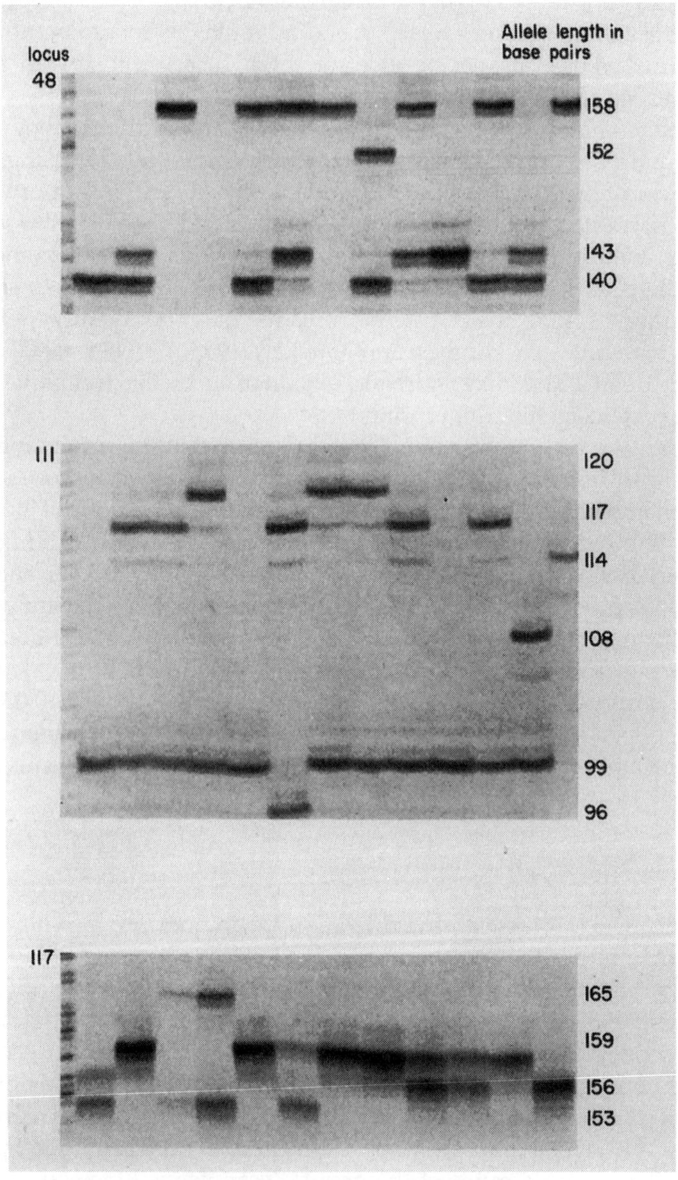

Figure 7.8 Highly polymorphic microsatellite DNA markers revealed following autoradiography. Sequencing gels showing three loci amplified from unrelated female social wasps, *Polistes annularis*. Part of a sequencing reaction is loaded in lane 1 as a size marker. Note the characteristic phenotypes at each locus. (From Hughes and Queller, 1993.)

ciating fastest. The cooling DNA solution is passed through a hydroxyapatite column which preferentially binds double-stranded DNA, and allows the single-stranded to pass through. Thus it retains the more numerous sequences in preference to the less abundant ones. The latter are then labelled with radioactive iodine or phosphorus and mixed with unlabelled DNA from the same or a different species. The combined DNAs are incubated for 120 hours at 60°C, and will form a duplex, one-half of which is labelled. The hybrid strands are then returned to a hydroxyapatite column, heated in increments of 2.5°C from 55 to 95°C and single-stranded DNA is eluted from the column at each temperature increment. The amount of radioactivity in the elutant is measured per temperature. The greater the genetic relatedness between reference and test species, the higher the homology between labelled and unlabelled DNA and the higher the melting point of the duplex.

In such a study on termites, incremental melting point curves were produced for species comparisons. From these data, a dendrogram was constructed of the species relatedness using Distance–Wagner methods (Nei, 1987; Avise, 1994), that indicated some genetic distance between two morphologically-identical *Zootermopsis nevadensis* (Hagen) populations (subsp. *nevadensis* and *nuttingi*), hitherto known to differ only in cuticular hydrocarbon patterns (Haverty *et al.*, 1988). The results of the hybridization experiments also provided independent confirmation of the conclusions derived from a study of the *Zootermopsis* cuticular hydrocarbon phenotypes using allozymes (Korman *et al.*, 1991). However, according to Broughton and Kistner (1991), the hybridization technique is generally less useful in studying very close phylogenetic relationships and is probably best applied at the generic and species levels.

7.7 MITOCHONDRIAL DNA (mtDNA) AND RFLP ANALYSIS OF POPULATION GENETIC STRUCTURE, GENE FLOW AND MATING SYSTEMS

Analysis of mitochondrial DNA (mtDNA) is now a very familiar tool in animal biology for several reasons. It is easy to separate and purify (due to its unusual buoyant density in CsCl density gradient centrifugation), has a high copy number and is ubiquitous within the animal kingdom. Also, mitochondrial DNA has conserved genes (37, specifying 22 tRNAs, 13 mRNAs and two rRNAs), a similar gene order (changes only occurring between phyla) and simple structure (no introns or repetitive sequences) (Wilson *et al.*, 1985). It is predominantly maternally transmitted and functionally haploid, although there is some evidence from backcross experiments that leakage of paternal mtDNA does occur in vertebrates (*Mus*), insects (*Drosophila*) and molluscs (*Mytilus edulis* L.) (Avise, 1991). Mitochondrial DNAs usually comprise around 15–20 kb of double-stranded, tightly packed DNA, the arrangement being very stable (Wilson *et al.*, 1985).

Maternal inheritance and the lack of recombination are features which make mtDNA less susceptible to interpopulation genetic homogenization, while at the same time, more susceptible to bottlenecks and founder effects than nuclear genes (Wilson *et al.*, 1985; De Salle and Templeton, 1998). The extensive nucleotide sequence divergence between conspecific populations and sibling species make it especially suitable for studies at this level of evolutionary divergence (Lansman *et al.*, 1981; Avise and Lansman, 1983; Brown, 1983).

Mitochondrial DNA is purified, restricted, Southern blotted and the fragments detected using specific probes (see Brown, 1980; Lansman *et al.*, 1981; Feinberg and Vogelstein, 1983, 1984; Sambrook *et al.*, 1989; Tegelstrom, 1992; White and Densmore, 1992 for technical details). Alternatively, polymorphic regions can be detected using primers that flank the polymorphic sequence allowing it to be amplified using PCR. This type of polymorphism analysis is used primarily in studies of population structure and movement (gene flow), intraspecific genetic variation and phylogenetics, as well as mating systems, although comparative studies involving mapping have also recently been attempted (ses below).

The use of mtDNA in entomological studies has been extensive. Its application depends on the genetics and ecology of the insect in question: for example, whether it shows a lack of enzyme variation (e.g. aphids, cicadas), thereby necessitating use of another marker, or has a mating structure (e.g. Hymenoptera) which makes it particularly amenable for examining the distribution of mtDNA variants (haplotypes) within and between nests/colonies. The orders studied to date include: Coleoptera (Azeredo-Espin *et al.*, 1991; Zehnider *et al.*, 1992; Vogler *et al.*, 1993a,b); Homoptera (Martin and Simon, 1990; Martinez *et al.*, 1992; Simon *et al.*, 1993); Orthoptera (Chapco *et al.*, 1992a, 1994); Diptera (De Salle *et al.*, 1987; Cariou *et al.*, 1990; Cockburn *et al.*, 1990; Kambhampati and Rai, 1991b; Taylor *et al.*, 1991; Sheppard *et al.*, 1992; Frohlich *et al.*, 1993; Narang *et al.*, 1993); Lepidoptera (Brower and Boyce, 1991) and Hymenoptera (Smith and Brown, 1990; Hall and Smith, 1991; Stille *et al.*, 1991; Garnery *et al.*, 1992; Stille and Stille, 1992). Studies thus largely involve insects of agricultural and medical/veterinary importance, although two such studies (on the monarch butterfly, *Danaus plexippus* L. (Brower and Boyce, 1991) and on tiger beetles, *Cicindela* spp. (Vogler *et al.*, 1993a,b)), concern insects of conservation interest. Space permits only some of these examples, mainly of agricultural interest, to be discussed in more detail.

Azeredo-Espin *et al.* (1991) examined population structure of the Colorado potato beetle (CPB), *Leptinotarsa decemlineata* (Say) in the USA. Three restriction enzymes (*Eco*RI and *Hpa*I and *Pst*I) were used that revealed polymorphisms which differentiated each of the four US populations tested. This finding prompted the further use of these markers to examine spread of the beetle in the US and elsewhere. In contrast, a previous study using allozymes (Jacobson and Hsiao, 1983) had shown variability between populations across the USA and Central America, yet was unable to unambiguously resolve CPB population structure.

In a similar study, Martinez *et al.* (1992) examined 35 clones of the bird cherry-oat aphid, *Rhopalosiphum padi* (L.) from separate, distant localities in Spain. Unlike the previous study, only one enzyme (*Rsa*I) detected restriction site polymorphism among *R. padi* clones. Preliminary diversity analysis of the variable data (Rand and Harrison, 1989) gave a total diversity index of 0.465, while the among-population index (G*st*; Nei, 1973) was 0.28.

Previous allozyme results for *R. padi* in Britain using two polymorphic loci (GOT and SORDH) showed a high degree of homogeneity of allele frequencies between geographically distant sites (Loxdale and Brookes, 1988). The RFLP data of Martinez *et al.* (1992) on the other hand, suggests some degree of population structuring and restriction of interpopulation gene flow. This may be due to the large geographical separation of samples (>400 km apart) which has led to allopatric differentiation resulting from the influences of such factors as founder effects, genetic drift and selection.

In another aphid study, Powers *et al.* (1989) examined mtDNA RFLPs in eight female lineages of the biotypes B, C, E and F of the greenbug, *Schizaphis graminum*. Five restriction enzymes were found to produce distinct and diagnostic inter biotype differences. The fragment patterns from greenbug biotypes were compared using the formula $F = 2N_{xy}/(N_x+N_y)$, where N_x = the number of fragments in genotype x, N_y in genotype y and N_{xy} the number of common fragments [analogous to the Lynch (1988, 1990, 1991), Similarity Index]. F values were then converted to estimates of percentage nucleotide sequence divergence, p (Nei and Li, 1979; equation 20). Comparison of mtDNA from sample biotypes of a laboratory culture (biotype F) showed more than one haplotype occurred in the population, while for all restriction enzymes, fragments found in biotype E were also found in biotype F. Later examination of isofemale lines of biotype F showed it possessed mitochondrial genomes distinct from the other three biotypes examined. Estimates of p in pair-wise comparison revealed biotype B versus C and B versus E had a sequence divergence of approximately 1%, and C versus E, nearly 2%. Such values suggest that a long period of time has elapsed since the biotypes shared a common ancestral mitochondrial genome, perhaps 3 to 6×10^5 years (Powers *et al.*, 1989).

Besides aphids, other Homoptera populations examined using mtDNA techniques include cicadas of the genus *Magicicada* (Martin and Simon, 1990; Simon *et al.*, 1993). In the former study, 13- and 17-year periodic cicadas (*M. septendecim* (L.) and *M. tredecim* Walsh and Riley) were collected from 16 localities throughout the mid-west and eastern USA. Two distinct mtDNA haplotype lineages were found (A and B) within each year class, which differed by 2.5%, based on fragment pattern and restriction mapping. The year–class lineages displayed latitudinally related spatial patterns (A north, B south, the boundary being around 33° North) as well as low within-taxon diversity (π) (π, approximately 0.0001 and 0.001 for lineages A and B respectively). Pattern similarity was calculated using the maximum likelihood method of Nei and Tajima (1983), involving within- and between-group genetic differentiation estimates, while genotype diversity was calculated according to Ball *et al.* (1988). Other statistics were also employed, including calculation of Gst, a measure of among-population heterogeneity (Nei, 1973) (see Martin and Simon, 1990 for details).

The low level of mtDNA variability found by Martin and Simon (1990), is mainly attributed to population bottlenecks in the Pleistocene period, followed by historically recent expansion into areas vacated by glaciation. The differences found in mtDNA variation between lineages were explained by their different biogeographic histories, again related to changes in the Pleistocene (Martin and Simon, 1990).

The likelihood of revealing polymorphisms can be increased if certain variable regions of mtDNA are investigated by PCR and sequencing (Frohlich *et al.*, 1993), using an approach essentially similar to that for rDNA discussed earlier, i.e. Scott *et al.* (1993). Frohlich *et al.* (1993) amplified, using specific primers designed from known *Drosophila yakuba* sequences, six mitochondrial tRNA genes of the Medfly, *C. capitata* (Ala, Arg, Asn, Ser, Glu and Phe) between genes for mitochondrial NADH dehydrogenases 3 and 5. Sequences obtained (582 b.p.) were compared with the same region of mtDNA from *D. yakuba*. The six tRNA sequences could be folded into cloverleaf shapes to give characteristic secondary structures; compared with *D. yakuba*, a number of substitutions were apparent, most occurring in loops, while of the few found in stems, none disrupted the secondary structure. Frohlich *et al.* (1993) suggest that this approach will enable population structure to be studied in species such as Medfly which show little detectable genetic variation using other methods, for example, allozymes (Kourti *et al.*, 1990).

The gypsy moth, *Lymantria dispar* (L.) is a major forest pest in eastern North America and was introduced accidentally from France in the

1860s. Moths are also thought to have been introduced into the north-west of North America between 1989 – 1991 from Russian grain ships. The species is known to be largely invariant for allozyme markers and a study by Harrison and Odell (1989) could not differentiate French and North American moths using RFLP maps; however, moths from Japan and China were distinct from North American moths as well as from each other. Bogdanowicz *et al.* (1993) have used mtDNA markers as a means of determining the origin of the introduced moths.

Although the dried specimens of the gypsy moths sent from the Pacific north-west of North America for analysis were badly degraded, small sections of the mtDNA molecule could be amplified by PCR and sequenced. The oligonucleotide primers used were designed based on published sequences for *D. yakuba* and *Apis mellifera* mtDNA, and allowed amplification of a set of overlapping fragments including parts of the cytochrome oxidase I (COI) and II (COII) genes (Bogdanowicz *et al.*, 1993). Sequence data for an approximately 180 bp region of the COI gene, showed that four haplotypes could be distinguished (Figure 7.9) and enabled moths of Asian origin (haplotypes C and D) to be distinguished from North American (A) or French (B) moths. These differences in haplotype were then used to demonstrate the presence of moths of Asian origin (haplotype C) in North America, probably as the result of a recent introduction via the Pacific north-west.

```
  1 AAGAAATTTA TCGTTTGATC AAATACCTTT ATTTGTTTGA AGAGTTGGAA

 51 TTACAGCTTT CCTTCTACTT TTATCTTTAC CTGTTTTAGC AGGTGCTATT

101 ACAATATTAT TAACTGACCG AAATTTAAAT ACATCCTTTT TTGACCCTGC

151 AGGAGGAGGG GATCCAATCC TTTACCAACA TTT


                                          *  *           *
Gypsy Moth Haplotype A (NA)            GGGGGAGATCCAATCCTTTATCAA
Gypsy Moth Haplotype B (FR)            GGAGGAGATCCAATCCTTTATCAA
Gypsy Moth Haplotype C (Asia)          GGAGGGGATCCAATCCTTTACCAA
Gypsy Moth Haplotype D (Asia)          GGAGGAGATCCAATCCTTTACCAA
```

Figure 7.9 Mitochondrial DNA RFLP polymorphisms (cytochrome oxidase I) revealed from four 'strains' (haplotypes) of the gypsy moth, *Lymantria dispar* using sequencing techniques. The 183 base pairs shown are homologous to positions 1986–2168 (sense strand) in *Drosophila yakuba* sequence (Clary and Wolstenholme, 1985). Nucleotides 155–178 are underlined, this region containing substitutions which distinguish the four haplotypes shown. Haplotypes A–D are identical outside this region. The *Bam*HI restriction site (GGATCC) is underlined for haplotype C. (From Bogdanowicz *et al*, 1993.)

Apart from these population studies, analysis of mtDNA has been applied to solve phylogenetic problems. Garnery *et al.* (1992) mapped restriction sites for the mtDNA of honeybees from 68 colonies from world-wide sources and belonging to 10 different *A. mellifera* subspecies. They also used specific primers and PCR to amplify several regions of the mtDNA molecule, which were subsequently sequenced; the mtDNA from these insects could be grouped into 19 classes (haplotypes). Calculation of genetic distances using both nucleotide substitution and sequence data (Nei and Li, 1979; Kimura, 1980), and the construction of phylogenetic trees using both maximum parsimony (Swofford, 1989) and distance methods (Saitou and Nei, 1987), revealed three clades, corresponding to three groups of populations with distinct geographical distributions (branch A, African subspecies; branch C, North Mediterranean subspecies; and branch M, West Mediterranean subspecies). These results partially confirmed previous hypotheses based on morphological and allozyme data, except that North African populations were assigned to branch A instead of M. The study supported the view that the Middle East is the centre of dispersion of the species, the three main branches separating about one million years ago.

In the ant, *Leptothorax acervorum* (F.), polymorphisms detected in mtDNA have enabled nest infrastructure to be investigated (Stille and Stille, 1992). Previously, Stille *et al.* (1991) had used allozymes to calculate relatedness between individuals within polygynous nests, but the rare occurrence of unrelated queens could not be detected due to the lack of suitable markers. Ants, like many other hymenopterans, have a low enzyme variability (Packer and Owen, 1992). Once individual queens could be distinguished genetically using mtDNA markers (13 haplotypes were discovered), it was found that four out of 27 nests investigated had queens of two different haplotypes (Stille and Stille, 1992). This may be an example of inquilinism, nest 'parasitism', but without detriment to the owner (Stille and Stille, 1992).

7.8 RANDOM AMPLIFIED POLYMORPHIC DNA–POLYMERASE CHAIN REACTION (RAPD–PCR)

Of the various molecular techniques currently employed in biological studies, RAPD–PCR has had the greatest impact in entomology. Certainly, the number of studies performed since this technique was developed 5 years ago (Welsh and McClelland, 1990; Williams *et al.*, 1990) is already larger than those using all other molecular methods combined. This is primarily because RAPD–PCR is faster and cheaper than many other techniques; for example, a result can be obtained within a day instead of the week or more required for conventional fingerprinting studies.

The basis of RAPD–PCR is the use of short, usually 10-bp primers and, as with other PCR-mediated amplifications such as that earlier discussed in relation to rDNA and mtDNA, a thermostable DNA polymerase (e.g. Taq DNA polymerase, extracted from the thermophilic bacterium, *Thermus aquaticus*), is used to exponentially amplify tiny (ng) amounts of template DNA in controlled temperature conditions using a thermocycler machine. For RAPDs, primers are chosen arbitrarily with no prior knowledge of whether they will reveal any reproducible polymorphisms. After the establishment of conditions to allow amplification (see later), the first stage of any RAPD study is the screening of primers in order to select one or more sequence(s) suitable in terms of (i) revealing differences, (ii) producing a few well defined bands and (iii) producing reproducible patterns.

The complexity of RAPD–PCR comes from the need to determine the temperature and duration of each of the PCR steps, as well as to balance the concentration of $MgCl_2$ and KCl, and optimize the pH of the reaction mix. All of these factors affect the quality and reproducibility of the end result, as does the quality of the template DNA. The lack of reproducibility with RAPD–PCR is usually overcome once the correct conditions have been accurately determined. Numerous papers deal with the technique and how to overcome the various problems associated with it (Arnheim *et al.*, 1990; Hadrys *et al.*, 1992; Hill and Stewart, 1992; Hoelzel and Green, 1992; Black, 1993b; Meunier and Grimont, 1993; Haymer, 1994). Micheli *et al.* (1994) provide in-depth explanations of the technique and Rollinson and Stothard (1994) review recent applications of RAPD markers in biology, including entomology.

In medical and veterinary entomology, RAPD techniques have been successfully applied (Adamson *et al.*, 1993) to distinguish two morphologically similar and sympatric Venezuelan sandfly species, *Lutzomyia spinicrassa* Morales-Alarcon *et al.* and *L. youngi* Feliciangeli and Murillo, both possible vectors of cutaneous and mucocutaneous leishmaniasis. Wilkerson *et al.* (1993) have similarly separated the two morphologically identical mosquitoes, *Anopheles gambiae* and *A. arabiensis*, while Apostol *et al.* (1993) have used RAPDs to study mating behaviour in the mosquito *Aedes aegyptii* (L.).

In agricultural entomology, RAPDs have been used to investigate taxonomic differences between different aphid species (Cenis *et al.*, 1993), including the genus *Diuraphis* (Robinson *et al.*, 1993). Genetic variation has also been examined between clones of the greenbug, *Schizaphis graminum*, pea aphid, *Acyrthosiphon pisum* (Harris), Russian wheat aphid, *Diuraphis noxia* (Mordvilko) and brown ambrosia aphid, *Uroleucon ambrosiae* (Thomas) (Black *et al.*, 1992) and within- and between-field populations of *D. noxia* (Puterka *et al.*, 1993; Robinson *et al.*, 1993). Puterka *et al.* (1993) compared samples of *D. noxia* from world-wide sources using both allozymes (23 loci) and RAPDs (seven primers, producing 69 polymorphic bands). Most populations, which were invariant for allozyme markers, especially in areas where the insect had been recently introduced – for example, the USA, Canada and France – were found to be more variable using RAPD markers. Cluster analysis showed there to be strong similarities between USA, Mexican, South African, French and Turkish populations, while most variation occurred in populations from the Middle-East and south Russia, where the species is holocyclic (with sexual phase) (Kiriac *et al.*, 1990) and from where it probably originated (Kovalev *et al.*, 1991). Lastly with aphids, Black *et al.* (1992) have detected RAPD banding patterns of hymenopterous endoparasitoids *Diaeretiella rapae* (McIntosh) and *Lysiphlebus testaceipes* (Cresson), superimposed upon the RAPD profile of the aphid host(s), showing the technique to have potential for assessing percentage parasitism in aphid field populations.

RAPD markers have proved valuable in the differentiation of coccinellid beetles (Roehrdanz and Flanders, 1993), melanopline and oedipodine species of grasshopper (Chapco *et al.*, 1992b) and in separation of the A and B biotypes of the whitefly, *Bemisia tabaci* (Gennadius), all 20 of the primers tested distinguishing the types (Gawel and Bartlett, 1993). In the last study, the technique gave similar patterns even when DNA was extracted from eggs or nymphal stages. Genetic divergence statistics showed that the A and B forms are no more similar to each other than they are to two other whitefly species – the bayberry whitefly, *Parabemisia myricae* Kumana and the banded-winged whitefly, *Trialeurodes abutilonea* Haldeman. Williams *et al.* (1994) have investigated the geographic origin of the Argentine stem weevil, *Listronotus bonariensis* (Kuschel) in New Zealand by comparing these with material from South America and Australia. They concluded that South America was the likely source of introduction.

In Hymenoptera, Roehrdanz *et al.* (1993) have successfully differentiated five species of aphid parasitoid using RAPDs, i.e. *Diaeretiella rapae, Aphidius matricariae* Haliday, *A. picipes* (Nees), *A. colemani* Viereck and *Lysiphlebus testaceipes,* while Vanlerberghe-Masutti (1994a) examined genetic variability in the egg parasitoid *Trichogramma evanescens* Westwood using a single primer and found extensive variation in 40 isofemale lines of this species (see also Vanlerberghe-Masutti, 1994b). In the honeybee, RAPDs have been used to reveal the genetics of inheritance (Hunt and Page, 1992). Here, primers were selected to examine the inheritance of RAPD alleles in the haploid drone progeny and daughter queens of mated pairs of bees. From this work, four types of polymorphism were detected: (i) band presence and absence and (ii) brightness, both inherited in Mendelian fashion and as dominant markers; (iii), fragment length variation, segregating as a 1:1 ratio in drone progeny; and (iv), a diploid-specific band, the result of heteroduplex formation of the DNA of alternative alleles in heterozygotes (Hunt and Page, 1992). In two out of the four cases of heteroduplex formation, the alternative alleles were seen as small fragment length polymorphisms, so that the resultant heterozygotes appeared as co-dominant markers. Other studies of parasitic hymenoptera have included species separation of the microhymenoptera, *Anaphes* spp. and *Trichogramma* spp. (Landry *et al.*, 1993) and the detection of polymorphism in *Trioxys pallidus* (Haliday) and *Diglyphus begini* Ashmead (Edwards and Hoy, 1993).

The interpretation of RAPDs is similar to that used in other molecular studies, although greater care is required in the selection of bands so as to avoid mis-identification. To this end, Clark and Lanigan (1993) discuss the prospects of estimating nucleotide divergence using RAPDs and provide nine criteria which should be satisfied when closely related taxa are examined, while Rollinson and Stothard (1994) briefly outline the problems of constructing phylogenetic trees using these markers, especially concerning the assignment of band homology. Lynch and Milligan (1994) describe the statistical analysis of data on population genetic structure in sexually reproducing species.

Apart from the study of insects of agricultural importance, RAPDs have been applied to the study of mating behaviour in the dragonflies, *Anax parthenope* Selys and *Orthetrum coerulescens* (F.) (Hadrys *et al.*, 1993) and reproduction success in communally breeding burying beetles, *Nicrophorus tomentosus* Weber (Scott and Williams, 1993).

In our work on RAPDs and aphids (H.D. Loxdale and C.P. Brookes, in preparation), certain primers allow good discrimination of field-collected clones of *Sitobion avenae*, although it is essential to standardize conditions, especially with respect to the type of Taq polymerase used (see also Schierwater and Ender, 1993). Primers must also be screened for pattern reproducibility. Following screening of a number of primers, we found three (OPF-03, 04 and 08; Operon Technologies, Inc., Alameda, California, USA) that revealed reproducible interclonal polymorphism in *S. avenae*, readily differentiating 11 out of 12 clones (Figure 7.10a–c). Surprisingly, some clones were consistently different in their banding pattern between apterous (wingless) and alate (winged) individuals using certain primers (see Figure 7.10 legend). This phenomenon is presently under further investigation (C.P. Brookes *et al.*, in preparation).

Further work on *S. avenae* has also demonstrated considerable genetic variation between populations overwintering on wheat and on cocksfoot grass, *Dactylis glomerata*, L. The temporal aspect of this variation is currently under investigation, but results obtained so far suggest that the heterogeneity found in the winter disappears during the summer, to

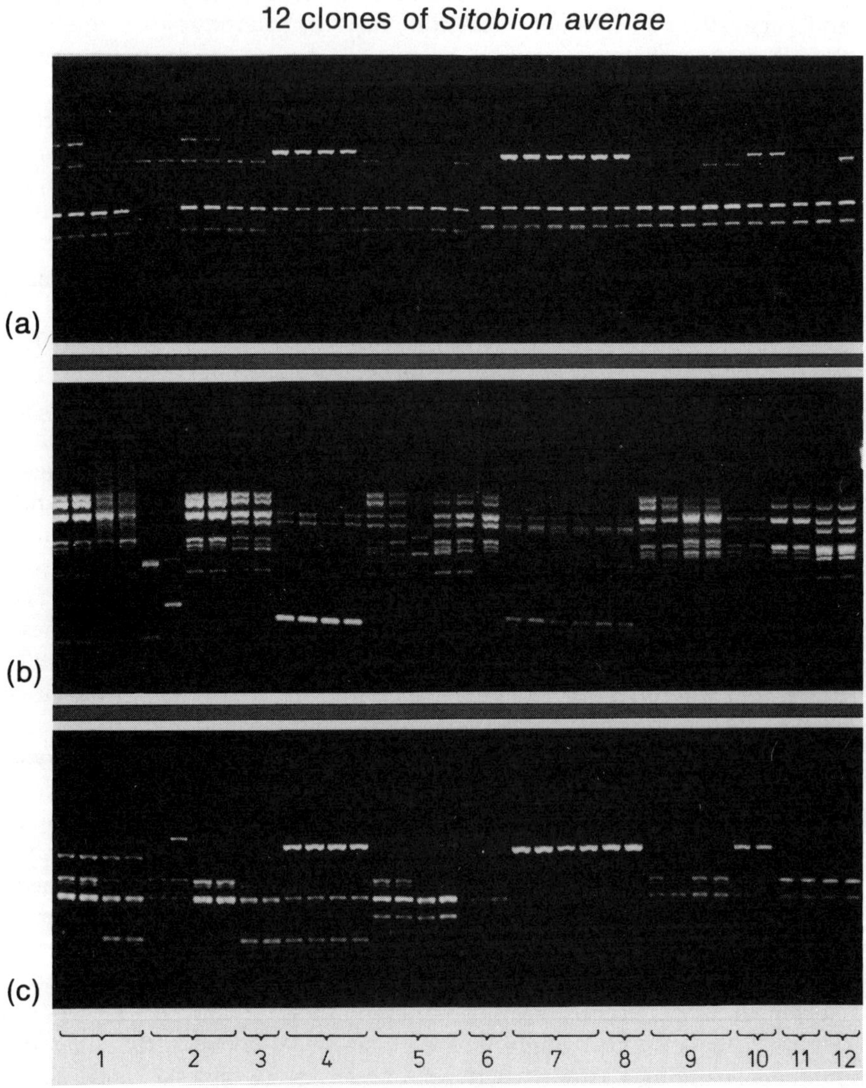

Figure 7.10 RAPD banding patterns ('fingerprints') from 12 clones of the grain aphid, *Sitobion avenae* collected in southern Britain and revealed using three arbitrary 10-base primers: (a), OPF-08; (b), OPF-04; and (c), OPF-03. For surveys of intraspecific variation, a range of primers are required to reveal the number of clones present in the population sample. Here, use of three primers revealed 11 clones, although a further 16 primers could not separate two 'sister' clones from the same parents (clones 7 and 8). In *S. avenae*, banding patterns between nymphal stages and adults were consistent per primer; however, using some

reform again the following winter (De Barro *et al.*, 1995b). The long-term stability of RAPD profiles in clonal *S. avenae* is also under investigation; so far, clones tested using several primers show continuous stability across several primers for 12 generations (P. De Barro, T.N. Sherratt and N. Maclean, unpublished data).

A variation on RAPD–PCR is arbitrarily primed-PCR (AP–PCR). The principle is the same as RAPD–PCR except that after the first round of amplification is completed, an aliquot is removed to a second PCR reaction mix containing ^{32}P-dCTP. Following a further round of amplification, the products are then separated on a polyacrylamide sequencing gel and the bands visualized using autoradiography instead of ethidium bromide and UV light. The advantages of AP–PCR over RAPD–PCR are that the increase in sensitivity allows very faint bands to be visualized, and when run on long sequencing gels, even slight changes in band length can be detected. However, for this to be meaningful, the results must first be shown to be highly reproducible. AP–PCR has been used in entomology (Fukatsu and Ishikawa, 1994) to study the gall-forming aphid, *Ceratovacuna nekoashi* (Sasaki). Here, even closely grouped galls were shown to be composed of single genotypes, indicating that inter-gall migration was a rare event.

7.9 CONCLUSIONS

Allozymes are very versatile multilocus genetic markers that, for many uses, will not be displaced easily by other molecular markers. But, as the present review shows, DNA techniques have a range of applications in studies of insect biology, especially phylogeny, clonal structure and mating systems. In population genetics, the number of samples required to produce data which can be compared statistically may limit studies of inter- or intrapopulation genetic variation and gene flow. With allozymes, a large throughput of samples assayed at many loci is readily obtained, especially with the cellulose acetate systems now available (Wynne *et al.*, 1992).

With DNA fingerprinting, due to the complexity and slow speed of data generation, it has not been generally possible to produce large amounts of data for population comparisons. Even when moderate amounts of data are available, complexity of the patterns precludes easy comparison of population parameters (Burke *et al.*, 1991; Lynch,

---

(*Contd*) clones and primers, banding pattern differences were found between alate (winged) and apterous (wingless) adult aphids (in the figure, where there are four aphids per clone, the first couplet are alatae, the second couplet apterae (H.D. Loxdale and C.P. Brookes, unpublished data)).

1991). One of the main problems is the assignment of the dominant bands to loci in multilocus fingerprints, especially in animals which cannot easily be crossed, such as parthenogenetically reproducing aphids. However, in aphids or other animals, which show very low levels of protein variation, DNA fingerprints provide extremely valuable information on population genetic structure.

The RAPDs technique is simpler and quicker than other fingerprinting techniques. Thus RAPDs can, to some degree, compete with allozymes as markers for large population samples, although most RAPD bands are dominant markers (Williams *et al.*, 1990), unlike allozymes or even RFLP markers, both of which are co-dominant. This effectively precludes assessment of homo- and heterozygote frequencies and the information derived from such studies (Workman and Niswander, 1970; Richardson *et al.*, 1986); see also Lynch and Milligan, 1994). However, locus-specific primers that target polymorphic microsatellites, as demonstrated by Queller *et al.* (1993) and Hughes and Queller (1993), may overcome this limitation.

All the PCR-based techniques use minute amounts of material, which is an advantage when working on extremely small insects such as *Trichogramma*, or certain life-cycle stages such as the egg stage of hymenopterous endoparasitoids, which cannot be detected by allozyme electrophoresis (Walton *et al.*, 1990a,b). In addition, the use of RAPDs or other PCR-mediated approaches to detect entomopathogenic fungal infections may also challenge allozyme electrophoresis, which in aphids at least, can only detect the fungi 3 days post-infection (Wilding *et al.*, 1993).

Other advantages of some DNA techniques, compared with allozymes, can be derived from the use of dried or ancient samples. These techniques include mtDNA (Wilson *et al.*, 1985), formal DNA fingerprint probes (Jeffreys *et al.*, 1991), microsatellite DNA markers (Hughes and Queller, 1993) and other PCR-mediated DNA techniques (Paabo, 1993). In contrast, allozymes require fresh or deep-frozen material, or material maintained in special preservatives such as glycerol-based solutions (Loxdale *et al.*, 1994). Examination of DNA from dried material opens up the possibility of obtaining long-term temporal data using museum material (e.g. Wilson *et al.*, 1985) or spatially and temporally collected insect material, such as that collected and stored by the Rothamsted Insect Survey which has operated a network of suction and light traps in the UK since the early 1960s (Woiwod, 1991).

In conclusion, the choice of DNA techniques versus allozymes depends very much on the question(s) to be answered. Allozymes are generally superior for studies such as the population genetics of sexually reproducing insects, whereas DNA techniques are superior for others, such as the analysis of mating and clonal structure of enzyme invariant insects. Presumably, DNA techniques have not yet attained their full potential as research tools, especially for population studies; new developments are certain to increase their use. We must state emphatically though that before launching into any 'molecular'-based investigation, it should first be ascertained whether allozymes are appropriate; there is no sense in using a 'hammer to crack a nut'!

Acknowledgements

We thank Drs Brian Kerry, Nicola von Mende, Lin Field, Alison Miller, David Bertioli, Joe Perry (Rothamsted) and two anonymous reviewers for their valuable criticisms of the manuscript and Drs Roger Blackman (Entomology Department, Natural History Museum, South Kensington, London) and Mike Wilson (Department of Zoology, National Museum of Wales, Cardiff) for help with taxonomic problems.

REFERENCES

Adamson, R.E., Ward, R.D., Feliciangeli, M.D. and Maingon, R. (1993) The application of random amplified polymorphic DNA for sandfly species identification. *Medical and Veterinary Entomology*, **7**, 203–7.

Apostol, B.L., Black, W.C., Miller, B.R., Reiter, P. and Beaty, B.J. (1993) Estimation of the number of full sibling families at an oviposition site using RAPD–PCR markers – applications to the mosquito *Aedes aegypti*. *Theoretical and Applied Genetics*, **86**, 991–1000.

Archmann, R., Heller, K.G. and Epplen, J.T. (1992) Last-male sperm precedence in the bushcricket *Poecilimon veluchianus* (Orthoptera, Tettigonioidea) demonstrated by DNA fingerprinting. *Molecular Ecology*, **1**, 47–54.

Armour, J.A.L., Povey, S., Jeremiah, S. and Jeffreys, A.J. (1990) Systematic cloning of human minisatellites from ordered array charomid libraries. *Genomics*, **8**, 501–12.

Arnheim, N., White, T. and Rainey, W.E. (1990) Application of PCR: organismal and population biology. *Bioscience*, **40**, 174–82

Atanassova, P., Brookes, C.P., Loxdale, H.D. and Powell, W. (submitted) Electrophoretic studies of five species of aphid parasitoids of the genus *Aphidius* Nees (Hymenoptera: Braconidae): Evidence of host adaptation and sympatric speciation. *Biological Journal of the Linnean Society*.

Avise, J.C. (1991) Matriarchal liberation. *Nature*, **352**, 192.

Avise, J.C. (1994) *Molecular Markers, Natural History and Evolution*, Chapman and Hall, New York.

Avise, J.C. and Lansman, R.A. (1983) Polymorphism of mitochondrial DNA in populations of higher animals, in *Evolution of Genes and Proteins*, (eds, M. Nei and R.K. Koehn), Sinauer Associates, Inc., Sunderland, Massachusetts, pp. 147–64.

Azeredo-Espin, A.M.L., Schroder, R.F.W., Huettel, M.D. and Sheppard, W.S. (1991) Mitochondrial DNA variation in geographic populations of Colorado potato beetle, *Leptinotarsa decemlineata* (Coleoptera: Chrysomelidae). *Experientia*, **47**, 483–5.

Ball, R.M., James, F.C., Freeman, S., Bermingham, E. and Avise, J.C. (1988) Phylogeographic population structure of red-winged blackbirds assessed by mitochondrial DNA. *Proceedings of the National Academy of Sciences, USA*, **5**, 1558–62.

Berlocher, S.H., McPheron, B.A, Feder, J.L. and Bush, G.L. (1993) Genetic differentiation at allozyme loci in the *Rhagoletis pomonella* (Diptera: Tephritidae) species complex. *Annals of the Entomological Society of America*, **86**, 716–27.

Birch, A.N.E., Fenton, B., Malloch, G., Jones, A.T., Phillips, M.S., Harrower, B.E., Woodford, J.A.T. and Catley, M.A. (1994) Ribosomal spacer length variability in the large raspberry aphid, *Amphorophora idaei* (Aphidinae: Macrosiphini). *Insect Molecular Biology*, **3**, 239–45.

Black, W.C., IV (1993a) Variation in the ribosomal RNA cistron among host-adapted races of an aphid (*Schizaphis graminum*). *Insect Molecular Biology*, **2**, 59–69.

Black, W.C., IV (1993b) PCR with arbitrary primers: approach with care. *Insect Molecular Biology*, **2**, 1–6.

Black, W.C., IV, Ferrari, J.A., Rai, K.S. and Sprenger, D. (1988) Breeding structure of a colonising species: *Aedes albopictus* in the United States. *Heredity*, **60**, 173–81.

Black, W.C., IV, McLain, D.K. and Rai, K.S. (1989) Patterns of variation in the rDNA cistron within and among world populations of a mosquito *Aedes albopictus* (Skuse). *Genetics*, **121**, 539–50.

Black, W.C., IV, DuTeau, N.M., Puterka, G.J., Nechols, J.R. and Pettorini, J.M. (1992) Use of the random amplified polymorphic DNA polymerase chain reaction (RAPD–PCR) to detect DNA polymorphisms in aphids (Homoptera: Aphididae). *Bulletin of Entomological Research*, **82**, 151–9.

Blackman, R.L. (1987) Reproduction, cytogenetics and development, in *Aphids, their Biology Natural Enemies and Control*, Vol. A. (eds A.K. Minks and P. Harrewijin), Elsevier Press, New York, pp. 163–95.

Blanchetot, A. (1991) Genetic relatedness in honeybees as established by DNA fingerprinting. *Journal of Heredity*, **82**, 391-6.

Blanchetot, A. (1992) DNA fingerprinting analysis in the solitary bee *Megachile rotundata*: variability and nest mate genetic relationships. *Genome*, **35**, 681–8.

Blanchetot, A. and Gooding, R.H. (1993) Genetic-analysis by DNA-fingerprinting in tsetse-fly genomes. *Insect Biochemistry and Molecular Biology*, **23**, 937–44.

Blanchetot, A. and Packer, L. (1992) Genetic variability in the social bee *Lasioglossum marginatum* and a cryptic undescribed sibling species, as detected by DNA fingerprinting and allozyme electrophoresis. *Insect Molecular Biology*, **1**, 89–97.

Bogdanowicz, S.M., Wallner, W.E., Bell, J., Odell, T.M. and Harrison, R.G. (1993) Asian gypsy moths (Lepidoptera: Lymantriidae) in North America: evidence from molecular data. *Annals of the Entomological Society of America*, **86**, 710–15.

Boudreaux, H.B. (1979) *Arthropod Phylogeny With Special Reference to Insects*, Wiley, New York.

Brookfield, J.F.Y. (1992) DNA fingerprinting in clonal organisms. *Molecular Ecology*, **1**, 21–6.

Broughton, R.E. and Kistner, D.H. (1991) A DNA hybridization study of the termite genus *Zootermopsis* (Isoptera: Termopsidae). *Sociobiology*, **19**, 15–40.

Brower, A.V.Z. and Boyce, T.M. (1991) Mitochondrial DNA variation in Monarch butterflies. *Evolution*, **45**, 1281–6.

Brown, W.M. (1980) Polymorphism in mitochondrial DNA of humans as revealed by restriction endonuclease analysis. *Proceedings of the National Academy of Sciences, USA*, **77**, 3605–9.

Brown, W.M. (1983) Evolution of animal mitochondrial DNA, in *Evolution of Genes and Proteins*, (eds M. Nei and R. K. Koehn), Sinauer Associates, Inc., Sunderland, Massachusetts., pp. 62–88.

Burke, T., Hanotte, O., Bruford, M.W. and Cairns, E. (1991) Multilocus and single locus minisatellite analysis in population biological studies, in *DNA Fingerprinting: Approaches and Applications*, (eds T. Burke, G. Dolf, A.J. Jeffreys and R. Wolff), Birkhauser, Basel, pp. 154–68.

Cariou, M.-L., Solignac, M., Monnerot, M. and David, J.R. (1990) Low allozyme and mtDNA variability in the island endemic species *Drosophila sechellia* (*D. melanogaster* complex). *Experientia*, **46**, 101–4.

Carmean, D., Kimsey, L.S. and Berbee, M.L. (1992) 18S rDNA sequences and the holometabolous insects. *Molecular Phylogenetics and Evolution*, **1**, 270–8.

Carvalho, G.R., Maclean, N., Wratten, S.D., Carter, R.E. and Thurston, J.P. (1991) Differentiation of aphid clones using DNA fingerprints from individual aphids. *Proceedings of the Royal Society, London, Series B*, **243**, 109–14.

Cenis, J.L., Perez, P. and Fereres, A. (1993) Identification of aphid (Homoptera: Aphididae) species and clones by random amplified polymorphic DNA. *Annals of the Entomological Society of America*, **86**, 545–50.

Chapco, W., Kelln, R.A. and McFadyen, D.A. (1992a) Intraspecific mitochondrial DNA variation in the migratory grasshopper, *Melanoplus sanguinipes. Heredity*, **69**, 547–57.

Chapco, W., Ashton., N.W., Martel, R.K.B. and Antonishyn, N. (1992b) A feasibility study of the use of random amplified polymorphic DNA in the population genetics and systematics of grasshoppers. *Genome*, **35**, 569–74.

Chapco, W., Kelln, R.A. and McFadyen, D.A. (1994) Mitochondrial DNA variation in North American *Melanopline* grasshoppers. *Heredity*, **72**, 1–9.

Clapp, J.P., McKee, R.A., Allen-Williams, L., Hopley, J.G. and Slater, R.J. (1993) Genomic subtractive hybridization to isolate species-specific DNA sequences in insects. *Insect Molecular Biology*, **1**, 133-8.

Clark, A.G. and Lanigan, C.M.S. (1993) Prospects for estimating nucleotide divergence with RAPDs. *Molecular Biology and Evolution*, **10**, 1096–111.

Clary, D.O. and Wolstenholme, D.R. (1985) The mitochondrial DNA molecule of *Drosophila yakuba*: nucleotide sequence, gene organization, and genetic code. *Journal of Molecular Evolution*, **22**, 252–71.

Cockburn, A.F., Mitchell, S.E. and Seawright, J.A. (1990) Cloning of the mitochondrial genome of *Anopheles quadrimaculatus*. *Archives of Insect Biochemistry and Physiology*, **14**, 31–6.

Colgan, D.J. (1986) Developmental changes in the isoenzymes controlling glycolysis in the acridine grasshopper, *Caledia captiva. Roux's Archives of Developmental Biology*, **195**, 197–201.

Collins, F.H., Mehafey, P.C., Rasmussen, M.O., Brandlinc-Bennett, A.D., Odera, J.S. and Finnerty, V. (1988) Comparison of DNA-probe and isoenzyme methods for differentiating *Anopheles gambiae* and *Anopheles arabiensis* (Diptera: Culicidae). *Journal of Medical Entomology*, **25**, 116–20.

Coluzzi, M. (1968) Chromosomi politenici delle cellule nutrici ovariche nel compresso *gambiae* del genera *Anopheles. Parasitologia*, **10**, 179–83.

Coluzzi, M. and Sabatini, A. (1967) Cytogenetic observations on species A and B of the *Anopheles gambiae* complex. *Parasitologia*, **9**, 73–88.

Daly, J. and Trowell, S. (1996) Biochemical approaches to the study of population genetics: the role of selection and gene flow in the evaluation of insecticide resistance, in *The Ecology of Agricultural Pests: Biochemical Approaches* (eds W.O.C. Symondson and J.E. Liddell), Chapman & Hall, London, pp. 73–92.

De Barro, P., Sherratt, T., Wratten, S and Maclean, N. (1994a) DNA fingerprinting of cereal aphids using $(GATA)_4$. *European Journal of Entomology*, **91**, 109–14.

De Barro, P.J., Sherratt, T.N., Carvalho, G.R., Maclean, N., Nicol, D. and Iyengar, A. (1994b) An analysis of secondary spread by putative clones of *Sitobion avenae* within a Hampshire wheat field using the multilocus $(GATA)_4$ probe. *Insect Molecular Biology*, **3**, 253–60.

De Barro, P.J., Sherratt, T.N., Carvalho, G.R., Maclean, N., Nicol, D. and Iyengar, A. (1995a) The use of the multilocus $(GATA)_4$ probe to investigate geographic and micro-geographic genetic differentiation in two aphid species over southern England. *Molecular Ecology*, **4**, 375–82.

De Barro, P.J., Sherratt, T.N., Brookes, C.P., David, O. and Maclean, N. (1995b) Spatial and temporal genetic variation in British field populations of the grain aphid, *Sitobion avenae* (F.) (Hemiptera: Aphididae) studied using RAPD–PCR. *Proceedings of the Royal Society, London, Series B*, **262**, 321–7.

DeSalle, R. and Templeton, A.R. (1988) Founder effects accelerate the rate of mitochondrial DNA evolution in Hawaiian *Drosophila*. *Evolution*, **42**, 1076–84.

DeSalle, R., Templeton, A., Mori, I., Pletscher, S. and Johnston, J.S. (1987) Temporal and spatial heterogeneity of mtDNA polymorphisms in natural populations of *Drosophila mercatorum*. *Genetics*, **116**, 215–23.

Devonshire, A.L. (1989) The role of electrophoresis in the biochemical detection of insecticide resistance in *Electrophoretic Studies on Agricultural Pests*, (eds H.D. Loxdale and J. den Hollander), Clarendon Press, Oxford, pp. 363–74.

Devonshire, A.L. and Field, L.M. (1991) Gene amplification and insecticide resistance. *Annual Review of Entomology*, **36**, 1–23.

Dorer, D.R. and Henikoff, S. (1994) Expansions of transgene repeats cause heterochromatin formation and gene silencing in *Drosophila*. *Cell*, **77**, 993–1002.

Dover, G.A. (1989) Victims or perpetrators of DNA turnover. *Nature*, **342**, 347–8.

Edwards, O.R. and Hoy, M. A. (1993) Polymorphism in two parasitoids detected using random amplified polymorphic DNA-polymerase chain-reaction. *Biological Control*, **3**, 243–57.

Eggers-Schumacher, H.A. (1987) Enzyme electrophoresis in biosystematics and taxonomy of aphids, in *Population Structure, Genetics and Taxonomy of Aphids and Thysanoptera*, (eds J. Holman, J. Pelikan, A.F.G. Dixon and L. Weismann), SPB Publishing, The Hague, The Netherlands, pp. 63–70.

Epplen, J.T. (1992) The methodology of multilocus DNA fingerprinting using radioactive or nonradioactive oligonucleotide probes specific for simple repeat motifs, in *Advances in Electrophoresis, Vol. 5*, (eds A. Chrambach, M.J. Dunn and B.J. Radola), VCH, Weinheim, New York, Basel, Cambridge, pp. 59–112.

Evans, J.D. (1993) Parentage analyses in ant colonies using simple sequence repeat loci. *Molecular Ecology*, **2**, 393–7.

Fang, Q., Black, W.C., IV, Blocker, H.D. and Whitcomb, R.F. (1993) A phylogeny of New World *Deltocephalus*-like leafhopper genera based on mitochondrial 16S ribosomal DNA sequences. *Molecular Phylogenetics and Evolution*, **2**, 119–31.

Feinberg, A.P. and Vogelstein, B. (1983) A technique for radiolabelling DNA restriction endonuclease fragments to high specific activity. *Analytical Biochemistry*, **132**, 6–13.

Feinberg, A.P. and Vogelstein, B. (1984) High specific activity labelling of DNA restriction endonuclease fragments. *Addendum Analytical Biochemistry*, **137**, 266.

Fenton, B., Birch, A.N,E., Malloch, G., Woodford, J.A.T. and Gonzalez, C. (1994) Molecular analysis of ribosomal DNA from the aphid *Amphorophora idaei* and an associated fungal organism. *Insect Molecular Biology*, **3**, 183–90.

ffrench-Constant, R.H., Devonshire, A.L. and White, R.P. (1988) Spontaneous loss and reselection of resistance in extremely resistant *Myzus persicae* (Sulzer). *Pesticide Biochemistry and Physiology*, **30**, 1–10.

Field, L.M. and Devonshire, A.L. (1992) Insecticide resistance by gene amplification in *Myzus persicae,* in *Resistance '91: Achievements and Developments in Combating Insecticide Resistance,* (eds I. Denholm, A.L. Devonshire and D.W. Hollomon), SCI, Elsevier Applied Science, London, New York, pp. 240–50.

Field, L.M., Devonshire, A.L. and Forde, B.G. (1988) Molecular evidence that insecticide resistance in peach-potato aphids (*Myzus persicae* Sulz.) results from amplification of an esterase gene. *Biochemical Journal*, **251**, 309–12.

Field, L.M., Devonshire, A.L., ffrench-Constant, R.H. and Forde, B.G. (1989) Changes in DNA methylation are associated with loss of insecticide resistance in the peach-potato aphid *Myzus persicae* (Sulz.). *FEBS Letters*, **243**, 323–7.

Field, L.M., Williamson, M.S., Moores, G.D. and Devonshire, A.L. (1993) Cloning and analysis of the esterase genes conferring insecticide resistance in the peach potato aphid *Myzus persicae* (Sulzer). *Biochemical Journal*, **294**, 569–74.

Field, L.M., Javed, N., Stribley, M.F. and Devonshire, A.L. (1994) The peach-potato aphid *Myzus persicae* and the tobacco aphid *Myzus nicotianae* have the same esterase-based mechanisms of insecticide resistance. *Insect Molecular Biology*, **3**, 143–8.

Frohlich, D.R., Robinson, A.S. and Wells, M.A. (1993) Mediterranean fruit fly, *Ceratitis capitata* (Weidemann), mitochondrial DNA: genes and secondary structures for six t-RNAs. *Insect Molecular Biology*, **1**, 165–9.

Fukatsu, T. and Ishikawa, H. (1994) Differentiation of aphid clones by arbitrary primed polymerase chain reaction (AP–PCR) DNA fingerprinting. *Molecular Ecology*, **3**, 187–92.

Futuyma, D.J. and McCafferty, S.S. (1990) Phylogeny and the evolution of host plant associations in the leaf beetle genus *Ophraella* (Coleoptera, Chrysomelidae). *Evolution*, **44**, 1885–913.

Gargiulo, G., Malva, C., Pennacchio, F. and Tremblay, E. (1988) Structure of *Aphidius* Nees (Hymenoptera, Braconidae) rDNA: a molecular tool in biosystematic research. *Bollettino Laboratorio di Entomologia Agraria "Filippo Silvestri"*, **45**, 203–19.

Garnery, L., Cornuet, J.M. and Solignac, M. (1992) Evolutionary history of the honey bee *Apis mellifera* inferred from mitochondrial DNA analysis. *Molecular Ecology*, **1**, 145–54.

Gawel, N.J. and Bartlett, A.C. (1993) Characterization of differences between whiteflies using RAPD–PCR. *Insect Molecular Biology*, **2**, 33–8.

Guldemond, J.A. and Eggers-Schumacher, H.A. (1989) Biosystematics of the genus *Cryptomyzus* (Homoptera: Aphididae): an electrophoretic analysis. *Zeitschrift zoologischer Systematik- und Evolutionsforschung*, **27**, 14–25.

Gutell, R.R. and Fox, G.E. (1988) A compilation of large subunit RNA sequences presented in a structural format. *Nucleic Acids Research*, **16**, r175–r269.

Hadrys, H., Balick, M. and Schierwater, B. (1992) Applications of random amplified polymorphic DNA (RAPD) in molecular ecology. *Molecular Ecology*, **1**, 55–63.

Hadrys, H., Schierwater, B., Dellaporta, S.L., DeSalle, R. and Buss, L.W. (1993) Determination of paternity in dragonflies by random amplified polymorphic DNA fingerprinting. *Molecular Ecology*, **2**, 79–87.

Hall, H.G. (1992a) Further characterization of nuclear DNA RFLP markers that distinguish African and European honeybees. *Archives of Insect Biochemistry and Physiology*, **19**, 163–75.

Hall, H.G. (1992b) DNA studies reveal processes involved in the spread of New World African honeybees. *Florida Entomologist*, **75**, 51–9.

Hall, H.G. and Smith, D.R. (1991) Distinguishing African and European honeybee matrilines using amplified mitochondrial DNA. *Proceedings of the National Academy of Sciences, USA*, **88**, 4548–52.

Hamby, R.K., Sims, L., Issel, L. and Zimmer, E.A. (1988) Direct ribosomal RNA sequencing: optimization of extraction and sequencing methods for work with higher plants. *Plant Molecular Biology Reporter*, **6**, 175–92.

Harrison, R.G. and Odell, T.M. (1989) Mitochondrial DNA as a tracer of gypsy moth origins, in *Proceedings, Lymantriidae: a comparison of features of New and Old World tussock moths*, (eds W.E. Wallner and K. McManus), Northeastern Forest Experiment Station, Broomall, PA, USA, pp. 265–73.

Hartl, D.L. (1980) *Principles of Population Genetics*, Sinauer Associates, Inc., Sunderland, Mass., USA.

Haverty, M.I., Page, M., Nelson, L.J. and Blomquist, G.J. (1988) Cuticular hydrocarbons of dampwood termites, *Zootermopsis*: intra- and intercolony variation and potential as taxonomic characters. *Journal of Chemical Ecology*, **14**, 1035–58.

Haymer, D.S. (1994) Arbitrary (RAPD) primer sequences used in insect studies. *Insect Molecular Biology*, **3**, 191–4.

Haymer, D.S., McInnis, D.O. and Arcangeli, L. (1992) Genetic variation between strains of the Mediterranean fruit fly, *Ceratitis capitata*, detected by DNA fingerprinting. *Genome*, **35**, 528–33.

Hemingway, J., Small, G.J., Lindsay, S.W. and Collins, F.H. (1996) Combined use of biochemical, immunological and molecular assays for infection, species identification and resistance detection in field populations of *Anopheles*, in *The Ecology of Agricultural Pests: Biochemical Approaches*, (eds W.O.C. Symondson and J.E. Liddell). Chapman & Hall, London, pp. 31–49.

Hennig, W. (1981) *Insect Phylogeny*, Wiley, New York.

Hill, P.J. and Stewart, G.S.A.B. (1992) The polymerase chain reaction in molecular and micro-biology. *Biotechnology and Genetic Engineering Reviews*, **10**, 343–77.

Hillis, D.M. and Davis, S.K. (1986) Evolution of ribosomal DNA: fifty million years of recorded history in the frog genus *Rana*. *Evolution*, **40**, 1275–88.

Hillis, D.M. and Dixon, M.T. (1991) Ribosomal DNA: molecular evolution and phylogenetic inference. *The Quarterly Review of Biology*, **66**, 411–53.

Hoelzel, A.R. and Green, A. (1992) Analysis of population level variation by sequencing PCR-amplified DNA, in *Molecular Genetic Analysis of Populations. A Practical Approach*, (ed. A.R. Hoelzel), IRL Press at Oxford University Press, Oxford, pp. 159–87.

Hughes, C.R. and Queller, D.C. (1993) Detection of highly polymorphic microsatellite loci in a species with little allozyme polymorphism. *Molecular Ecology*, **2**, 131–7.

Hunt, G.J. and Page, R.E., Jr. (1992) Patterns of inheritance with RAPD molecular markers reveal novel types of polymorphism in the honey bee. *Theoretical and Applied Genetics*, **85**, 15–20.

Hunt, J.A. and Carson, H.L. (1983) Evolutionary relationships of four species of Hawaiian *Drosophila* as measured by DNA reassociation. *Genetics*, **104**, 353–64.

Hunt, J.A., Hall, T.J. and Britten, R.J. (1981) Evolutionary distances in Hawaiian *Drosophila* measured by DNA reassociation. *Journal of Molecular Evolution*, **17**, 361–7.

Jacobson, J.W. and Hsiao, T.H. (1983) Isozyme variation between geographic populations of the Colorado potato beetle, *Leptinotarsa decemlineata* (Coleoptera: Chrysomelidae). *Annals of the Entomological Society of America*, **76**, 162–6.

Jeffreys, A.J., Wilson, V. and Thein, S.W. (1985a) Hypervariable 'minisatellite' regions in human DNA. *Nature*, **314**, 67–73.

Jeffreys, A.J., Wilson, V. and Thein, S.W. (1985b) Individual-specific 'fingerprints' of human DNA. *Nature*, **316**, 76–9.

Jeffreys, A.J., Royle, N.J., Patel, I. *et al.* (1991) Principles and recent advances in human DNA fingerprinting, in *DNA Fingerprinting: Approaches and Applications*, (eds T. Burke, G. Dolf, A.J. Jeffreys and R. Wolff), Birkhauser, Basel, pp. 1–19.

Kambhampati, S. and Rai, K.S. (1991a) Temporal variation in the ribosomal DNA nontranscribed spacer of *Aedes albopictus* (Diptera: Culicidae). *Genome*, **34**, 293–7.

Kambhampati, S. and Rai, K.S. (1991b) Mitochondrial DNA variation within and among populations of the mosquito *Aedes albopictus*. *Genome*, **34**, 288–92.

Kambhampati, S., Black, W.C., IV, Rai, K.S. and Sprenger, D. (1990) Temporal variation in the genetic structure of a colonising species: *Aedes albopictus* in the United States. *Heredity*, **64**, 281–7.

Karpen, G.H. (1994) Position–effect variegation and the new biology of heterochromatin. *Current Opinions in Genetics and Development*, **4**, 281–91.

Kimura, M. (1980) A simple method for estimating evolutionary rates of base substitutions through comparative studies of nucleotide sequences. *Journal of Molecular Evolution*, **16**, 11–20.

Kirby, L.T. (1992) *DNA Fingerprinting. An Introduction*, W.H. Freeman & Co, New York.

Kiriac, I., Gruber, F., Poprawski, T., Halbert, S. and Elberson, L. (1990) Occurrence of sexual morphs of Russian wheat aphid, *Diuraphis noxia* (Homoptera: Aphididae), in several locations in the Soviet Union and the northwestern United States. *Proceedings of the Entomological Society of Washington*, **92**, 544–7.

Korman, A.K., Pashley, D.P., Haverty, M.I. and LaFage, J.P. (1991) Allozymic relationships among cuticular hydrocarbon phenotypes of dampwood termites, *Zootermopsis* sp. (Isoptera: Termopsidae). *Annals of the Entomological Society of America*, **84**, 1–9.

Kourti, A., Loukas, M. and Economopolous, A.P. (1990) Population genetics of Medfly. I. Enzyme polymorphisms in natural populations and routes of dispersion of the fly, in *Genetic Sexing of the Mediterranean Fruit Fly*, FAO/IAEA Research Co-ordination Meeting, Colymbari, Crete, 16–21 September, 1988, p. 7.

Kovalev, O.V., Poprawski, T.J., Stekolshchikov, A.V., Vereshchagina, A.B. and Gandrabur, S.A. (1991) *Diuraphis* Aizenberg (Hom., Aphididae): key to apterous viviparous females and review of Russian language literature on the natural history of *Diuraphis noxia* (Kurdjumov, 1913). *Journal of Applied Entomology*, **112**, 425–36.

Landry, B.S., Dextraze, L. and Boivin, G. (1993) Random amplified polymorphic DNA markers for DNA fingerprinting and genetic-variability assessment of minute parasitic wasp species (Hymenoptera, Mymaridae and Trichogrammatidae) used in biological-control programs of phytophagous insects. *Genome*, **36**, 580–7.

Lansman, R.A., Shade, R.O., Shapira, J.F. and Avise, J.C. (1981) The use of restriction endonucleases to measure mitochondrial DNA sequence relatedness in natural populations. *Journal of Molecular Evolution*, **17**, 214–26.

Loxdale, H.D. (1994) Isozyme and protein profiles of insects of agricultural and horticultural importance, in *Identification and Characterization of Pest Organisms*, (ed. D.L. Hawksworth), CAB International, Wallingford, UK, pp. 337–75.

Loxdale, H.D. and Brookes, C.P. (1988) Electrophoretic study of enzymes from cereal aphid populations. V. Spatial and temporal genetic similarity of holocyclic populations of the bird-cherry oat aphid, *Rhopalosiphum padi* (L.) (Hemiptera: Aphididae), in Britain. *Bulletin of Entomological Research*, **78**, 241–9.

Loxdale, H.D. and Brookes, C.P. (1990) Genetic stability within and restricted migration (gene flow) between local populations of the blackberry-grain aphid *Sitobion fragariae* in south-east England. *Journal of Animal Ecology*, **59**, 495–512.

Loxdale, H.D. and den Hollander, J. (eds) (1989) *Electrophoretic Studies on Agricultural Pests*, Systematics Association Special Volume No. 39, Clarendon Press, Oxford.

Loxdale, H.D., Tarr, I.J., Weber, C.P., Brookes, C.P., Digby, P.G.N. and Castañera, P. (1985) Electrophoretic study of enzymes from cereal aphid populations. III. Spatial and temporal genetic variation of populations of *Sitobion avenae* (F.) (Hemiptera: Aphididae). *Bulletin of Entomological Research*, **75**, 121–41.

Loxdale, H.D., Brookes, C.P., Woiwod, I.P. and Harrington, R. (1994) Another solution for preserving activity and integrity? (Letter). *Trends in Ecology and Evolution*, **9**, 441.

Lupoli, R., Irwin, M.E. and Vossbrinck, C.R. (1990) A ribosomal DNA probe to distinguish populations of *Rhopalosiphum maidis* (Homoptera: Aphididae). *Annals of Applied Biology*, **117**, 3–8.

Lynch, M. (1988) Estimation of relatedness by DNA fingerprinting. *Molecular Biology and Evolution*, **5**, 584–99.

Lynch, M. (1990) The similarity index and DNA fingerprinting. *Molecular Biology and Evolution*, **7**, 478–84.

Lynch, M. (1991) Analysis of population genetic structure by DNA fingerprinting, in *DNA Fingerprinting: Approaches and Applications*, (eds T. Burke, G. Dolf, A.J. Jeffreys and R. Wolff), Birkhauser, Basel, pp. 113–26.

Lynch, M. and Milligan, B.G. (1994) Analysis of population genetic structure with RAPD markers. *Molecular Ecology*, **3**, 91–9.

Machado, M.F.P.S. and Contel, E.P.B. (1991) Glycerol-3-phosphate dehydrogenase isozyme variation in adult Meliponids (Hymenoptera: Apidae). *Biochemical Genetics*, **29**, 593–600.

Martin, A. and Simon, C. (1990) Differing levels of among-population divergence in the mitochondrial DNA of periodical cicadas related to historical biogeography. *Evolution*, **44**, 1066–80.

Martinez, D., Moya, A., Latorre, A. and Fereres, A. (1992) Mitochondrial DNA variation in *Rhopalosiphum padi* (Homoptera: Aphididae) populations from four Spanish localities. *Annals of the Entomological Society of America*, **85**, 241–6.

May, B. and Holbrook, F.R. (1978) Absence of genetic variability in the green peach aphid, *Myzus persicae* (Hemiptera: Aphidiae). *Annals of the Entomological Society of America*, **71**, 809–12.

McLain, D.K., Rai, K.S. and Fraser, M.J. (1986) Interspecific variation in the abundance of highly repeated DNA sequences in the *Aedes scutellaris* (Diptera: Culicidae) subgroup. *Annals of the Entomological Society of America*, **79**, 784–91.

Metcalf, R.A., Marlin, J.C. and Whitt, G.S. (1975) Low levels of genetic heterozygosity in Hymenoptera. *Nature*, **257**, 792–4.

Meunier, J. R. and Grimont, P.A.D. (1993) Factors affecting reproducibility of random amplified polymorphic DNA-fingerprinting. *Research In Microbiology*, **144**, 373–9.

Micheli, M.R., Beva, R., Pascale, E. and D'Ambrosio (1994) Reproducible DNA fingerprinting with the random amplified polymorphic DNA (RAPD) method. *Nucleic Acids Research*, **22**, 1921–2.

Miles, S.J. (1978) Enzymatic variation in the *Anopheles gambiae* Giles group of species (Diptera: Culicidae). *Bulletin of Entomological Research*, **68**, 85–96.

Miles, S.J. (1979) A biochemical key to adult members of the *Anopheles gambiae* group of species (Diptera: Culicidae). *Journal of Medical Entomology*, **15**, 297–9.

Moritz, R.F.A. and Haberl, M. (1994) Lack of meiotic recombination in thelytocous parthenogenesis of laying workers of *Apis mellifera-capensis* (the cape honeybee). *Heredity*, **73**, 98–102.

Mouches, C., Pauplin, Y., Agarwal, M. *et al.* (1990) Characterization of amplification core and esterase BI gene responsible for insecticide resistance in *Culex. Proceedings of the National Academy of Sciences, USA*, **87**, 2574–8.

Mueller, U.G., Eickwort, G.C. and Aquadro, C.F. (1994) DNA-fingerprinting analysis of parent–offspring conflict in a bee. *Proceedings of the National Academy of Sciences, USA*, **91**, 5143–7.

Muralidharan, K. and Hall, H.G. (1990) Prevalence of African DNA RFLP alleles in neotropical African honeybees. *Archives of Insect Biochemistry and Physiology*, **15**, 229–36.

Murray, R.A., Solomon, M.G. and Fitzgerald, J.D. (1989) The use of electrophoresis for determining patterns of predation in arthropods, in *Electrophoretic Studies on Agricultural Pests*, (eds H.D. Loxdale and J. den Hollander), Systematics Association, Special Volume No. 39, Clarendon Press, Oxford, pp. 467–83.

Narang, S.K., Klein, T.A., Perera, O.P., Bento Lima, J. and Toda Tang, A. (1993) Genetic evidence for the existence of cryptic species in the *Anopheles albitarsis* complex in Brazil: allozymes and mitochondrial DNA restriction fragment length polymorphisms. *Biochemical Genetics*, **31**, 97–112.

Nei, M. (1973) Analysis of gene diversity in subdivided populations. *Proceedings of the National Academy of Sciences, USA*, **20**, 3321–3.

Nei, M. (1975) *Molecular Population Genetics and Evolution*, North Holland Publishing Co., Amsterdam.

Nei, M. (1987) *Molecular Evolutionary Genetics*, Columbia University Press, New York.

Nei, M. and Li, W.H. (1979) Mathematical model for studying genetic variation in terms of restriction endonucleases. *Proceedings of the National Academy of Sciences, USA*, **76**, 5269–73.

Nei, M. and Tajima, F. (1983) Maximum likelihood estimation of the number of nucleotide substitutions for restriction sites data. *Genetics*, **105**, 207–16.

Oman, P.W. (1949) The Nearctic leafhoppers (Homoptera: Cicadellidae). A generic classification and checklist. *Memoirs of the Entomological Society of Washington*, **3**, 1–253.

Paabo, S. (1993) Ancient DNA. *Scientific American*, November, 60–6.

Packer, L. and Owen, R.E. (1990) Allozyme variation, linkage disequilibrium and diploid male production in a primitively social bee *Augochlorella striata* (Hymenoptera: Halictidae) *Heredity*, **65**, 241–8.

Packer, L. and Owen, R.E. (1992) Variable enzyme systems in the Hymenoptera. *Biochemical Systematics and Ecology*, **20**, 1–7.

Panyim, S., Yasothornsrikul, S. and Baimai, V. (1988) Species-specific DNA sequences from the *Anopheles dirus* complex – a potential for efficient identification of isomorphic species, in *Biosystematics of Haematophagous Insects*, (ed. M.W. Service), Systematics Association, Special Volume No. 37, Clarendon Press, Oxford, pp. 193–202.

Pashley, D.P., McPheron, B. and Zimmer, E.A. (1993) Systematics of holometabolous insect orders based on 18S ribosomal RNA. *Molecular Phylogenetics and Evolution*, **2**, 132–42.

Pelandakis, M. and Solignac, M. (1993) Molecular phylogeny of *Drosophila* based on ribosomal RNA sequences. *Journal of Molecular Evolution*, **37**, 525–43.

Pennacchio, F., Digilio, M.C., Tremblay, E. and Tranfaglia, A. (1994) Host recognition and acceptance behaviour in two aphid parasitoid species: *Aphidius ervi* and *Aphidius microlophii* (Hymenoptera: Braconidae). *Bulletin of Entomological Research*, **84**, 57–64.

Pfennig, D.W. and Reeve, H.K. (1993) Nepotism in a solitary wasp as revealed by DNA fingerprinting. *Evolution*, **47**, 700–4.

Pinto, J.D., Kazmer, D.J., Platner, G.R. and Sassaman, C.A. (1992) Taxonomy of the *Trichogramma minutum* complex (Hymenoptera: Trichogrammatidae): allozymic variation and its relationship to reproductive and geographic data. *Annals of the Entomological Society of America*, **85**, 413–22.

Post, R.J. and Crampton, J.M. (1988) The taxonomic use of variation in repetitive DNA sequences in the *Simulium damnosum* complex, in *Biosystematics of Haematophagous Insects*, (ed. M.W. Service), Systematics Association, Special Volume No. 37, Clarendon Press, Oxford, pp. 245–55.

Powell, W. and Walton, M.P. (1989). The use of electrophoresis in the study of hymenopteran parasitoids of agricultural pests, in *Electrophoretic Studies on Agricultural Pests*, (eds H.D. Loxdale and J. den Hollander), Systematics Association, Special Volume No. 39, Clarendon Press, Oxford, pp. 443–65.

Powers, T.O., Jensen, S.G., Kindler, S.D, Stryker, C.J. and Sandall, L.J. (1989) Mitochondrial divergence among greenbug (Homoptera: Aphididae) biotypes. *Annals of the Entomological Society of America*, **82**, 298–302.

Puterka, G.J., Black, W.C., IV, Steiner, W.M. and Burton, R.L. (1993) Genetic variation and phylogenetic relationships among worldwide collections of the Russian wheat aphid, *Diuraphis noxia* (Mordvilko), inferred from allozyme and RAPD–PCR markers. *Heredity*, **70**, 604–18.

Queller, D.C., Strassmann, J.E. and Hughes, C. (1993) Microsatellites and kinship. *Trends in Ecology and Evolution*, **8**, 285–8.

Rand, D.M. and Harrison, R.G. (1989) Molecular population genetics of mitochondrial DNA size variation in crickets. *Genetics*, **121**, 551–69.

Raymond, M., Callaghan, A., Fort, P. and Pasteur, N. (1991) Worldwide migration of amplified insecticide resistance genes in mosquitoes. *Nature*, **350**, 151–3.

Richardson, B.J., Baverstock, P.R. and Adams, M. (1986) *Allozyme Electrophoresis. A Handbook for Animal Systematics and Population Studies*, Academic Press, London.

Robinson, J., Fischer, M. and Hoisington, D. (1993) Molecular characterization of *Diuraphis* spp. using random amplified polymorphic DNA. *Southwestern Entomologist*, **18**, 121–7.

Roehrdanz, R.L. and Flanders, R.V. (1993) Detection of DNA polymorphisms in predatory Coccinellids using polymerase chain reaction and arbitrary primers (RAPD–PCR). *Entomophaga*, **38**, 479–91.

Roehrdanz, R.L., Reed, D.K. and Burton, R.L. (1993) Use of polymerase chain reaction and arbitrary primers to distinguish laboratory-raised colonies of parasitic hymenoptera. *Biological Control*, **3**, 199–206.

Rollinson, D. and Stothard, J.R. (1994) Identification of pests and pathogens by random amplification of polymorphic DNA (RAPDs), in *Identification and Characterization of Pest Organisms*, (ed. D.L. Hawksworth), CAB International, Wallingford, UK, pp. 447–59.

Saccheri, I.J. and Bruford, M.W. (1993) DNA fingerprinting in a butterfly, *Bicyclus anynana* (Satyridae). *Journal of Heredity*, **84**, 195–200.

Saitou, N. and Nei, M. (1987) The neighbor-joining method: a new method for reconstructing phylogenetic trees. *Molecular Biology and Evolution*, **4**, 406–25.

Sambrook, J., Fritsch E.F. and Maniatis, T. (1989) *Molecular Cloning: A Laboratory Manual*, 2nd edn, Cold Spring Harbor Laboratory Press, New York.

Schafer, R., Zischler, H., Birsner, U., Becker, A. and Epplen, J. (1988) Optimized oligonucleotide probes for DNA fingerprinting. *Electrophoresis*, **9**, 369–74.

Schierwater, B. and Ender, A. (1993) Different thermostable DNA polymerases may amplify different RAPD products. *Nucleic Acids Research*, **21**, 4647–8.

Schmitz, J. and Moritz, R.F.A. (1994) Sequence analysis of the D1 and D2 regions of 28S rDNA in the hornet (*Vespa crabro*) (Hymenoptera, Vespinae). *Insect Molecular Biology*, **3**, 273–7.

Scott, J.A., Brogdon, W.G. and Collins, F.H. (1993) Identification of single specimens of the *Anopheles gambiae* complex by the polymerase chain reaction. *American Journal of Tropical Medicine and Hygiene*, **49**, 520-9.

Scott, M. P. and Williams, S.M. (1993) Comparative reproductive success of communally breeding burying beetles as assessed by PCR with randomly amplified polymorphic DNA. *Proceedings of the National Academy of Sciences, USA*, **90**, 2242–5.

Sheppard, W.S., Steck, G.S. and McPheron, B.A. (1992) Geographic populations of the medfly may be differentiated by mitochondrial DNA variation. *Experientia*, **48**, 1010–13.

Shufran, K.A. and Wilde, G.E. (1994) Clonal diversity in overwintering populations of *Schizaphis graminum* (Homoptera: Aphididae). *Bulletin of Entomological Research*, **84**, 105–14.

Shufran, K.A., Black, W.C., IV and Margolies, D.C. (1991) DNA fingerprinting to study spatial and temporal distributions of an aphid, *Schizaphis graminum* (Homoptera: Aphididae). *Bulletin of Entomological Research*, **81**, 303–13.

Shufran, K.A., Margolies, D.C. and Black, W.C., IV (1992) Variation between biotype E clones of *Schizaphis graminum* (Homoptera: Aphididae). *Bulletin of Entomological Research*, **82**, 407–16.

Sibley, C.G. and Ahlquist, J.E. (1983) Phylogeny and classification of birds based on the data of DNA hybridization, in *Current Ornithology*, Volume 1, (ed. R.F. Johnson), Plenum Press, New York, pp. 245–91.

Sibley, C.G. and Ahlquist, J.E. (1984) The phylogeny of the hominoid primates as indicated by DNA–DNA hybridization. *Journal of Molecular Evolution*, **20**, 2–15.

Sibley, C.G. and Ahlquist, J.E. (1986) Reconstructing bird phylogeny by comparing DNAs. *Scientific American*, February, 68–78.

Sibley, C.G., Ahlquist, J.E. and Sheldon, F.H. (1987) DNA hybridization and avian phylogenetics, in *Evolutionary Biology*, Volume 21, (eds M.K. Hecht, B. Wallace and G.T. Prance), Plenum Press, New York, pp. 97–125.

Simon, C., McIntosh, C. and Deniega, J. (1993) Standard restriction fragment analysis of the mitochondrial genome is not sensitive enough for phylogenetic analysis or identification of 17-year periodical cicada broods (Hemiptera:Cicadidae): the potential for a new technique. *Annals of the Entomological Society of America*, **86**, 228–38.

Slatkin, M. (1985) Gene flow in natural populations. *Annual Review of Ecology and Systematics*, **16**, 393–430.

Slatkin, M. (1987) Gene flow and the geographic structure of natural populations. *Science*, **236**, 787–92.

Smith, D.R. and Brown, W.M. (1990) Restriction endonuclease cleavage site and length polymorphisms in mitochondrial DNA of *Apis mellifera mellifera* and *A.m. carnica* (Hymenoptera: Apidae). *Annals of the Entomological Society of America*, **83**, 81–8.

Solomon, M.G., Fitzgerald, J.D. and Murray, R.A. (1996) Electrophoretic approaches to predator–prey interactions, in *The Ecology of Agricultural Pests: Biochemical Approaches*, (eds W.O.C. Symondson and J.E. Liddell). Chapman & Hall, London, pp. 457–68.

Spicer, G.S. (1992) Reevaluation of the phylogeny of the *Drosophila virilis* species group (Diptera: Drosophilidae). *Annals of the Entomological Society of America*, **85**, 11–25.

Stephan, W. and Cho, S. (1994) Possible role of natural selection in the formation of tandem-repeat noncoding DNA. *Genetics*, **136**, 333–441.

Stille, M. and Stille, B. (1992) Intra- and inter-nest variation in mitochondrial DNA in the polygynous ant *Leptothorax acervorum* (Hymenoptera; Formicidae). *Insectes Sociaux*, **39**, 335–40.

Stille, M., Stille, B. and Douwes, P. (1991) Polygyny, relatedness and nestfounding in the polygynous myrmicine ant *Leptothorax acervorum* (Hymenoptera; Formicidae). *Behavioural Ecology and Sociobiology*, **28**, 91–6.

Swofford, D.L. (1989) *Phylogenetic analysis using parsimony (PAUP)*, Release 3. Illinois Natural History Survey, Champaign, Ill., USA.

Symondson, W.O.C. and Liddell, J.E. (1993) A monoclonal antibody for the detection of Arionid slug remains in Carabid predators. *Biological Control*, **3**, 207–14.

Symondson, W.O.C. and Liddell, J,E. (1996) Polyclonal, monoclonal and engineered antibodies to investigate the role of predation in slug population dynamics, in *The Ecology of Agricultural Pests: Biochemical Approaches*, (eds W.O.C. Symondson and J.E. Liddell), Chapman & Hall, London, pp. 323–45.

Taylor, D.B., Hammack, L. and Roehrdanz, R.L. (1991) Reproductive compatibility and mitochondrial DNA restriction site analysis of New World screwworm, *Cochliomyia hominivorax*, from North Africa and Central America. *Medical and Veterinary Entomology*, **5**, 145–51.

Tegelstrom, H. (1992) Detection of mitochondrial DNA fragments, in *Molecular Genetic Analysis of Populations. A Practical Approach*, (ed. A.R. Hoelzel), IRL Press at Oxford University Press, Oxford, pp. 89–113.

Tomiuk, J. and Wöhrmann, K. (1983) Enzyme polymorphism and taxonomy of aphid species. *Zeitschrift zoologischer Systematik- und Evolutionsforschung*, **21**, 266–74.

Traut, W., Epplen, J.T., Weichenhan, D. and Rohwedel, J. (1992) Inheritance and mutation of hypervariable $(GATA)_n$ microsatellite loci in a moth, *Ephestia kuehniella. Genome*, **35**, 659-66.

Tretjk, A.P., Ryskov, A.P., Sevastyanova, G.A., Filippovich, Y.B. and Strunnikov, V.A. (1992) DNA fingerprints of *Bombyx mori* 1 – testing of genotypic variability of parthenogenetic strains. *FEBS Letters*, **303**, 258–60.

Vanlerberghe-Masutti, F. (1994a) Detection of genetic variability in *Trichogramma* populations using molecular markers, in *Norwegian Journal of Agricultural Sciences, Supplement No. 16, Insect Parasitoids: Biology and Ecology*, (eds Eline B. Hågvar and Trond Hofsvang), pp. 171–6.

Vanlerberghe-Masutti, F. (1994b) Molecular identification and phylogeny of parasitic wasp species (Hymenoptera: Trichogrammatidae) by mitochondrial DNA RFLP and RAPD markers. *Insect Molecular Biology*, **3**, 229–37.

Vogler, A.P., Knisley, C.B., Glueck, S.B., Hill, J.M. and DeSalle, R. (1993a) Using molecular and ecological data to diagnose endangered populations of the puritan tiger beetle *Cicindela puritana. Molecular Ecology*, **2**, 375–83.

Vogler, A.P., De Salle, R., Assman, T., Knisley, C.B. and Schultz, T.D. (1993b) Molecular population genetics of the endangered tiger beetle *Cicindela dorsalis* (Coleoptera: Cicindelidae). *Annals of the Entomological Society of America*, **86**, 142–52.

Vossbrinck, C.R. and Friedman, S. (1989) A 28S ribosomal RNA phylogeny of certain cyclorrhaphous Diptera based upon a hypervariable region. *Systematic Entomology*, **14**, 417–31.

Walton, M.P., Loxdale, H.D. and Allen-Williams, L. (1990a) Electrophoretic 'keys' for the identification of aphidiid parasitoids (Hymenoptera: Braconidae: Aphelinidae) of *Sitobion avenae* (F.) (Hemiptera: Aphididae). *Biological Journal of the Linnean Society*, **40**, 333–46.

Walton, M.P., Powell, W., Loxdale, H.D. and Allen-Williams, L. (1990b) Electrophoresis as a tool for estimating levels of hymenopterous parasitism in field populations of the cereal aphid, *Sitobion avenae. Entomologia experimentalis et applicata*, **54**, 271–9.

Weising, K., Ramser, J., Kaemmer, D., Kahl, G. and Epplen, J.T. (1991a) Oligonucleotide fingerprinting in plants and fungi, in *DNA Fingerprinting: Approaches and Applications* (eds T. Burke, G. Dolf, A.J. Jeffreys and R. Wolf), Birkhauser Verlag Basel, Switzerland, pp. 312–29.

Weising, K., Beyerman, B., Ramser, J. and Kahl, G. (1991b) Plant DNA fingerprinting with radioactive and digoxigenated nucleotide probes complementary to simple repetitive DNA sequences. *Electrophoresis*, **12**, 159–69.

Welsh, J. and McClelland, M. (1990) Finger printing genomes using PCR with arbitrary primers. *Nucleic Acids Research*, **18**, 7213–18.

Wetton, J.H., Carter, R.E., Parkin, D.T. and Walters D. (1987) Demographic study of a wild house sparrow population by DNA fingerprinting. *Nature*, **327**, 147–9.

White, P. S. and Densmore, L.D. (1992) Mitochondrial DNA isolation, in *Molecular Genetic Analysis of Populations. A Practical Approach*, (ed. A.R. Hoelzel), IRL Press at Oxford University Press, Oxford, pp. 29–58.

Wilding, N., Mardell, S.K., Brookes, C.P. and Loxdale, H.D. (1993) The use of polyacrylamide gel electrophoresis of enzymes to identify entomophthoralean fungi in aphid hosts. *Journal of Invertebrate Pathology*, **62**, 268–72.

Wilkerson, R.C., Parsons, T.J., Albright, D.G., Klein, T.A. and Braun, M.J. (1993) Random amplified polymorphic DNA (RAPD) markers readily distinguish cryptic mosquito species (Diptera; Culicidae: *Anopheles*). *Insect Molecular Biology*, **1**, 205–11.

Williams, C.L., Goldson, S.L., Baird, D.B. and Bullock, D.W. (1994) Geographical origin of an introduced insect pest, *Listronotus bonariensis* (Kuschel), determined by RAPD analysis. *Heredity*, **72**, 412–19.

Williams, J.G.K., Kubelik, A.R., Livak, K.J., Rafalski, J.A. and Tingey, S.V. (1990) DNA polymorphisms amplified by arbitrary primers are useful as genetic markers. *Nucleic Acids Research*, **18**, 6531–5.

Wilson, A.C., Cann, R.L., Carr, S.M. *et al.* (1985) Mitochondrial DNA and two perspectives on evolutionary genetics. *Biological Journal of the Linnean Society*, **26**, 375–400.

Woiwod, I.P. (1991) The ecological importance of long-term synoptic monitoring, in *The Ecology of Temperate Cereal Fields*, (eds L.G. Firbank, N. Carter, J.F. Darbyshire and G.R. Potts), Blackwell Scientific Publications, Oxford, pp. 275–304.

Workman, P.L. and Niswander, J.D. (1970) Population studies of southwestern indian tribes.II. Local genetic differentiation in the Papago. *American Journal of Human Genetics*, **22**, 24–49.

Wyman, A.R. and White, R. (1980) A highly polymorphic locus in human DNA. *Proceedings of the National Academy of Sciences, USA*, **77**, 6754–8.

Wynne, I.R., Loxdale, H.D. and Brookes, C.P. (1992) Use of a cellulose acetate system for allozyme electrophoresis, in *Genes in Ecology*, (eds R.J. Berry, T.J. Crawford and G.M. Hewitt), Blackwell Scientific Publications, Oxford, pp. 494–9.

Wynne, I.R., Howard, J.J., Loxdale, H.D. and Brookes, C.P. (1994) Population genetic structure during aestivation in the sycamore aphid *Drepanosiphum platanoidis* (Hemiptera: Drepanosiphidae). *European Journal of Entomology*, **91**, 375–83.

Yang-jiang Lu, Adang, M.J., Isenhour, D.J. and Kochert, G.D. (1992) RFLP analysis of genetic variation in North American populations of the fall armyworm moth *Spodoptera frugiperda* (Lepidoptera: Noctuidae). *Molecular Ecology*, **1**, 199–207.

Zeh, D.W., Zeh, J.A., Coffroth, M.A. and Bermingham, E. (1992) Population-specific fingerprints in a neotropical pseudoscorpion *(Cordylochernes scorpioides)*. *Heredity*, **69**, 201–8.

Zeh, D.W., May, C.A., Coffroth, M.A. and Bermingham, E. (1993) *MboI* and *macrohaltica* – quality of DNA fingerprints is strongly enzyme-dependent in an insect (Coleoptera). *Molecular Ecology*, **2**, 61–3.

Zehnider, G.W., Sandall, L., Tisler, A.M. and Powers, T.O. (1992) Mitochondrial DNA diversity among 17 geographic populations of *Leptinotarsa decemlineata* (Coleoptera: Chrysomelidae). *Annals of the Entomological Society of America*, **85**, 234–40.

Zraket, C.A., Barth, J.L., Heckel, D.G. and Abbott, A.G. (1990) Genetic linkage mapping with restriction fragment length polymorphisms in the tobacco budworm, *Heliothis virescens*, in *Molecular Insect Science*, (eds H.H. Hagedorn *et al.*), Plenum Press, New York, pp. 13–20.

8

Location of resistance to *Brevicoryne brassicae* in wild brassica species

R.A. Cole and T. Fenning

ABSTRACT

Resistance to colonization by the cabbage aphid *Brevicoryne brassicae* (L.), a form of antibiosis, has been identified in the wild *Brassica* species *B. fruticulosa* (Cyrillo) and *B. spinescens* (Pomel) in glasshouse and field experiments. The mechanism of resistance is thought to be located in the sieve element as normal phloem feeding was quickly terminated by withdrawal of the stylets. A lectin, with a chitin-binding domain and chitinase activity, is present in significant quantities in the wild brassica species but at low levels in cultivated cabbage *B. oleracea* (L.). The lectin was isolated and its amino acid composition found to be similar to other chitin-binding lectins with insecticidal and antifungal activities. Brassica lectin and the closely related agglutinin from nettle, the gene for which has already been isolated, show significant insecticidal activity when presented to *B. brassicae* in chemically defined synthetic diets. Immunocytochemistry showed the lectin to be located primarily in specific granular cells of both *B. fruticulosa* and *B. spinescens* but absent in *B. oleracea*. Homology was detected, in a Southern blot, between the nettle lectin gene (*uda*) and DNA from *B. fruticulosa*.

8.1 INTRODUCTION

Chemical ecology can be defined as the study of chemical interactions between organisms and their environment. Pests of agricultural crops are as old as agriculture itself; estimates of the losses due to pests and

The Ecology of Agricultural Pests, Edited by W.O.C. Symondson and J.E. Liddell.
Published in 1996 by Chapman & Hall, London. ISBN 0 412 62190 8

diseases total 14% of the total yield. In present-day agricultural monocultures, there is increasing selection pressure on insects to adapt to crops, but effective protection by insecticides has until recently reduced the urgency for the production of plant varieties with resistance to insects. As agricultural crops have been selected for yield and form under reduced selection pressure by insects, many modern crop varieties have little pest resistance. In fact, as the production of secondary plant chemicals associated with resistance requires an expenditure of energy by the plant, which may yield less as a consequence, selection for crop improvement has been actively against crop resistance. Thus, many cultivated plant species show less resistance than their wild relatives to phytophagous pests (de Ponti *et al.*, 1981). By understanding the complex chemical aspects of insect–host plants relationships, researchers aim to accelerate the production of new resistant varieties, utilizing a wide variety of resistance mechanisms.

8.2 MECHANISMS OF RESISTANCE

In order to produce insect-resistant crops it is important to understand the underlying mechanisms of host plant resistance. The importance of determining the chemical composition of tissues at the source of feeding rather than in the entire plant cannot be over-emphasized. This involves determining precise differences in behaviour of insects on resistant and susceptible varieties. Techniques, such as electronic monitoring of aphid feeding behaviour (Tjallingii, 1990) allow the location of resistance to be determined. Only then can hypotheses be proposed to explain the mechanism and compounds involved in resistance. Further advances using genetic manipulation to transfer host plant resistance can only be achieved reliably with an understanding of the resistance mechanism and the primary gene products, such as proteins or modification of biosynthetic enzymes, involved in the synthesis of secondary plant compounds.

Natural defence mechanisms which depend on a primary gene product for effect are among those currently suitable for producing insect-resistant, transgenic crops by genetic engineering. One such mechanism involves the proteinase inhibitors from cowpea which are anti-metabolic to a wide range of insect pests (Gatehouse and Hilder, 1988). Ryan and co-workers (Ryan, 1983) demonstrated that proteinase inhibitors are induced systemically either by insect feeding or mechanical wounding, More recently Alarcon and Malone (1994) have shown that caterpillars feeding on tomato leaves generate sufficient damage to disperse elicitors throughout the xylem, through mass flow associated with hydraulic signals, and systemically induce proteinase inhibitors within 1 hour of the damage. α-Amylase inhibitors are protein inhibitors found mainly in seeds and which are toxic to mammals and a wide range of insects (Deponte *et al.*, 1976); fortunately, inhibitors have been found which are specific for the insect enzyme only. Lectins are plant proteins which have been identified as having insecticidal properties (Gatehouse *et al.*, 1984; Heusing *et al.*, 1991), particularly against Coleoptera and Lepidoptera. Genes encoding pea lectin have been transferred successfully to other plants, where preliminary results indicate enhanced levels of resistance to Coleopteran and Lepidopteran pests (Gatehouse *et al.*, 1992). One drawback of this mode of resistance is the need for the insect to consume leaf material in order for a lethal dose of toxin to be acquired; this in turn can result in a reduction of plant quality or the transmission of plant diseases. Also, as resistance requires consumption of leaf material, proteinase inhibitors are not effective against phloem-feeding insects, feeding directly on free amino acids and sucrose in the phloem sieve element. Recently, the toxicity of a mannose-binding lectin from *Galanthus nivalis* (snowdrop) (GNA) to some Homopteran pests including rice planthoppers *Nilaparvata lugens* (Stål) and rice leafhoppers *Nephotettix cinciteps* (Uhler) (Powell *et al.*, 1993), has been demonstrated. The insecticidal activity of a chitin-binding lectin, wheatgerm aggluttin (WGA) (Powell *et al.*, 1993), has been demonstrated against brown planthopper when incorporated into diets at a concentration of 0.1% WGA, but had little effect on pea aphid, *Acyrthosiphon pisum* (Harris), (Rahbe and Febvay, 1993), at a concentration of 0.05% WGA. To control Homopteran pests, such as aphids and planthoppers, an insecticidal gene must be expressed in the phloem tissue and the gene product be present in the phloem sap. Using the rice sucrose synthase-1 (RSs1) promoter Shi *et al.* (1994) directed the tissue-specific expression of the snowdrop lectin GNA gene in the phloem tissue, and GNA was subsequently detected in phloem tissue and aphid honeydew. This chapter describes research over the past 4 years at Horticulture Research International, from the discovery of aphid-resistant wild *Brassica* species, to the identification of a defence mechanism dependent on a primary gene product, and the possible use in genetically transformed crops expressing the resistance gene.

8.3 LOCATING THE RESISTANCE MECHANISM TO *B. brassicae* IN WILD BRASSICA SPECIES

The cabbage aphid (*Brevicoryne brassicae* L.) continues to increase in importance as a pest of horticultural brassicas in most temperate regions. Currently, *B. brassicae* control depends on insecticides, but the production of resistant cultivars is under investigation. Resistance to colonization by *B. brassicae*, a form of antibiosis, has been identified in the wild *Brassica* species *B. fruticulosa* (Cyrillo) and *B. spinescens* (Pomel)

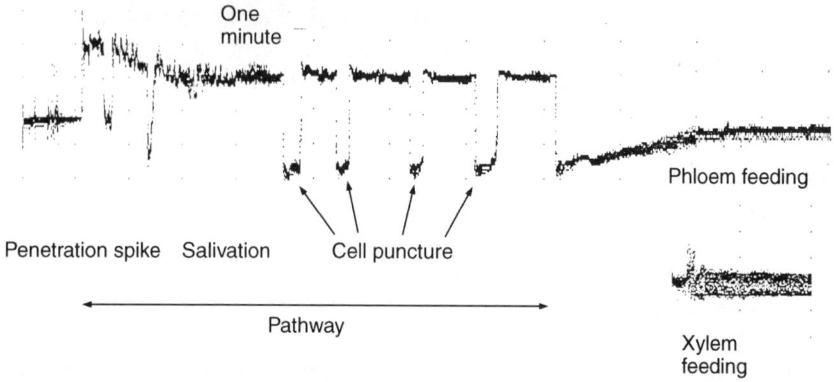

Figure 8.1 Electronic monitoring of *Brevicoryne brassicae* on *Brassica* sp., showing different feeding activities.

in glasshouse and field experiments (Singh *et al.*, 1994). Although several resistance mechanisms have been suggested for host plant–aphid interactions (Dreyer and Campbell, 1984; Weibull *et al.*, 1986; Cole *et al.*, 1993) the mechanisms of resistance to *B. brassicae* were unknown. MacLean and Kinsey (1964) were the first to monitor aphid feeding behaviour electronically using an AC method which was modified subsequently to include a DC circuit. DC electronic recording of aphid feeding behaviour classifies voltage fluctuations, comprising distinct patterns, according to the activities of the aphids and the positions of the aphid stylets within the plant tissue (Figure 8.1) (Tjallingii, 1985, 1990). These different feeding activities can be summarized as follows:

- Non-probing (NP) aphids on the leaf without stylet penetration or electrical stylet contact
- Probe, a period of stylet penetration
- Pathway, patterns A (initial penetration), B (salivation), and C (stylet pathway) were regarded as one pattern representing stylet pathway
- Cell punctures, short intracellular stylet penetrations during pathway
- E_1, sieve element salivation
- E_2, passive phloem ingestion
- G, xylem feeding

Using this detailed information on the location and behaviour of aphid stylets, electronic monitoring has been successful in determining the location of resistance mechanisms in brassica and lettuce cultivars (Cole and Riggal, 1992; Cole *et al.*, 1993; van Helden and Tjallingii, 1993). The probing behaviour of *B. brassicae*, on leaves of resistant *B.*

fruticulosa and *B. spinescens* and on a susceptible *B. oleracea* var. capitata (L.), was monitored electronically to determine the mechanisms of resistance. *B. brassicae* was not deterred initially from accepting all three *Brassica* species as host plants, penetration of the leaf cuticle occurring in less than 5 minutes on all three plant species (Cole, 1994a). On leaves of all species, after less than 10 initial short probes, a penetration continued towards the phloem vessels, as indicated by the increased number of cell punctures as the stylets manoeuvred around the closely packed cells of the vascular bundle (Cole *et al.*, 1993). The stylets usually continued to penetrate to the phloem where pattern E_2 was recorded within 3 hours of the cabbage aphid being placed on the leaf. Penetration of the phloem sieve element was characterized by a short (<3 min) E_1 pattern, assigned to salivation into the sieve element in preparation for passive phloem ingestion. This signal in the susceptible *B. oleracea* preceded a long sustained E_2 pattern often lasting for several hours (Figure 8.2). On resistant *Brassica* species a normal E_1 signal was recorded which was quickly terminated by withdrawal of the stylets, often with an increase in the amplitude of the E_1 pattern (Figure 8.2). The major difference recorded between *B. brassicae* feeding on the susceptible *B. oleracea* and the resistant *Brassica* species, *B. fruticulosa* and *B.*

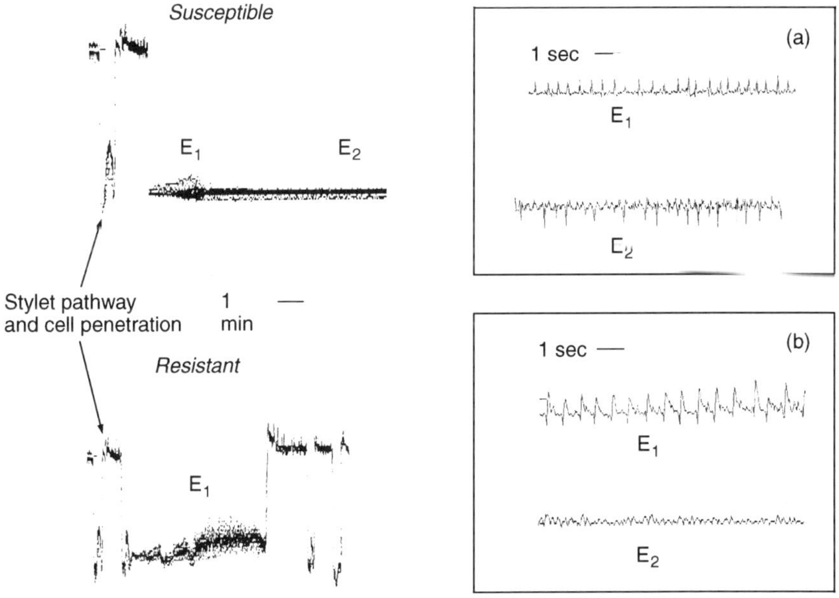

Figure 8.2 Sieve element penetrations during feeding behaviour of *B. brassica* on *Brassica* sp.

Table 8.1 Mean % duration and number of probing events recorded electronically for *Brevicoryne brassicae* on *Brassica* species over the total feeding period

Species	*Mean % of total duration*				*No. of*	
	Non-probing	*Pathway*	*Phloem ingestion*		*Probes*	*Cell punctures*
			E_1	E_2		
B. spinescens	22	64	3	11	39	219
B. fruticulosa	15	69	5	10	40	346
B. oleracea	9	31	2	56	21	133
LSD ($P<0.05$)	10.8	36.6	2.8	11.2	19.2	82.0

spinescens, was a significant reduction in the duration of passive phloem ingestion (Table 8.1). That is, the aphids penetrated the phloem, but were unable to continue sustained phloem ingestion. Passive food uptake was not initiated, as confirmed by the lack of honeydew droplets around the feeding site on the resistant species while a number of droplets were still visible in the vicinity of aphids feeding on *B. oleracea*. This difference in feeding behaviour suggests that the resistance mechanism is located in the phloem sieve element as either a chemical deterrent, lack of stimulant or the production of a mechanical barrier. Further results indicated that although glucosinolate profiles, the characteristic secondary compounds found in brassica species, were different in the wild brassica species, inclusion of these compounds did not reduce feeding from chemically defined synthetic diets (Cole *et al.*, 1993). It was possible that the passage of the stylets towards the vascular bundle or the actual penetration of the phloem sieve element could release elicitors which would induce a hypersensitive response, causing a blockage of the phloem sieve pores by callose or P-proteins which include chitin binding lectins (Read and Northcote, 1983).

The mechanism of resistance appears to be very similar to the resistance of lettuce lines to *Nasonovia ribis-nigri* (Moseley) which was originally demonstrated in the wild *Lactuca* species *L. virosa* (L.) (Eenink *el al.*, 1982; van Helden and Tjallingii, 1993). It is possible that the mechanism of resistance was common to other aphid–plant interactions and perhaps dependent on the location of a primary product, such as a protein, for effect. If this were so, the mechanism could be transferred to other plant species by cloning the gene to enhance resistance to a number of pests.

8.4 ISOLATION OF A CHITIN-BINDING LECTIN, WITH INSECTICIDAL ACTIVITY TO THE CABBAGE APHID *B. brassicae*

Phloem sap is generally poor in proteins ($<100\ \mu g/ml$) but many lectins are constitutive in the sieve tubes of plant species (Sabnis and Hart, 1978; Read and Northcote, 1983; Bostwick *el al.*, 1993), providing a defence against attacks by bacteria, fungi and perhaps by phloem-feeding insects. These lectins have been shown to have a chitin, *N*-acetylglucosamine (GlcNAc)-binding domain (Chrispeels and Raikhel, 1991).

Both wild brassica species *B. fruticulosa* and *B. spinescens* were shown to contain significant concentrations of a constitutive chitin-binding lectin BL (brassica lectin), which was present in lower concentrations in the cultivated *B. oleracea* (Cole, 1994b). The lectins migrated as a major band with a M_r of 14 500 under denaturing conditions, after SDS–PAGE electrophoresis at 10% and pH 8.8 (Figure 8.3(a)). Chitin-binding lectins have been found in such diverse species as wheat (*Triticum aestivum* L.) WGA, rice (*Oryza sativa* L.), RL, and stinging nettle (*Urtica dioica* L.) in UDA (Chrispeels and Raikhel, 1991). Figure 8.3(b) shows the chitinase activity of the lectins BL and WGA and UDA run under native conditions, detected as dark zones in the glycol chitin overlay when stained with Calcofluor white M2R. BL, WGA and UDA did not migrate in SDS–PAGE gels containing 0.01% glycol chitin because of the affinity

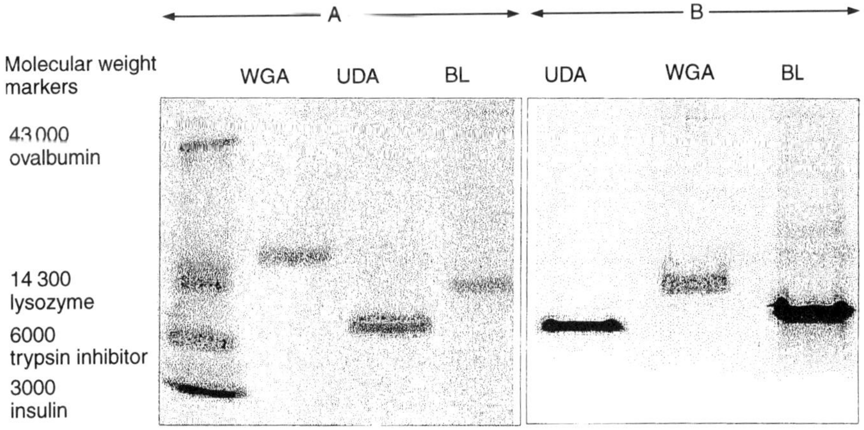

Figure 8.3 (a) SDS–PAGE of wheatgerm agglutinin (WGA), *Urtica dioica* agglutinin (UDA) and brassica lectin (BL) in 10% polyacrylamide gel at pH 8.8 stained with Coomassie blue. (b) Chitinase activity in 10% native polyacrylamide gels overlaid with 0.05% glycol chitin/agar and stained with Calcofluor white M2R.

for chitin. Because the lectins from the three brassica species had the same mobility after SDS–PAGE electrophoresis they were assumed to represent the same constitutive lectin and further analysis was continued on the extract from *B. spinescens*. The brassica lectin, in similarity to other chitin-binding lectins, contained a high concentration of asparagine/aspartic acid, serine, glycine and leucine and low concentrations of cysteine and methionine (Cole, 1994b).

8.5 INSECTICIDAL EFFECTS OF LECTINS IN CHEMICALLY DEFINED ARTIFICIAL DIETS

Brassica lectin and the closely related agglutinin from wheatgerm and nettle showed insecticidal activity when present in chemically defined synthetic diets (Figure 8.4) (Cole, 1994b). Some 40% of *B. brassicae* died on the control diet while all aphids died on a diet containing 2.5 mg/ml of each lectin. Reduced feeding was indicated by a lack of honeydew production when the lectin was present at this high concentration (2.5 mg/ml) but this was not measured. The mean percentage mortality rate of *B. brassicae* on diets with UDA, WGA and BL at 8 days was significantly higher than the control (Figure 8.4(a–c)) and there was a quadratic, almost linear, relationship with log dose (Figure 8.4(d)). BL and UDA were not significantly different but were more effective than WGA at reducing aphid survival. The curves in Figure 8.4 may reflect two mechanisms of mortality; initially an anti-feedant activity with a rapid mortality up to 50% in the concentration range <25 μg/ml where aphids were deterred from feeding, and died from starvation, while at the higher concentration of 2.5 mg/ml mortality was linked to the toxicity of the lectin. Subsequent experiments have shown that other proteins such as soybean trypsin inhibitor – which show antinutritive effects against Coleoptera and Lepidoptera – can improve survival of *B. brassicae* on artificial diets (R.A. Cole, unpublished data), The differing toxicities of plant protective proteins on various insect orders may reflect different modes of feeding behaviour or the different mechanisms of actions of the proteins against various insect orders. The concentration of lectin in the artificial diet needed to produce an insecticidal effect would appear rather high in comparison with the concentration of protein detected in phloem sap (<100 μg/ml) (Rahbe and Febvay, 1993). However, the concentration detected by phloem exudation may not represent the concentration at the point of stylet penetration into the phloem sieve element. Aphid stylets have been shown to penetrate the cells around the vascular bundles (Cole *el al.*, 1993) during passage to the phloem and may contact relatively high concentrations of lectin sequestered in vacuoles (Chrispeels and Raikhel, 1991) or cell damage during stylet penetration may induce the synthesis

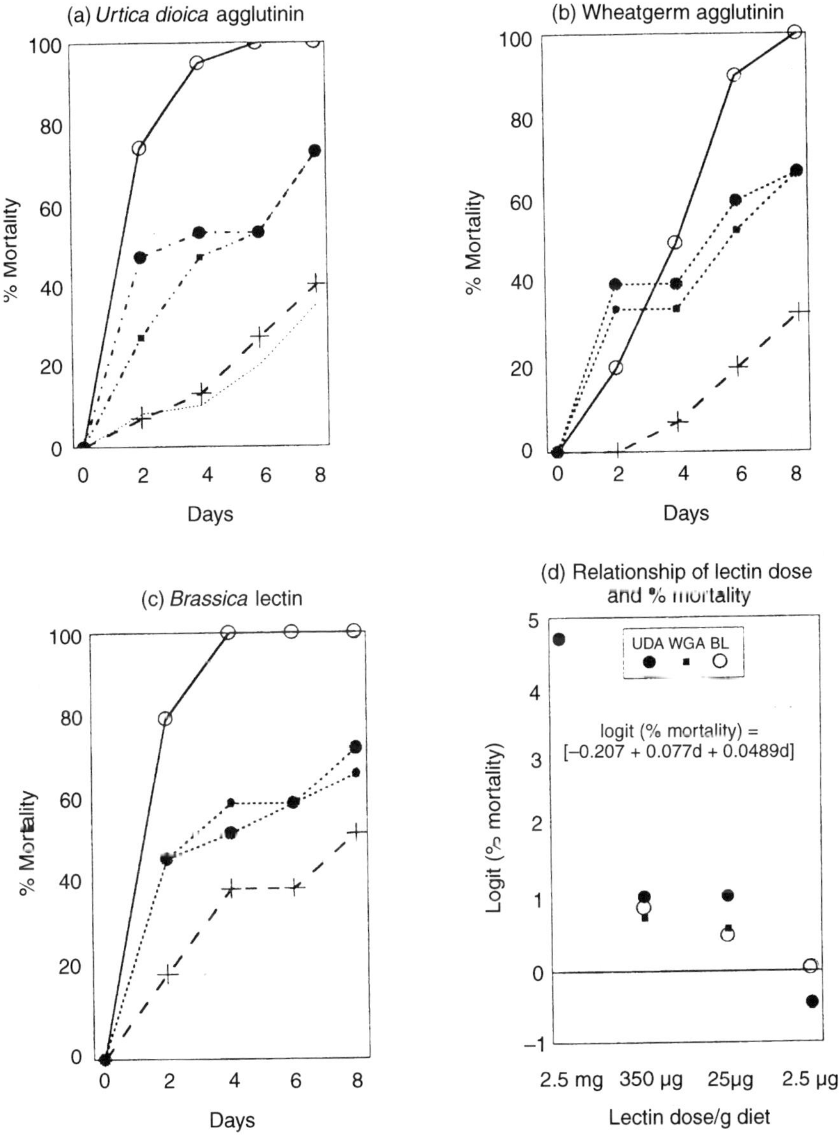

Figure 8.4 Effect of dietary lectin on mortality rate (%) of *B. brassicae*. ○, 2.5 mg; ●, 250 µg; ■, 25 µg; +, 2.5 µg; ... indicates control.

of higher concentrations of lectins. A pathogen-induced chitinase, with a chitin-binding domain of M_r 30 000 has been isolated from rape (*B. napus* L. ssp. *olifera*) seedlings, after less than 24 hours exposure to *Phoma lingam* Desm (Rasmussen *et al.*, 1992). Protease inhibitors which are involved in other forms of plant defence, begin to be induced systemically in response to insect attack 12 hours after damage occurred (Alarcon and Malone, 1994), with maximal expression taking 48–64 hours (Stanford *et al.*, 1990), but this is not rapid enough to provide the constitutive protection that *B. fruticulosa* has against aphids initiating passive phloem ingestion after penetration of the sieve element.

8.6 IMMUNOLOCALIZATION OF BRASSICA LECTIN IN *Brassica* SPECIES

Since its introduction, immunological detection has become a standard tool in histological research. To determine the location of brassica lectin within *Brassica* species, polyclonal rabbit anti-BL antisera was raised to purified lectin from *B. spinescens*. Antiserum at 100-fold dilution was applied to dewaxed sections with skimmed milk as a blocking agent. After thorough washing with phosphate buffered saline (PBS) the anti-rabbit IgG FITC conjugate was added. Using UV-epifluorescence microscopy, brassica lectin was detected in leaf tissue of the wild *Brassica* species, but not in *B. oleracea*.

Stylets of *B. brassica* were cut by radiocautery (high-frequency radio waves), using electronic monitoring to determine the location and behaviour of the aphid stylets during phloem penetration in all three species. Paraffin wax sections were prepared as before for immunolocalization of BL in wild brassica leaves. Enhanced fluorescence of the anti-rabbit IgG FITC conjugate was detected in phloem tissue of *B. fruticulosa* penetrated by the stylets but this was not observed in *B. oleracea*. This may indicate that other chitin-binding proteins may be produced in the wild species in response to plant damage, in addition to those constitutively present. It might be reasonable therefore, to expect a family of lectin defence genes to be present in *B. fruticulosa* and *B. spinescens*, inducible independently or constitutively expressed as appropriate.

8.7 DETECTION OF HOMOLOGY IN A SOUTHERN BLOT, BETWEEN THE NETTLE LECTIN GENE (*uda*) AND DNA FROM *B. fruticulosa*

Plants contain or produce a number of proteins as a response to stress or damage, termed pathogenesis-related (PR) proteins. Many PR proteins from widely diverse families possess a conserved 43-amino acid domain and all other lectins with insecticidal activity have been shown to possess this domain (Chrispeels and Raikhel, 1991). If the chitin-binding domain of the brassica lectin has homology to the nettle lectin (UDA) as seems likely, then the *uda* gene which has already been isolated (Chrispeels and Raikhel, 1991) could be utilized as a probe in a Southern blot to enable isolation of the brassica lectin gene(s).

Potentially, the probe could be of use in brassica breeding programmes for determining the level of aphid resistance of a new variety without lengthy and laborious aphid trials. In the case of other horticultural crops which may lack such lectin genes altogether, the *uda* or brassica lectin genes could be transferred by genetic transformation to increase their resistance to aphid or fungal attack.

8.8 CONCLUSIONS

Mirelman *et al.* (1975) suggested that chitin-binding lectins can disturb the balance of growth in fungi by binding or cross-linking chitin chains, but the mode of action of chitin-binding lectins on phloem-feeding insects is not known. The insecticidal activity may be due to binding to the chitinous stylets enclosing the food and salivary canals and extending to the insect's foregut (Ponsen, 1987),or alternatively the lectin may bind to cell surface receptors in the gut and disrupt signalling molecules regulating cell growth there (de Olivera *et al.*, 1988).

The particularly small size of the brassica lectin (M_r 14 500) may render it effective against phloem-feeding pests as it would be translocated more easily through the phloem than larger proteins. A phloem-translocated lectin would be ideally located to respond to invasion of the sugar-rich sieve element by the aphid stylet (Smith *et al.*, 1987). For resistance to be effective in plants which have not previously been exposed to Homopteran pests, lectin genes must be expressed constitutively at sufficient levels in or near to the phloem prior to stylet penetration into the sieve elements.

The use of genetically engineered crops expressing plant-derived insect resistance genes is now seen as a potential means of producing crops with significantly enhanced levels of resistance. This may not seem so important for crop species such as brassicas where suitable resistance mechanisms have been identified in the genetic resources available to the plant breeder. Conventional plant breeding techniques may encounter great difficulties in transferring the lectin-based resistance to horticultural crops as resistance has only been demonstrated in distantly related wild brassica species. However, the identification of the genes involved could usefully act as markers for aphid resistance. As single gene products, lectins are suitable for genetic engineering into other crop plants which lack them.

In transgenic tobacco plants the snowdrop lectin was expressed at only low levels under the CaMV 35S promoter, but still caused a significant reduction in aphid populations, suggesting the presence of at least some GNA protein in the phloem sap (Shi *et al.*, 1994). The use of more suitable promoters, such as the rice sucrose synthase-1 promoter (Shi *et al.*, 1994), or perhaps those for the brassica lectin gene(s) themselves, may enable expression of a variety of proteins with insecticidal properties in the correct cellular location for sustained resistance to attack by phloem-feeding insects in transgenic plants. In summary, it is only by understanding the location and basis of pest-resistance mechanisms at the molecular level, especially in the wild relatives of crop plants, that the entomologist, plant breeder and genetic engineer will be able to produce the new varieties needed for tomorrow's reduced input agriculture.

Acknowledgements

This work was funded by the Ministry of Acriculture, Fisheries and Food. Acknowledgements also to Dr W. Broekaert for supplying samples of the UDA protein, to Dr N. Raikhel for supplying the *uda* gene and Dr N. Lyons for polyclonal antibodies to the brassica lectin.

REFERENCES

Alarcon, J.-J. and Malone, M. (1994) Substantial hydraulic signals are triggered by leaf-biting insects in tomato. *Journal of Experimental Botany*, **45**, 953–7.

Bostwick, D.E., Dannenhoffer, J.M., Skaggs, M.I., Lister, R.M., Larkin, B.A. and Thompson, G.A. (1993) Pumpkin phloem lectin genes are specifically expressed in companion cells. *Plant Cell*, **4**, 1539–48.

Chrispeels, J.M. and Raikhel, V.J. (1991) Lectins, lectin genes, and their role in plant defense. *The Plant Cell*, **3**, 1–9.

Cole, R.A. (1994a) Locating a resistance mechanism to the cabbage aphid in two wild Brassicas. *Entomologia Experimentia et Applicata*, **71**, 23–31.

Cole, R.A. (1994b) Isolation of a chitin-binding lectin, with insecticidal activity in chemically-defined synthetic diets, from two wild brassica species with resistance to cabbage aphid *Brevicoryne brassicae*. *Entomologia Experimentia et Applicata*, **72**, 181–7.

Cole, R.A. and Riggall, W.E. (1992) Pleiotropic effects of genes in glossy *Brassica oleracea* resistant to *Brevicoryne brassica*, in *Proceedings 8th International Symposium on Insect–Plant Relationships*, (eds S.B.J. Menken, J.H. Visser and P. Harrewijn), Kluwer Academic Press, Dordrecht, pp. 313–15.

Cole, R.A., Riggall, W.E. and Morgan, A. (1993) Electronically monitored feeding behaviour of the lettuce root aphid (*Pemphigus bursarius*) on resistant and susceptible lettuce varieties. *Entomologia Experimentia et Applicata*, **68**, 179–86.

Deponte, R., Parlamenti, R., Petrucci, T., Silano, V. and Tomasi, M. (1976) Albumin α-amylase inhibitor families from wheat flour. *Cereal Chemistry*, **53**, 805–20.

de Ponti, O.M.B. (1981) Conserving the natural resistance to insects by the proper use of the pesticide umberella, in *Genetic Resources and Plant-breeding for Resistance to Disease, Pests and Abiotic Environmental Conditions 1X Eucarpia Congress*, Leningrad, September 1980, pp. 236–48.

Dreyer, D.L. and Campbell, B.C. (1984) Chemical basis of host-plant resistance to aphids. *Plant Cell Environment*, **10**, 35.

Eenink, A.H., Groenwold, R. and Dieleman, F.L. (1982) Resistance of lettuce (*Lactuca*) to the leaf aphid *Nasonovia ribis-nigri* 1: Transfer of resistance from *L. virosa* to *L. sativa* by interspecific crosses and selection of resistant breeding lines. *Euphytica*, **31**, 291–300.

Gatehouse, A.M.R. and Hilder, V.A. (1988) Introduction of genes conferring insect resistance, in *Proceedings of Brighton Crop Protection Conference*, Vol. 49, Lavenham Press Ltd, Lavenham, Suffolk, UK, pp. 1234–54.

Gatehouse, A.M.R., Dewey, F.M., Dove, J. Fenton, K.A. and Pusztai, A. (1984) Effect of seed lectin from *Phaseolus vulgaris* on the development of larvae of *Callosobruchus maculatus*; mechanism of toxicity. *Journal of the Science of Food and Agriculture*, **47**, 373–80.

Gatehouse, A.M.R., Hilder, V.A., Powell, K., Boulter, D. and Gatehouse, J.A. (1992) Potential of plant-derived genes in the genetic manipulation of crops for insect resistance, in *Proceedings 8th International Symposium on Insect–Plant Relationships*, (eds S.B.J. Menken, J.H. Visser and P. Harrewijn), Kluwer Academic Publishers, Dordrecht, pp. 221–34.

Heusing, J.E., Murdock, L.L. and Shade, R.E. (1991) Rice and stinging nettle lectins: insecticidal activity similar to wheat germ agglutinin. *Phytochemistry*, **30**, 3565–8.

MacLean, D.L. and Kinsey, M.G. (1964) A technique for electronically recording aphid feeding behaviour and salivation. *Nature*, **256**, 1358–9.

Mirelman, D., Galun, E., Sharom, N. and Lotan, R. (1975) Inhibition of fungal growth by wheat germ agglutinin. *Nature*, **256**, 414.

de Olivera, J.T.A., Puszai, A.J. and Grant, G. (1988) Changes in organs and tissues induced by feeding of purified kidney bean (*Phaseolus vulgaris*) lectins *Nutrition Research*, **8**, 943–7.

Ponsen, M.B. (1987) Alimentary tract, in *Aphids their Biology, Natural Enemies and Control*, (eds A.K. Minks and P. Harrewijn), Elsevier, Amsterdam, pp. 79–97.

Powell, K.S., Gatehouse, A.M.R., Hilder, V.A. and Gatehouse, J.A. (1993) Anti-metabolic effects of plant lectins and plant and fungal enzymes on the nymphal stages of two important rice pests, *Nilaparvata lugens* and *Nephotettix cinciteps*. *Entomologia Experimentia et Applicata*, **66**, 119–26.

Rahbe, Y. and Febvay, G. (1993) Protein toxicity to aphids: an in vitro test on *Acyrthosiphon*. *Entomologia Experimentia et Applicata*, **67**, 149–60.

Rasmussen, U., Giese, H. and Mikkelsen, J.D. (1992) Induction and purification of chitinase in *Brassica napus* infected with *Phoma lingam*. *Planta*, **187**, 328–34.

Read, S.M. and Northcote, D.H. (1983) Chemical and immunological similarities between the phloem proteins of three genera of the Cucurbitaceae. *Planta*, **158**, 119–27.

Ryan, C.A. (1983) Insect-induced chemical signals regulating natural plant protection responses, in *Variable Plants and Herbivores in Natural and Managed Systems*, (eds R.F. Denno and M.S. McClure), Academic Press, New York, pp. 43–60.

Sabnis, D.D. and Hart, J.W. (1978) The isolation and some properties of a lectin (haemagglutinin) from Cucurbita phloem exudate. *Planta*, **142**, 97–101.

Shi, Y., Wang, M.B., Powell, K.S., Van Damme, E., Hilder, V.A., Gatehouse, A.M.R., Boulter, D. and Gatehouse, J.A. (1994) Use of the rice sucrose synthase-1 promoter to direct phloem-specific expression of β-glucuronidase and snowdrop lectin genes in transgenic tobacco plants. *Journal of Experimental Botany*, **45**, 623–31.

Singh, R., Ellis, P.R. and Pink, D.A.C. (1994) Sources of resistance to cabbage aphid *Brevicoryne brassicae* in the wild *Brassica* species, *B. fruticulosa* and *B. spinescens*. *Annals of Applied Biology*, **125**, 457–65.

Smith, L.M., Sabnis, D.D. and Johnson, R.P.C. (1987) Immunochemical localisation of phloem lectins from *Cucurbita maxima* using peroxide and colloidal-gold labels. *Planta*, **170**, 461–70.

Stanford, A.C., Northcote, D.H. and Bevan, M.W. (1990) Spatial and temporal patterns of wound-induced genes in potato. *The EMBO Journal*, **9**, 593–603.

Tjallingii, W.F. (1985) Electrical nature of recorded signals during stylet penetration by aphids. *Entomologia Experimentia et Applicata*, **38**, 177–86.

Tjallingii, W.F. (1990) Continuous recording of stylet penetration activities by aphids, in *Aphid–Plant Genotype Interactions*, (eds R.K. Campbell and R.D. Eikenberg), Elsevier, Amsterdam, pp. 89–99.

van Helden, M. and Tjallingii, W.F. (1993) The resistance of lettuce (*Lactuca sativa* L.) to *Nasonovia ribisnigri* (Homoptera Aphididae): the use of electrical penetration graphs to locate the resistance factor(s) in the plant. *Entomologia Experimentia et Applicata*, **68**, 269–78.

Weibull, J., Brishammar, S. and Petterson, J. (1986) Amino acid analysis of phloem sap from oats and barley a combination of aphid stylet excision and high performance liquid chromatography. *Entomologia Experimentalis et Applicata*, **42**, 27–30.

9

The use of DNA markers in population genetics and ecological studies of the Desert Locust *Schistocerca gregaria* (Orthoptera: Acrididae)

D-X. Zhang and G.M. Hewitt

ABSTRACT

The Desert Locust *Schistocerca gregaria* is a major agricultural pest, affecting more than one-fifth of the total land surface of the world. Few genetic studies have been carried out, though a fuller understanding of its population biology is necessary for the eventual control of the plague. In this study, nuclear and mitochondrial (mt) DNA markers have been used to investigate the genetic diversity, population structure and dynamics of Desert Locust populations on a large geographic scale, using samples collected from populations across its range (Pakistan, Saudi Arabia, Eritrea, Sudan, Niger and Mauritania). Analysis of DNA sequence data shows that genetic diversity widely exists in *S. gregaria* populations, and the pattern of mtDNA diversity is different from that of nuclear DNA. Two distinct genetic lineages were identified from both the nuclear and mtDNA data. One nuclear lineage seems to be under-represented in some populations. The large genetic distance observed between the two nuclear lineages suggests that they represent an ancient genetic split in *S. gregaria* populations. It has been estimated that these two lineages have been separated for about one million years, assuming a substitution rate of 30×10^{-9} per site per year. A hypothesis of 'divergence by isolation' is proposed to explain such observations,

The Ecology of Agricultural Pests, Edited by W.O.C. Symondson and J.E. Liddell.
Published in 1996 by Chapman & Hall, London. ISBN 0 412 62190 8

considering the sensitivity of the Desert Locust to environmental changes. The present distribution pattern of the two lineages in *S. gregaria* populations can be explained by migration and population hybridization under two hypothetical models, which suggest that the present widespread populations are the consequence of subsequent expansion after the restoration of living conditions for the desert locust. The suitability of various PCR-based techniques and different DNA markers for population biological studies are also discussed.

9.1 INTRODUCTION

The Desert Locust (*Schistocerca gregaria*, Forskål) is a large grasshopper which belongs to the family Acrididae of the Orthoptera. It has been one of the most important agricultural pests since the earliest recorded times (a plague caused by the Desert Locust occurred at about 1300 BC recorded by the ancient Egyptians). The Desert Locust has a vast invasion area of about 29 million km^2 ranging across Africa to Western Asia (Centre for Overseas Pest Research, 1982; Steedman, 1990). During recession periods, the Desert Locust is in its solitary phase, and swarms and hopper bands are rare. However, with suitable weather conditions involving widespread heavy and prolonged rains in several successive seasonal breeding areas, recession populations can rapidly increase both in number and in size, forming swarms. Swarms of the Desert Locust in the gregarious phase are highly mobile and cause serious damage over a very wide range of wild and cultivated plants.

Great efforts have been made to control these Desert Locust plagues, mainly by chemical control over gregarious populations. However, solitary locusts, for practical and economical reasons, are not suitable targets for chemical control (Steedman, 1990). Therefore, they continue to multiply under suitable conditions, migrate, concentrate, and finally develop into new swarms. Consequently, solitary locusts serve as a permanent source of swarms. Furthermore, the use of insecticides in chemical control can lead to environmental pollution, reduces biodiversity in areas far from the original sites of application, and eventually produces severe ecological effects. Biological control by natural enemies may be more environment-friendly, but unfortunately it is not an effective method in practice due to a number of limiting factors (Steedman, 1990).

Only methods able to control both swarming and solitary locusts efficiently would eventually resolve the problem. This necessitates a fuller understanding of the population biology of the Desert Locust, including population genetic structure and dynamics, gene flow between populations, evolution and geographic origin of swarms of which we are still largely ignorant. Modern DNA and polymerase chain reaction (PCR) technologies provide powerful tools to address such questions. In this paper, we discuss the possibility of using various molecular techniques for population genetic and ecological studies of agricultural pests, and present results obtained by applying some of the techniques to the Desert Locust.

9.2 MOLECULAR MARKERS AND PCR-BASED DNA TECHNOLOGY: TECHNICAL CONSIDERATIONS

The recently developed PCR-based DNA technology is both very sensitive and efficient. It can start from trace amounts of initial materials and partly degraded DNA samples, which greatly facilitates sample collection and allows the use of museum specimens or historic relics in the analysis. It also makes possible the rapid analysis of a large number of samples at the individual level. A number of PCR-based techniques have been increasingly used to detect genetic polymorphism, viz: RAPD (random amplified polymorphic DNA), PCR–SSCP (single-strand conformation polymorphism of PCR products), microsatellite analysis, PCR–RFLP (restriction fragment length polymorphism of PCR-amplified DNA), and DNA sequencing. Their applicability depends largely on the organism studied and the objective of the study.

9.2.1 RAPD technique

The RAPD reaction employs an arbitrary oligonucleotide primer (often a 10-base oligonucleotide) to amplify small inverted repeats scattered throughout the genome (Welsh and McClelland, 1990; Williams *et al.*, 1990). Often, several discrete DNA products result which are individual-specific. RAPD primers are commercially available, the techniques are easy to learn and apply and RAPD appears to be the simplest method to reveal genetic polymorphism in populations. However, RAPDs are very sensitive to minute changes in reaction conditions, and the sources of *Taq* polymerase, DNA extraction protocols, and the type of DNA thermal cycler all affect RAPD results (for reviews, see Hadrys *et al.*, 1992; Black, 1993; Penner *et al.*, 1993; Williams *et al.*, 1993). Consequently, it is regularly emphasized that stringent standardization should be carried out before they can be used reliably and consistently. Moreover, RAPD markers are amplified anonymous bands; thus, bands that apparently comigrate in different individuals may not be identical in size, and size-identical co-migrating bands may not be homologous; potentially, this introduces artefacts. It is also important to note that only high-molecular weight DNA should be subject to RAPD analysis. This may be a limiting factor for some organisms for which sample collection and preservation are problematic.

9.2.2 SSCP technique

SSCP is a molecular scanning method for rapid detection of nucleotide changes present within DNA molecules (Orita *et al.*, 1989). This technique is based on the principle that single-stranded DNA molecules take up sequence-specific secondary structures under non-denaturing conditions (called conformers). Molecules differing by as little as one single base substitution may form different conformers. These different conformers migrate at different rates in a non-denaturing separation gel and can thus be distinguished. PCR-based SSCP has been widely used in medicine for diagnostic detection of single-base mutations. However, it has been reported that SSCP has limited sensitivity to detect single base substitution (Sheffield *et al.*, 1993; Vidal-Puig and Moller, 1994), that is, it cannot detect all mutation changes, particularly if very short or large DNA fragments are investigated (Sheffield *et al.*, 1993). Multiple analyses using different and delicately controlled conditions are necessary to detect all mutations. Multiple mutations in a DNA molecule are more likely to change its single-strand conformation and are easier to detect by SSCP. However, any mutations failing to change the secondary structure cannot be detected; and it is not known whether bands with identical mobility shift could always represent identical mutations of the target molecules. The sensitivity of this technique and difficulty for analysing large fragments could be the limiting factors for their use in molecular population studies.

9.2.3 Microsatellite analysis

Microsatellite sequences (simple sequence repeats) consist of tandem repeats of short nucleotide units, each generally less than 6 bp in length, such as $(CA)_n$ or $(AAT)_n$. Such sequences are abundant in eukaryote genomes. These sequences are highly polymorphic probably due to high mutability of the repeat during replication. The flanking regions of these repeats are in general unique sequences and appear to be conserved in closely related species. Primers can be designed in these flanking regions for PCR isolation of microsatellite sequence. Therefore, they could well serve as good molecular markers for ecological and population studies (for reviews, see Tautz, 1989; Bruford and Wayne, 1993; Queller *et al.*, 1993). The degree of variability of microsatellite markers at the population level, whether it is just right or too much, is currently being investigated in a few organisms. However, to set up this technique for a new project, considerable work by an experienced person is needed to construct a partial genomic library, screen it with simple repeat oligos, and perform DNA sequence analysis, primer design and testing; all of which are quite time consuming. Such a technical prerequisite may limit its use in some laboratories.

9.2.4 PCR–RFLP analysis

This is a simple, economic and efficient method for detecting polymorphism, provided that good markers have already been found. However, a considerable amount of effort and expense is first required to identify highly polymorphic DNA markers, especially for single-copy nuclear ones; in general, a genomic library is needed for isolation of such markers (see Karl and Avise, 1993, for more detail). Second, the variability revealed by PCR–RFLP analysis may not be enough for addressing questions of interest, particularly if the differences are small and the organisms are highly migratory or frequently inbred. Nevertheless, with the invention of 'long PCR' technology (Barnes, 1994; Cheng *et al.*, 1994), which allows us to amplify DNA fragments of more than 10 kb, PCR–RFLP will certainly become one of the best methods to study population genetic variation.

9.2.5 DNA sequencing

The methods described above can only detect a fraction of variation present in DNA as markers, and thus only provide limited information. DNA sequence analysis allows one to reveal all genetic variation present in a DNA fragment, and thereby gives the highest resolution without ambiguity. The strategy for this technique is to amplify DNA markers of interest by PCR, then determine their nucleotide sequence from amplified products. DNA sequence data obtained are then analysed by various computer programs to generate a phenogram or phylogenetic tree. In animals, mitochondrial DNA and single-copy nuclear DNA (scnDNA) have been widely used for this practice (for reviews, see Avise *et al.*, 1987; Moritz *et al.*, 1987; Harrison, 1989; Simon, 1991 for mtDNA; Karl and Avise, 1993 for scnDNA). For mtDNA, conserved (universal) primers for PCR amplification and DNA sequencing have been identified in both vertebrates and insects (Kocher *et al.*, 1989; Simon *et al.*, 1991; Lunt *et al.*, in press; R.G. Harrison, personal communication). For single-copy nuclear DNA, it seems more difficult to develop universal primers for different organisms as compared with mtDNA. In general, a particular marker must be found and individual primers designed for it. The single-copy regions most likely to be highly polymorphic at a population level are nuclear non-coding regions, including introns, 5′ non-coding region and 3′ non-coding region of a gene, and intergenic regions along the chromosomes.

This is because these regions are more likely to be selectively neutral, and thereby have a faster rate of evolution.

DNA sequence analysis is more laborious and expensive than the other techniques discussed earlier, but provides the most precise information and the highest resolution. Nevertheless, it is usually a necessary step leading to the other simple, effective and inexpensive techniques. For example, once a highly variable DNA marker has been found by sequence analysis, the position of polymorphic sites and the restriction enzyme polymorphic sites can easily be worked out, and PCR–RFLP or SSCP methods can then be rapidly employed to screen a large number of samples for polymorphism.

Despite the mitochondrial genome coding for some 37 genes, mtDNA represents only a single allele. Both the maternal inheritance and the haploid nature of mtDNA reduce its effective population size, and thus its diversity in a population is much less sensitive to immigration and more sensitive to population bottlenecks (see later for examples). This is because only female immigrants are likely to contribute their mitochondrial genetic diversity to the recipient population; the size of populations going through bottlenecks is greatly reduced, and as the effective population size of mtDNA is only one-quarter that of nuclear DNA, mtDNA diversity will consequently be four-fold reduced compared with nuclear DNA diversity. Most importantly, genetic diversity revealed by mtDNA analysis does not necessarily represent the genetic diversity of nuclear DNA, and the large nuclear genome contributes most of the genetic diversity of an organism. Therefore, analysis of mtDNA diversity alone is not really enough except for some particular applications (such as study of species origin).

9.3 A CASE STUDY OF *S. gregaria*

In the case of the Desert Locust *S. gregaria*, since it is a highly migratory insect, population mixing may have occurred frequently. So, highly polymorphic DNA markers are required, which should be able to reveal both intra- and inter-population variation and allow us to address questions such as gene flow between populations and phylogeography of the insect across its range. RAPD analysis is not suitable in *S. gregaria*, as only non-degraded DNA can be used; and we have decided to develop microsatellite markers, mtDNA and single-copy nuclear DNA markers for this study. DNA sequence analysis of PCR-amplified markers was used to detect genetic variation in populations. Here, we summarize the main results and conclusions obtained from DNA sequence data from mitochondrial and single-copy nuclear markers. Detailed descriptions, and discussion will be published elsewhere (D-X. Zhang and G.M. Hewitt, unpublished data).

The Desert Locust samples have been collected from populations across its wide range, i.e. from Pakistan, Saudi Arabia, Eritrea, Sudan, Niger and Mauritania in the past 4 years. Some samples from Saudi Arabia were collected in 1971 during a plague and conserved in alcohol. DNA extracted from these 23-year-old samples works as well as from recently collected samples.

9.3.1 Population structure and genetic diversity

The nuclear marker used in this study (Sgnm1) is a non-coding region of an *Antennapedia*-class homeobox gene which is involved in neurogenesis and early development (Akam M., EMBL access number X73982). Partial sequence of this gene has been determined by Dawson (1989). We have designed primers to amplify a 643-bp fragment located in the presumptive 3′ non-coding region of this gene. PCR amplification and direct sequencing indicate that this 643-kb fragment is highly polymorphic. Internal primers for PCR and sequencing were then designed to increase specificity and facilitate subsequent analysis. The 290-bp fragment amplified by internal primer N694 (5′-GTTCCTTCGTTCCCAAATGTCTC-3′) and N969 (5′-TGAGGCAGAGGAGGGAACAAG-3′) was used as a diagnostic marker (named Sgnm1) to analyse different populations. Forty-four individuals from six populations (Pakistan, Saudi Arabia, Eritrea, Sudan, Niger and Mauritania) have been analysed. A large number of individuals are heterozygous for Sgnm1 marker, as two different types of DNA sequences were detected from the amplified PCR product. DNA sequence data reveal 33 haplotype variants of Sgnm1 in the 44 individuals, produced by 20 polymorphic sites. In no case were more than two haplotypes observed in any individual, showing that Sgnm1 is almost certainly a single-copy nuclear sequence.

The mitochondrial marker used in this study is the central part of the control region (also known as the A+T-rich region), which is highly variable. The strategy of analysis is as follows: the whole mitochondrial control region was first amplified using the primers J1 and J6 which are located in the flanking small rRNA gene and $tRNA^{ile}$ gene respectively (Zhang *et al.*, 1995). Internal primers G1, G2, G3 (Zhang *et al.*, 1995), G5 and L424 (D-X. Zhang and G.M. Hewitt, unpublished data) are then used to sequence the central highly variable region (about 450 bp). Fifty-six individuals from the above-mentioned six populations have been analysed, including the 44 individuals analysed using the nuclear marker Sgnm1 (see above). Seventeen mtDNA types have been identified from DNA sequence data.

Phylogenetic analyses of these data have been carried out using distance, parsimony and maximum likelihood methods (Programs Fitch/Neighbor, Dnapars and Dnaml respectively, PHYLIP 3.5; Felsen-

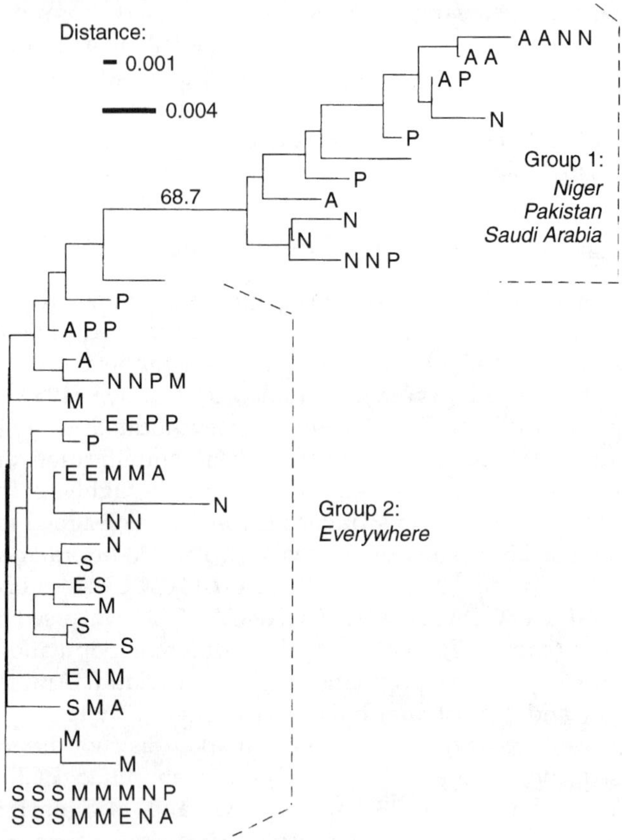

Figure 9.1 Unrooted distance tree produced by the program Neighbor (PHYLIP 3.5) from nuclear DNA (Sgnm1) sequence data, showing two distinct genetic groups. P, A, E, S, N and M represent samples from Pakistan, Saudi Arabia, Eritrea, Sudan, Niger and Mauritania respectively. Branch lengths are approximately scaled to DNA distance. The percentage bootstrap proportion for the branch leading to a distant cluster is 68.7.

stein, 1993). All three methods produce very similar phylogenetic trees. Figure 9.1 shows an unrooted distance tree generated by the program Neighbor81 in the PHYLIP package (V3.5) from nuclear DNA sequence data; Figure 9.2 shows another unrooted neighbor-joining distance tree produced from mtDNA sequence data. These phylogenetic trees were produced from a DNA distance matrix computed directly from sequence data using the above mentioned program (DNA distance is a measure of nucleotide divergence between pairs of DNA sequences; refer to PHYLIP documentation for details).

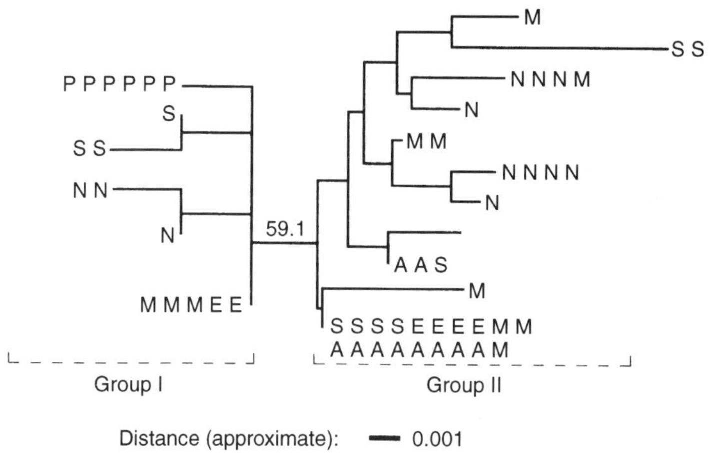

Figure 9.2 Unrooted distances tree produced by the program Neighbor (PHYLIP 3.5) from mtDNA sequence data, showing two distinctive groups. P, A, E, S, N and M represent samples from Pakistan, Saudi Arabia, Eritrea, Sudan, Niger and Mauritania respectively. Branch lengths are approximately scaled to DNA distance. The percentage bootstrap proportion for the branch leading to a distant cluster is 59.1.

Nuclear DNA sequence data clearly show the existence of two distinct genetic groups (lineages) (Figure 9.1). Group 1 consists only of samples from Pakistan, Saudi Arabia and Niger; group 2 contains samples from everywhere (all six populations studied, Pakistan, Saudi Arabia, Eritrea, Sudan, Niger and Mauritania). Bootstrap analysis (see PHYLIP documentation for details) was used to estimate the robustness of the phylogenetic tree, giving a proportion of 68.7%, thereby supporting strongly the presence of two distinct lineages (Figure 9.1). The evolutionary distance between the two groups is large and significant, suggesting that such a split is ancient (see section 9.3.3 for an estimation of the timescale). It is clear that the nuclear DNA diversity in Pakistan, Saudi Arabia and Niger populations is much higher than in other populations, as reflected by their containing both lineages. Statistical tests (Kst) using a permutation-based method (Hudson *et al.*, 1992) support such observations and show that Pakistan, Saudi Arabia and Niger populations are genetically differentiated from Eritrea, Sudan, and Mauritania populations at 1% or less level of significance. The two main lineages are genetically differentiated from each other at less than 1% level of significance.

There also appear to be two main genetic groups for mtDNA (Figure 9.2). Group I contains samples from all but the Saudi Arabia populations, and group II consists of samples from all but the Pakistan populations. The bootstrap proportion for the branch leading to a distant cluster is 59.1% (Figure 9.2). The mtDNA diversity within populations is quite different; for example, samples from the Pakistan population are all identical, samples from Saudi Arabia show very little diversity, the rest (especially Sudan and Niger) show a relatively high mtDNA diversity. The same permutation based statistical test used for the nuclear data (Hudson *et al.*, 1992) shows that Pakistan and Niger populations are genetically significantly differentiated from the other populations.

Both the nuclear and mitochondrial markers are able to detect genetic variation within and between populations. However, the nuclear marker Sgnm1 reveals a much higher genetic diversity than the mtDNA marker. This could be partly because *S. gregaria* is highly migratory and populations may frequently go through bottlenecks (from swarming populations to recession populations), which tends preferentially to reduce mtDNA diversity. It is worth noting that in vertebrates (especially in mammals) mtDNA appears to evolve much faster than nuclear DNA (see Simon, 1991, for a recent review). The pattern of mtDNA diversity is also different from that of nuclear DNA; thus, the Saudi Arabia and Pakistan populations both have a very high nuclear DNA diversity but show little or no mtDNA diversity. The Niger population is the only one showing a high genetic diversity for both nuclear DNA and mtDNA

9.3.2 Gene flow between populations

A large number of individuals are heterozygous for the nuclear marker Sgnm1, having two different DNA sequences for this marker in their diploid genome. Such widespread genetic heterozygosity is probably the consequence of population hybridization due to migration. Figure 9.3 shows that several individuals from Pakistan, Saudi Arabia and Niger populations have two different haplotypes belonging to each of the two distinct genetic groups. Within each group such heterozygosities are also observed (Figure 9.3). This is a genetical reflection of the migratory habit of the Desert Locust.

No individuals from Eritrea, Sudan and Mauritania populations possess highly diverged Sgnm1 haplotypes in their genome. This may be due to sampling error (though 19 individuals from six different locations have been analysed); otherwise it might suggest that gene flow between the two ancient once-isolated genetic pools (see later) had different consequences in the different populations, or has become much weaker, or simply that frequent swarming in these areas had homogenized populations.

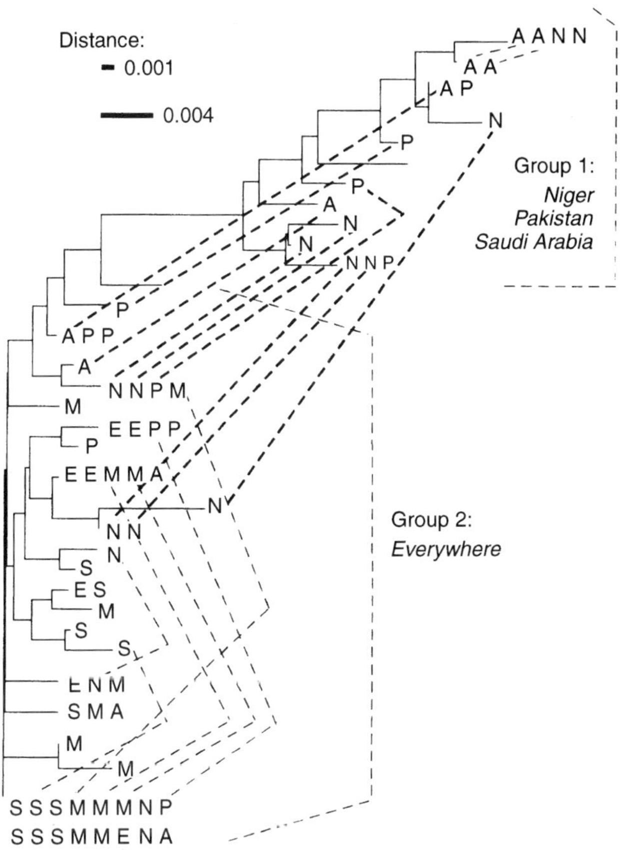

Figure 9.3 Gene flow between populations revealed by the DNA sequence data of the nuclear marker Sgnm1. The two Sgnm1 haplotypes of heterozygous individuals are connected by dashed lines. Heavy dashed lines thus indicate gene flow between the two distinct genetic groups; light dashed lines indicate gene flow within each group. See Figure 9.1 for other information.

9.3.3 Population dynamics and evolution of the Desert Locust populations

It is evident that more samples need to be analysed before firm conclusions can be drawn about the geographic distribution and evolutionary history of the Desert Locust populations. However, the present data do clearly indicate a significant genetic split and provide important evidence for the understanding of the evolution of locust populations. The Desert Locust is highly migratory. The existence of two distinct genetic groups, revealed by both nuclear and mitochondrial DNA sequence data, suggests that there was once a long-term isolation between two main population communities in the evolutionary history of the species. For the nuclear marker Sgnm1, the observed maximum nucleotide divergence between lineages is about 5.8% (estimated using Kimura's two parameter method). Sharp and Li (1989) have estimated that the silent substitution rate of *Drosophila* nuclear genes is about 16×10^{-9} per site per year. Assuming a similar molecular clock for *S. gregaria* and considering that the substitution rate of the 3′ non-coding region of a nuclear gene may be higher than that of the silent sites (Li *et al.*, 1985), we may take a value of 30×10^{-9} as the substitution rate for the nuclear marker *Sgnm1*. This would mean the two lineages of *Sgnm1* in *S. gregaria* diverged about 1 million years ago. It seems hard to explain a divergence on such a time-scale by just random drift (e.g. lineage sorting) without isolation.

The Desert Locust probably had a much smaller and limited distribution area in its history as a result of different weather conditions. The present-day widespread populations are the consequence of subsequent expansion after an extension of suitable living conditions.

In order to attempt to explain such genetic data by evolutionary processes, two initial hypotheses can be proposed. The first is the 'isolation and secondary contact of two refugial populations' model (Figure 9.4(a)). From the nuclear DNA sequence data, it seems likely that there once existed two isolated refugial population communities in the species history of the Desert Locust (assuming one was in Pakistan, i.e. Group 1 types in Figure 9.1, and another in the western Red Sea coast Sudan/Eritrea, i.e. Group 2 types). As suitable living conditions were restored, the Pakistan refugial population migrated to the west across Western Asia to Africa; the western Red Sea coast refugial population migrated to the west to other parts of Africa and to the east across Saudi Arabia to Pakistan/India. Individuals from the two refugial populations hybridized where they mixed. This could explain the high nuclear DNA diversity observed in Pakistan/Saudi Arabia populations. The absence of Group 1 Sgmn1 types in Eritrea/Sudan populations could be explained by random drift and dilution effect, since the native populations here represent a refugial population (though sampling error cannot be excluded). High genetic diversity in the Niger population could be the result of the mixing of immigrants from both refugial populations. The Mauritania population consists only of Group 2 types, which might have resulted from the fixation of immigrants from the Red Sea coast refugial population.

This model is consistent with the mtDNA data which show two distinct genetic groups and samples from Pakistan appear in only one of the groups. It agrees also with the long-term observation that Red Sea coast and the Pakistan/India boundary are two of the most impor-

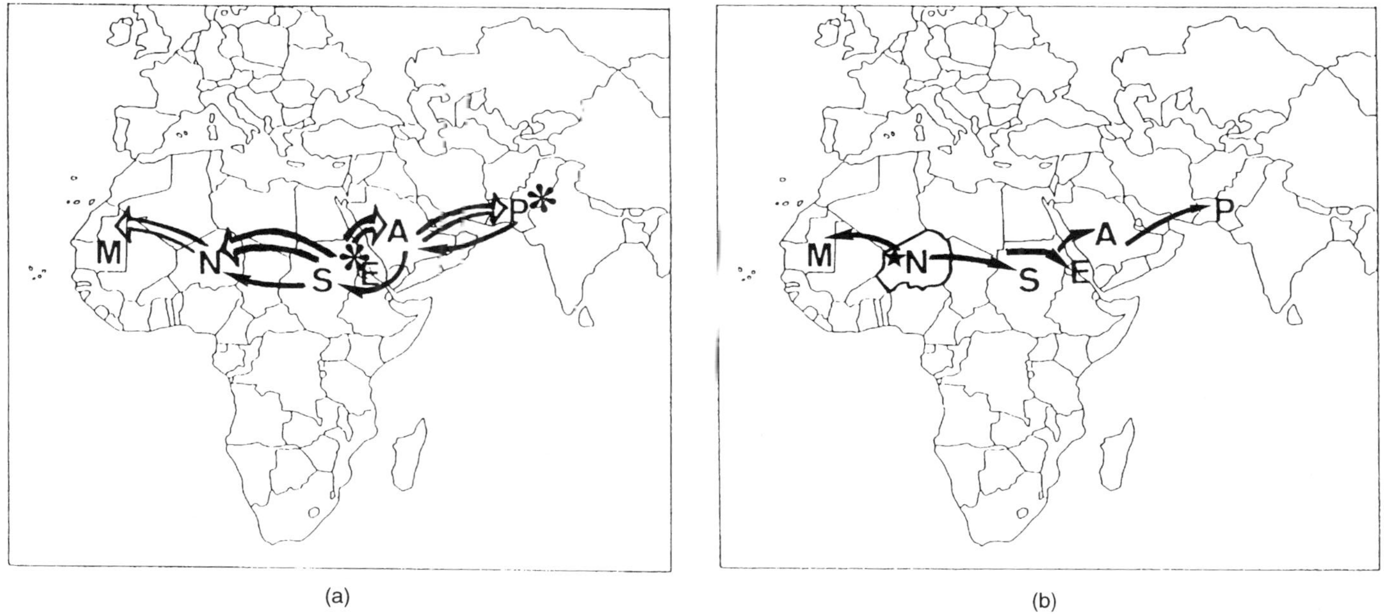

Figure 9.4 Maps schematizing the two hypothetical models on evolution of the Desert Locust populations. (a) The 'isolation and secondary contact of two refugial populations' model. (b) The 'constriction and expansion of one refugial population' model. Asterisks in the figures indicate the hypothetical refugial areas. M, N, S, E, A and P represent Mauritania, Niger, Sudan, Eritrea, Saudi Arabia and Pakistan, respectively.

tant breeding and gregariation areas of the Desert Locust during recession (Steedman, 1990).

The second hypothesis is the 'constriction and expansion of one refugial population' model (Figure 9.4(b)). This proposes that the present locust populations have not evolved directly from two refugial populations, but from a single, already diverged refugial population. As Niger is the only population showing a high genetic diversity for both nuclear and mitochondrial DNA (see earlier), it might represent an ancestral population. The ancestral population was itself a refugial population formed by constriction of already highly diverged populations. The present-day Desert Locust might have survived in Niger and spread out to other places by subsequent migrations after the restoration of living conditions.

Step-by-step migration from Niger to Sudan/Eritrea, then to Saudi Arabia, and finally to Indo-Pakistan could explain the genetic data obtained. Only a few migrants may have reached Indo-Pakistan as a consequence of such step-wise migration across the large geographic distance from Niger. These migrants would act as sequential 'founder populations', resulting in a relatively homogenized mtDNA. The same might have happened to the Saudi Arabia population. Nuclear DNA diversity in these populations could have survived in the founder population itself since it is less affected by small population size. Furthermore, subsequent immigrants may have little effect on mtDNA diversity also because the effective population size of mtDNA is much smaller than that of nuclear DNA (see section 9.2.5), and because the migrants often are composed mainly of male locusts in swarming populations (according to Popov, 1954).

The absence of Group 1 nuclear types in Sudan/Eritrea and Mauritania populations could be due to frequent swarming in these areas which may tend to homogenize populations (again, sampling error cannot be excluded).

The exact geographic sites involved in the above two models are not critical for these hypotheses and may need to be modified in the light of new data being obtained. What is important are the suggestions themselves. Also, these models are clearly quite speculative and need more evidence to support them. Nevertheless they are suggestive and may promote understanding of possible genetic and ecological processes involved in the evolution of the Desert Locust populations.

9.3.4 Alternative explanations and necessary caution in interpretating molecular data

DNA sequence data provide records of evolutionary genetic changes with the most resolution, from which DNA trees and hopefully composite trees can be constructed, allowing the reconstruction of the evolution history of populations and species. However, sampling errors (including both individual sampling error and DNA sequence sampling error) in such data may lead to misinterpretation, especially in molecular population genetics and ecological studies. Thus, the conclusions reached earlier are preliminary, being sensitive to several possible sampling errors. Consequently, alternative explanations exist regarding different genetic diversity observed in various populations, the distribution of different genetic lineages identified and the hypotheses about the evolution of the Desert Locust populations. For example, the absence of the nuclear lineage 1 from Eritrea, Sudan and Mauritania samples could simply be due to sampling error or local extinction, rather than to its real absence in these regions; the reduced mtDNA diversity observed in some localities (Pakistan and Saudi Arabia) may result from a female line sampled from small populations. Some genetic types may have a low frequency in some populations; therefore, analysis of small size samples is less likely to reveal their presence.

The regions of DNA chosen as markers could also bias the conclusions and make interpretation diffficult, particularly if they are not representative and only one or few different markers are employed. For instance, observed mtDNA lineages could be explained as a result of lineage sorting if no nuclear data were available, and no firmer conclusions could be made about the evolution of locust populations. In our case, the nuclear data independently reveal a divergence on a million-year time-scale and, though this does not exclude possible lineage sorting of mtDNA in populations, it supports our hypothesis for isolation as the main cause of divergence in the Desert Locust. In population biological studies employing DNA markers, it is crucial to carry out a large systematic study using several independent DNA markers and representative samples to produce a reliable answer.

9.4 CONCLUDING REMARKS

DNA sequence data demonstrate that both mitochondrial and nuclear DNA markers can be used to investigate population structure and genetic diversity of the Desert Locust, and when they are employed together a clearer answer is obtained. Our data reveal an ancient genetic split of the Desert Locust population, suggesting that there was once a long-term isolation between two main population communities in the evolutionary history of the species probably due to a period of unsuitable weather conditions during which the Desert Locust had a much smaller and more limited distribution range. The present-day widespread populations are the consequence of subsequent expansion after the restoration of suitable living conditions for locusts. Two initial models have been proposed to explain this evolutionary process based on nuclear and rnitochondrial DNA sequence data and field observation data. One is the 'isolation and secondary contact of two refugial populations' model; the other is the 'constriction and expansion of one refugial population' model. Both models assume that the data obtained here are representative, and that little sampling error is involved.

The conclusions obtained in this study are also suggestive for locust control. First, genetic diversity exists in every population studied and two deeply diverged nuclear types have been identified. This might be of interest for chemical control in terms of drug resistance in the Desert Locust. Second, the relatively homogenized genetic compositions observed in Sudan/Eritrea and Mauritania populations suggest that these areas are frequent swarming sites. If this relationship is confirmed by future study, it would clearly be useful in developing a control strategy. Third, a long period of divergence in Desert Locust history is demonstrated by DNA data, very probably caused by the isolation of two main population communities as a result of environmental changes. This is of particular importance for understanding how the Desert Locust populations have evolved to their present state, and the two proposed models are strongly suggestive of how genetical data may clarify ecological control of the Desert Locust.

However, it is important to note that the conclusions reached here are preliminary ones, for the following reasons: (i) some sampling error is involved due to the limited number analysed and possibly unrepresentative collection, both of which could bias the observed data; (ii) only two different DNA markers are used here: clearly a larger systematic study using several independent DNA markers would provide a more reliable answer. Therefore, further investigation using different nuclear markers and more representative samples is needed to test the assumptions and hypotheses proposed in this study.

Acknowledgements

We are grateful to M. Abdallahi, Y. Ashour, S. Bamofleh, H.D. Brown, S.E.L. Bashir, D. Batt, C.F. Dewhurst, H. Dobson, K. Ibrahim, N. Jago, G. Popov, C. Prior, J. Roffey, S. Simpson, R. Skaf and D. Zelazny for collecting and providing *S. gregaria* samples, and K. Bradley and M.F. Khan for technical help. We thank very much K. Ibrahim for helpful discussion and an anonymous reviewer for valuable comments on the manuscript. This work was supported by a grant from the United Nations Development Program (UNDP).

REFERENCES

Avise, J.C., Arnold, J., Ball, R.M., Bermingham, E., Lamb, T., Neigel, J.E., Reeb, C.A. and Saunders, N.C. (1987) Intraspecific phylogeography: The mitochondrial DNA bridge between population genetics and systematics. *Annual Review of Ecology and Systematics*, **18**, 489–522.

Barnes, W.M. (1994) PCR amplification of up to 35-kb DNA with high fidelity and high yield from λ bacteriophage templates. *Proceedings of the National Academy of Sciences, USA*, **91**, 2216–20.

Black, W.C. (1993) PCR with arbitrary primers: approach with care. *Insect Molecular Biology*, **2**, 1–6.

Bruford, M.W. and Wayne, R.K. (1993) Microsatellites and their application to population genetic studies. *Current Opinion in Genetics and Dvelopment*, **3**, 939–43.

Centre for Overseas Pest Research (1982) *The Locust and Grasshopper Agricultural Manual*, Centre for Overseas Pest Research, London, pp. 305–16.

Cheng, S., Fockler, C., Barnes, W.M. and Higuchi, R. (1994) Effective amplification of long targets from cloned inserts and human genomic DNA. *Proceedings of the National Academy of Sciences, USA*, **91**, 5695–9.

Dawson, I.A. (1989) Characterization of homeobox containing clones from the grasshopper *Schistocerca gregaria*. PhD Thesis, University of Cambridge, Cambridge, UK.

Felsenstein, J. (1993) PHYLIP 3.5 (Phylogeny Inference Package), version 3.5. Distributed by the author, University of Washington, Seattle.

Hadrys, H., Balick, M. and Schierwater, B. (1992) Applications of random amplified polymorphic DNA (RAPD) in molecular ecology. *Molecular Ecology*, **1**, 55–63.

Harrison, R.G. (1989) Animal mitochondrial DNA as a genetic marker in population and evolutionary biology. *Trends in Ecology and Evolution*, **4**, 6–11.

Hudson, R.R., Boos, D.D. and Kaplan, N.L. (1992) A statistical test for detecting geographic subdivision. *Molecular Biology and Evolution*, **9**, 138–51.

Karl, S.A. and Avise, J.C. (1993) PCR-based assays of Mendelian polymorphisms from anonymous single-copy nuclear DNA: techniques and applications for population genetics. *Molecular Biology and Evolution*, **10**, 342–61.

Kocher, T.D., Thomas, W.K, Meyer, A., Edwards, S.V., Pääbo, S., Villablanca, F.X. and Wilson, A.C. (1989) Dynamics of mitochondrial DNA sequence evolution in animals: amplification and sequencing with conserved primers. *Proceedings of the National Academy of Sciences, USA*, **86**, 6196–200.

Li, W.-H., Luo, C.-C. and Wu, C.-I. (1985) Evolution of DNA sequences, in *Molecular Evolutionary Genetics*, (ed. R.J. MacIntyre), Plenum, New York, pp 1–94.

Lunt, D.H., Zhang, D.X., Szymura, J.M. and Hewitt, G.M. (in press) The insect Cytochrome Oxidase I gene: evolutionary patterns and conserved primers for phylogenetic studies. *Insect Molecular Biology*.

Moritz, C., Dowling, T.E. and Brown, W.M. (1987) Evolution of animal mitochondrial DNA: relevance for population biology and systematics. *Annual Review of Ecology and Systematics*, **18**, 269–92.

Orita, M., Iwahana, H., Kanazawa, H., Hayashi, K. and Sekiya, T. (1989) Detection of polymorphisms of human DNA by gel electrophoresis as single-strand conformation polymorphisms. *Proceedings of the National Academy of Sciences, USA*, **86**, 2766–70.

Penner, G.A., Bush, A., Wise, R. *et al.* (1993) Reproducibility of random amplified polymorphic DNA (RAPD) analysis among laboratories. *PCR Methods and Applications*, **2**, 341–5.

Popov, G. (1954) Notes on the behaviour of swarms of the Desert Locust (*Schistocerca gregaria* Forskål) during oviposition in Iran. *Transactions of the Royal Entomological Society of London*, **105**, 65–77.

Queller, D.C., Strassmann, J.E. and Hughes, C.R. (1993) Microsatellites and kinship. *Trends in Ecology and Evolution*, **8**, 285–8.

Sharp, P.M. and Li, W.H. (1989) On the rate of DNA sequence evolution in *Drosophila. Journal of Molecular Evolution*, **28**, 398–402.

Sheffield, V.C., Beck, J.S., Kwitek, A.E., Sandstrom, D.W. and Stone, E.M. (1993) The sensitivity of single-strand conformation polymorphism analysis for the detection of single base substitutions. *Genomics*, **16**, 325–32.

Simon, C. (1991) Molecular systematics at species boundary: exploiting conserved and variable regions of the mitochondrial genome of animals via direct sequencing from amplified DNA, in *Molecular Techniques in Taxonomy*, (eds G.M. Hewitt, A.W.B. Johnston and J.P.W. Young), Springer-Verlag, Berlin, pp. 33–71.

Simon, C., Franke, A. and A. Martin (1991) The Polymerase Chain Reaction: DNA extraction and amplification, in *Molecular Techniques in Taxonomy*, (eds G.M. Hewitt, A.W.B. Johnston and J.P.W. Young), Springer-Verlag, Berlin, pp. 329–55.

Steedman, A. (ed.) (1990) *Locust Handbook*, 3rd edn, Natural Resources Institute, Chatham.

Tautz, D. (1989) Hypervariability of simple sequences as a general source for polymorphic DNA markers. *Nucleic Acids Research*, **17**, 6463–71.

Vidal-Puig, A. and Moller, D.E. (1994) Comparative sensitivity of alternative single-strand comformation polymorphism (SSCP) methods. *BioTechniques*, **17**, 490–6.

Welsh, J. and McClelland, M. (1990) Fingerprinting genomes using PCR with arbitrary primers. *Nucleic Acids Research*, **18**, 7213–19.

Williams, J.G.K., Kubelik, A.R., Livak, K.J., Rafalski, J.A. and Tingey, S.V. (1990) DNA polymorphisms amplified by arbitrary primers are useful genetic markers. *Nucleic Acids Research*, **18**, 6531–5.

Williams, J.G.K, Hanafey, M.K., Rafalski, J.A. and Tingey, S.V. (1993) Genetic analysis using random amplified polymorphic DNA markers. *Methods in Enzymology*, **218**, 704–40.

Zhang, D.X., Szymura, J.M. and Hewitt, G.M. (1995) Evolution and structural conservation of the control region of insect mitochondrial DNA. *Journal of Molecular Evolution*, **40**, 382–91.

10

The use of DNA analysis and the polymerase chain reaction in the study of introduced pests in New Zealand

K.F. Armstrong and S.D. Wratten

ABSTRACT

Many of the most damaging insect pests in New Zealand are non-native. Questions pertinent to effective pest management, such as their spatial and temporal distributions or accurate identification at the species and population level, can now be addressed more effectively with the recent advances in molecular biology techniques. This review focuses on those techniques based on the polymerase chain reaction (PCR) that can provide molecular markers for species with no previous gene sequence information. Discussion of the use of universal and random primers is illustrated with our own application of them to several areas of applied ecology. We have amplified nuclear ribosomal DNA in fruit fly (Tephritidae) with universal PCR primers and used restriction site analysis of the product to distinguish species. This shows promise as a diagnostic technique to identify Pacific Island and Australian species considered a potential threat to the New Zealand export fruit industry. PCR–RAPD analysis has been used in several population studies involving a range of spatial and evolutionary scales. These random genetic markers provided supporting information for the South American geographic origin of New Zealand populations of the pasture pest Argentine stem weevil, *Listronotus bonariensis* (Kuschel). They have also contributed fundamental information on the within-paddock clonal variation of New Zealand populations of the white-fringed weevil,

The Ecology of Agricultural Pests, Edited by W.O.C. Symondson and J.E. Liddell.
Published in 1996 by Chapman & Hall, London. ISBN 0 412 62190 8

Graphognathus leucoloma (Boheman). A final example is given in an investigation of their suitability to determine paternity in a population dynamics study of possums.

10.1 INTRODUCTION

New Zealand is characterized by a high level of endemism in its invertebrate fauna. However, many of the insect communities are depauperate, resulting in some niches being unfilled. Consequently, there is high likelihood of establishment by introduced species, and indeed many of the most damaging insect pests in New Zealand are non-native. Species of aphids, weevils, thrips and moths are good examples. In most cases, the source of the introduction is not known, nor is the genetic constitution of the colonists in relation to that of the original populations; although morphological and allozyme differences have been used world-wide to address these questions, these methods often do not discriminate enough. Modern molecular techniques can, however, separate isomorphic species, sub-populations, 'clones', 'biotypes', etc. Examination of spatial and temporal variation of populations at the DNA level enables us to assess relative 'fitness' and possibly forecast likely persistence of some introductions. Analyses of this type can give data which are critical to our understanding of the factors affecting rates and adaptive potential of genetic change in pests. This in turn may help us explain the successes or failures of invading pest populations, the development of insecticide resistance in pests, as well as more fundamental ideas concerning the diversity and dynamics of introduced pests and the success or otherwise of their introduced biocontrol agents.

In other cases, the origin of some of New Zealand's introduced pests is fairly well known, in terms of date of arrival/source/multiple or single arrival points/dates, etc. For instance, the rose-grain aphid (*Metopolophium dirhodum* Wlk.) is known to have arrived in 1982 (Farrell and Stufkens, 1992) but its immediate source, number of arriving insects and their clonal composition are unknown. It was a yield-reducing pest in the 1980s, but release of the parasitoid *Aphidius rhopalosiphi* De Stefani Perez from Europe appears to have kept the aphid's populations below damage thresholds (Farrell and Stufkens, 1990). In contrast, the origins of a second cereal aphid in New Zealand, *Rhopalosiphum padi* L., are unknown. In Europe, *A. rhopalosiphi* parasitises *R. padi* and can reduce its populations (Wratten and Powell, 1991). However, in New Zealand the parasitoid exerts a negligible influence on the aphid's numbers. This implicates the genetic matching of parasitoid biotypes to the aphid's clones. In the absence of a knowledge of the clonal genetics of these aphid pests in New Zealand, the reasons for their success or failure remain matters for speculation.

10.1.1 A perspective for DNA analysis in the study of pest species

The existence of heterogeneity in natural populations determines the success of a given population, or components of it, in relation to environmental variables such as weather, habitat, host and natural enemies (Jones, 1980; Wallner, 1987). For a long time behavioural, morphological and cytogenetic markers were the only means of assessing genetic polymorphism in populations. In insects these were largely restricted to laboratory stocks or model systems such as *Drosophila*. Undoubtedly there is a requirement for behavioural observation and phenotypic assessment in both phylogenetic and population studies (Heuttel, 1979), but often this does not give the resolution required. The use of protein markers for allozyme analysis was a popular approach in population studies for many years (Menken, 1989). However this technique is limiting, with some species revealing useful levels of variation, such as *Drosophila*, and others showing very little, such as those within the Hymenoptera (Graur, 1985) and Aphididae (Black *et al.*, 1992). The use of polymorphic markers at the DNA level can provide the ultimate genetic resolution and contribute supporting and/or new information on genetic variability. This can be illustrated in studies addressing the dubious assignment of biotype status. An example is that of Perring *et al.* (1993) in their investigation of the sweet-potato whitefly, *Bemisia tabaci* (Gennadius). Whitefly types A and B were morphologically indistinguishable and had been considered strains of the same species. Strain B was introduced and, in contrast to strain A, was responsible for 'over half a billion dollars damage to the US agricultural production in 1991'. Using DNA analysis (PCR–RAPD; see section 10.4) to complement mating studies and isoenzyme analysis, the results strongly indicated that the two 'types' were in fact two distinct species. Recognition of this would be critical for determining an appropriate pest management strategy which could be compromised if differences regarding, for example, life-cycles, host plant relationship, associated parasitoids, etc. were not known. Shufran *et al.* (1992) were able to resolve biotypes of the greenbug, *Schizaphis graminum* Rodani, into clones. Biotypes A–E were characterized phenotypically by their ability to kill or damage particular plant species or cultivars. Biotype E was characterized by its ability to damage wheat that contains the Amigo gene which confers resistance in the host plant against biotype C. Based on a survey of 24 enzymatic loci and four phenotypic variables, clones of biotype E were indistinguishable. However, DNA analysis based on the size of the intergenic spacer region of ribosomal DNA revealed that of the 61 clones tested, 56 were unique. Assuming that no one aphid clone is equally fit on all hosts or in all environments (Service and Lenski, 1982) then care must be taken when characterizing biotypes for which that variation is not associated with variation in other traits – in this case, possibly a spectrum of virulence within the biotype. The sensitivity of DNA markers to detect variation in previously indistinguishable aphid clones was also demonstrated by Carvalho *et al.* (1991) using a derivative of an RNA probe for DNA fingerprint analysis.

As a consequence of widely differing sequence organization, mutation rates and selective pressures, different regions within the genome display a heterogeneity of evolutionary rates (Figure 10.1). In addition, there is a variety of molecular techniques that are informative at different evolutionary scales (Table 10.1). Thus, comparative studies at the DNA level can accommodate applications at a range of taxonomic scales. Examples range from the need for biosystematics of pest organisms in sustainable agriculture (Hawksworth, 1994) to the understanding of adaptive mechanisms such as those responsible for the evolution of insecticide resistance (McKenzie and Batterham, 1994). Recent advances in molecular genetics have led to the development of many new techniques useful to the ecologist. Even a complete dearth of gene sequence data for the organism of interest need no longer be a barrier to obtaining information at the DNA level. This review focuses on techniques using the polymerase chain reaction (PCR) which are of benefit to the ecologist. Genetic markers, over a range of taxonomic scales, can

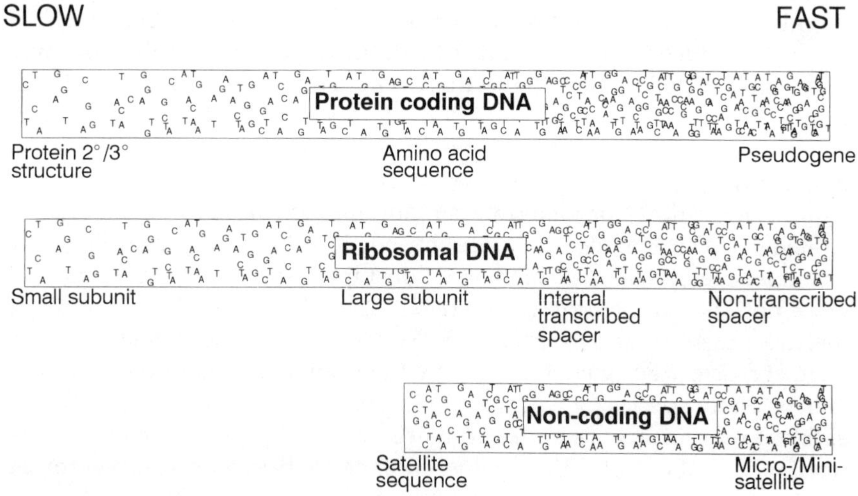

Figure 10.1 Relative rates of DNA sequence evolution. (Redrawn from Burke *et al* ., 1992.)

Table 10.1 Molecular methods providing informative genetic markers at different evolutionary levels

Genetic level	*DNA–DNA hybridization*	*Restriction site/fragment length*				*DNA sequencing*
		rDNA	*mtDNA*	*scnDNA*	*VNTR loci*	
Genetic identity	–	–	*	*	**	*
Parentage	–	–	*	**	**	*
Conspecific population	–	–	**	**	*	*
Closely related species	*	**	*	*	–	*
Intergenic	**	**	–	–	–	**
Interfamily	*	*	–	–	–	**

*, Useful; **, very useful.
mtDNA, mitochondrial DNA; rDNA, ribosomal DNA; scnDNA, single-copy nuclear DNA; VNTR, variable number of tandem repeats.

now be obtained for use with species, typical of agricultural pests, that are currently poorly defined at the DNA level.

10.1.2 The polymerase chain reaction (PCR)

Automation of the PCR (Mullis and Faloona, 1987) has revolutionized techniques in molecular biology. The principle of the reaction and functions of the component reagents have been presented in depth elsewhere (e.g. Taylor, 1993), as has the breadth of its application (e.g. Innis *et al.*, 1990; McPherson *et al.*, 1991; Erlich and Arnheim, 1992). In brief, the basis of the reaction is an enzyme-catalysed synthesis of small regions (typically <3000 base pairs (bp)) of DNA template. The region(s) amplified is defined by a pair of primers (typically 10–30 bases of synthetic oligonucleotide) whose sequences are complementary to those flanking a specific or anonymous piece of DNA. A heat-stable enzyme, usually *Taq* polymerase, catalyses the extension of these primers along the length of DNA by joining nucleosides (dinucleotriphosphates, dNTPs) complementary to the target sequence. One reaction cycle consists of three temperature-dependent components: denaturation of the DNA template at 95°C, annealing of the primers to complementary sites on the single stranded template at a temperature usually between 35–65°C and then extension of the primers with the dNTPs and DNA polymerase at 72°C. Each cycle effectively synthesises an exact copy of the original template, though the fidelity of the enzyme may introduce errors. For *Taq* polymerase one incorrect base may be placed in every 500 000 (McPherson *et al.*, 1991). The DNA is amplified geometrically such that, theoretically, over 20 cycles the original DNA template is increased a million-fold. Thus, amplification of extremely small (pmol) amounts of DNA can result in enough PCR product to: (i) visualize directly on an electrophoresis gel without the need for time-consuming and expensive radioactive or non-radioactive labelling techniques; and (ii) use for subsequent manipulations such as digestion with restriction enzymes (Rolfs *et al.*, 1992) or sequence analysis (e.g. Thomas and Kocher, 1993).

The apparent simplicity of the PCR technique (Taylor, 1993) does, however, belie a very complex reaction (Sardelli, 1993). Careful optimization (section 10.2.3) is recommended as target amplification is sensitive to PCR reagent concentrations (Muralidharan and Wakeland, 1993) and components of cellular and DNA preparation (Rolfs *et al.*, 1992). Despite this, however, the speed and technically undemanding nature of the process makes it an attractive starting point for many studies, and as with any new technology the more it is applied the more quickly it can be understood and refined.

The principal requirement for a PCR reaction is access to sequence(s) from which to design primers that flank the DNA region of interest. For many applications, primer sequences of 20–30 bases can be designed from existing species-specific gene sequence data. However, often very little or nothing is known of the gene sequence data for a species of ecological or agricultural interest. In this situation there are two possibilities: one is the use of 'universal' or 'versatile' primers and another is the use of random-sequence primers. This review focuses on the use of these primers in applying the PCR to studies of agricultural pests.

10.2 GENERAL METHODOLOGICAL CONSIDERATIONS

10.2.1 Sample collection and preservation

A major attraction of PCR-based DNA analysis for ecological application is that it is economical in the amount (mg) of tissue required. Providing there are a few intact copies of the region of interest (normally around 500 to 3000 bp long), then even badly degraded DNA is sufficient for many purposes. Thus, in contrast to the samples required for protein analysis or DNA fingerprinting, benefits to the ecologist are that relatively poor quality or precious samples are suitable for DNA analysis. Consequently, field collection is feasible without the need for freezing facilities and dried or alcohol-preserved samples stored at room temperature are satisfactory (Smith *et al.*, 1987; Bramwell and Burns, 1988; Kocher *et al.*, 1989; Oakenfull, 1994). One should be aware, however, that fixation procedures involving excessive drying of tissue, such as long-term storage in absolute ethanol or air, could encourage irreversible binding of some proteins to DNA. This in turn may inhibit subsequent enzymatic manipulations such as polymerase amplification or restriction digests (J. Hickford and D. Blunt, personal communication). Our own experience with long-term storage of vertebrate and invertebrate tissue in alcohol, however, has not presented any problems with subsequent PCR amplification using either specific (section 10.3.1) or random (section 10.4.1(b)) primers.

Nonetheless, just as the sensitivity of the technique permits minimum care with the target sample, caution should be taken with respect to contamination of that sample with, for example, parasites or microorganisms. This is especially important when using universal or random primers, which of course are designed to be compatible with a large range of species (sections 10.3 and 10.4, respectively). Unless preliminary experiments can show that either the PCR primers are highly specific for the target organism, or that the concentration of target DNA significantly outweighs that of the contaminating DNA, precautions may need to be considered. In our work at Lincoln on introduced aphid pests we rear all field-collected aphids individually in the laboratory and exclude those containing parasitoid (Hymenoptera: Aphididae) larvae as parasitism becomes apparent. This is also done by collaborators in the UK (De Barro *et al.*, 1994). Alternatively, dissection may be required to remove contaminating tissue, or the use of life-stages that are not vulnerable to parasitism could be screened. The latter has been a consideration in our studies of white-fringed weevil; adults have been used preferentially over larvae due to the common parasitism by nematodes in the latter.

10.2.2 DNA extraction

For PCR applications DNA may be obtained from a variety of sources including other than fresh, frozen or ethanol preserved specimens (section 10.2.1). For example, crude tissue digestions such as that described for the scrapings obtained from butterfly wings (Rose *et al.*, 1994), protein extracts intended for allozyme analysis (Tan and Orrego, 1992) or whole-blood samples (Mercier *et al.*, 1990). Extraction can be achieved by a variety of methods. The 'cleanest' DNA of high molecular weight (ca. 50 kilobases (kb)) is preferable as it limits problems with subsequent molecular manipulation. Phenol extraction (Sambrook *et al.*, 1989) is relatively gentle and results in high-molecular weight DNA: several repeat extractions remove all protein and any likely inhibitors of the PCR (Rolfs *et al.*, 1992). However, the use of toxic chemicals for organic extraction is cumbersome when dealing with a large number of samples. Also a substantial proportion of DNA can partition into the phenol (Tower, 1991) so the technique may be unsuitable for very small samples. For population studies, alternative non-toxic methods, such as the salting-out of cellular protein from tissue digests using a saturated sodium chloride solution, can be considered (Miller *et al.*, 1988; Montgomery and Sise, 1990). Few manipulations are involved in this case so high-molecular weight DNA can be obtained. We routinely use this high-salt method of extraction in our laboratory (including the weevil and possum projects described later) and achieve high-molecular weight DNA (>40 kb) and absorbance ratios (A_{260}/A_{280}) of 1.5–1.9. Another method of obtaining 'clean' DNA is the use of commercial kits based on the specific binding of DNA to a matrix. For good quality DNA this may compromise molecular weight by shearing long strands, but for very small or highly degraded samples this can be a 'safe' alternative method to organic or high-salt extraction. We routinely use the Prep-a-gene kit (Bio-Rad Laboratories) for poor quality fruit fly specimens and obtain DNA of 10–20 kb, which is more than adequate for the majority of PCR applications.

10.2.3 PCR conditions and optimization

There are many publications available that discuss manipulation of the PCR components and method (e.g. Erlich, 1989; Innis *et al.*, 1990; Mullis *et al.*, 1994) and this will not be considered in depth here. Standard procedure for our own work described in this review included all reactions set up in a laminar flow cabinet, from aliquoted reagents and with dedicated equipment, to reduce the chance of contamination. 25 μl reactions were composed of *Taq* polymerase buffer (10 mM Tris–HCl; 2.5 mM $MgCl_2$; 50 mM KCl; pH 8.3), 1 unit *Taq* polymerase (Boehringer-Mannheim), 100 μM each dNTP, 5 pmol each primer, 0.5–25 ng DNA and overlaid with 20 μl oil. All reactions were carried out on a Perkin–Elmer Cetus 480 thermal cycler and employed the hot start technique (D'Aquila *et al.*, 1991). The cycling regime consisted of an initial denaturation step of 95°C for 5 min followed by addition of the *Taq* polymerase (2 μl of a diluted stock) at 72°C and then either (i) 35 cycles of 1 min 95°C, 1 min 50°C and 2 min 72°C with universal primers, or (ii) 40 cycles of 30 s 95°C, 30 s 35°C and 1 min 72°C for randomly amplified DNA, and then a final extension step at 72°C for 5 min. Any modifications to this protocol are included in the appropriate sections. Mindful of the sensitivity of the technique, a negative control (without DNA template) was included with all reactions to monitor for contamination. Preliminary experiments with a new species/primer combination also included a positive control tube with known template giving an expected product.

Optimization of the PCR components is a worthwhile preliminary exercise and examples from our own work illustrate this. We have found PCR with universal primers (section 10.3.1) was very tolerant of DNA concentration and a 100-fold difference (ca. 0.5–50 ng per reaction) could be amplified from different species. In contrast, optimal DNA concentration for amplification with random primers was found to be highly species-dependent: possum DNA could be amplified using 10–50 ng per reaction (25 ng was used as standard), but the same random primers would not amplify white-fringed weevil DNA at greater than 5 ng (2.5 ng per reaction was used as standard). Similarly, in an aphid/parasitoid study (unpublished) concentrations of 15–40 ng of aphid DNA were amplifiable, but with the same random primer parasitoid DNA needed to be <2 ng. During initial reactions with new species/primer combinations, the concentration of $MgCl_2$ and annealing temperature were optimized to produce positive amplification with minimal amount of non-specific products. With random primers an annealing temperature outside the 35–37°C range is often not considered but, in our experience for some species/primer combinations, temperatures up to 48°C can be useful.

The choice of primers, universal or random, for use with a new species is determined largely by trial and error. Different combinations of universal primers may need to be tested before whole or part of a gene/region can be amplified. This will depend on the occurrence of polymorphisms within the primer regions. Evaluation of random primers can involve the screening of many (often 20–200, e.g. Shoemaker *et al.*, 1994) single primer sequences to determine which produce clear, reproducible bands on an electrophoresis gel. Informative banding patterns will depend on the application (refer to section 10.2.5). Further information about these two groups of primers, and on those specifically used in our own projects (Table 10.2), is given in sections 10.3 and 10.4.

10.2.4 Visualization of amplified products

Much of the decision about the usefulness of a PCR technique for a particular study can depend on the physical separation and visualization of the PCR products. Choosing an appropriate separation media and staining technique may involve a balance between being satisfied that PCR products of similar size can be sufficiently distinguished, such that variation is not overlooked (Pedley *et al.*, 1988), and avoiding too much 'noise'. Several methods are available which offer a range of resolution. The most popular method, using separation by agarose gel electrophoresis, staining with ethidium bromide and visualization over UV (Sambrook *et al.*, 1989) is relatively cheap, quick and simple. Greater

Table 10.2 Primers useful for PCR amplification in the New Zealand studies of this review

Species	*Target*	*Primer*	*Sequence*	*Source*
Fruit fly	18S	18SA	AACCTGGTTGATCCTGCCAGT	Sogin (1990)*
		18SB	TGATCCTTCTGCAGGTTCACCTAC	
	18S/ITS1/ 5.8s/ITS2	ITS4	TCCTCCGCTTA(G/T)T(T/G)ATATGC	Stiller and Waaland, 1993
White-fringed weevil	Random	OPA-01	CAGGCCCTTC	Operon Technologies Inc., Alameda, CA 94501, USA
		OPA-02	TGCCGAGCTG	
		OPA-07	GAAACGGGTG	
		OPB-01	GTTTCGCTCC	
		OPB-07	GGTGACGCAG	
Possum	Random	OPA-09	GGGTAACGCC	Operon Technologies Inc., Alameda, CA 94501, USA
		OPA-10	GTGATCGCAG	
		OPB-08	GTCCACACGG	
		OPB-15	GGAGGGTGTT	
		OPK-03	CCAGCTTAGG	
		OPK-16	AGCCAGCGAA	

*Primers without polylinkers.

resolution can be achieved with polyacrylamide gel electrophoresis (PAGE) and better sensitivity with silver stain (Caetano-Anollés *et al.*, 1991) or radioactive/non-radioactive labels. Application of high-resolution techniques used in combination with PCR, including denaturing gradient gel electrophoresis (DGGE), temperature gradient gel electrophoresis (TGGE) and single-strand conformational polymorphism (SSCP) have been reviewed by Lessa and Applebaum (1993). For our own work (see sections 10.3 and 10.4), separation with agarose (SeaKem LE, 1–2%) and ethidium stain was sufficient for most applications, but the use of PAGE (PhastSystem, Amrad Pharmacia) in combination with silver stain significantly improved the use of RAPDs for detecting variation between individual possums (section 10.4.1(b)).

10.2.5 Scoring and analysis of PCR products

In the case of PCR with universal primers, presence of a single band on an electrophoresis gel is used as an initial indication of positive amplification of a specific product. For comparative studies further analysis of the product, via restriction enzyme digestion or sequencing, is required. Details of these techniques can be found in texts such as Sambrook *et al.* (1989) and are not pursued here. In the case of RAPDS, PCR products are commonly scored subjectively as presence/absence of electrophoretic bands with respect to molecular weight standards and recorded in a 1/0 format. A map of relative mobility values can be generated, much the same as for allozyme analysis, but the presence of internal standard or common bands are required. However, intensity of bands has also been shown to be a useful genetic marker by Hunt and Page (1992) who illustrated that brightness was polymorphic and heritable in a RAPD analysis of *Apis mellifera* (L). Levitan and Grosberg (1993) also found this trait to be transmitted in Mendelian fashion in the marine hydrozoan *Hydractinia symbiolongicarpus*. Intensity of bands can also, however, be an artifact of the PCR and produced as a result of very minor differences between reactions (Ellsworth *et al.*, 1993). The assumption that fragments of randomly amplified DNA with the same electrophoretic mobility are the same, i.e. show sequence homology, can be an important limitation for some applications such as phylogeny reconstruction or familial relationships (Black, 1993). Homology may have to be confirmed by probing with a known PCR product (Hanboonsong, 1994) or sequence, or by comparison of sequences of excised bands. In many population studies, however, reproducibility of the product for the same DNA sample and individuals of the same group can be a useful and reliable genetic marker. Positive controls included at the time of amplification and comparison on the gel with previously amplified products can reduce ambiguity.

Different aspects of RAPD patterns can be considered informative for different applications. Conserved bands can be used as diagnostic for a population or species (Kambhampati *et al.*, 1992). Multiple banding patterns may contain several polymorphic loci useful for monitoring gene flow between populations. High-resolution applications, such as studies of familial relationships, can require a large number of polymorphic loci. These can be generated by combining the results of amplifications with different primers (Shoemaker *et al.*, 1994). Some studies have set criteria for selecting bands suitable for analysis, such as the proximity of adjacent bands, frequency of bands, linkage to sex genes and intensity of the fragment band on the gel (Ballinger-Crabtree *et al.*, 1992). Phylogenetic analysis of RAPD genetic markers can be achieved, much the same as for RFLP, allozyme or morphological data and uses both distance and parsimony methods (Black, 1993), for which computer packages such as PAUP (Swofford, 1992) and PHYLIP (Felsenstein, 1989) are available. The relative merits of different analyses for use at different taxonomic scales have been considered in several publications (Ballinger-Crabtree *et al.*, 1992; Kambhampati *et al.*, 1992; Black, 1993). For example, nearest-neighbour discriminant (non-parametric, multivariate) analysis may be useful to classify individuals according to species but not to population. Cluster analysis is a common method of summarizing relatedness between species or populations, but different methods of estimating genetic distance/similarity (e.g. similarity index, percent match, Nei's unbiased genetic distance analyses) may be more, or less, appropriate for subsequent transformation. Cluster analysis has also been useful in resolving genealogical relationships in multiple offspring breeding systems (Levitan and Grosberg, 1993).

10.3 PCR WITH UNIVERSAL PRIMERS

The sequences of universal primers are based on those of highly conserved regions of DNA, determined by comparing sequence data from an evolutionarily diverse group of organisms. They flank relatively variable stretches of DNA, of which there are many throughout the genome, characterized by differing rates of nucleotide substitution (Figure 10.1). Those evolving slowly are useful for comparing distantly related taxa and those evolving rapidly are appropriate for closely related taxa. Factors to consider for choosing a region appropriate for a particular systematic level have recently been discussed at length by Brower and DeSalle (1994). Historically, high-copy genes have been popular targets, such as those coding for mitochondrial and nuclear ribosomal RNA. RNA sequencing became feasible before DNA sequencing and, in eukaryotic cells, both gene regions are present in high copy numbers lending them well to the traditional detection by radioactive probes. These attributes, together with their versatility in terms of subregions within them evolving at different rates (Avise, 1994), have resulted in a substantial body of literature and sequence data (accessible through databases, e.g. GENBANK [Bilofsky and Burks, 1988]) for these targets in a diverse range of species.

In addition to the attributes mentioned above mitochondrial DNA (mtDNA) is maternally (and therefore clonally) inherited without recombination. A single copy is present in each mitochondrion, therefore its 'high copy' status is a reflection of the number of mitochondria per cell which will be different for different tissues. Care is required in the choice of mitochondrial gene for a particular application, as the same gene may not show the same flexibility in all organisms (Crozier, 1990; Meyer, 1994). Nonetheless, availability of universal primers makes it relatively easy to screen for informative mtDNA regions in genetically obscure species. Kocher *et al.* (1989) designed some of the earlier universal primers to amplify the regions coding for cytochrome b, 12S rRNA and the mtDNA control region. Their versatility was illustrated by amplification of homologous segments of mtDNA in over 100 different species of mammals, birds, amphibians, fish and invertebrates. Another popular set of primers were designed by White *et al.* (1990) from conserved sequences in several species of fungi; these were also shown to be useful in species of algae, plants and vertebrates. More recently, an extensive set of universal primers for mtDNA has been published by, and is available from, Simon *et al.* (1994). Emphasis is on their application to insect studies, though many have shown potential for use with a wider range of taxa. Even without access to known versatile sequences/primers, useful primers can be derived from data of distantly related species. This approach has been used by Xiong and Kocher (1991) who used primers designed from the 16S rRNA gene of *Drosophila yakuba* Burla mtDNA to amplify the same in black flies (Diptera: Simulidae) and distinguish seven morphologically indistinguishable species by sequence analysis.

Nuclear ribosomal RNA is coded for by a multigene family consisting of many copies within each cell. In contrast to mtDNA, these copies form tandem repeats found at one or very few locations on the chromosome (Lewin, 1990). In insects ribosomal DNA (rDNA) is present as 200–500 copies per nucleus (Finnerty and Collins, 1988). The repeat of an rDNA unit includes three regions coding for the 18S, 5.8S and 28S ribosomal subunits, two internal transcribed spacer regions (ITS1 and ITS2) joining them, plus an external transcribed spacer. This cluster is flanked by a non-transcribed spacer (NTS), or intergenic spacer (IGS) region (Figure 10.2). Avise (1994, pp. 75–8) has discussed the attributes and potential complexity of nuclear rDNA as a genetic marker arising from the heterogeneity of molecular mechanisms controlling evolution of the repeat elements. Hence, some parts of the subunit sequences are strongly conserved, even among distantly related species, while others are highly variable among species within a family (Vossbrinck and Friedmann, 1989). The ITS regions evolve more quickly, and the NTS even more so, demonstrating intra-individual and intra-specific polymorphism that make it unsuitable for macroevolutionary comparisons. This range of sequence variability for nuclear rDNA has been illustrated concisely by Allard and Honeycutt (1991) from restriction site and sequence data of rodents (Figure 10.2). A similar example occurs in insects where extensive and continuous variation was found in the NTS of the tiger mosquito *Aedes albopictus* (Skuse) (Diptera: Culicidae), but no variation was found in the coding regions of the 18S and 28S subunits (Black *et al.*, 1989). Thus, the most appropriate region to study

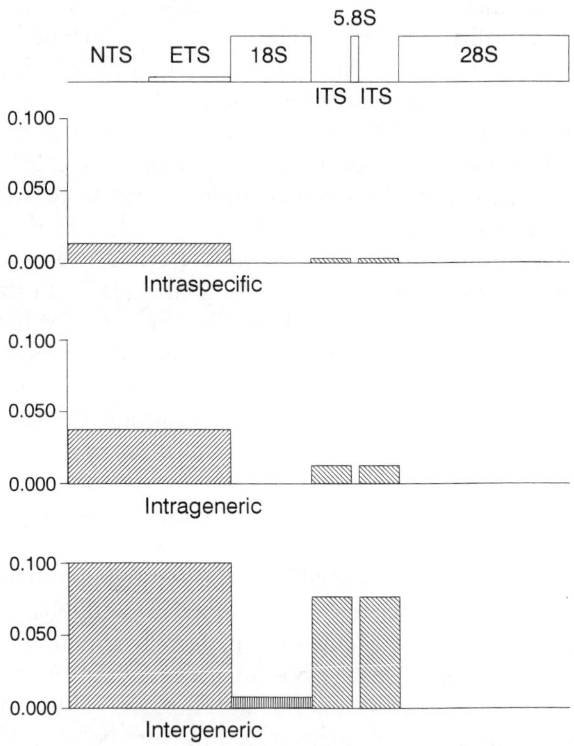

Figure 10.2 Relative levels of variation for sub-regions of the nuclear rDNA multigene family compared at three taxonomic levels. Data represent the number of nucleotide substitutions per nucleotide site in rodent spp. determined from pairwise comparisons of 86 restriction sites. ETS, external transcribed spacer; ITS, internal transcribed spacer; NTS, non-transcribed spacer. (Redrawn from Allard and Honeycutt, 1991.)

depends on the application. At the microevolutionary level, there are some interesting illustrations in studies of insects. Shufran *et al.* (1991) used the intergenic spacer as a molecular fingerprinting probe to distinguish clones of the greenbug aphid, *Schizaphis graminum* (Rondani), which had previously shown no allozyme polymorphism. Also a heterospecific probe, cloned from the tiger mosquito (*A. albopictus*), was used by Lupoli *et al.* (1990) to distinguish clones of the corn leaf aphid, *Rhopalosiphum maidis* (Fitch). In that study, probe hybridization was dependent on highly conserved regions of the two species, from different families. Location of this region in restriction digests of the genomic DNA enabled detection of the adjacent, highly variable NTS region and subsequent resolution of the aphid clones. Universal primers for PCR have, however, made nuclear rDNA much more accessible to study. A versatile set of such primers have been designed by Sogin (1990) and White *et al.* (1990). Both sets were designed from sequence data of several fungi, but are compatible with a diverse range of organisms. The whole of the gene coding for the 18S has been amplified in organisms as diverse as algae (Stiller and Waaland, 1993), yeast (Sogin, 1990), fungi (Chen *et al.*, 1992) and mammals (White *et al.*, 1990). We have also shown these primers to be useful in our own studies with insects (section 10.3.1; Table 10.2).

Heterozygosity of single-copy nuclear (scn) genes can also be useful in comparative studies. These targets were not readily accessible before PCR but have recently been considered as candidates for the design of universal primers. Slade *et al.* (1993) constructed a set of primers based on sequences of the relatively conserved exon regions of specific scn genes compared for several mammals, chicken and *Drosophila*. These were shown to prime across the potentially more variable intron (specifically the histone *H2AF* and myoglobin) or hypervariable exon (specifically the MHC *DQA*) regions of several other species. Subsequent restriction site analysis of the amplification products showed these genes to be a useful source of population-specific genetic markers. The authors discuss this system, with respect to simplicity and flexibility, for developing an 'ideal' protocol for population-level analysis. The use of scn genes for population analysis has been reviewed recently by Brower and DeSalle (1994) who also discuss the factors to consider in the design of universal primers for this type of DNA target.

10.3.1 Application of universal primers to the study of agricultural pests in New Zealand

We have applied the use of PCR towards identification of fruit fly species (Tephritidae: *Bactrocera*), in particular, those species that are native to the Australasian and Oceanian regions, and thus are a potential threat to New Zealand. New Zealand is currently the only major fruit-producing country that is free of fruit flies and their establishment there, even for a short time, would have serious consequences for the export horticultural industry. The high pest status of fruit flies has led to a significant commitment by the New Zealand Ministry of Agriculture and Fisheries to put national surveillance and eradication systems in place (Cowley and Frampton, 1989; Stephenson, 1989). There are over 4600 described species; of these, only one-third are considered pests, though they infest a wide range of fruit hosts. Some species are extremely polyphagous (e.g. Mediterranean fruit fly, *Ceratitis capitata* (Weidmann), has been reared from over 250 different hosts), though most are associated with relatively few hosts (e.g. 10–50 fruits) (White and Elson-Harris, 1992). Different species would be considered to carry different levels of 'risk' in terms of their likely establishment. Species such as *Bactrocera passiflorae* (Froggatt) originating from a broad range of climates in Australia (overlapping with that of New Zealand) or *B. tryoni* (Froggatt) and *B. xanthodes* (Broun) originating from tropical climates in Australia and Fiji are extremely polyphagous and therefore would have a high likelihood of colonisation in New Zealand in the summer months. Other species such as *B. cucumis* (French), also from tropical regions, have a much narrower host range and therefore would be considered lower risk. Correct identification of the species causing an infestation is a critical step in progressing toward an eradication program (Armstrong and Couey, 1989). However, fruit fly larvae are difficult to identify to genus, let alone species level, at any but the third instar and for many that are also part of a species complex, reliable larval characters have not been described (White and Elson-Harris, 1992). In the case of an infestation consisting of immature life-stages, normal procedure would be to rear specimens through to imagos. This could take weeks and is not always successful. Thus, in order that the most appropriate and efficient course of action can be taken, a more rapid and reliable method of species identification is needed that can ideally use minimal material (single specimens) from any stage of the life-cycle and perhaps not suitable for morphological classification.

There are several recent studies using molecular techniques for the taxonomic classification of Tephritidae spp. The higher taxonomic classification of Tephritidae has been improved using comparative sequence data for the nuclear 18S and 28S rDNA regions (Han and McPheron, 1994). Probes designed to detect repetitive regions of DNA have been used to distinguish three Hawaiian Tephritid species, the Mediterranean fruit fly, *C. capitata*, the oriental fruit fly, *Bactrocera dorsalis* (Hendel) and the melon fly, *Bactrocera cucurbitae* (Coquillett) (Haymer *et al.*, 1994). Other techniques using the PCR have addressed intra-specific population variation using primers based on species-specific microsatellite sequences (Robson *et al.*, 1994), primers based on species-specific mtDNA sequences (Steck *et al.*, 1994) and random primers (Haymer and McInnis, 1993; He and Haymer, 1994). None of these techniques is suitable to adapt as a diagnostic method for the range of species of interest to New Zealand.

To address the issue of Tephritidae speciation, we have investigated the use of PCR with universal primers to amplify rDNA for restriction fragment length polymorphism (RFLP) DNA analysis. Four species, belonging to three subgenera, have been studied: *B. (Bactrocera) tryoni, B. (Bactrocera) passiflorae, B. (Notodacus) xanthodes* and *B. (Austrodacus) cucumis*. DNA was extracted, as described in section 10.2.2, from several (ca. 10) individual, ethanol-preserved, larval specimens of each species. The universal primers, listed in Table 10.1, were used to amplify the rDNA 18S and combined 18S/ITS/5.8S regions. Using primers 18SA and 18SB, PCR conditions are given in section 10.2.3, except that optimal $MgCl_2$ was determined as 1.5 mM and ca. 0.5–5 ng DNA was used routinely. A single fragment (ca. 2 kb) was amplified from all samples (Figure 10.3(A)). A 4-μl sample of the PCR product in an 18 μl restriction digest (>5 units enzyme) revealed diagnostic patterns for the three subgenera (Figure 10.4(A)); digestion with *Rsa*I distinguished *B. xanthodes* and with *Hinf*I distinguishes *B. cucumis* from *B. passiflorae* and *B. tryoni*. The latter two species could not be separated using this method with the 18 restriction enzymes tried so far (14 of which cut the PCR product). This is perhaps not surprising as polymorphism between closely related species, such as those belonging to

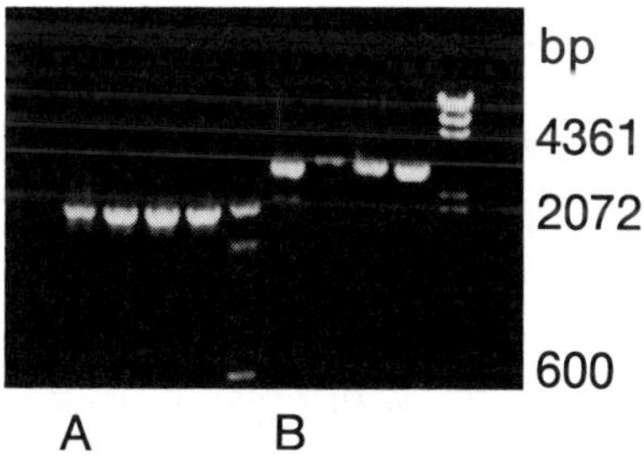

Figure 10.3 PCR amplification of fruit fly rDNA regions (A) 18S and (B) 18S/ITS1/5.8S/ITS2. Lanes 1–4 and 6–9 are *Bactrocera cucumis; B. tryoni; B. passiflorae; B. xanthodes* in order. Lanes 5 and 10 are molecular weight standards (100 bp ladder Gibco-BRL and λ*Hind*III, Boehringer-Mannheim, respectively). Submerged gel electrophoresis was carried out at 80 V for 1 h through 1.0% agarose (SeaKem LE) gel in 0.5 × TBE buffer (45 mM Tris-borate, 1 mM EDTA) containing 0.08 μg/ml ethidium bromide.

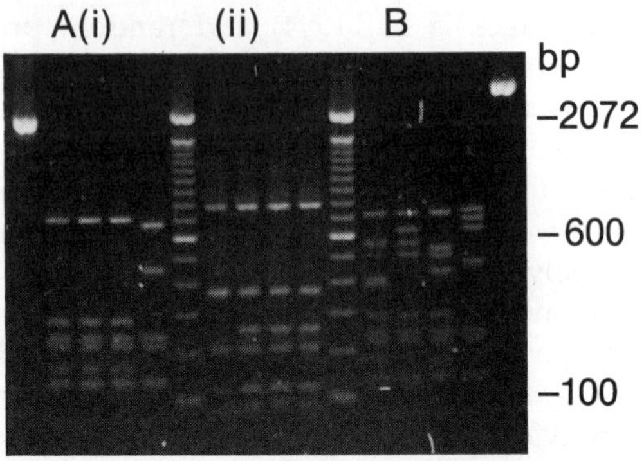

Figure 10.4 Restriction digests of fruit fly rDNA regions (A) 18S with (i) *Rsa* I and (ii) *Hinf* I; (B) 18S/ITS1/5.8S/ITS2 with *Rsa* I. Lane 1 is the undigested 18S PCR product and lane 16 is undigested 18S/ITS1/5.8S/ITS2 PCR product. Lanes 2–5, 7–10 and 12–15 are *Bactrocera cucumis; B. tryoni; B. passiflorae; B. xanthodes* in order. Lanes 6 and 11 are molecular weight standards (100 bp ladder, Gibco-BRL). Electrophoresis details are as for Figure 10.3.

the same sub-genus, would be rare within the rDNA 18S. Confronting this problem we aimed to amplify a more variable region using primers 18SA and ITS4. PCR amplification employed a different polymerase enzyme, Expand Hi Fi (Boehringer-Mannheim), which has a greater fidelity for amplifying larger targets and its exonuclease activity can improve amplification with primers for which there may be some mismatch with the priming site. Amplification conditions were as recommended by the manufacturer. A ca. 3 kb fragment (Figure 10.3B) was consistently produced. Characteristic, faint non-specific products were also amplified for *B. cucumis* which may, with further investigation, be diagnostic; this did not interfere with interpretation of restriction digests of the main product. Close examination indicated that the PCR products were not the same size across the species. This may be due to differences in the length of the ITS regions and could also be species-specific. Digestion of the PCR product with *Rsa* I, however, did show variation within the *Bactrocera* subgenus and all four species could be readily distinguished (Figure 10.4B).

Clearly this technique is useful in distinguishing the four species screened so far. It also shows promise as a diagnostic technique for this application; the entire procedure from DNA extraction to restriction profile is technically undemanding and takes less than 8 hours. Further work, such as sequence analysis, is required to confirm these PCR products as being the intended rDNA loci and more species are to be tested. In our laboratory the SSU primers A and B also amplify a similar, single ca. 2-kb fragment in aphid (*Metapolophium dirhodum* Walk.), fish (*Galaxia* spp.) (L. Djikstra, personal communication) and an introduced parasitoid wasp *Microctonus hyperodae* Loan (C. Williams, personal communication).

10.4 RANDOMLY AMPLIFIED POLYMORPHIC DNA (RAPD) PCR

DNA amplification using randomly sequenced primers varies from the classical PCR in several ways. The primers are short (ca. 10 bases), can be used singly (Welsh and McClelland, 1990; Williams *et al.*, 1990) as well as in pairs (Welsh and McClelland, 1991; Micheli *et al.*, 1993) and relatively non-stringent annealing conditions are employed. Products are of unknown sequence from anonymous regions of the genome. For successful amplification, the 3-prime ends of annealed primers must face each other and be less than 3 kb apart for *Taq* polymerase to synthesize the region between them. When just one primer is used in the reaction the annealing sites must be exact, or similar, inverted repeats. Inverted and direct repeats occur frequently in the genome and are known to undergo rapid changes in size and sequence through unequal crossing over, gene conversion and slippage replication of nucleotides (Levinson and Gutman, 1987). Inverted repeats, however, are not confined to repetitive DNA, and unique regions can also be amplified (Williams *et al.*, 1990). Thus, it is not surprising that RAPDs commonly reveal genetic polymorphism and can provide molecular markers over a range of taxonomic scales, especially in species which were previously inaccessible at the molecular level. The independent segregation of RAPD markers has been illustrated in honey bees (*Apis mellifera*, L.) with PCR products present in the 1:1 ratio expected for a haplodiploid species (Hunt and Page, 1992). Mendelian inheritance was also shown by Kazan *et al.* (1992) to be common in inter-specific crosses of plants belonging to the legume genus *Stylosanthes*, and likely mechanisms of non-Mendelian segregation were also discussed for some RAPD products.

While the use of RAPDs can be an attractive starting point for population analysis, a major limitation can be the deceptive level of allelic information due to the 'dominant' appearance of the markers (Milligan and McMurry, 1993). Hence, in diploid tissue, homozygotes (two copies of an amplifiable region of DNA or gene) are indistinguishable from heterozygotes (one copy of an amplifiable region of DNA or gene) and therefore it is impossible directly to estimate allele frequencies. In addition, the assumption that absence of a band represents a homozygous null genotype can be inaccurate if the null allele constitutes multiple, rather than single, alleles (Levitan and Grosberg, 1993). For many studies this can be overcome by obtaining many markers to increase statistical power. Lewis and Snow (1992) considered the implications of dominance in the application of RAPDs to paternity analysis. In a theoretical presentation they suggested that one would need to score more than 50 RAPD loci for each offspring. They concluded that, while co-dominant systems such as RFLP and allozyme analysis are generally more efficient for assignment of paternity, RAPDs would be useful in species lacking sufficient polymorphism by these traditional methods. In addition, ameliorating factors in the search for a large number of markers are being able to score several markers from one primer, the speed and automation of the technique and the ability to 'pick and choose' loci of value. Similarly, Levitan and Grosberg (1993) recommended scoring a large number of polymorphisms in their study of parentage in a multiple-offspring system to minimize the confounding effects of mutation, non-Mendelian inheritance and PCR artifacts. They also give an eloquent discussion on the statistical analysis of RAPDs for such a system. If simple or monomorphic RAPD patterns are amplified by some primers, another means of generating markers can be digestion of the PCR product(s) with restriction enzymes. This may reveal co-dominant characters as in traditional RFLP analysis (Hedrick, 1992; Lewis and Snow, 1992). An alternative approach is the application of RAPDs as phenotypic rather than allelic markers. For population studies, this overcomes the assumptions such as Hardy–Weinburg equilibrium that have to be made (Haig *et al.*, 1994).

There are also some potential technical problems worthy of note that may result in 'non-genetic variation' in RAPDs. Low stringency of the reaction conditions leaves the amplification process more vulnerable to competition for priming sites, especially during the first few cycles (Ruano *et al.*, 1991). Thus, minor differences in reaction conditions can affect the probability of amplification of some products which, consequently, may seem to be absent (McClelland and Welsh, 1994). Other factors such as primer–template DNA ratio, annealing temperature and magnesium concentration, have been highlighted by Ellsworth *et al.* (1993). Assuming that these parameters can be adequately controlled, then it should be possible to achieve reproducibility across different DNA extractions and thermal cyclers (Bucci and Menozzi, 1993). Finally, in addition to all the above, the pure complexity of the electrophoretic banding patterns can be difficult to interpret, though this is a common problem with all multiple-product techniques such as DNA fingerprinting (Lewin, 1989).

Despite the potential shortcomings of the technique (Black, 1993) the use of RAPDs has been greeted with enthusiasm and, to its testament, a diverse range of applications are now cited in the literature. Examples include species identification (Kambhampati *et al.*, 1992; Megnegneau *et al.*, 1993; Perring *et al.*, 1993), biotype variation (Black *et al.*, 1992), geographic population variation (Ballinger-Crabtree *et al.*, 1992; Haymer and McInnis, 1993), differentiation of conspecific populations (Kambhampati *et al.*, 1992), differentiation of clones (Fukatsu and Ishikawa, 1994) and reproductive success via parentage analysis (Hadrys *et al.*, 1993; Scott and Williams, 1993). RAPD data have also demonstrated strong agreement with data derived by other techniques. For example, Puterka *et al.* (1992) obtained almost perfect agreement of allozyme and RAPD data in a phylogenetic analysis of Russian wheat aphid, *Diuraphis noxia* (Mordvilko). Similarly, Megnegneau *et al.* (1993) found the classification of black *Aspergillus* and related fungal species using RAPD markers to be completely compatible with those based on RFLP and allozyme data.

10.4.1 Application of RAPDs in the study of agricultural pests in New Zealand

(a) Weevils

Of the 1546 weevil species in New Zealand, 50 are introduced (Watt, 1982). The large majority of those would be considered pests attacking a range of vegetation including ornamentals, berry fruits, root vegetables and pasture crops (Scott, 1984). RAPDs have recently been used for population analysis of two pest species. Both are native to South America: the Argentine stem weevil (ASW), *Listronotus bonariensis* (Kuschel), and the white-fringed weevil (WFW), *Graphognathus leucoloma* (Boheman). They are thought to have been introduced into New Zealand and Australia with European trade a matter of decades ago (Todd, 1964; Goldson and Emberson, 1980). With the advent of livestock farming they have quickly become established as pests of improved graminaceous pasture crops (Scott, 1984, Prestidge *et al.*, 1991). Like most soil pests these species are difficult to control, especially since the gradual banning of residual insecticides in New Zealand over the past 20 years. To assist in the development of alternative control strategies, fundamental knowledge of the genetic constitution of the New Zealand 'population' is desirable. This information can then underpin related issues such as the geographical origin of the colonizing populations of weevil and of candidate parasitoids with potential for biological control and also the behavioural and ecological interactions between them. Modern molecular methods, as outlined above, now allow us to explore the genetic basis of any metapopulation (*sensu* Levin, 1974) of introduced pests in New Zealand. The PCR–RAPD technique was employed in separate studies of the two weevil species for which different levels and mechanisms of genetic mixing were likely. Different spatial scales were targeted; between-countries for ASW and within-paddock for WFW. The two species also have different reproductive biology. ASW is sexually outbreeding, winged and has a generation time in New Zealand of around 6–12 months. In contrast, WFW is clonal, completely parthenogenic with no known incidence of males (East, 1980), flightless and has a generation time in New Zealand of 18–24 months.

The RAPD analysis of ASW, recently published by Williams *et al.* (1994), aimed to identify the likely geographic origin of the New Zealand population. This information, together with that from behavioural, ecological and bioassay studies, will serve to assist in the search for appropriate candidates for biological control. Populations of ASW from five geographically distinct locations around New Zealand, one locality in South Australia and nine climatically diverse regions in South America were compared. Two random primers gave useful population profiles, presenting largely homogeneous patterns for individuals within populations but detecting variation between the geographic groups. The results were analysed by both canonical discriminant function analysis and cluster analysis and supported the initial visual impressions. The populations in South America are distinct from each other, suggesting that these populations are to some extent genetically isolated. This was consistent with the geographical and climatic barriers that separate them. Within New Zealand, however, the populations throughout the country generated very similar profiles which in turn were similar to those of the Australian population. The New Zealand and Australian profiles closely resembled that from the east coast of South America (Uruguay and Buenos Aires) and showed much less similarity with the populations from the other regions in South America. The conclusion therefore was that the most likely source of the populations sampled in New Zealand and Australian was the Rio de la Plata region of South America. Independent of this molecular study, has been the release of the parasitoid *M. hyperodae*. This species was collected from a range of South American sites. Rates of parasitism in New Zealand have reached 70%, so it is useful to characterize, albeit retrospectively, the genetics and origin of the successful parasitoid strain.

The second study is investigating the within-New Zealand variation in WFW. No parasitoids have been released against this weevil, so the opportunity exists to characterize the clone, or clones, present in New Zealand pastures before selection of the most appropriate South American parasitoid(s). To date, only two geographic locations have been analysed, both in the North Island about 20 km apart and each sampled within a 20 m^2 area. Of 60 primers screened six were chosen (Table 10.2) for population analysis, on the basis that they each amplified several strong products.

All six primers revealed polymorphism in one population (Ruakura, n = 80) and effectively detecting two consistent groups. In contrast, the same primers did not show any variation in the second population (Wellsford, n = 65) (Figure 10.5). This initial result is compatible with the very narrow margin for genetic mixing expected on the basis of the biology/reproductive isolation of this species. DNA analysis of three other geographically distinct populations has yet to be done. In addition populations from several locations along the length of Chile are to be screened in the same manner and a comparison of the native endemic population variation made with that of the introduced population in New Zealand that may have arisen from very few original individuals.

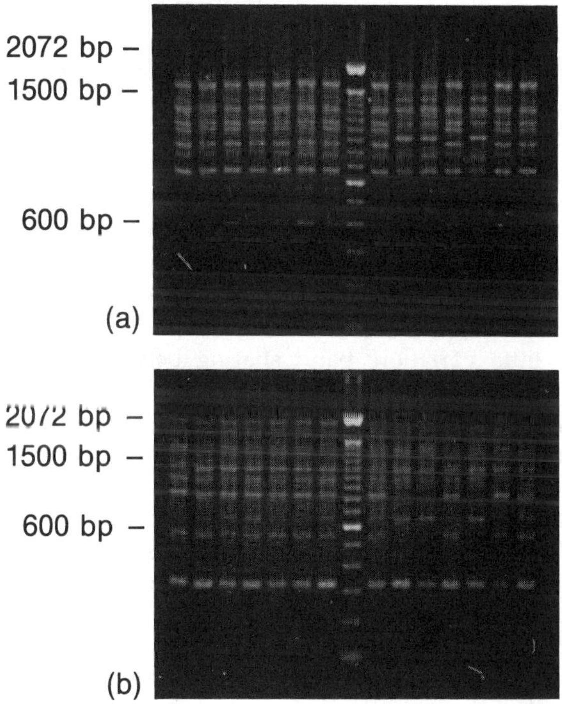

Figure 10.5 PCR–RAPD profiles of samples from two white-fringed weevil populations. Lanes 1–7 Ruakura; lanes 9–15 Wellsford; lane 8 is a molecular weight standard (100 bp, Gibco-BRL). Examples are from amplification with random primer (a) OPA-01 and (b) OPA-07. Electrophoresis details are as for Figure 10.3, except a 1.5% agarose gel was used.

(b) Possum

In addition to an abundance of introduced invertebrate pest species, New Zealand is also endowed with a number of introduced vertebrate pests of which the common brushtail possum, *Trichosurus vulpecula* (Kerr), is currently the most economically significant. The founder population of a few hundred individuals were originally liberated into New Zealand from Australia in 1858 to establish a fur trade. This is the only species in New Zealand and is now a major pest of native and plantation forests. It is also responsible for damage and economic loss in improved pastures and various farm and horticultural crops (reviewed by Cowan, 1990). Consequently, considerable effort has been dedicated to various, but as yet unsuccessful, eradication programmes. More recently, the possum has been identified as a reservoir of bovine tuberculosis (Tb), increasing its pest status yet again. This has led to studies focusing on the population dynamics and social structure of wild populations in an effort to understand aspects of disease epidemiology such as the clumped nature of bovine Tb prevalence (Barlow, 1991), the spread of potential sexually transmitted biological control agents (Bomford, 1990) and natural population regulation (Efford, 1991). A long-term live-trapping study is underway to underpin spatial and temporal population studies such as these. However, due to male-biased dispersal, male relatedness is difficult to monitor and so molecular techniques for paternity testing have been considered. The multi-locus DNA fingerprinting technique of Jeffreys *et al.* (1985) is currently being used by G.L. Dick and P.E. Cowan (personal communication). However, the probe/restriction enzyme combinations tested so far have detected too little variation; band sharing between unrelated animals was <0.6 compared with <0.26 for humans. Even sequential probing of each sample with several probes did not lead to satisfactory exclusion of potential fathers. This could reflect a genetic bottleneck that may have occurred following the original introduction of animals, or be a function of the technique.

We have investigated the use of RAPDs as an alternative method. This technique has been applied to parentage analysis in a variety of mating systems, including broods of communally breeding burying beetles (Scott *et al.*, 1992), free-spawning colonial marine hydrozoans (Levitan and Grosberg, 1993) and inheritance in human families (Caetano-Anollés *et al.*, 1991). Reidy *et al.* (1992) assessed the use of RAPDs for assigning paternity in a troop of baboons (*Papio cynocephalus ursinus*) and concluded the technique to be unsuitable due to the high frequency of non-parental bands in known pedigrees. In contrast, however, Scott *et al.* (1992) suggested that this need not preclude their use for parentage analysis if (i) all bands are reproducible, (ii) only bands unique to one of the putative parents are used and (iii) the probability of false inclusion of an adult as a parent should be calculated from frequencies of the specific variable bands in the population. We have screened 60 random primers with DNA extracted from ethanol-preserved possum ear tissue (section 10.2.2). Of 32 potential primers producing multiple banding patterns, six have so far been further tested based on strong banding and polymorphism. Polyacrylamide gel electrophoresis and silver staining were used to improve resolution and optimize detection of variation. An example using primer B8 is given in Figure 10.6. Using DNA from a wild, forest possum population 40 individuals were screened with these primers which together produced 23 polymorphic loci. A pairwise analysis, recommended for use with dominant markers such as RAPDs (Black, 1993), has been employed where every band position scored by the individuals in the population tested is considered. Shared presence and shared absence of a band are both considered similarity between individuals. This initial study indicated a relatively low range of 0.33–45 band sharing between individuals. However, this could be lower in reality as not all family relationships within the population were known. The analysis was repeated for eight captive family groups. Within-family (joey/true father) band sharing was 0.59 to 0.78 while between-unrelated individuals was <0.38. This indicated a lower level of band sharing than was suggested by multilocus fingerprinting. While promising, this is still insufficient to exclude paternity. Further analysis is being undertaken to confirm or reject the potential of the RAPD technique for the purpose

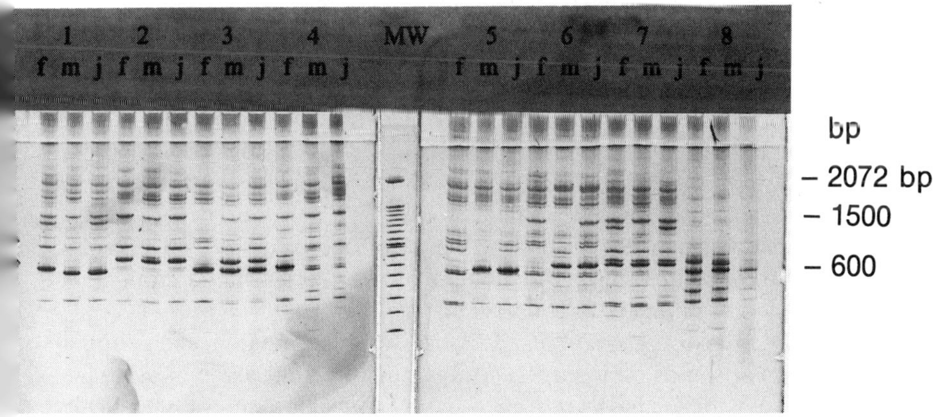

Figure 10.6 PCR–RAPD profiles of possum families #1–#8 using random primer OPB-08. Father, f; Mother, m; Joey, j. Lane 13 is a molecular weight standard (100 bp, Gibco-BRL). Electrophoresis was carried out on an 8–25% gradient polyacrylamide gel followed by silver staining as recommended for PhastSystem (Pharmacia-LKB).

of paternity testing in a sexually outbreeding, single offspring family animal such as possum.

10.5 CONCLUSIONS AND FUTURE PROSPECTS

The examples above illustrate how we now have access to a versatile range of molecular markers for use with species in which we have no previous gene sequence data. Browsing of any issue of the journal *Molecular Ecology* reveals the diversity of applications of modern molecular techniques to ecological questions. The examples below illustrate some of the invertebrate orientated questions to be addressed in our Lincoln laboratory in the near future, in collaboration with other laboratories in New Zealand and Europe.

1. Clonal biology
   - (a) Spatial and temporal variation in the clonal composition of invertebrate populations. Examples of applications of this include the dynamics of pest forecasting, host plant relationships, ecological fitness on a range of plant hosts, pesticide resistance and its maintenance within populations, selection of appropriate imported biological control agents, etc.
   - (b) Clonal stability through time. This is highly relevant to the genetical basis of the development of insecticide resistance and to the breakdown of host plant resistance by new 'biotypes' of aphids and other organisms. Biocontrol failures may also have their basis in genetic changes in the clonal host.
2. Sexually-out breeding and social insects
   - (a) Selection of appropriate biocontrol agents.
   - (b) Genetic diversity in space and time.
   - (c) Evolutionarily rapid acquisition of a new host plant range such as in the western flower thrips, *Frankliniella* sp. (Thysanoptera: Thripidae), and the glasshouse whitefly, *Trialeurodes vaporariorum* (Westwood), in New Zealand.
   - (d) Genetic diversity within and between bee colonies as a potential marker to determine the number of nests 'servicing' a crop.
3. Rare insects of amenity/conservation value
   - (a) The role of geographical isolation within previous metapopulations (islands, landscape features, uncultivated areas in an arable or pastoral landscape, etc.) in the development of genetically distinct sub-populations.
   - (b) The prospects and limitations involved in 'restoration ecology' in which invertebrates are translocated from one site to a newly created one, often without attention to the genetic origin of the introduced invertebrate.

These, and a host of other 'basic science' and applied questions await the efforts of research teams which comprise ecologists and molecular biologists working together. The Lincoln group is using long-term New Zealand government funding to begin to answer questions such as these in invertebrate molecular ecology.

Acknowledgements

We are grateful to the Plant Protection Research Unit, Lincoln University and the Vice Chancellor of Lincoln University for sponsoring K.F. Armstrong to present this work at the symposium on *Ecology of Agricultural Pests: Biochemical Approaches*, Cardiff, 21–23 September, 1994. The studies presented here were supported by funding from Ministry of Agriculture and Fisheries (NZ), the Foundation for Research Science and Technology (PGSF) and Lincoln University Research Committee Special Projects Fund (for K.F.A). We would like to thank S. Hardwick and R. Prestidge (New Zealand Pastoral Agriculture Research Institute Ltd) for supplying white-fringed weevil samples, A. Allwood (Regional Fruit Fly Project, South Pacific Commission) for providing fruit fly samples and P. Cowan (Landcare Research New Zealand Ltd) for kindly providing support and tissue samples to K.F.A. for the possum study. C. Vink kindly provided the line drawings. Thanks also to C. Williams for the discussion of *L. bonariensis* RAPD analysis.

REFERENCES

Allard, M.W. and Honeycutt, R.L. (1991) Ribosomal DNA variation within and between species of rodents, with emphasis on the genus *Onychomys*. *Molecular Biology and Evolution*, **8**, 71–84.

Armstrong, J.W. and Couey, H.M. (1989) Fumigation, heat and cold, in *Fruit Flies: Their Biology, Natural Enemies and Control*, Vol. 3B, (eds A. Robinson and G. Hooper), Elsevier, New York, pp. 411–24.

Avise, J.C. (1994) *Molecular Markers, Natural History and Evolution*, Chapman & Hall, London.

Ballinger-Crabtree, M.E., Black, W.C. VI and Miller, B.R. (1992) Use of genetic polymorphisms may be detected by the random-amplified polymorphic DNA polymerase chain reaction (RAPD–PCR) for differentiation and identification of *Aedes aegypti* subspecies and populations. *American Journal of Tropical Medicine and Hygiene*, **47**, 893–901.

Barlow, N.D. (1991) A spatially aggregated disease/host model for bovine Tb in New Zealand possum populations. *Journal of Applied Ecology*, **28**, 777–93.

Bilofsky, H.S. and Burks, C. (1988) The GENBANK genetic sequence data bank. *Nucleic Acids Research*, **16**, 1861–4.

Black, W.C. (1993) PCR with arbitrary primers: approach with care. *Insect Molecular Biology*, **2**, 1–6.

Black, W.C., McLain, D.K. and Rai, K.S. (1989) Patterns of variation in the RDNA cistron within and among world populations of a mosquito, *Aedes albopictus* (Skuse). *Genetics*, **121**, 539–50.

Black, W.C., DuTeau, N.M., Puterka, G.J., Nechols, J.R. and Pettorini, J.M. (1992) Use of the random amplified polymorphic DNA polymerase chain reaction (RAPD–PCR) to detect DNA polymorphisms in aphids (Homoptera: Aphididae). *Bulletin of Entomological Research*, **82**, 151–9.

Bomford, M. (1990) A role for fertility control in wildlife management? *Bureau of Rural Resources Bulletin*. No. 7, Canberra, ACT.

Bramwell, N.H. and Burns, B.F. (1988) The effects of fixative type and fixation time on the quantity and quality of extractable DNA for hybridisation studies on lymphoid tissue. *Experimental Haematology*, **16**, 730–2.

Brower, A.V.Z. and DeSalle, R. (1994) Practical and theoretical considerations for choice of a DNA sequence region in insect molecular systematics, with a short review of published studies using nuclear gene regions. *Annals of the Entomological Society of America*, **87**, 702–16.

Bucci, G. and Menozzi, P. (1993) Segregation analysis of random amplified polymorphic DNA (RAPD) markers in *Picea abies* Karst. *Molecular Biology*, **2**, 227–32.

Burke, T., Rainey, W.E. and White, T.J. (1992) Molecular variation and ecological problems, in *Genes in Ecology*, (eds R.J. Berry, T.J. Crawford and G.M. Hewitt), Blackwell Scientific Publications, Oxford, pp. 229–54.

Caetano-Anollés, G. and Bassam, B.J. (1993) DNA amplification fingerprinting using arbitrary oligonucleotide primers. *Applied Biochemistry and Biotechnology*, **42**, 189–200.

Caetano-Anollés, G., Bassam, B.J. and Gresshof P.M. (1991) DNA amplification fingerprinting using very short arbitrary oligonucteotide primers. *BioTechnology*, **9**, 553–6.

Carvalho, G.R., Maclean, N., Wratten, S.D., Carter, R.E. and Thurston, J.P. (1991) Differentiation of aphid clones using DNA fingerprints from individual aphids. *Proceedings of the Royal Society London, B*, **243**, 109–14.

Chen, W., Hoy, J.W. and Schneider, R.W. (1992) Species-specific polymorphisms in transcribed ribosomal DNA of five *Pythium* species. *Experimental Mycology*, **16**, 22–34.

Cowan, P.E. (1990) Family Phalangeridae, in *The Handbook of New Zealand Mammals*, (ed. C.M. King), Oxford University Press, Auckland, pp. 68–98.

Cowley, J.M. and Frampton, E.R. (1989) Significance of and monitoring for alien fruit flies (Tephritidae) in *Proceedings of the 42nd New Zealand Weed and Pest Control Conference, New Plymouth*, New Zealand Weed & Pest Society Inc.

Crozier, R.H. (1990) From population genetics to phylogeny: uses and limits of mitochondrial DNA. *Australian Systematic Botany*, **3**, 111–24.

D'Aquila, R.T., Bechtel, L.J., Vidler, J.A., Eron, J.J., Gorezyca, P. and Kaplan, J.C. (1991) Maximising sensitivity and specificity of PCR by preamplification heating. *Nucleic Acids Research*, **19**, 3749.

De Barro, P., Sherratt, T., Wratten, S. and Maclean, N. (1994) DNA fingerprinting of cereal aphids using $(GATA)_4$. *European Journal of Entomology*, **91**, 109–14.

East, R. (1980) Sampling white-fringed weevil (*Graphognathus leucoloma*) populations. *New Zealand Journal of Agricultural Research*, **23**, 581–7.

Efford, M.G. (1991) A review of possum dispersal. DSIR Land Resources Scientific Report No. 23, 70 p.

Ellsworth, D.L., Rittenhouse, K.D. and Honeycutt, R.L. (1993) Artifactual variation in randomly amplified polymorphic DNA banding patterns. *BioTechniques*, **14**, 214–17.

Erlich, H.A. (1989) *PCR Technology: Principles and Applications for DNA Amplification*, Stockton Press, New York.

Erlich, H.A. and Arnheim, N. (1992) Genetic analysis using the polymerase chain reaction. *Annual Review of Genetics*, **26**, 479–506.

Farrell, J.A. and Stufkens, M.W. (1990) The impact of *Aphidius rhopalosiphi* (Hymenoptera: Aphididae) on populations of the rose grain aphid (*Metopolophium dirhodum*) (Hemiptera: Aphididae) on cereals in Canterbury, New Zealand. *Bulletin of Entomological Research*, **80**, 377–83.

Farrell, J.A. and Stufkens, M.W. (1992) Cereal aphid flights and barley yellow dwarf virus infection of cereals in Canterbury, New Zealand. *New Zealand Journal of Crop and Horticultural Science*, **20**, 407–12.

Felsenstein, J. (1989). PHYLIP – phylogenetic inferences package (version 3.2). *Cladistics*, **5**, 164–6.

Finnerty, V. and Collins, F.H. (1988) Ribosomal DNA probes for identification of member species of the *Anopheles gambiae* complex. *Florida Entomologist*, **71**, 288–94.

Fukatsu, T. and Ishikawa, H. (1994) Differentiation of aphid clones by arbitrarily primed polymerase chain reaction (AP–PCR) DNA fingerprinting. *Molecular Ecology*, **3**, 187–92.

Goldson, S.L. and Emberson, R.M. (1980) Relict diapause in an introduced weevil in New Zealand. *Nature*, **286**, 489–90.

Graur, D. (1985) Gene diversity in Hymenoptera. *Evolution*, **39**, 190–9.

Hadrys, H., Schierwater, B., Dellaporta, S.L., DeSalle, R. and Buss, L.W. (1993) Determination of paternity in dragonflies by random amplified polymorphic DNA fingerprinting. *Molecular Ecology*, **2**, 79–87.

Haig, S.M., Rhymer, J.M. and Heckel (1994) Population differentiation in randomly amplified polymorphic DNA of red-cockaded woodpeckers *Pioides borealis*. *Molecular Ecology*, **3**, 581–94.

Han, H.-Y. and McPheron, B.A. (1994) Molecular approach to the higher classification of Tephritidae using nuclear ribosomal DNA sequences. *Fourth International Symposium on Fruit Flies of Economic Importance*, Sand Key, Florida, June 5–10.

Hanboonsong Y. (1994) A comparative phenetic and cladistic analysis of the genus *Holcaspis* Chaudoir (Coleoptera: Carabidae). PhD Thesis, Lincoln University, NZ.

Hawksworth, D.L. (ed.) (1994) *The Identification and Characterization of Pest Organisms*, CAB International, Wallingford, UK.

Haymer, D.S. and McInnis, D.O. (1993) Resolution of populations of the Mediterranean fruit fly at the DNA level using random primers for the polymerase chain reaction. *Genome*, **37**, 244–8.

Haymer, D.S., Tanaka, T. and Teramae, C. (1994) DNA probes can be used to discriminate between Tephritid species at all stages of the life cycle (Diptera: Tephritidae). *Entomological Society of America*, **87**, 741–6.

He, M. and Haymer, D. (1994) Genetic relationships and genetic origin of medfly populations determined by RAPD PCR. *Fourth International Symposium on Fruit Flies of Economic Importance*, Sand Key, Florida, June 5–10.

Hedrick, P. (1992) Shooting the RAPDs. *Nature*, **355**, 679–80.

Heuttel, M.D. (1979) Genetic approaches to basic problems in insect behaviour and ecology, in *Genetics in Relation to Insect Management*, (eds M.A. Hoy and J.J. McKelvey), The Rockefeller Foundation, pp. 161–9.

Hunt, G.J. and Page, R.E. (1992) Patterns of inheritance with RAPD molecular markers reveal novel types of polymorphism in the honeybee. *Theoretical and Applied Genetics*, **85**, 15–20.

Innis, M.A., Gelfand, D.H. and Sninsky, J.J. (eds) (1990) *PCR Protocols: A Guide to Methods and Applications*, Academic Press Inc., San Diego.

Jeffreys, A.J., Wilson, V. and Thein, S.L. (1985) Hypervariable 'minisatellite' regions in human DNA. *Nature*, **314**, 67.

Jones, J. (1980) Can genes choose habitats? *Nature*, **286**, 757–8.

Kambhampati, S. Black, W.C. and Rai, K.S. (1992) Random amplified polymorphic DNA of mosquito species and populations (Diptera: Culicidae): techniques, statistical analysis and applications. *Entomological Society of America*, **29**, 939–45.

Kazan K., Manners, J.M. and Cameron D.F. (1992) Inheritance of random amplified polymorphic DNA markers in an interspecific cross in the genus *Stylosanthes. Genome*, **36**, 50–6.

Kocher, T.D., Thomas, W.K., Meyer, A., Edwards, S.V., Pääbo, Villablanca, F.X. and Wilson, A.C. (1989) Dynamics of mitochondrial DNA evolution in animals: Amplification and sequencing with conserved primers. *Proceedings of the National Academy of Science, USA*, **86**, 6196–200.

Lessa, E.P. and Applebaum, G. (1993) Screening techniques for detecting allelic variation in DNA sequences. *Molecular Ecology*, **2**, 119–29.

Levin, S.A. (1974) Dispersion and population interactions. *American Naturalist*, **104**, 413–23.

Levinson, G. and Gutman, G.A. (1987) Slipped-strand mispairing: a major mechanism for DNA sequence evolution. *Molecular Biology and Evolution*, **4**, 203–21.

Levitan, D.R. and Grosberg, R.K. (1993) The analysis of paternity and maternity in the marine hydrozoan *Hydractinion symbiolongicarpus* using randomly amplified polymorphic DNA (RAPD) markers. *Molecular Ecology*, **2**, 315–26.

Lewin, R. (1989) Limits to DNA fingerprinting. *Science*, **243**, 1549–51.

Lewin, B. (1990) *Genes IV*, Oxford University Press.

Lewis, P.O. and Snow, A.A. (1992) Deterministic paternity exclusion using RAPD markers. *Molecular Ecology*, **1**, 155–60.

Lupoli, R., Irwin, M.E. and Vossbrinck, C.R. (1990) A ribosomal DNA probe to distinguish populations of *Rhopalosiphum maidis* (Homoptera: Aphididae). *Annals of Applied Biology*, **117**, 3–8.

McClelland, M. and Welsh, J. (1994) DNA fingerprinting by arbitrarily primed PCR. *PCR Methods and Applications*, **4**, 559–65.

McKenzie, J.A. and Batterham, P. (1994) The genetic, molecular and phenotypic consequences of selection for insecticide resistance. *Trends in Ecology and Evolution*, **9**, 166-9.

McPherson, M.J., Quirke, P. and Taylor, G.R. (1991) *PCR – A Practical Approach*, IRL Press at Oxford University Press.

Megnegneau, B., Debets, F. and Hockstra, R.F. (1993) Genetic variability and relatedness in the complex group of black aspergilli based on random amplification of polymorphic DNA. *Current Genetics*, **23**, 323–9.

Menken, S.B.J. (1989) Electrophoretic studies on geographic populations, host races, and sibling species of insect pests, in *Electrophoretic Studies on Agricultural Pests*, (eds H.D. Loxdale and J.D. Hollander) Systematics Association, Special Volume 39, Clarendon Press, Oxford, pp. 181–202.

Mercier, B., Gaucher, C., Feugeas, O. and Mazurier, C. (1990) Direct PCR from whole blood, without DNA extraction. *Nucleic Acids Research*, **28**, 5908.

Meyer, A. (1994) Shortcomings of the cytochrome b gene as a molecular marker. *Trends in Ecology and Evolution*, **9**, 278–80.

Micheli, M.R., Bova, R., Calissano, P. and D'Ambrosio, E. (1993) Randomly amplified polymorphic DNA fingerprinting using combinations of oligonucleotide primers. *BioTechniques*, **15**, 388–90.

Miller, S.A., Dykes, D.D. and Polesky, H.F. (1988) A simple salting out procedure for extracting DNA from human nucleated cells. *Nucleic Acids Research*, **16**, 1215.

Milligan, B.G. and McMurry, C.K. (1993) Dominant vs. codominant genetic markers in the estimation of male mating success. *Molecular Ecology*, **2**, 275–83.

Montgomery, G.W. and Sise, J.A. (1990) Extraction of DNA from sheep white blood cells. *New Zealand Journal of Agricultural Research*, **33**, 437–41.

Mullis, K.B. and Faloona, F.A. (1987) Specific synthesis of DNA *in vitro* via a polymerase-catalyzed chain reaction. *Methods in Enzymology*, **155**, 335–50.

Mullis, K.B., Ferre, F. and Gibbs, R.A. (eds) (1994) *The Polymerase Chain Reaction*, Birkhauser, Boston.

Muralidharan, K. and Wakeland, E.K. (1993) Concentration of primer and template qualitatively affects products in random-amplified polymorphic DNA PCR. *BioTechniques*, **14**, 362–4.

Oakenfull, E.A. (1994) Vodka, meths and DNA. *Trends in Ecology and Evolution*, **9**, 26.

Pedley, S., Mohamed, M.E.H. and Mertens, P.P.C. (1988) Analysis of genome segments from six different isolates of bluetongue virus using RNA–RNA hybridisation: a generalised coding assignment for bluetongue viruses. *Virus Research*, **10**, 381–90.

Perring, T.M., Cooper, A.D., Rodriguez, R.J., Farrar, C.A. and Bellows, T.S. (1993) Identification of a whitefly species by genomic and behavioral studies. *Science*, **259**, 74–7.

Prestidge, R.A., Barker, G.M. and Pottinger, R.P. (1991). The economic costs of Argentine stem weevil in pastures in New Zealand. *Proceedings of the 44th New Zealand Weed and Pest Control Conference*, New Zealand Weed & Pest Society Inc., 165–70.

Puterka, G.J., Bird, J.D. and Broton, R.L. (1992) Biotypic variation in a worldwide collection of Russian wheat aphid (Homoptera: Aphididae). *Journal of Economic Entomology*, **85**, 1497–506.

Reidy, M.F., Hamilton, W.J. and Aquadro, C.F. (1992) Excess of non-parental bands in offspring from known primate pedigrees assayed using RAPD-PCR. *Nucleic Acids Research*, **20**, 918.

Robson, M., Osborne, R., Green, E., Sved, J., Meats, A. and Frommer, M. (1994) A demography of *Bactrocera tryoni* in Australia: use of microsatellites to define the source of outbreaks in fly-free zones. *Fourth International Symposium on Fruit Flies of Economic Importance*, Sand Key, Florida, June 5–10.

Rolfs, A., Schuller, I., Finckh, U. and Weber-Rolfs, I. (eds) (1992) Substances affecting PCR: inhibition or enhancement, in *PCR: Clinical Diagnostics and Research*, Springer Verlag, Berlin, pp. 51–60.

Rose, O.C., Brookes, M.I. and Mallet, J.L.B. (1994) A quick and simple nonlethal method for extracting DNA from butterfly wings. *Molecular Ecology*, **3**, 275.

Ruano, G., Brash, D.E. and Kidd, K.K. (1991) PCR: The first few cycles. *Amplifications*, **7**, 1–4.

Sambrook, J., Fritsch, E.F. and Maniatis, T. (1989) *Molecular Cloning – A Laboratory Manual*, Cold Spring Harbor Laboratory Press, Cold Spring Harbor, NY.

Sardelli, A.D. (1993) Plateau effect – understanding PCR limitations. *Amplifications*, **9**, 1–5.

Scott, M.P. and Williams, S.M. (1993) Comparative reproductive success of communally breeding burying beetles as assessed by PCR with randomly amplified polymorphic DNA. *Proceedings of the National Academy of Science, USA*, **90**, 2242–5.

Scott, M.P., Haymes, K.M. and Williams, S.M. (1992) Parentage analysis using RAPD PCR. *Nucleic Acids Research*, **20**, 5493.

Scott, R.R. (1984) *New Zealand Pest and Beneficial Insects*, Lincoln University College of Agriculture, Canterbury, New Zealand.

Service, P.M. and Lenski, R.E. (1982) Aphid genotypes, plant genotypes and genetic diversity: a demographic analysis of experimental data. *Evolution*, **36**, 1276–82.

Shoemaker, D.D., Ross, K.G. and Arnold, M.L. (1994) Development of RAPD markers in two introduced fire ants, *Solenopsis invicta* and *S. richteri* and their application to the study of a hybrid zone. *Molecular Ecology*, **3**, 531–9.

Shufran, K.A., Black, W.C. and Margolies, D.C. (1991) DNA fingerprinting to study spatial and temporal distributions of an aphid, *Schizaphis graminum* (Homoptera: Aphididae). *Bulletin of Entomological Research*, **81**, 303–13.

Shufran, K.A., Margolies, D.C. and Black, W.C. (1992) Variation between biotype E clones of *Schizaphis graminum* (Homoptera: Aphididae). *Bulletin of Entomological Research*, **82**, 407–16.

Simon, C., Francesco, F., Beckenbach, A., Crespi, B., Liu, H. and Flook, P. (1994) Evolution, weighting, and phylogenetic utility of mitochondrial gene sequences and a compilation of conserved polymerase chain reaction primers. *Annals of the Entomological Society of American*, **87**, 651–701.

Slade, R.W., Moritz, C., Heideman, A. and Hale, P.T. (1993) Rapid assessment of single-copy nuclear DNA variation in diverse species. *Molecular Ecology*, **2**, 359–73.

Smith, L.J., Braylon, R.C., Nutkis, J.E., Edmundson, K.B., Downing, J.R. and Wakeland, E.K. (1987) Extraction of cellular DNA from human cells and tissues fixed in ethanol. *Analytical Biochemistry*, **160**, 135–8.

Sogin, M.L. (1990) Amplification of ribosomal RNA genes for molecular evolution studies, in *PCR Protocols: A Guide to Methods and Applications*, (eds M.A. Innis, D.H. Gelfand, J.J. Sninsky and T.J. White) Academic Press Inc., San Diego, pp. 307–426.

Steck, G.J., Gasparich, G.E., Han, H-Y, McPheron, B.A. and Sheppard, W.S. (1994) Distribution of mitochondrial DNA haplotypes among *Ceratitis capitata* populations worldwide. *Fourth International Symposium on Fruit Flies of Economic Importance*, Sand Key, Florida, June 5–10.

Stephenson, B.P. (1989) Procedures for fruit fly (Tephritidae) eradication in New Zealand. *Proceedings of the 42nd New Zealand Weed and Pest Control Conference, New Plymouth*, New Zealand Weed and Pest Control Society Inc.

Stiller, J.W. and Waaland, J.R. (1993) Molecular analysis reveals cryptic diversity in *Porphyra* (Rhodophyta). *Journal of Phycology*, **29**, 506–17.

Swofford, D.L. (1992). *PAUP: Phylogenetic analysis using parsimony, version 3.0*, Illinois Natural History Survey, Champaign.

Tan, A.-M. and Orrego, C. (1992) DNA stabilisation and amplification from museum collections of extracts originally intended for allozyme analysis. *Molecular Ecology*, **1**, 195–7.

Taylor, G.R. (1993) Polymerase chain reaction: basic principles and automation, in *PCR*, (eds M.J. McPherson, P. Quirke and G.R. Taylor), Oxford University Press, Oxford, pp. 1–13.

Thomas, K.T. and Kocher, T.D. (1993) Sequence of polymerase chain reaction amplified DNAs. *Methods in Enzymology*, **224**, 391–9.

Todd, D.H. (1964) Biology and control of whitefringed weevil. *Proceedings of the 17th New Zealand Weed and Pest Control Conference*, New Zealand Weed and Pest Control Society Inc. pp. 125–9.

Tower, P. (1991) Purification of DNA, in *Essential Molecular Biology, Vol. I, A Practical Approach*, (ed. T.A. Brown), IRL Press.

Vossbrinck, C.R. and Friedman, S. (1989) A 28S ribosomal RNA phylogeny of certain cyclorrhaphous Diptera based upon a hypervariable region. *Systematic Entomology*, **14**, 417–31.

Wallner, W.E. (1987) Factors affecting insect population dynamics: differences between outbreak and non-outbreak species. *Annual Review of Entomology*, **32**, 317–40.

Watt, J.C. (1982) New Zealand beetles. *New Zealand Entomologist*, **7**, 213–21.

Welsh, J. and McClelland, M. (1990) Fingerprinting genomes using PCR with arbitrary primers. *Nucleic Acids Research*, **18**, 7213–18.

Welsh, J. and McClelland, M. (1991) Genomic fingerprinting using arbitrarily primed PCR and a matrix of pairwise combinations of primers. *Nucleic Acids Research*, **19**, 5275–9.

White, I.M. and Elson-Harris, M.M. (1992) *Fruit Flies of Economic Significance: Their Identification and Bionomics*, CAB International, Wallingford.

White, T.J., Bruns, T., Lee, S. and Taylor, J. (1990) Amplification and direct sequencing of fungal ribosomal RNA genes for phylogenetics, in *PCR Protocols: A Guide to Methods and Applications*, (eds M.A. Innis, D.H. Gelfand, J.J. Sninsky and T.J. White), Academic Press Inc., San Diego, pp. 315-22.

Williams, C.L., Goldson, S.L., Baird, D.B. and Bullock, D.W. (1994) Geographical origin of an introduced insect pest, *Listronotus bonariensis* (Kuschel), determined by RAPD analysis. *Heredity*, **72**, 412 19.

Williams, J.G.K., Kubelik, A.R., Livak, K.J., Rafalski, J.A. and Tingey, S.V. (1990) DNA polymorphisms amplified by arbitrary primers are useful as genetic markers. *Nucleic Acids Research*, **18**, 6531–5

Wratten, S.D. and Powell, W. (1991) Cereal aphids and their natural enemies, in *The Ecology of Temperate Cereal Aphids*, (eds L.G. Firbank, N. Carter and G.R. Potts), Blackwell Scientific Publications, Oxford.

Xiong, B. and Kocher, T.D. (1991) Comparison of mitochondrial DNA sequences of seven morphospecies of black flies (Diptera: Simuliidae), *Genome*, **34**, 306–11

11

Serological analysis of arthropod predation: past, present and future

M.H. Greenstone

ABSTRACT

This history of serological analysis of arthropod predation is reviewed, emphasizing the evolution of technologies for antigen preparation, antibody production, and immunoassay. The first recorded serological gut analyses used crude homogenates as antigen, rabbit antisera as antibody source, and tube precipitin tests as immunoassays. In the succeeding 48 years antigen preparations have been refined to the levels of total homogenate protein, proteins partially purified by two-dimensional electrophoresis, and finally to biochemically characterized purified individual proteins. Antibodies have included absorbed as well as crude rabbit antisera, rabbit immunoglobulin, affinity-purified rabbit and mouse antisera, and mouse monoclonal antibodies. The tube precipitin test has been eclipsed by Ouchterlony and other gel precipitin tests, latex agglutination, passive haemagglutination inhibition, radioimmunoassay, ELISA and immunodot assays.

As a result of these efforts we are now able routinely to achieve high sensitivity and ecologically useful levels of specificity, and effective technical approaches have been sifted from among the many choices. It transpires that antigens do not need to be highly purified to yield specific assays, so simple homogenates are used with even higher frequency now than in past decades. ELISA, which is simple once optimized, has become the assay of choice.

The Ecology of Agricultural Pests, Edited by W.O.C. Symondson and J.E. Liddell.
Published in 1996 by Chapman & Hall, London. ISBN 0 412 62190 8

Unlike antigen production and assay technologies, antibody production has not become simpler. The dominant technology, hybridoma or monoclonal antibody production, is expensive and complex. New technologies promise to make monoclonal antibody production simpler and less expensive, and it may be possible in the future to engineer antibodies or antibody-like molecules with desired specificity from scratch. In the meantime, affinity chromatography of antisera can also produce high levels of specificity, and may be a viable economical alternative.

Gut analysis technologies other than serological methods may be appropriate in some circumstances. A new technology, the production of nucleic acid probes, will probably be easier to master than monoclonal antibody technology, but we need to know more about rates of digestion of specific nucleotide sequences in the gut before making further predictions about their usefulness.

11.1 INTRODUCTION

Arthropod predators are the most abundant and consistently present insect natural enemies in most agroecosystems. Nevertheless, we know very little about their impact on pest populations. Most predator species are cryptic, feed infrequently and, with a few notable exceptions, do not consume hard parts of their prey. This makes arthropod predation one of the most difficult ecological interactions to study. Ecologists have used various approaches to obtain arthropod predation data. Methods which identify prey molecules in the guts of predators are least disruptive of ecosystem processes. One of these, serological analysis, i.e. the use of vertebrate antibodies to detect prey antigens, has been the subject of intense experimentation and adaptation for almost 50 years.

Like other users of applied immunology, insect ecologists make two demands on serological assays. The first is for specificity, because predator guts may contain not only target antigens, but also antigens from other, related prey species in the ecosystem. The second is for high sensitivity, because the target antigens will often be found in exceedingly small amounts. We have pursued these goals by trial and error, adopting technologies made available in other fields, notably clinical diagnostics.

Progress was slow until recently, with levels of sensitivity and specificity falling far short of what was needed to produce ecologically useful data. In order to understand why this was so, and why we have finally reached the stage where we can expect routinely to meet our needs, it is useful to review the technologies of antibody production and immunoassay.

11.2 ANTIBODY PRODUCTION

11.2.1 Conventional antiserum production

Insect ecologists have used serological methods since the 1920s. In most cases we have employed so-called conventional antisera, also known as polyclonal antibodies. This means that following a course of antigen injections (initial immunization followed by 'booster' injections), a sample of the vertebrate host's blood is taken and allowed to clot, and the supernatant serum is used directly in the assay.

Suppose we wish to produce antibodies to study predation on larvae of a particular moth species. A conventional immunization protocol, and the way we used to think about antibody production, is illustrated in Figure 11.1. We begin by extracting the antigen from the target species (Species 1 in Figure 11.1) We think of the antigen as a large molecule (typically a protein although carbohydrates also are usually very antigenic) having a unique shape (shown here as hexagonal) that will differentiate it from similar antigens (Figure 11.1(a)). The antigen is injected into the vertebrate host (Figure 11.1(b)), typically a rabbit. After several booster injections and the passage of sufficient time, antibodies

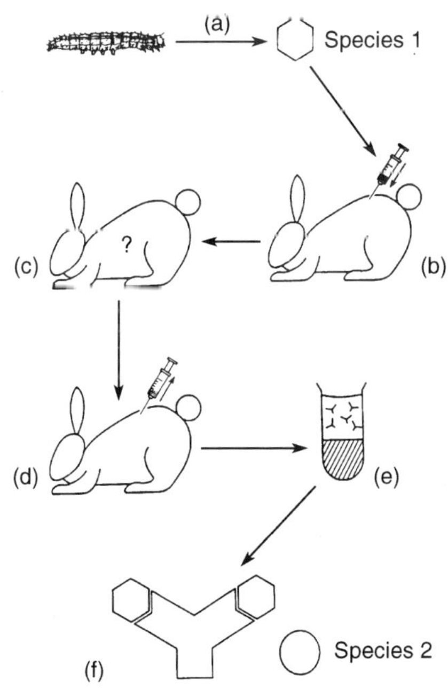

Figure 11.1 Polyclonal antibody (antiserum) production. See text for explanation.

are produced within the rabbit by processes which are beyond the scope of this chapter (see Hood *et al.*, 1978) (Figure 11.1(c)). Next, we bleed the host (Figure 11.1(d)) and allow the blood to clot (Figure 11.1(e)). The resulting antibodies are mostly Y-shaped bipolar molecules, with identical antibody-combining sites on the ends of their arms. These combining sites have shapes which are complementary to that of the immunizing antigen; this is what enables them to recognize the target antigen with specificity (Figure 11.1(f)). The other species in the system, even close relatives (Species 2 in Figure 11.1), would have antigens of different shapes (circular in this hypothetical example), which would not be recognized by the antibodies' combining sites.

Early investigators referred to this antibody mixture as 'specific antiserum,' exhibiting a faith which was seldom justified by the results. By opening up the 'black box' which is the vertebrate host's immune system (Figure 11.1(c)), we can begin to understand why antisera produced in this manner have seldom exhibited the desired specificity.

11.2.2 The vertebrate immune response and the origins of antibody heterogeneity

We now know that the specificity of antigens derives not from the shapes of antigen molecules, but from the shapes of regions of molecules, referred to as antigenic determinants or epitopes (Hood *et al.*, 1978). In Figure 11.2(a) the antigen molecule is shown as bearing three determinants, which are circular, triangular and square (an antigen may bear fewer or many more than three determinants, and a given determinant may be present as multiple copies). When the antigen is injected into the vertebrate host (Figure 11.2(b); mice are shown for consistency with the rest of the figure but the essential details are the same in the rabbit or any other vertebrate) three pre-existing clones of lymphocytes in the immune system are activated (Figure 11.2(c)), each producing copies of an antibody specific to one of the three determinants. As in Figure 11.1, the specificity of each antibody is indicated by the shape of its two identical combining sites, which matches that of one of the determinants (Figure 11.2(d)). The antibodies are secreted into the host's blood stream; after the usual booster immunizations the investigator takes a blood sample, allows it to clot, and removes the serum (Figure 11.2(e,f).

The specificity of this antiserum is evaluated by immunoassay against antigens from target and non-target prey species. When the immunizing antigen from Species 1 (the target) is combined with the antiserum a positive result is obtained (denoted by the asterisk in Figure 11.2(g)) because the antiserum contains antibodies directed against all three determinants. When we perform the assay against the antigen from closely related non-target Species 2, we still obtain a positive result

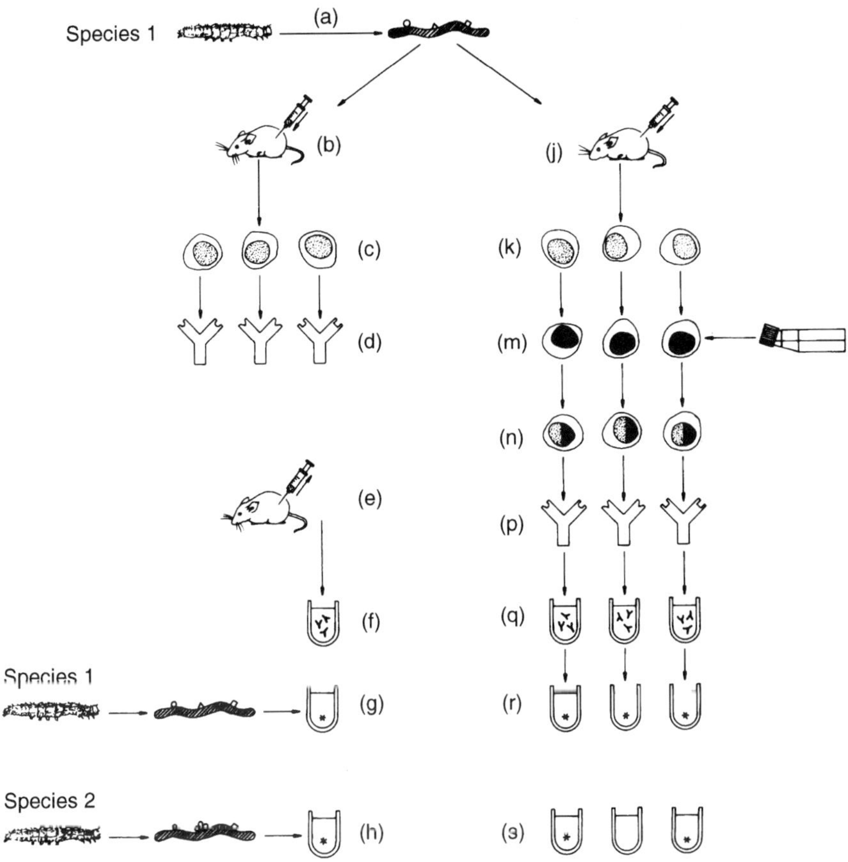

Figure 11.2 Polyclonal (left) and MAb (right) production compared. See text for explanation.

(Figure 11.2(h)), because the two species share a common ancestor and therefore share some determinants (Greenstone *et al.*, 1991); they are shown as sharing the terminal circular and square determinants. Because two of the antibodies in the antiserum recognize the shared determinants, our antiserum cross-reacts; it lacks specificity and produces false positives. Now we can see why antisera are also referred to as 'polyclonal' antibody mixtures, because they are produced by many clones of lymphocytes.

11.3 SOLVING THE SPECIFICITY PROBLEM

Analysis of Figure 11.2(a–h) shows that the cross-reactivity derives from two sources: both species share some antigenic determinants, and the antibodies to all determinants are commingled in the vertebrate host's blood. Attacks on the specificity problem have proceeded along two fronts: (i) refined preparations of antigen; and (ii) purification of the resulting antisera, first by isolating the antibodies in the serum, later by isolating the antibody-secreting lymphocyte clones themselves.

11.3.1 Antigen purification

The first refinement in antigen preparation was precipitation of total protein in the homogenate preparation (Dempster, 1963), an approach used in about one-sixth of the literature reviewed here (Table 11.1). The next approach was the use of a well-defined and easily obtained fraction of body fluid, haemolymph (Ashby, 1974; Greenstone, 1977; Greenstone and Morgan, 1989; Stuart and Greenstone, 1990; Symondson and Liddell, 1993a). It is unlikely that either of these approaches has been particularly efficacious in the attack on specificity (see below). However a third approach, isolation of target-specific protein fractions by two-dimensional (2-D) immunoelectrophoresis (Fichter and Stephen, 1979; Miller, 1979, 1981; Ragsdale, 1980), led to the first stage-specific assay for predator gut analysis (Ragsdale *et al.*, 1981). The most sophisticated level of antigen purification has been to use a single, well-characterized protein (Symondson and Liddell, 1993b; Greenstone and Trowell, 1994). However, this has not always led to high levels of specificity (see later).

11.3.2 Antibody purification

(a) Isolating antibodies from the serum

Slightly more than one-third of the papers reviewed here made use of crude rabbit antisera; about one-tenth of the papers used just the immunoglobulin fraction precipitated from the antiserum. The earliest attempt to reduce antibody heterogeneity and thereby reduce cross-reactivity (Dempster, 1960) was by absorption, i.e. by adding to the antiserum, antigens from cross-reacting insect species and thereby precipitating out the offending antibodies. While this sometimes worked it invariably reduced the already low sensitivity to homologous (target) antigen, often to unusable levels (see Boreham and Ohiagu, 1978 for a review). The most powerful method available for antibody purification is affinity-chromatography, in which antiserum is passed over a column to which target antigen has been covalently linked, allowing non-target antigens to pass through. The pure antigen-specific antibodies are then eluted from the column (Cuatrecasas, 1970). This method has been used only once in predation research (Schoof *et al.*, 1986).

In addition to these active methods of antibody purification, fortuitous gains in specificity were also made with the adoption of more sensitive assays, permitting the 'titreing-out' of cross-reacting antibodies, which apparently are often of lower concentration than homologous ones (Greenstone, 1979a).

(b) Isolating antibody-producing lymphocytes

If the individual antibody-producing lymphocytes could be isolated and propagated *in vitro*, then the specific antibodies they produce could be collected in pure form. Unfortunately long-term culture of healthy lymphocytes producing antibodies of defined specificity has never been achieved, and the only lymphocytes that will grow indefinitely in culture, the naturally occurring or experimentally induced tumours known as myelomas, produce antibodies of random specificity.

Lymphocyte hybridomas

While performing basic research on the genetic basis of antibody specificity, Köhler and Milstein (1975) developed a technique for fusing myeloma cells with healthy lymphocytes from immunized animals. The new cell was dubbed a 'hybridoma' to indicate the hybrid nature of its genome, which contains the information both to produce antibody of defined specificity and to grow indefinitely as a tumour in culture. With such cells it is possible to isolate the production of a single, elicited antibody *in vitro*. The immunization is performed as before, and in our example the same three clones are activated in the host's immune system (Figure 11.2(j,k)). These cells are then fused with myeloma cells from a pre-existing culture (Figure 11.2(m)) to produce three hybridomas (Figure 11.2(n)), each of which can be grown separately in culture to produce the same three antibodies as previously (Figure 11.2(p)). Now, however, these so-called 'monoclonal antibodies' (MAbs), each from the tissue culture supernatant of a different hybridoma clone, can be isolated in separate tubes (Figure 11.2(q)). When these antibody preparations are used in immunoassays, all three recognize the target species (Figure 11.2(r)), but only two of them recognize the heterologous species (Figure 11.2(s)); the antibody in the central tube is directed against the triangular determinant, which is unique to the target species. This antibody is completely species-specific (Figure 11.2(s)).

The procedures involved in the production of MAbs have been well known for more than 15 years (Köhler and Milstein, 1975; see appendices in Kennett *et al.*, 1980). Mice are immunized (Figure 11.3(a)) according to standard protocols (Lenz and Greenstone, 1988). Four days after a final intravenous booster immunization, the mice are killed and splenectomized (Figure 11.3(b)). Viable lymphocytes are freed from the

Table 11.1 Summary of investigations employing serological methods for arthropod predator gut analysis. COE, cross-over electrophoresis; EIA, ELISA; dab, double-antibody; das, double-antibody sandwich; Hmgnt, homogenate; HmProt, homogenate or haemolymph total protein; Hmlmph, haemolymph; ind, indirect; ID, immunodot; IEF, isolectric focussing; LtxAgg, latex agglutination; MsMAb, mouse MAb; MsPcAff, affinity-purified mouse polyclonal; NR, not reported; Oak, Oakley–Fulthorpe; Ouch, Ouchterlony; Prot2DE, protein purified by two-dimensional immunoelectrophoresis; PurProt, purified single protein; RabPc, rabbit polyclonal antiserum; RabPc-ab, RabPc purified by absorption; RabIg, rabbit immunoglobulin; RabPcAff, affinity-purified rabbit polyclonal; RIA, radioimmunoassay; RID, radial immunodiffusion; Tptn, tube precipitin. Sensitivities in dilutions refer to dilutions of a standardized antigen preparation (normal typeface) or dilutions of a target prey individual (**bold face**). See text for explanation.

Target	*Predator*	*Antigen*	*Antibody*	*Assay*	*Specificity*	*Sensitivity*	*Reference*
Anopheles quadrimaculatus	Insecta	Hmgnt	RabPc	Tptn	Order	1:4320	Brooke and Proske, 1946
Malacosoma disstria	*Zelus exsanguis*	Hmgnt	RabPc	Tptn	Family	1:12 500	Hall *et al.*, 1953
						1:20 000	
Phytodecta olivacea	Insecta, Acarina	Hmgnt	RabPc-ab	Tptn	Family	1:8000	Dempster, 1960
						1:20 000	
Eroischia brassicae	Carabidae	Hmgnt	RabPc	Tptn	NR	1:8000	Coaker and Williams, 1963
	Staphylinidae					**1:1**	
Arytaina genistae	*Anthocorus* spp.	HmProt	RabPc-ab	Tptn	Genus	NR	Dempster, 1963
A. spartiophila							
Acyrthosiphon spartii							
Heterocordylus tibialis							
Asciodema obsoletum							
Orthotylus adenocarpi							
O. virescens							
Chesias legatella							
Sarothamnus scoparius							
Choristoneura fumiferana	Araneae	Hmgnt	RabPc	Tptn	Order	**1:1**	Loughton *et al.*, 1963
Asellus meridianus	*Dendrocoelum lacteum*	HmProt	RabPc	Oak	Family	1:512	Young *et al.*, 1964
Pieris rapae	Insecta, Arachnida	HmProt	RabPc	Tptn	Family	NR	Dempster, 1967

Conomelus anceps	Insecta, Arachnida	Hmgnt	RabPc	Tptn	Family	1:3000	Rothschild, 1966
Operophtera brumata	Coleoptera	Hmgnt	RabIg	Ouch	Order	NR	Frank, 1967a
Sitona regensteinensis	Coleoptera	Hmgnt	RabPc	Tptn	NR	NR	Danthanarayana, 1969
Tricladida	*Pyrrhosoma nymphuls*	HmProt	RabPc-ab	Oak	Genus	**1:32**	Davies, 1969
Crustacea, Insecta Annelida, Pelecypoda	*Dugesia tigrina*	Hmgnt	RabPc	Ouch	Order	1:16 000	Pickavance, 1970
Leptocorisa oratorius	Insecta, Araneae	HmProt	RabPc	Tptn	NR	1:900	Rothschild, 1970
Amphipoda	*Dugesia* spp.	HmProt	RabPc	Ouch	Class	NR	Reynoldson and Davies, 1970
Gastropoda Oligochaeta							
Eurygaster integriceps	Insecta	Hmgnt	RabPc-ab	Ouch	Family	**1:250**	Sergeeva, 1970a,b
Philoscia muscorum	*Sorex* spp. *Apodemus sylvaticus* Insecta, Arachnida	Hmgnt	RabPc	Ouch	Order	NR	Sutton, 1970
Tyria jacobaeae	Insecta, Isopoda, Arachnida	HmProt	RabPc-ab	Tptn	Species	NR	Dempster, 1971
Euxoa ochrogaster	Carabidae, Staphylinidae	Hmgnt	RabIg	Ouch	NR	NR	Frank, 1971
Tryporyza spp. *Chilo* spp. *Sesamia inferens*	Insecta, Arachnida	HmProt	RabPc	Tptn	Family	1:4000	Rothschild, 1971
Agrotis ipsilon	Carabidae	Hmgnt	RabPc	Ouch	Class	1:1280 **1:640**	Lund and Turpin, 1977
Rhopalosiphum padi	Insecta, Arachnida	Hmgnt	RabPc	IEF/Ouch	Species	**1:1**	Petterson, 1972
Epiphyas postvittana	Insecta, Araneae	Hmgnt	RabPc	Tptn	Family	NR	MacLellan, 1973
Aedes contans	Insecta, Arachnida Tricladida	HmProt	RabPc-ab	Tptn	Family	1:4000 **1:64**	Service, 1973a
Anopheles gambiae	Insecta, Araneae Amphibia	HmProt	RabPc	Tptn	Family	1:4000 **1:128**	Service, 1973b
Agriolimax reticulatus	Coleoptera, Opiliones	Hmgnt	RabPc	Ouch	NR	NR	Tod, 1973

Table 11.1 continued

Target	*Predator*	*Antigen*	*Antibody*	*Assay*	*Specificity*	*Sensitivity*	*Reference*
Daphnia spp. *Chironomus* spp. Tubificidae	*Phaenocora typhlops*	Hmgnt	RabPc	Ouch	Family	NR	Young, 1973
Pieris rapae	Insecta, Arachnida	Hmlmph	RabPc	Tptn	NR	1:1600	Ashby, 1974
Tubificidae Gastropoda Isopoda Amphipoda	*Dugesia polychroa*	Hmgnt	RabPc	Ouch	Order	1:16	Boddington and Mettrick, 1974
Ixodes ricinus	NR	Hmgnt	RabPc	COE	NR	1:10	Healy and Cross, 1975
Simulium damnosum	Insecta, Pices	HmProt	RabPc-ab	Tptn	Family	1:4000 **1:64**	Service and Lyle, 1975
Aphididae	Insecta, Araneae	NR	RabPc	Ouch	Family	1:2560	Vickerman and Sunderland, 1975
Gammarus Plecoptera Trichoptera Oligochaeta Ephemeroptera	*Polycelis felina* *Crenobia alpine*	HmProt	RabPc	Ouch	Order	NR	Lock and Reynoldson, 1976
Tubifex tubifex *Asellus meridianus* *Chironomus* spp.	*Planaria torva*	Hmgnt	PabPc	Ouch	Order	NR	Reynoldson and Sefton, 1976
Aedes dorsalis *Ephydra riparia* *Trichocorixa reticulata*	*Pardosa ramulosa*	Hmlmph	RabPc	PHI	Genus	1:96 000	Greenstone, 1977
Agrotis ipsilon	*Petrostichus* spp.	Hmgnt	RabPc	Ouch	Class	1:1280 **1:640**	Lund and Turpin, 1977

Dryocosmus kuriphilus	Araneae	Hmgnt	RabPc	Tptn	Order	1:32 **1:160**	Nakamura and Nakamura, 1977
Hypera postica	*Harpalus pennsylvanicus*	Hmgnt	RabPc	Ouch	Family	NR	Mollet and Armbrust, 1978
Acyrthosiphon pisum	*Coccinella septempunctata*	Hmgnt	RabIg	LtxAg	Family	1:80	Ohiagu and Boreham, 1978
Lygus lineolaris	*Coleomegilla maculata* Hemiptera	NR	RabPc-ab	Ouch	NR	1:1280	Whalon and Parker, 1978
Dendroctonus frontalis	NR	Prot2DE	RabPc	Ouch COE EIAdas	Genus	1:10 1:80 1:2560	Miller, 1979, 1981
Isopoda	Insecta, Arachnida, Chilopoda	Hmgnt	RabPc-ab	Pptn	Species	1:520 **1:50**	Sunderland and Sutton, 1980
Orygia pseudotsugata	*Podisus maculiventris*	Prot2DE	RabIg	EIAdas	Family	1:260 000	Fichter and Stephen, 1981
Noctuidae	Carabidae, Hemiptera	Hmgnt	RabPc	Pptn	NR	1:3200	Gardner *et al.*, 1981
Drosophila spp. *Oncopeltus fasciatus*	*Pardosa sternalis*	Hmgnt	RabPc-ab	Ouch RID	Order	1:800	McIver, 1981
Nezara viridula	Insecta, Araneae	Prot2DE	RabIg	EIAdas	Stage	7.6ng	Ragsdale, 1980; Ragsdale *et al.*, 1981
Tortricidae	Carabidae	Hmgnt	RabPc-ab	COE	Order	1:1024	Allen and Hagley, 1982
Helicoverpa zea	Carabidae	Hmgnt	RabPc	Ouch	Family	1:1024	Lesiewicz *et al.*, 1982
Aphididae	Carabidae	Hmgnt	RabPc	Ouch	NR	NR	Hance and Gregoire-Wibo, 1983
Isopoda, Diplopoda,	Coleoptera	Hmgnt	RabPc	Pptn	NR	NR	Dennison and Hodkinson, 1983
Acarina, Araneae, Collembola, Diptera, Nematoda, Lumbricidae, Enchytraeidae, Carrion, Fungi							
Aphididae	Carabidae, *Forficula auricularia*	Hmgnt	RabIg	EIAdas	Family	**1:1000**	Crook and Sunderland, 1984

Table 11.1 continued

Target	*Predator*	*Antigen*	*Antibody*	*Assay*	*Specificity*	*Sensitivity*	*Reference*
Acyrthosiphon	Insecta, Arachnida	Hmgnt	RabPc	Pptn	Family	1:32	Leathwick and Winterbourn, 1984
Inopus rubriceps	Elateridae spp.	HmProt	RabPc	COE	Order	1:128 438 ng **1:1700**	Doane *et al.*, 1985
Ostrinia nubilalis	*Poecilus cupreus*	Hmgnt	RabIg	EIAdas	NR	NR	Lövei *et al.*, 1985
Plutella xylostella	NR	Hmgnt	RabPc-ab	Ouch	Species	NR	Nemoto *et al.*, 1985
Mansonia dyari	*Toxorhynchites rutilus*	Hmgnt	RabPc	RIA	Genus	1:5210	Bailey *et al.*, 1986
Musca vetustissima	Coleoptera	Hmgnt	RabPc	COE	Family	1:9000	Calver *et al.*, 1986
Culex tarsalis	Hemiptera, Coenagrionidae	Hmgnt	RabPcAff MsPcAff	EIAdab	Genus	2.5 ng	Schoof *et al.*, 1986
Rhopalosiphum padi	*Bembidion lampros*	Hmgnt	RabIg	EIAdas	Genus	100 μg	Chiverton, 1987
Aphis fabae	*Pterostichus melanarius*	Hmgnt	RabIg	EIAind	Order	NR	Hance and Renier, 1987
Pieris rapae	Coleoptera	Hmgnt	RabPc	EIAdas	Family	NR	Kapuge *et al.*, 1987
Sitobion avenae	*Bembidion lampros*	Hmgnt	RabIg	EIAdas	NR	NR	Lovei *et al.*, 1987
Aphididae	Insecta, Araneae	Hmgnt	RabPc	EIAdas	Family	**1:100**	Sunderland *et al.*, 1987
Helicoverpa zea	*Podisus maculiventris*	Hmlmph	MsMAb	EIAdas	Instar	NR	Greenstone and Morgan, 1989
Sitobion avenae	Coleoptera, Linyphiidae	Hmgnt	RabIg	EIAdas	NR	NR	Sopp and Sunderland, 1989
Rhagoletis pomonella	Insecta, Araneae	Hmgnt	RabPc	COE	NR	NR	Allen and Hagley, 1990
Lymantria dispar	Carabidae, Arachnida, Chilopoda	Hmgnt	RabPc	EIAdas	Order	NR	DuDevoir and Reeves, 1990
Helicoverpa zea	*Phidippus audax* *Podisus maculiventris*	Hmlmph	MsMAb	IDdas	Instar	16 ng	Stuart and Greenstone, 1990
Lygus hesperus	NR	Hmgnt	MsMAb	EIAind	Stage	NR	Hagler *et al.*, 1991
Crustacea, Odonata, Gastropoda,	*Ranatra montezuma*	Hmgnt	RabPc-ab	Oak	Class	NR	Blinn *et al.*, 1993

Oligochaeta							
Bemisia tabaci	NR	Hmgnt	MsMAb	EIAind	Stage	NR	Hagler *et al.*, 1993
Lopidea nigridea	Insecta, Araneae	Hmgnt	RabPc	EIAdAb	Genus	12.5 ng	McIver and Tempelis, 1993
Coquillettia insignis							
Periplaneta americana	*Urodacus* spp.	Hmgnt	RabPc	RID	NR	100µg/ml	Quinlan *et al.*, 1993
Arion ater	*Abax parallelepipidus* *Pterostichus madidus* *Phalangium opilio*	Hmlmph	RabPc-ab	EIAind	Subclass	0.86 ng	Symondson and Liddell, 1993a
Arion ater	*Abax parallelepipidus*	PurProt	MsMAb	EIAind	Genus	3.4 ng	Symondson and Liddell, 1993b
Deroceras reticulatum	*Pterostichus melanarius*	Hmgnt	MsMAb	EIAind	Species	7.3 ng	Symondson and Liddell, 1993c, 1995
Pectinophora gossypiella	*Collops vittatus* *Orius tristicolor* *Geocoris punctipes* *Chrysoperla carnea*	Hmgnt	MsMAb	EIAind	Stage	**1:1**	Hagler *et al.*, 1994
Heliothis virescens	*Podisus maculiventris*	PurProt	MsMAb	IDind	SuF/Stage	16 ng **1:1**	Greenstone and Trowell, 1994
Helicoverpa zea	NT	Hmgnt	MsMAb	IDind	Stage	125 ng **1:1**	Greenstone, 1995

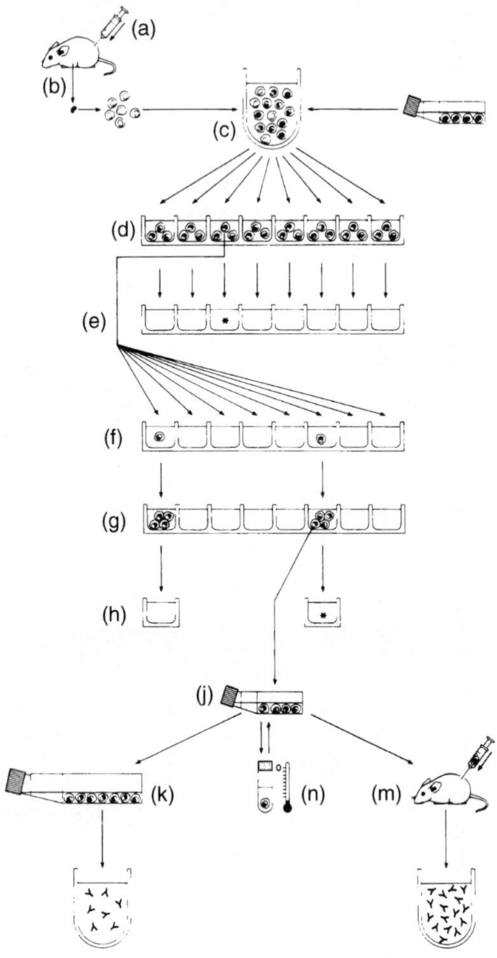

Figure 11.3 Production, screening, cloning, and storage of hybridoma cell lines. See text for explanation.

spleen by maceration and combined in the presence of polyethylene glycol (Gefter *et al.*, 1977) with myeloma cells from culture (Figure 11.3(c)). The cells are grown in micro-well tissue culture plates in a selective medium (Littlefield, 1964) which allows only the hybrids to live (Figure 11.3(d)), the myeloma having been previously selected for enzyme deficiency (Milstein, 1981) and the healthy lymphocytes possessing the enzyme but unable to grow in culture. The antibody-containing supernatants of these cell cultures are subsequently assayed (Figure 11.3(e)) for their ability to recognize the immunizing antigen, and cells from any positive cultures are cloned (Figure 11.3(f)) by limiting dilution (McKearn, 1980), that is, 'plated out' to such a dilute suspension that the majority of wells of the culture plate will receive no cells while 20–30% of them will receive a single cell, which will give rise to a clone (Figure 11.3(g)). Supernatants from these clones are assayed again (Figure 11.3(h)) to ensure that they are still producing antibody which recognizes the immunizing antigen. At this point they are also screened against all relevant heterologous antigens to determine their specificity. After at least one more cycle of cloning and assaying (not shown), stable cell lines of appropriate specificity are grown in larger cultures (Figure 11.3(j)). These cells may then follow one of three routes. They may be grown in mass culture to provide a stock of supernatant antibody for use in assays (Figure 11.3(k)). Alternatively, they may be injected into mice of the same strain to establish soft tumours (Köhler and Milstein, 1975) which secrete antibodies at 10–1000 times higher concentration than in cell culture medium (Figure 11.3(m)). Finally, once the investigator has sufficient antibody on hand for immediate needs, the cells can be frozen and held indefinitely until more antibody is needed, whereupon they can be thawed and grown again in culture (Figure 11.3(n)).

MAb technology yields the highest specificity obtainable from the vertebrate immune system. The investigator is assured an unlimited supply of pure, uniform antibody which recognizes a single antigenic determinant. Furthermore, because antibody selection is performed at the level of the hybridoma clone, the crudest preparations can be used for immunization, provided that small amounts of target and heterologous antigens are available for screening assays (Milstein, 1981; Lenz and Greenstone, 1988).

11.4 IMMUNOASSAY TECHNOLOGIES

Once we have an antibody preparation that recognizes target antigen with the required specificity, we combine it with preparations of predators to see whether target antigens are present in the gut. Since antigens and antibodies are microscopic molecules, we require some mechanism to make the recognition of antigen by antibody detectable; this mechanism is the immunoassay.

11.4.1 Precipitin tests

The first immunoassays used in predation research were precipitin tests, which are performed by combining the soluble fractions of predator preparations and a solution of the antibody. Homologous antigen and antibody combine, forming a visible precipitate. Because large amounts of both antigen and antibody are required in order to form a large enough precipitate to be visible, the precipitin test is inherently insensitive (Greenstone, 1979a). The first precipitin tests used by entomologists were tube precipitin tests, performed first in serological tubes and later in microcapillary tubes. Subsequently various other formats, in which the antibody and antigen interact in gels, were introduced. The most popular of these was the Ouchterlony test (first used for predation research by Pickavance, 1970), which had the advantage that it was able to distinguish homologous and cross-reacting heterologous antigens (Eisen, 1974). Other gel precipitin formats include the Oakley–Fulthorpe test (first used for predation research by Davies, 1969), cross-over electrophoresis, also known as immunoelectro-osmophoresis (first used for predation research by Healy and Cross, 1975), and a quantitative variant, radial immunodiffusion (McIver, 1981). Although gel precipitin assays have been more popular than tube precipitin tests, they do not differ more than a few-fold in sensitivity (Boreham and Ohiagu, 1978), and have been largely overtaken by solid-phase assays.

11.4.2 Agglutination assays

A second family of assays uses macroscopic particles to make the antigen–antibody interaction visible. In the latex agglutination assay, latex beads are coated with antibody; when antigen-containing predator extract is added, the antigen molecules bridge antibody molecules on adjacent latex beads, causing them to clump. As reported by Ohiagu and Boreham (1978), this assay was fast and simple but not very sensitive. In the passive haemagglutination inhibition (PHI) assay, sheep red blood cells are coated with antigen, and the smallest concentration of antibody required to maintain them in suspension in microtitre plates is determined. The assay is performed by incubating this antibody solution with predator extracts and then adding the coated cells: target antigen in predator extracts will have bound the antibody, thereby allowing the cells to fall out of suspension (Greenstone, 1979a). Though extremely sensitive, PHI was never widely adopted, due to the availability of equally sensitive and simpler ELISA assays.

11.4.3 Solid-phase assays

Solid-phase assays are so called because the reactions involve reagents which are bound to solid surfaces. There are three common formats for solid-phase assays (Figure 11.4). In the double antibody sandwich format, the first reagent bound on to the surface is the antibody. Following a period of incubation, any unbound antibody is removed by extensive washing. Now the predator extract is added; if it contains homologous antigen, it will be bound by the attached antibody. Incuba-

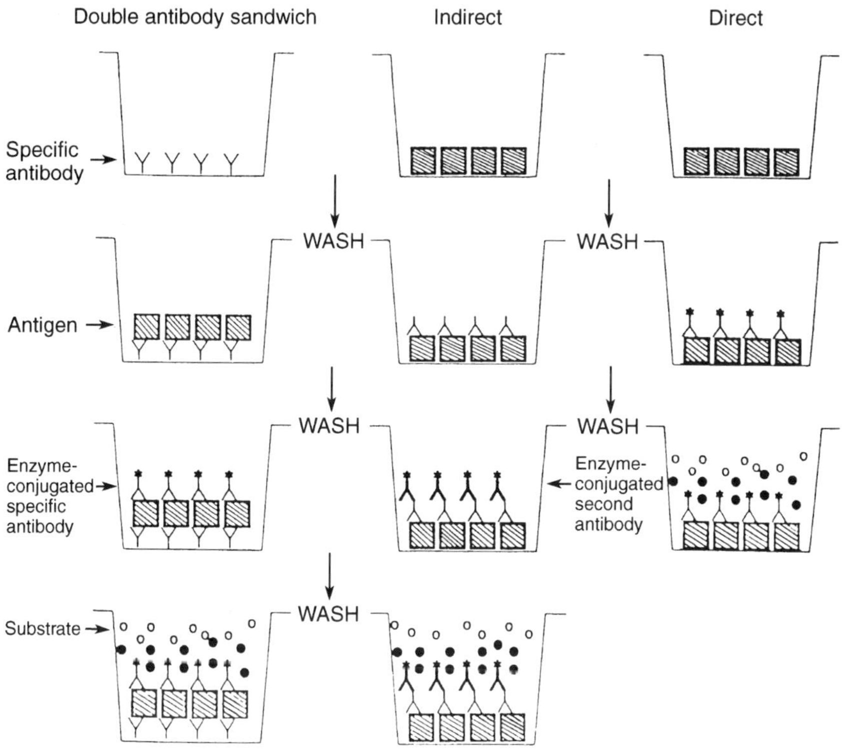

Figure 11.4 Common formats for solid phase immunoassays. See text for explanation.

tion and washing are repeated and a conjugate, which is another preparation of the same antibody, or of a different antibody to a second determinant of the same antigen, this time with a label attached, is added. In most cases the label is an enzyme (but see later). After incubation and washing, the final reagent is added; if the penultimate reagent was an enzyme-conjugated antibody, then the ultimate reagent will be a substrate which, when hydrolysed by the enzyme, will yield a coloured product, signifying antibody–antigen recognition.

In the indirect format, the first reagent is the predator extract, and the second is the antibody. Following this, a labelled second antibody to immunoglobulins from the host species which produced the first antibody is added. For example, if the first antibody is a mouse MAb, then the second may be antibody to mouse immunoglobulin produced in sheep. Finally, in the case of the enzyme–substrate assays, the substrate is added and the dye development is evaluated. Advantages of the indirect format are that second antibody conjugates are commercially available, and they may be more sensitive.

In the direct format, the first reagent is the antigen (predator extract), followed by the labelled first antibody. This shortens the assay by at least one step, but the investigator must prepare the conjugate.

There are other variations on these formats. For example the second antibody conjugate label may be biotin, to be followed by an enzyme–avidin complex (Greenstone and Trowell, 1994). Although this adds a step, it boosts sensitivity because several biotin molecules may be conjugated to one antibody molecule and several antibody molecules to each avidin molecule. In the double antibody format, the antigen is sandwiched between two antibodies to target antigen, raised in different vertebrate host species, followed by a commercially produced conjugate to the second specific antibody (Schoof *et al.*, 1986).

(a) ELISA

Enzyme-linked immunosorbent assays (ELISAs) were first employed in predation research in 1981 by Miller and by Ragsdale and coworkers. In this case the solid surface is polystyrene, and the assay is usually performed in moulded, 96-well plates. The substrate produces a soluble dye upon hydrolysis, and the assay is usually scored with a plate reader, which is a multi-channel spectrophotometer that reads the absorbance at the peak wavelength for the dye in question. ELISA takes just a few hours, is fool-proof once optimized and routinely produces sensitivities in the 1–50 ng range. Furthermore, the 96-well format has been standardized and rapid liquid-handling and washing apparatus is available. A disadvantage is the expense of the plate reader, a minimum £1300 (US$2000). The plate reader is essential, because there is always some dye development, even in a known negative ELISA, so that the absorbance must be read and a statistical (or arbitrary quantitative) decision made on how to score the result (Stuart and Greenstone, 1990; Fenlon and Sopp, 1991; Symondson and Liddell, 1993a).

ELISA is the assay of choice in predation research today. Although most investigators have used double antibody sandwich assays, the indirect format is becoming more popular. Early literature suggested that the indirect format is about an order of magnitude more sensitive than the double antibody sandwich (Greenstone, 1983a), and this is confirmed by the quantitative data in Table 11.1.

(b) Immunodot

In 1990, Stuart and Greenstone adapted for predation research an assay employing a nylon or nitrocellulose membrane as the solid phase, and enzyme substrates which yield particulate dyes that are deposited on the membrane and can be read by eye (Stott, 1989). When the assay is properly optimized, known negatives do not produce any dye development, so that a quantitative decision to distinguish positives from negatives does not need to be made. The expensive plate reader is thereby eliminated. Although the original assay, which was a double antibody sandwich, requires a membrane holder (about £260; US$400), an indirect version eliminates the need for even that piece of apparatus (Greenstone and Trowell, 1994); it also requires only 1 μl of predator extract for each test. Sensitivity of the immunodot is comparable with that of ELISA, and specificity is not compromised (Stuart and Greenstone, 1990).

Like ELISAs, immunodots can be quantified (Hagler *et al.*, 1995). Nevertheless, knowing the quantity of antigen recognized in a positive assay does not lead straightforwardly to an estimate of the number of prey consumed per predator, and some predation models derive quantitative predation rates from qualitative gut assay data (see later).

(c) Radioimmunoassay

Bailey *et al.* (1986) used an indirect radioimmunoassay (RIA) performed on 96-well polyvinyl plates. Protein-A conjugated to a radionuclide took the place of an enzyme-conjugated second antibody, and the assay was read by cutting out and placing each well in a glass tube and reading the bound radioactivity with a gamma counter. RIA is sensitive, but the availability of simpler and less expensive assays, and difficulties in using radionuclides and ultimately disposing of them, have doomed this technology for our purposes.

11.5 HISTORICAL OVERVIEW OF SEROLOGICAL GUT ANALYSIS

Serological analysis has been used by ecologists to study a myriad of arthropod-interspecific interactions, beginning with mosquito–vertebrate host relationships, in a publication (Bull and King, 1923) that was also credited with providing the inspiration for the first serological analysis of arthropod predation (Brooke and Proske, 1946). Information on the technologies used in the development and application of serological gut analysis from 1946 to the present has been tabulated (Table 11.1). In order to be included in the analysis, a publication had to describe assays designed either to detect arthropod prey or to detect prey in arthropod predator guts (or both), identify the nature of the antigen, antibody preparation, and assay used, and record its sensitivity and specificity. Because the ultimate goal was to compare the efficacies of different approaches, papers that lacked data in two or more categories were excluded. Also, papers employing antibodies that had already been characterized in earlier papers were excluded, so as not to bias the data. In a few cases more than one publication was required in order to get all of the information for one example. These criteria yielded a total of 93 publications (Table 11.1).

A few notes on some of the categories are in order. Target organisms are recorded as the lowest taxonomic level given by the investigator. For example, in most cases antigens were derived from single species, but in others they were from multi-species pools (e.g. for aphids compare Petterson, 1972 and Vickerman and Sunderland, 1975). Predators were usually identified to genus and species, but where they were identified only to higher levels (e.g. 'spiders') or include long multi-species lists, the lowest encompassing taxonomic level has been given. Specificity is the most difficult category to score, since investigators do not usually make close comparisons if, say, congeners or members of the same subfamily or family are not apt to be encountered in the ecosystem under study. Therefore, the lowest taxonomic level not excluded by the investigator's tests of cross-reactivity has been given. Sensitivity has, unfortunately, been reported in three different ways: as the highest dilution of the investigator's 'standardized' antigen preparation, as the highest detectable dilution of a target individual, and as the smallest mass of antigen detectable.

It is obvious from inspection of Table 11.1 that we can produce antibodies against any prey item, invertebrate or vertebrate (Dennison and Hodkinson, 1983), and use it to detect prey in the gut of any predator, invertebrate or vertebrate. Progress in the evolution of technology is more easily appreciated by graphical presentation. Figure 11.5 shows the frequency of use of the options in the three component technologies of serological gut analysis in the pioneering phase (1946–1964) and each of the subsequent three decades leading to the present. Antigen preparations remained relatively crude through 1974, comprising only whole homogenate, homogenate protein, and haemolymph. The decade 1975–1984 saw the most strenuous efforts in antigen preparation, crowned by three studies using proteins purified by 2-D immunoelectrophoresis (Fichter and Stephen, 1979; Miller, 1979; Ragsdale, 1980). In the decade now ending, biochemically characterized individual proteins have been used twice (Symondson and Liddell, 1993b; Greenstone and Trowell, 1994), but 2-D immunoelectrophoresis, though effective, has passed out of use, probably because crude preparations have been found to be adequate for most purposes (see later).

Antibody production was dominated by rabbit polyclonal antisera, crude or absorbed, or sometimes precipitated rabbit immunoglobulin, until the present decade, when mouse MAbs have been used increasingly. There was also one use of affinity-purified antisera (Schoof *et al.*, 1986).

Immunoassay technology has undergone the most exploration and evolution of the three component technologies. The tube precipitin test has disappeared entirely; the Ouchterlony test appears to be on the way out but other gel precipitin tests are still in use. Agglutination assays have also disappeared (but see later), as has radioimmunoassay. ELISAs have emerged as the most popular format.

The connection between technology choices and the level of specificity achieved is explored in Figure 11.6. Specificities ranging from instar to class have been reported (Table 11.1). In order to make this analysis comprehensive, levels of specificity have been arbitrarily divided into two classes, genus or lower and subfamily or higher. In practice these will often translate, respectively, into ecologically useful and ecologically inadequate levels for many of the agroecosystems in which we work. Options in each of the three technology families have also been combined in ways which do not appear to mask the virtues and deficiencies of each.

All antigens prepared by 2-D immunoelectrophoresis have produced ecologically useful results, but only one of the single well-characterized protein preparations did so (Symondson and Liddell, 1993b). While most uses of essentially unfractionated antigen preparations (comprising crude homogenate, haemolymph, and homogenate and haemolymph total protein) have led to poor specificity, most of the cases of ecologically useful assays have also employed them, which perhaps explains their continuing domination into the present decade (Figure 11.5).

Antibody preparation has been dominated by rabbit polyclonals. Most of these have not been associated with ecologically useful specificity as defined here. In most cases where they have done so, they were improved by absorption (Dempster, 1963; Davies, 1969) or raised against antigens selected by 2-D immunoelectrophoresis (Miller, 1979; Ragsdale, 1980). On the other hand, most assays achieving ecologically useful specificity employed MAbs (Greenstone and Morgan, 1989; Hagler *et al.*, 1991, 1993, 1994; Symondson and Liddell, 1993b, 1993c; Greenstone, 1995).

Ecologically useful levels of specificity have been achieved by immunoassays of all three families. However, most of them involved solid phase (ELISA and immunodot) assays.

Sensitivities in terms of mass of antigen detectable were not reported until sensitive assays appeared (Ragsdale *et al.*, 1981). Almost all ELISA and immunodot assays have detectabilities ranging from 1 to 20 ng (Table 11.1). In terms of the proportion of a target individual which can be detected, where the target is a larva or pupa it often ranges from one-fiftieth to one part in several thousand (Table 11.1); investigators working on egg predation have usually been content to report that a

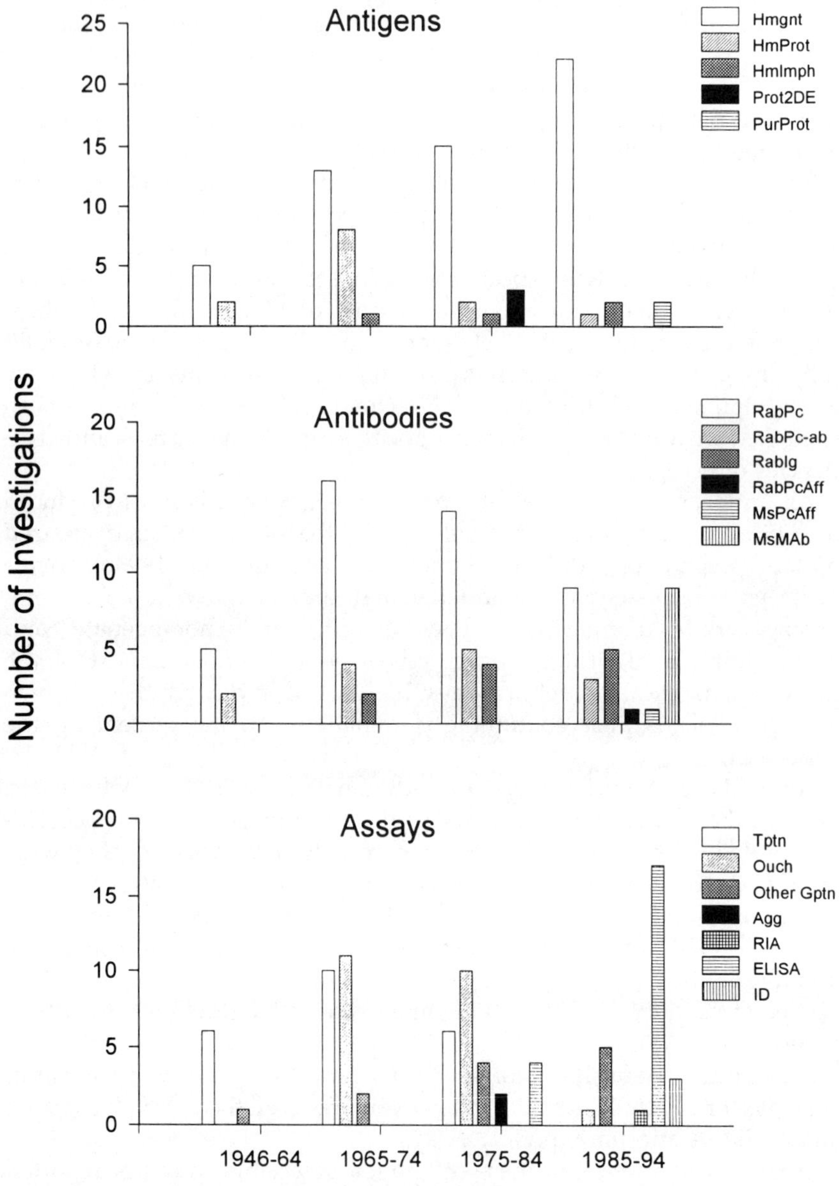

Figure 11.5 Chronology of progress in antigen preparation (top), antibody production (middle), and immunoassay technologies (bottom). ELISA, enzyme-linked immunosorbent assay; Hmgnt, homogenate; Hmlmph, haemolymph; HmProt, homogenate or haemolymph total protein; ID, immunodot; MsPcAff, affinity-purified mouse polyclonal; MsMAb, mouse MAb; Ouch, Ouchterlony; Other Gptn, other gel precipitin assays; Prot2DE, protein purified by two-

single egg is detectable (Hagler *et al.*, 1994; Greenstone and Trowell, 1994; Sigsgaard, 1996, this volume).

11.6 THE STATE-OF-THE-ART IN SEROLOGICAL ARTHROPOD GUT ANALYSIS

Serological gut analysis is a mature technology from which we can now expect to achieve levels of sensitivity and specificity required for ecological research. On the basis of the foregoing analysis, the following protocol can be recommended for achieving a sensitive, specific analysis for any system: (i) immunize with a crude homogenate of the target species; (ii) use MAbs, not polyclonal antisera; and (iii) use an ELISA or immunodot assay.

Some will argue that while MAb production may be the surest route to antibody specificity, it is expensive, time-consuming, and requires training and institutional support available only to a few investigators. This is unfortunately correct, which is why investigators should be urged to consider an alternative: affinity-chromatography of antisera. It is much less expensive and time consuming than MAb technology, and while it requires some skill, it has produced very high levels of specificity in this and other areas of insect ecology (Tempelis and Rodrick, 1972; Schoof *et al.*, 1986).

I also suggest that while ELISA and immunodot are excellent immunoassay formats, latex agglutination technology may be an attractive alternative. Latex bead technology has undergone a renaissance in the last 10 years, and various sensitive, rapid assay formats are now available (Bangs, 1990) and worthy of exploration.

11.7 THE FUTURE OF SEROLOGICAL GUT ANALYSIS

Improvements in serological gut analysis will depend in the future – as they have in the past – on technological innovations in applied immunology generally. A particularly exciting new area is the introduction of genes encoding the variable antibody region into bacteria by bacteriophage vectors (Symondson and Liddell, 1993c; Engelhardt *et al.*, 1994; George, 1994; Liddell and Symondson, 1996, this volume). Since the genes are cloned directly from lymphocytes rather than hybridoma cells, the considerable expense and complexity of hybridoma technology can be avoided.

---

dimensional immunoelectrophoresis; PurProt, purified single protein; RabPc, rabbit polyclonal antiserum; RabPc-ab, RabPc purified by absorption; RabIg, rabbit immunoglobulin; RabPcAff, affinity-purified rabbit polyclonal; RIA, radioimmunoassay; Tptn, tube precipitin.

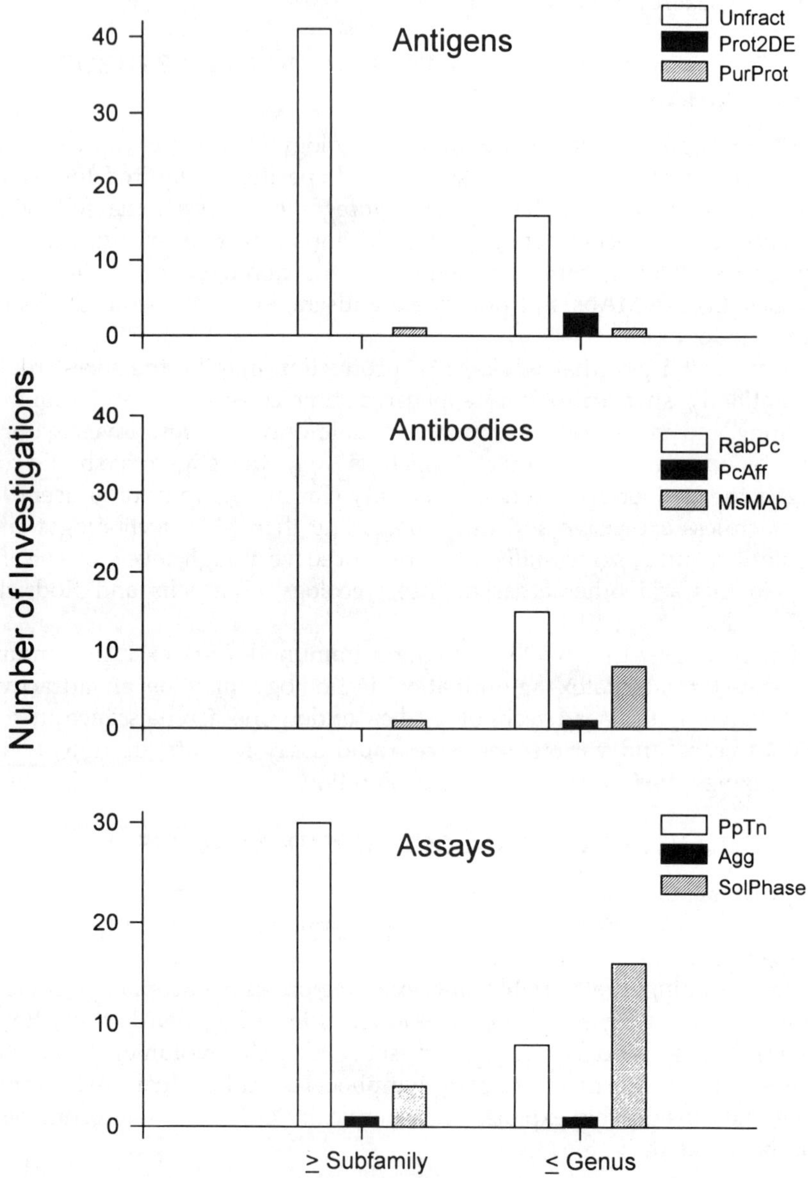

Figure 11.6 Levels of taxonomic specificity associated with different antigen preparation (top), antibody preparation (middle), and immunoassay (bottom) technologies. Agg, agglutinating assays; MsMAb, mouse MAb; PcAff, affinity-purified polyclonal; Pptn, tube and gel precipitin assays; Prot2DE, protein purified by two-dimensional immunoelectrophoresis; PurProt, purified single

There is a great deal of research going on into the nature of antigenic determinants, antibody combining sites, and antigen–antibody recognition mechanisms (Paterson, 1992; Rini *et al.*, 1992; Lang *et al.*, 1994), and prototype antibody-like molecules have already been synthesised (Saragovi *et al.*, 1991; Haynes *et al.*, 1994). The time may therefore not be far off when target-specific antigenic determinants can be identified and antibody-like molecules which recognize them prepared from scratch.

11.8 QUANTITATION OF SEROLOGICAL GUT DATA

The transformation of assay data into ecologically useful per-capita estimates of the number of prey eaten per unit time remains an unsolved technical problem in serological gut analysis. The exquisite quantifiability of assays like ELISA is misleading, for while they enable us quite accurately to estimate the mass of antigen represented by a predator assay of a particular absorbance, we cannot know how many prey individuals are represented by that positive assay, or when they were consumed. Clearly, a large mass of antigen could represent a large prey item, a prey item that had been extensively preyed upon, several small prey items, a very recent feeding, or any combination thereof, and conversely for a small mass of antigen (see also Sunderland, 1996, this volume, for a sobering discussion of additional possible confounding variables).

The simplest approach is to assume that a positive represents a single prey individual. When this assumption can be defended, useful data can be derived (Dempster, 1960), but in many predator–prey systems – especially given the sensitivity of the assays now in use – it is usually incorrect. When a positive is apt to represent more than one prey individual, a simple and elegant approach is to assume that the number of prey contained in any given predator gut is a Poisson variate. Then the proportion of negatives can be used as the zero class to calculate the mean number of prey individuals per gut (Nakamura and Nakamura, 1977; Greenstone, 1979b; Lister *et al.*, 1987). Sopp *et al.* (1992) developed a more elaborate model employing data on the quantity of antigen present and previously determined digestive rates. Estimates from their model gave good agreement with predation rate data obtained independently in an arena experiment. The Poisson

protein; RabPc, Rabbit polyclonal, absorbed or unabsorbed; SolPhase, solid phase assays; Unfract, homogenate, haemolymph, or protein precipitated from homogenate or haemolymph.

model fits the actual data much less well, but almost all predators in the experiment were positive for the target prey, which not only violates the assumptions of the model (Greenstone, 1979b) but is rarely the case for most predator-prey systems which have been studied in the field (Coaker and Williams, 1963; Loughton *et al.*, 1963; Young *et al.*, 1964; Rothschild, 1966; Dempster, 1967, 1971; Frank, 1967a,b, 1971; Danthanarayana, 1969; Sutton, 1970; MacClellan, 1973; Service, 1973a,b; Tod, 1973; Ashby, 1974; Service and Lyle, 1975; Vickerman and Sunderland, 1975; Lund and Turpin, 1977; Nakamura and Nakamura, 1977; Greenstone, 1979b; Sunderland and Sutton, 1980; Gardner *et al.*, 1981; Ragsdale *et al.*, 1981; Lesiewicz *et al.*, 1982; Dennison and Hodkinson, 1983; Leathwick and Winterbourn, 1984; Doane *et al.*, 1985; Calver *et al.*, 1986; Chiverton, 1987; Sunderland *et al.*, 1987; Kapuge *et al.*, 1987; Allen and Hagley, 1990; DuDevoir and Reeves, 1990; Hagler *et al.*, 1992; Blinn *et al.*, 1993; Symondson and Liddell, 1993d; Hagler and Naranjo, 1994a,b). The experiment should be repeated under conditions where not all prey are consumed and live non-target prey are also available.

It may prove more difficult to obtain a useful expression of the detectability interval. Several investigators have studied the rate at which antigen mass decreases in the gut. With the exception of a single linear example (Fichter and Stephen, 1981), antigen masses have always been found to decline exponentially (Lövei *et al.*, 1985, 1987; Sopp and Sunderland, 1989; Symondson and Liddell, 1993e, 1995). Since exponentially declining masses technically never become undetectable, it may be necessary to establish an arbitrarily low threshold for a detectable mass (Sopp and Sunderland, 1989).

An alternative approach is to ask the question, 'For how long can a single meal be detected?' This has only been done twice, once by Fichter and Stephen (1981), who found detectability to be linear, and once by Greenstone and Hunt (1993), who found it to be exponential. Here again an exponential relationship makes it impossible to determine a strict limit to detectability, and an arbitrary limit, such as five half-lives (encompassing almost 97% of all prey individuals consumed), might be appropriate.

These two ways of looking at detectability are subtly different and may have different uses. Expressions defining the rate of antigen mass disappearance may be useful for back-calculating to the mass ingested (Sopp and Sunderland, 1989), while those defining the loss of detectability of individual meals may be useful for providing a time unit for models employing Poisson-based estimates of number consumed (Nakamura and Nakamura, 1977; Greenstone, 1979b; Lister *et al.*, 1987).

An unsurprising but critical fact which has been reinforced by studies of digestive rate is that different predator species may vary several-fold in digestive rates and consequently in rates of loss of detectable antigen

(Rothschild, 1966; McIver, 1981; Ragsdale *et al.*, 1981; Greenstone, 1983b; Sunderland *et al.*, 1987; Symondson and Liddell, 1993e). This means that in systems involving more than one predator species, one must weight assay data to determine the relative contributions of each predator species to prey mortality rates.

As we continue to work on the quantitation problem, we should conduct our experiments under conditions as much like those in the field as possible. For example, trials should be run under field fluctuating temperatures (Greenstone and Hunt, 1993; Sigsgaard, 1996, this volume), and predators should be fed non-target prey species at normal consumption rates following the initial target species feeding (Fichter and Stephen, 1981; Lövei *et al.*, 1987, 1990; Greenstone and Hunt, 1993; Symondson and Liddell, 1996).

11.8 ALTERNATIVE TECHNOLOGIES FOR GUT ANALYSIS

A number of alternatives to serological methods have been used for gut analysis. They include dissection (restricted to predators with chewing mouthparts), the use of radioactive or non-radioactive tracers, paper chromatography (for pigmented prey), and enzyme electrophoresis (Solomon *et al.*, 1996, this volume). A thorough review of this extensive literature is beyond the scope of the present paper. Fortunately, Sunderland has published a very comprehensive recent review (1988), which includes comments on the strengths and weaknesses of each method.

An approach which Sunderland (1988) suggested – but which apparently has not yet been proven – is the detection of prey-specific nucleic acid sequences in a predator's gut (research is in progress in this area for a nematode–bacteria predation system [M.A. Banks and N.D. Martinez, personal communication]). The polymerase chain reaction (PCR) has already been employed in vector–pathogen diagnosis (Howe *et al.*, 1992; Vodkin *et al.*, 1995) and bloodmeal identification (Coulson *et al.*, 1990; Tobolewski *et al.*, 1992; Gokool *et al.*, 1993). However, one must electrophorese the PCR products, and make an interpretation of the patterns derived. An exciting alternative is using non-radioactive nucleic acid probes, directed against repeated DNA or RNA sequences, for use in rapid dot-blot tests (Johnson *et al.*, 1992; Hegedus and Khachatourians, 1993; Hill and Crampton, 1994). It is possible to develop such probes without prior knowledge of nucleic acid sequences in the species of interest, by creating and screening genomic or cDNA libraries (Cockburn, 1990). An appealing aspect of nucleic acid technologies is that, while contemporary, they are beyond the pioneering phase, and many procedures can be performed using commercially available kits. They are, therefore, apt to be a lot easier to master than MAb technology. Since DNA is found in all stages of the prey, specificity below the species level would not be possible with DNA probes. However, RNA probes could be made, restoring the possibility of stage- and instar-specificity.

Using nucleic acid probes for gut analysis will only be useful if specific nucleotide sequences have a reasonable half-life in the gut. Although this has not been determined for predators, in an analogous case of blood feeding, host DNA was detectable in mosquito guts for 10–20 hours, depending upon storage conditions (Coulson *et al.*, 1990).

11.9 FIELD APPLICATIONS OF SEROLOGICAL GUT ANALYSIS

A thorough review of the uses to which serological gut analysis has been put in field research is beyond the scope of this chapter. Nevertheless, it is useful to see what sorts of things have been accomplished. One conclusion which may be drawn is that serological analysis has made it possible to identify, rapidly, the remains of specific prey species in the guts of large samples of predators. Samples have often numbered into the thousands (e.g. Dempster, 1960; Coaker and Williams, 1963; Loughton *et al.*, 1963; Danthanarayana, 1969; MacLellan, 1973; Greenstone, 1979b; Sunderland and Sutton, 1980; Allen and Hagley, 1990; Hagler and Naranjo, 1944b). Some studies have shown that only a restricted set of the many polyphagous predators found in the habitat may feed on a given species, and that they are not necessarily the most abundant ones (Dempster, 1960; Rothschild, 1966; Service, 1973a; Service and Lyle, 1975; Sunderland *et al.*, 1978; Sunderland and Sutton, 1980; Leathwick and Winterbourne, 1984; DuDevoir and Reeves, 1990). Such information is important if chemical insecticides must be used, since predator species vary in their susceptibility to insecticides, and chemicals that are deleterious to predators that do not feed on the target pest might still be used. These data also show that the notion that polyphagous predators will take prey more or less in proportion to their availability is not always correct. Serological studies have also shown that the importance of different predators in a given field situation may vary seasonally (Rothschild, 1966; Sunderland *et al.*, 1987; Ragsdale *et al.*, 1981; Symondson and Liddell, 1993d) and with time of day (Vickerman and Sunderland, 1975; Blinn *et al.*, 1993).

A glimpse of the possibilities which will be opened up when we solve the quantitation problem is given by Dempster's classic paper (1960) on the predators of the broom beetle, *Phytodecta olivacea* Forster. A combination of long feeding intervals and short detectability intervals made the simplifying assumption that a serological positive indicates a single prey item tenable. Combining meticulous prey and predator population sampling with the results of more than 11 000 predator gut analyses, Dempster was able to show that the disappearance of eggs and larvae, estimated by independent life-table estimates, was largely due to arthropod predation.

Finally, serological studies have shed light on various ecological phenomena of broader interest to community ecologists, such as habitat use (Reynoldson and Sefton, 1976; Greenstone, 1983b), resource partitioning (Reynoldson and Davies, 1970; Lock and Reynoldson, 1976), prey selection (Greenstone, 1979b; Blinn *et al.*, 1993; McIver and Tempelis, 1993) and cannibalism (Sigsgaard, 1996, this volume).

11.10 CONCLUSION

Serological arthropod predator gut analysis is a mature and proven technology. It is now possible, using techniques developed over the last five decades, to develop antibodies of high specificity, and use them in simple, sensitive assays to detect prey remains in any predator–prey system. Quantitation of the resulting data to determine predator-induced prey mortality rates remains an elusive goal, but one on which substantial progress is being made.

Acknowledgements

I am deeply grateful to Eryl Liddell and William Symondson for the opportunity to participate in this symposium. I thank Clyde Morgan for technical support, Susan Winewriter for preparing Figures 11.1–3, and Howard Frank, James McIver, Melissa Stuart, William Symondson and two anonymous reviewers for comments on an early draft of the manuscript.

REFERENCES

Allen, W.R. and Hagley, A.C. (1982) Evaluation of immunoelectroosmophoresis on cellulose acetate for assessing predation of Lepidoptera (Tortricidae) by Coleoptera (Carabidae) species. *Canadian Entomologist*, **114**, 1047–54.

Allen, W.R. and Hagley, A.C. (1990) Epigeal arthropods as predators of mature larvae and pupae of the apple maggot (Diptera: Tephritidae). *Environmental Entomology*, **19**, 309–12.

Ashby, J.W. (1974) A study of arthropod predation of *Pieris rapae* using serological and exclusion techniques. *Journal of Applied Ecology*, **11**, 419–26.

Bailey, D.L., Choate, A.L. and Lawman, M.J.P. (1986) A rapid radioimmunoassay for the detection of *Mansonia* antigen (Diptera: Culicidae): its potential use as a sensitive method for studying predator–prey relationships. *Bulletin of Entomological Research*, **76**, 141–50.

Bangs, L.B. (1990) Latex immunossays. *Journal of Clinical Immunoassay*, **13**, 127–31.

Blinn, D.W., Runck, C. and Davies, R.W. (1993) The impact of prey behaviour and prey density on the foraging ecology of *Ranatra montezuma* (Heteroptera): a serological examination. *Canadian Journal of Zoology*, **71**, 387–91.

Boddington, M.J. and Mettrick, D.F. (1974) The distribution, abundance, feeding habits, and population biology of the immigrant triclad *Dugesia polychroa* (Platyhelminthes: Turbellaria) in Toronto Harbour, Canada. *Journal of Animal Ecology*, **43**, 681–99.

Boreham, P.F.L. and Ohiagu, C.E. (1978) The use of serology in evaluating prey–predator relationships: a review. *Bulletin of Entomological Research*, **68**, 171–94.

Brooke, M.M. and Proske, H.O. (1946) Precipitin test for determining natural insect predators of immature mosquitoes. *Journal of the National Malaria Society*, **5**, 45–56.

Bull, C.G. and King, W.V. (1923) The identification of blood meals of mosquitoes by means of the precipitin test. *American Journal of Hygiene*, **3**, 491–6.

Calver, M.C., Mathiessen, J.N., Hall, G.P., Bradley, J.S. and Lillywhite, J.H. (1986) Immunological determination of predators of the bush fly, *Musca vetustissima* (Diptera: Muscidae), in south-western Australia. *Bulletin of Entomological Research*, **76**, 133–9.

Chiverton, P.A. (1987) Predation of *Rhopalosiphum padi* (Homoptera: Aphididae) by polyphagous predatory arthropods during the aphids' pre-peak period in spring barley. *Annals of Applied Biology*, **111**, 257–69.

Coaker, T.H. and Williams, D.A. (1963) The importance of some Carabidae and Staphylinidae as predators of the cabbage root fly, *Erioischia brassicae* (Bouché). *Entomologia Experimentalis et Applicata*, **6**, 156–64.

Cockburn, A.F. (1990) A simple and rapid technique for identification of large numbers of individual mosquitoes using DNA hybridization. *Archives of Insect Biochemistry and Physiology*, **14**, 191–9.

Coulson, R.M.R., Curtis, C.F., Ready, P.D., Hill, N. and Smith, D.F. (1990) Amplification and analysis of human DNA present in mosquito bloodmeals. *Medical and Veterinary Entomology*, **4**, 357–66.

Crook, N.E. and Sunderland, K.D. (1984) Detection of aphid remains in predatory insects and spiders by ELISA. *Annals of Applied Biology*, **105**, 413–22.

Cuatrecasas, P. (1970) Protein purification by affinity chromatography. Derivatizations of agarose and polyacrylamide beads. *Journal of Biological Chemistry*, **245**, 3059–65.

Davies, R.W. (1969) The production of antisera for detecting specific triclad antigens in the guts contents of predators. *Oikos*, **20**, 248–60.

Danthanarayana W. (1969) Population dynamics of the weevil *Sitona regensteinensis* (Hbst.) on broom. *Journal of Animal Ecology*, **38**, 1–18.

Dempster, J.P. (1960) Quantitative study of the predators on the eggs and larvae of the broom beetle, *Phytodecta olivacea* Forster, using the precipitin test. *Journal of Animal Ecology*, **29**, 149–67.

Dempster, J.P. (1963) The natural prey of three species of *Anthocoris* (Heteroptera: Anthocoridae) living on broom (*Sarothamnus scoparius* L.). *Entomologia Experimentalis et Applicata*, **7**, 149–54.

Dempster, J.P. (1967) The control of *Pieris rapae* with DDT. I. The natural mortality of the young stages of *Pieris*. *Journal of Animal Ecology*, **4**, 485–500.

Dempster, J.P. (1971) The population ecology of the cinnabar moth, *Tyria Jacobeae* L. (Lepidoptera, Arctiidae). *Oecologia*, **7**, 26–67.

Dennison, D.F. and Hodkinson, I.D. (1983) Structure of the predatory beetle community in a woodland soil ecosystem. I. Prey selection. *Pedobiologia*, **25**, 109–15.

Doane, J.F., Scotti, P.D., Sutherland, O.R.W. and Pottinger, R.P. (1985) Serological identification of wireworm and staphylinid predators of the Australian soldier fly (*Inopus rubriceps*) and wireworm feeding on plant and animal food. *Entomologia Experimentalis et Applicata*, **38**, 65–72.

DuDevoir, D.S. and Reeves, M.R. (1990) Feeding activity of carabid beetles and spiders on gypsy moth larvae (Lepidoptera: Lymantriidae) at high density prey populations. *Journal of Entomological Science*, **25**, 341–56.

Eisen, H.N. (1974) *Immunology*, Harper & Row, Hagerstown, MD.

Engelhardt, O., Grabherr, R., Himmler, G. and Rüker, F. (1994) Two-step cloning of antibody variable domains in a phage display vector. *BioTechniques*, **17**, 44–6.

Fichter, B.L. and Stephen, W.P. (1979) Selection and use of host-specific antigens. *Miscellaneous Publications of the Entomological Society of America*, **11**, 25–33.

Fichter, B.L. and Stephen, W.P. (1981) Time related decay in prey antigens ingested by the predator *Podisus maculiventris* (Hemiptera, Pentatomidae) as detected by ELISA. *Oecologia*, **51**, 404–7.

Fenlon, J.S. and Sopp, P.I. (1991) Some statistical considerations in the determination of thresholds in ELISA. *Annals of Applied Biology*, **119**, 177–89.

Frank, J.H. (1967a) A serological method used in the investigation of the predators of the pupal stage of the winter moth, *Operophtera brumata* (L.) (Hydromeniidae). *Quaestiones Entomologicae*, **3**, 95–105.

Frank, J.H. (1967b) The effect of pupal predators on a population of winter moth, *Operophtera brumata* (L.) (Hydromeniidae). *Journal of Animal Ecology*, **36**, 611–20.

Frank, J.H. (1971) Carabidae (Coleoptera) as predators of the red-backed cutworm (Lepidoptera: Noctuidae) in central Alberta. *Canadian Entomologist*, **103**, 1039–44.

Gardner, W.A., Shepard, M. and Noblet, R. (1981) Precipitin test for examining predator–prey interactions in soybean fields. *Canadian Entomologist*, **113**, 365–9.

Gefter, M.L., Margulies, D.H. and Scharff, M.D. (1977) A simple method for polyethylene glycol-promoted hybridization of mouse myeloma cells. *Somatic Cell Genetics*, **3**, 231–6.

George, A.J.T. (1994) Antibody engineering *Endeavour*, **18**, 27–31.

Gokool, S., Curtis, C.F. and Smith, D.F. (1993) Analysis of mosquito bloodmeals by DNA profiling. *Medical and Veterinary Entomology*, **7**, 208–16.

Greenstone, M.H. (1977) A passive haemagglutination inhibition assay for the identification of stomach contents of invertebrate predators. *Journal of Applied Ecology*, **14**, 457–64.

Greenstone, M.H. (1979a) Passive hemagglutination inhibition: a powerful new tool for field studies of entomophagous predators. *Miscellaneous Publications of the Entomological Society of America*, **11**, 69–78.

Greenstone, M.H. (1979b) Spider feeding behaviour optimises dietary essential amino acid composition. *Nature*, **181**, 501–3.

Greenstone, M.H. (1983a) An enzyme-linked immunosorbent assay for the *Amblyospora* of *Culex salinarius* (Microspora: Amblyosporidae). *Journal of Invertebrate Pathology*, **14**, 250–5.

Greenstone, M.H. (1983b) Site-specificity and site-tenacity in a wolf spider: a serological dietary analysis. *Oecologia*, **56**, 79–83.

Greenstone, M.H. (1995) Bollworm or budworm? Squashblot immunoassay distinguishes eggs of *Helicoverpa zea* and *Heliothis virescens* (Lepidoptera: Noctuidae). *Journal of Economic Entomology*, **88**, 213–18.

Greenstone, M.H. and Morgan, C.E. (1989) Predation on *Heliothis zea* (Lepidoptera: Noctuidae): an instar-specific ELISA for stomach analysis. *Annals of the Entomological Society of America*, **82**, 45–9.

Greenstone, M.H. and Hunt, J.H. (1993) Determination of prey antigen half-life in *Polistes metricus* using a monoclonal antibody-based immunodot assay. *Entomologia Experimentalis et Applicata*, **68**, 1–7.

Greenstone, M.H. and Trowell, S.C. (1994) Arthropod predation: a simplified immunodot format for predator gut analysis. *Annals of Entomological Society of America*, **87**, 214–17.

Greenstone, M.H., Stuart, M.K. and Haunerland, N.H. (1991) Using monoclonal antibodies for phylogenetic analysis: an example from the Heliothinae (Lepidoptera: Noctuidae). *Annals of the Entomological Society of America*, **84**, 457–64.

Hagler, J.R. and Naranjo, S.E. (1994a) Qualitative survey of two coleopteran predators of *Bemisia tabaci* (Homoptera: Aleyrodidae) and *Pectinophora gossypiella* (Lepidoptera: Gelechiidae) using a multiple prey gut content ELISA. *Environmental Entomology*, **23**, 193–7.

Hagler, J.R. and Naranjo, S.E. (1994b) Determining the frequency of heteropteran predation on sweet potato whitefly and pink bollworm using multiple ELISAs. *Entomologia Experimentalis et Applicata*, **72**, 59–66.

Hagler, J.R., Cohen, A.C., Enriquez, F.J. and Bradley-Dunlop, D. (1991) An egg-specific monoclonal antibody to *Lygus hesperus*. *Biological Control*, **1**, 75–80.

Hagler, J.R., Cohen, A.C., Bradley-Dunlop, D. and Enriquez, F.J. (1992) Field evaluation of predation on *Lygus hesperus* using a pest-specific monoclonal antibody. *Environmental Entomology*, **21**, 896–900.

Hagler, J.R., Brower, A.G., Tu, Z., Byrne, D.N., Bradley-Dunlop, D. and Enriquez, F.J. (1993) Development of a monoclonal antibody to detect predation of the sweetpotato whitefly, *Bemisia tabaci*. *Entomologia Experimentalis et Applicata*, **68**, 231–6.

Hagler, J.R., Naranjo, S.E., Bradley-Dunlop, D., Enriquez, F.J. and Henneberry, T.J. (1994) A monoclonal antibody to pink bollworm (Lepidoptera: Gelechiidae) egg antigen: a tool for predator gut analysis. *Annals of the Entomological Society of America*, **87**, 85–90.

Hagler, J.R., Buchmann, S.L. and Hagler, D.A. (1995) A simple method to quantify dot blots for predator gut content analyses. *Journal of Entomological Science*, **30**, 95–8.

Hall, R.R., Downe, A.E.R., MacLellan, C.R. and West, A.S. (1953) Evaluation of insect predator-prey relationships by precipitin test studies. *Mosquito News*, **13**, 199–204.

Hance, T. and Grégoire-Wibo, C. (1983) Étude du régime alimentair des Carabides par voie sérologique, in *New Trends in Soil Biology*, (eds Ph. Lebrun *et al.*) Université Catholique de Louvain, Louvain-La-Neuve, Belgium, pp. 620–2.

Hance, T. and Renier, L. (1987) An ELISA technique for the study of the food of carabids. *Acta Phytopathologica et Entomologicae Hungaricae*, **22**, 363–8.

Haynes, M.R., Stura, E.A., Hilvert, D. and Wilson, I.A. (1994) Routes to catalysis: structure of a catalytic antibody and comparison with its natural counterpart. *Science*, **263**, 646–52.

Healy, J.A. and Cross, T.F. (1975) Immunoelectroosmophoresis for serological identification of predators of the sheep tick *Ixodes ricinus*. *Oikos*, **26**, 97–101.

Hegedus, D.D. and Khachatourians, G.G. (1993) Construction of cloned DNA probes for the specific detection of the entomopathogenic fungus *Beauveria bassiana* in grasshoppers. *Journal of Invertebrate Pathology*, **62**, 233–40.

Hill, S.M. and Crampton, J.M. (1994) DNA-based methods for the identification of insect vectors. *Annals of Tropical Medicine and Parasitology*, **88**, 227–50.

Hood, L.E., Weissman, I.L. and Wood, W.B. (1978) *Immunology*, Benjamin Cummings, Menlo Park.

Howe, D.K., Vodkin, M.H., Novak, R.J., Mitchell, C.J. and McLaughlin, G.L. (1992) Detection of St. Louis encephalitis virus in mosquitoes by use of the polymerase chain reaction. *Journal of the American Mosquito Control Association*, **8**, 333–5.

Johnson, D.W., Cockburn, A.F. and Seawright, J.A. (1992) Quick blots and non-radioactive detection of DNA probes for the identification of mosquitoes. *Journal of the American Mosquito Control Association*, **8**, 231–6.

Kapuge, S.H., Danthanarayana, W. and Hoogenraad, N. (1987) Immunological investigation of prey–predator relationships for *Pieris rapae* (L.) (Lepidoptera: Pieridae). *Bulletin of Entomological Research*, **77**, 247–54.

Kennett, R.H., McKearn, T.J. and Bechtols, K.B. (eds) (1980) *Monoclonal Antibodies*, Plenum, New York.

Köhler, G., and Milstein, C. (1975) Continuous culture of fused cells secreting antibody of predefined specificity. *Nature*, **256**, 495–7.

Lang, E., Szendrei, G.I., Lee, V.M.-Y. and Otvos, L. Jr (1994) Spectroscopic evidence that monoclonal antibodies recognize the dominant conformation of medium-sized synthetic peptide. *Journal of Immunological Methods*, **170**, 103–15.

Leathwick, D.M. and Winterbourn, M.J. (1984) Arthropod predation on aphids in a lucerne crop. *New Zealand Entomologist*, **8**, 75–80.

Lenz, C.J. and Greenstone, M.H. (1988) Production of a monoclonal antibody to the arylphorin of *Heliothis zea*. *Archives of Insect Biochemistry and Physiology*, **9**, 167–77.

Lesiewicz, D.S., Lesiewicz, J.L., Bradley, J.R., Jr. and Van Duyn, J.W. (1982) Serological determination of carabid (Coleoptera: Adephaga) predation of corn earworm (Lepidoptera: Noctuidae) in field corn. *Environmental Entomology*, **11**, 1183–6.

Liddell, J.E. and Symondson, W.O.C. (1996) The potential of combinatorial antibody libraries in pest/predator relationship studies, in *The Ecology of Agricultural Pests: Biochemical Approaches* (eds W.O.C. Symondson and J.E. Liddell), Chapman & Hall, London, pp. 347–66.

Lister, A., Usher, M.B. and Block, W. (1987) Description and quantification of field attack rates by predatory mites: an example using an electrophoresis method with a species of Antarctic mite. *Oecologia*, **72**, 185–91.

Littlefield, J.W. (1964) Selection of hybrids from matings of fibroblasts in vitro and their presumed recombinants. *Science*, **145**, 709–10.

Lock, M.A. and Reynoldson, T.B. (1976) The role of interspecific competition in the distribution of two stream dwelling triclads, *Crenobia alpina* (Dana) and *Polycelis felina* (Dalyell), in North Wales. *Journal of Animal Ecology*, **45**, 581–92.

Loughton, B.G., Derry, C. and West, A.S. (1963) Spiders and the spruce budworm. *Memoirs of the Entomological Society of Canada*, **31**, 249–68.

Lövei, G.L., Monostori, E. and Andó, I. (1985) Digestion rate in relation to starvation in the larva of a carabid predator, *Poecilus cupreus*. *Entomologica Experimentalis et Applicata*, **37**, 123–7.

Lövei, G.L., Sopp, P. and Sunderland, K.D. (1987) The effect of mixed feeding on the digestion of the carabid *Bembidion lampros*. *Acta Phytopathologica et Entomologica Hungaricae*, **22**, 403–7.

Lövei, G.L., Sopp, P. and Sunderland, K.D. (1990) Digestion rate in relation to alternative feeding in three species of polyphagous predators. *Ecological Entomology*, **15**, 293–300.

Lund, R.D. and Turpin, F.T. (1977) Serological investigation of black cutworm larval consumption by ground beetles. *Annals of the Entomological Society of America*, **70**, 322–4.

MacLellan, C.R. (1973) Natural enemies of the light brown apple moth, *Epiphyas postvittana*, in the Australian Capital Territory. *Canadian Entomologist*, **105**, 681–700.

McIver, J.D. (1981) An examination of the utility of the precipitin test for evaluation of arthropod predator-prey relationships. *Canadian Entomologist*, **113**, 213–22.

McIver, J.D. and Tempelis, C.H. (1993) The arthropod predators of anti-mimetic and aposematic prey: a serological analysis. *Ecological Entomology*, **18**, 218–22.

McKearn, T.J. (1980) Cloning of hybridoma cells by limiting dilution in fluid phase, in *Monoclonal Antibodies* (eds R.H. Kennett, T.J. McKearn and K.B. Bechtols), Plenum, New York.

Miller, M.C. (1979) Preparatory immunodiffusion for production of specific anti-adult southern pine beetle serum. *Annals of the Entomological Society of America*, **72**, 820–5.

Miller, M.C. (1981) Evaluation of enzyme-linked immunosorbent assay of narrow- and broad-spectrum anti-adult southern pine beetle serum. *Annals of the Entomological Society of America*, **74**, 279–82.

Milstein, C. (1981) Monoclonal antibodies from hybrid myelomas. *Proceedings of the Royal Society of London (Biology)*, **211**, 393.

Mollet, J.A. and E.J. Armbrust (1978) Age specific serological identification of adult stages of alfalfa weevil, *Hypera postica. Annals of the Entomological Society of America*, **71**, 207–11.

Nakamura, M. and Nakamura, K. (1977) Population dynamics of the chestnut gall wasp, *Dryocosmus kuriphilus* Yasumatsu (Hymenoptera: Cynipidae). V. Estimation of the effect of predation by spiders on the mortality of imaginal wasps based on the precipitin test. *Oecologia*, **27**, 97–116.

Nemoto, H., Sekijima, Y., Fujikura, Y., Kiritani, K., and Shibukawa, S. (1985) Application of an immunological method for the identification of predators of the diamond-back moth, *Plutella xylostella* (L.) (Lepidoptera: Yponomeutidae). *Japanese Journal of Applied Entomology and Zoology*, **29**, 61–6.

Ohiagu, C.E. and Boreham, P.F.L. (1978) A simple field test for evaluating insect prey-predator relationships. *Entomologia Experimentalis et Applicata*, **23**, 40–7.

Paterson, Y. (1992) Mapping antibody binding sites on protein antigens. *Nature*, **356**, 456–7.

Pettersson, J. (1972) Technical description of a serological method for quantitative predator efficiency studies on *Rhopalosiphum padi* (L.). *Swedish Journal of Agricultural Research*, **2**, 65–9.

Pickavance, J.R. (1970) A new approach to the immunological analysis of invertebrate diets. *Journal of Animal Ecology*, **39**, 715–24.

Quinlan, T.G., Calver, M.C. and Smith, G.T. (1993) Immunological determination of digestive rates in the syntopic scorpions *Urodacus armatus* Pocock and *Urodacus novaehollandiae* Peters. *Oecologia*, **95**, 459–62.

Ragsdale, D.W. (1980) The development and use of an enzyme linked immunosorbent assay (ELISA) for the quantitative assessment of predation of *Nezara viridula* (L.) within a soybean agroecosystem. PhD Dissertation, Louisiana State University.

Ragsdale, D.W., Larson, A.D. and Newsom, L.D. (1981) Quantitative assessment of the predators of *Nezara viridula* eggs and nymphs within a soybean agroecosystem using an ELISA. *Environmental Entomology*, **10**, 402–5.

Reynoldson, T.B. and Davies, R.W. (1970) Food niche and coexistence in lake-dwelling triclads. *Journal of Animal Ecology*, **39**, 599–617.

Reynoldson, T.B. and Sefton, A.D. (1976) The food of *Planaria torva* (Müller) (Turbellaria-Tricladida) a laboratory and field study. *Freshwater Biology*, **6**, 383–93.

Rini, J.M., Schulze-Gahmen, U. and Wilson, I.A. (1992) Structural evidence for induced fit as a mechanism for antibody–antigen recognition. *Science*, **255**, 959–65.

Rothschild, G.H.L. (1966) A study of a natural population of *Conomelus anceps*

(Germar) (Homoptera: Delphacidae) including observations on predation using the precipitin test. *Journal of Animal Ecology*, **35**, 413–34.

Rothschild, G.H.L. (1970) Observations on the ecology of the rice-ear bug *Leptocorisa oratorius* (F.) (Hemiptera: Alydidae) in Sarawak (Malaysian Borneo). *Journal of Applied Ecology*, **7**, 147–67.

Rothschild, G.H.L. (1971) The biology and ecology of rice-stem borers in Sarawak (Malaysian Borneo). *Journal of Applied Ecology*, **8**, 287–322.

Saragovi, H., Fitzpatrick, D., Raktabutr, A., Nakanishi, H., Kahn, M. and Greene, M.I. (1991) Design and synthesis of a mimetic from an antibody complementarity-determining region. *Science*, **253**, 792–5.

Schoof, D.D., Palchick, S. and Tempelis, C.H. (1986) Evaluation of predator–prey relationships using an enzyme immunoassay. *Annals of the Entomological Society of America*, **79**, 91–5.

Sergeeva, T.K. (1970a) A method for obtaining an antiserum to the proteins of *Eurygaster integriceps* for studying its predators. *Zoologicheskii Zhurnal*, **49**, 440–4.

Sergeeva, T.K. (1970b) Fixation and identification of proteins of *Eurygaster integriceps* in stomachs of predaceous arthropods. *Zoologicheskii Zhurnal*, **49**, 1715–20.

Service, M.W. (1973a) Study of the natural predators of *Aedes cantans* (Meigen) using the precipitin test. *Journal of Medical Entomology*, **10**, 503–10.

Service, M.W. (1973b) Mortalities of the larvae of the *Anopheles gambiae* Giles complex and detection of predators by the precipitin test. *Bulletin of Entomological Research*, **62**, 359–69.

Service, M.W. and Lyle, P.T.W. (1975) Detection of the predators of *Simulium damnosum* by the precipitin test. *Annals of Tropical Medicine and Parasitology*, **69**, 105–8.

Sigsgaard, L. (1996) Serological analysis of predators of *Helicoverpa armigera* Hübner (Lepidoptera: Noctuidae) eggs in sorghum-pigeonpea intercropping at ICRISAT, India: a preliminary field study, in *The Ecology of Agricultural Pests: Biochemical Approaches* (eds W.O.C. Symondson and J.E. Liddell), Chapman & Hall, London, pp. 367–81.

Solomon, M.G., Fitzgerald, J.D. and Murray, R.A. (1996) Electrophoretic approaches to predator–prey interactions, in *The Ecology of Agricultural Pests: Biochemical Approaches* (eds W.O.C. Symondson and J.E. Liddell), Chapman & Hall, London, pp. 457–68.

Sopp, P.I. and Sunderland, K.D. (1989) Some factors affecting the detection period of aphid remains in predators using ELISA. *Entomologia Experimentalis et Applicata*, **51**, 11–20.

Sopp, P.I., Sunderland, K.D., Fenlon, J.S. and Wratten, S.D. (1992) An improved quantitative method for estimating invertebrate predation in the field using an enzyme-linked immunosorbent assay (ELISA). *Journal of Applied Ecology*, **79**, 295–302.

Stott, D.I. (1989) Immunoblotting and dot blotting. *Journal of Immunological Methods*, **119**, 153–87.

Stuart, M.K. and Greenstone, M.H. (1990) Beyond ELISA: a rapid, sensitive, specific immunodot assay for identification of predator stomach contents. *Annals of the Entomological Society of America*, **83**, 1101–7.

Sunderland, K.D. (1988) Quantitative methods for detecting invertebrate predation occurring in the field. *Annals of Applied Biology*, **112**, 201–24.

Sunderland, K.D. (1996) Progress in quantifying predation using antibody techniques, in *The Ecology of Agricultural Pests: Biochemical Approaches* (eds W.O.C. Symondson and J.E. Liddell), Chapman & Hall, London, pp. 419–55.

Sunderland, K.D. and Sutton, S.L. (1980) A serological study of arthropod predation on woodlice in a dune grassland ecosystem. *Journal of Animal Ecology*, **49**, 987–1004.

Sunderland, K.D., Crook, N.E., Stacey, D.L. and Fuller, B.J. (1987) A study of feeding by polyphagous predators on cereal aphids using ELISA and gut dissection. *Journal of Applied Ecology*, **24**, 907–33.

Sutton, S.L. (1970) Predation on woodlice: an investigation using the precipitin test. *Entomologia Experimentalis et Applicata*, **13**, 279–85.

Symondson, W.O.C. and Liddell, J.E. (1993a) The development and characterization of an anti-haemolymph antiserum for the detection of mollusc remains within carabid beetles. *Biocontrol Science and Technology*, **3**, 261–75.

Symondson, W.O.C. and Liddell, J.E. (1993b) A monoclonal antibody for the detection of arionid slug remains in carabid predators. *Biological Control*, **3**, 207–14.

Symondson, W.O.C. and Liddell, J.E. (1993c) The detection of predator–mollusc interactions using advanced antibody technologies. *Annales, A.N.P.P. Third International Conference of Pests in Agriculture*, Montpelier, Association Nationale de Protection des Plantes, Paris, pp. 417–24.

Symondson, W.O.C. and Liddell, J.E. (1993d) The detection of predation by *Abax parallelepipedus* and *Pterostichus madidus* (Coleoptera: Carabidae) on Mollusca using a quantitative ELISA. *Bulletin of Entomological Research*, **83**, 641–7.

Symondson, W.O.C. and Liddell, J.E. (1993e) Differential antigen decay rates during digestion of molluscan prey by carabid predators. *Entomologica Experimentalis et Applicata*, **69**, 277–87.

Symondson, W.O.C. and Liddell, J.E. (1995) Decay rates for slug antigens within the carabid predator *Pterostichus melanarius* monitored with a monoclonal antibody. *Entomologia Experimentalis et Applicata*, **75**, 245–50.

Tempelis, C.H. and Rodrick, M.L. (1972) Passive hemagglutination inhibition technique for the identification of arthropod blood meals. *American Journal of Tropical Medicine and Hygiene*, **21**, 238–45.

Tobolewski, J., Kaliszewski, M.J., Colwell, R.K. and Oliver, J.H. (1992) Detection and identification of mammalian DNA from the gut of museum specimens of ticks. *Journal of Medical Entomology*, **29**, 1049–51.

Tod, M.E. (1973) Notes on beetle predators of molluscs. *Entomologist*, **106**, 196–201.

Vickerman, G.P. and Sunderland, K.D. (1975) Arthropods on cereal crops: nocturnal activity, vertical distribution and aphid predation. *Journal of Applied Ecology*, **12**, 755–66.

Vodkin, M.H., Streit, T., Mitchell, C.J., McGlaughlin, G.L. and Novak, R.J. (1995) PCR-based detection of arboviral RNA from mosquitoes homogenized in detergent. *Biotechniques*, **17**, 114–16.

Whalon, M.E. and Parker, B.L. (1978) Immunological identification of tarnished plant bug predators. *Annals of the Entomological Society of America*, **71**, 453–6.

Young, J.O. (1973) The prey and predators of *Phaenocora typhlops* (Vejdovsky) (Turbellaria: Neorhabdocoela) living in a small pond. *Journal of Animal Ecology*, **42**, 637–43.

Young, J.O., Morris, I.G. and Reynoldson, T.B. (1964) A serological study of *Asellus* in the diet of lake-dwelling triclads. *Archives of Hydrobiology*, **60**, 366–73.

12

Serological diagnosis of parasitism: a monoclonal antibody-based immunodot assay for *Microplitis croceipes* (Hymenoptera: Braconidae)

M.K. Stuart and M.H. Greenstone

ABSTRACT

A monoclonal antibody-based immunodot assay was developed that detects second instars of the endoparasitoid *Microplitis croceipes* (Cresson) within homogenates of parasitized *Helicoverpa zea* (Boddie). The monoclonal antibody did not cross-react with *Cotesia marginiventris* (Cresson), another braconid endoparasitoid of *H. zea* whose larvae resemble those of *M. croceipes*. The specificity of the immunodot assay was 100% and its sensitivity ranged from 84.9% to 94.3% in blind studies. Binding of the monoclonal antibody to *M. croceipes* was detected with a peroxidase-conjugated second antibody specific for mouse IgG. Three different reagents were evaluated for development of the nitrocellulose membranes. Of these, 4-chloro-1-naphthol/3,3′-diaminobenzidine tetrahydrochloride was preferred to 3-amino-9-ethylcarbazole and a chemiluminescent reagent because it was easy to use and to interpret by eye.

12.1 INTRODUCTION

Insect endoparasitoids may inflict heavy mortality on insect pest populations. For example, more than 30 species of hymenopterous and dipter-

The Ecology of Agricultural Pests, Edited by W.O.C. Symondson and J.E. Liddell.
Published in 1996 by Chapman & Hall, London. ISBN 0 412 62190 8

ous egg and larval parasitoids have been reared from the cotton bollworm, *Helicoverpa zea* (Boddie) (Lepidoptera: Noctuidae), and numerous quantitative field observational and experimental studies have conclusively demonstrated their potential for reducing bollworm populations (Quaintance and Brues, 1905; Phillips and Barber, 1933; Oatman, 1966; Nuenzing, 1969; Martin *et al.*, 1976; Ridgway *et al.*, 1977; Burleigh and Farmer, 1978; Caron and Bradley, 1978; Pair *et al.*, 1982; Mueller and Phillips, 1983; Hopper and King, 1984; Soteres *et al.*, 1984; Stadelbacher *et al.*, 1984; Puterka *et al.*, 1985; Lewis and Gross, 1989). Unfortunately, the technologies available for detecting parasitism preclude the sort of continuous monitoring which would make knowledge of parasitism rates available routinely to pest management consultants.

Although X-ray examination enabled rapid and accurate diagnosis of parasitism by a large pupal endoparasitoid of the imported cabbageworm (Biever and Boldt, 1970), it does not provide sufficient resolution for distinguishing small, closely related larval parasitoids such as are commonly found in many host species. Electrophoresis permits species-specific identification of endoparasitoids (Wool *et al.*, 1978; Tomiuk *et al.*, 1979; Castañera *et al.*, 1985; Powell and Walton, 1989; Walton *et al.*, 1990), but it requires expensive equipment and special skills to perform and interpret.

The only two methods which are used routinely for determining whether or not a larva contains an internal parasitoid are holding for adult emergence and dissection.

12.1.1 Holding

Holding for adult emergence entails a great deal of expense, time and effort, as well as space and facilities for holding and feeding larvae individually. Furthermore, because there is a lag between the time at which larvae are collected and that at which the incidence of parasitism of those larvae becomes known, the results are usually obtained too late to be of practical value to a consultant or grower who would use these data in making management decisions based upon projected host larval populations in relation to economic thresholds (Hopper and King, 1984). Also, holding larvae to determine rates of parasitism leads to inaccurate estimates, because some proportion of held larvae invariably die of other causes before it can be determined whether parasitoids would have emerged had they lived. These deaths are ascribed to disease (Hamm *et al.*, 1983), injury, or 'unknown causes.' The losses to analysable data due to unexplained larval death vary from moderate (Bottrell *et al.*, 1968; Lewis and Brazzel, 1968) to 'substantial' (Mueller and Phillips, 1983) to 'astronomical' (802 of 994 larvae collected by Eger *et al.*, 1982). It cannot be safely assumed that the proportions of parasitized larvae are the same in the unexplained as in the explained deaths. Therefore, the loss of data represents more than just a reduction in sample sizes; it means that estimates of rates of parasitism contain biases which are unknown in direction and magnitude.

12.1.2 Dissection

Dissection provides immediate information about whether or not a larva is parasitized; however, it is tedious and requires special skills. Where several parasitoid species attack a host, identification can be especially challenging. For example, in most of the United States, the principal parasitoids of *H. zea* are *Microplitis croceipes* (Cresson) (Hymenoptera: Braconidae) and *Cotesia marginiventris* (Cresson) (Hymenoptera: Braconidae), two parasitic wasps whose larvae are morphologically similar. Here we describe a simple, rapid, species-specific monoclonal antibody-based assay that distinguishes *M. croceipes* second instars from other parasitoids of *H. zea*.

12.2 EXPERIMENTAL APPROACH

12.2.1 Insects

M. croceipes and *C. marginiventris* were reared on host larvae from colonies maintained in continuous culture at the Biological Control of Insects Research Laboratory (BCIRL) in Columbia, Missouri, using established rearing methods (Berger, 1963; Ignoffo, 1965; Ignoffo and Boening, 1970). Groups of 25–35 *H. zea* late second to early third instars were exposed overnight to three mated adult female *M. croceipes* in 200-ml ice-cream cartons with plastic Petri dish lids as covers; diet was provided for host larvae and honey was provided for the parasitoids. Following oviposition, larvae were held at 25°C and a light : dark (L:D) cycle of 16:8 hours. Representative samples of larvae were dissected daily to determine parasitoid instar. Groups of 50–75 neonates or late first instars of *H. zea* or *Spodoptera frugiperda* (J.E. Smith) (Lepidoptera: Noctuidae) were exposed overnight to four or five adult female *C. marginiventris*; the rest of the rearing protocol was the same as for *M. croceipes*.

12.2.2 Monoclonal antibody production and screening

Young female BALB/c mice were immunized according to an established protocol (Stuart and Burkholder, 1991) with wholebody homogenates of *M. croceipes* second instars dissected from *H. zea* larvae. Spleen lymphocytes from the immunized mice were fused to non-secreting Sp2/O-Ag14 myeloma cells with polyethylene glycol (Van Deusen,

1983). Supernatants from hybridomas selected in HAT medium (Van Deusen, 1983) were screened for antibody activity against *M. croceipes* second instars by an indirect ELISA (Lenz and Greenstone, 1988). Supernatants yielding absorbances at least three-fold higher than myeloma-conditioned medium negative controls were further assayed by ELISA for cross-reactivity with additional insect species and life-stages.

Cell lines displaying appropriate specificities were cloned by the limiting dilution method (Campbell, 1984). Monoclonal antibodies (MAbs) from cloned lines were produced in serum-free medium (Carlsson *et al.*, 1985), concentrated by ultrafiltration through an Amicon YM 100 membrane (Amicon Corp., Danvers, Massachusetts, USA), and further purified by precipitation of contaminating proteins with Biocryl BPA-1000 (TosoHaas, Philadelphia, Pennsylvania, USA). Protein content of the final MAb preparations was determined using a commercial dye concentrate (Bio-Rad Laboratories, Hercules, California, USA) before storage of aliquots at –85°C.

12.2.3 Antigen characterization

Parasitoid antigens were characterized by immunoblot analysis and Coomassie blue staining of electrophoretic gels. Isoelectric focusing (IEF) was performed as described by Lenz and Greenstone (1988). The methods of Stuart *et al.* (1994) were used during immunoblotting and electrophoresis of native and denaturing polyacrylamide gels.

The onset and duration of antigen expression in *M. croceipes* were determined by indirect immunodot analysis of first, second and third instar parasitoids dissected from *H. zea*. *M. croceipes* larvae collected at 24-hour intervals, from day 1 post-hatch (first instar) through day 7 post-hatch (early second instar), were pooled according to the day of collection and stored in distilled water at –20°C until analysed. Dissected mid- and late-second instar parasitoids were frozen individually, as were third instars that had emerged from the host. Early, mid-, and late-second instar parasitoids were determined according to criteria given by Lewis (1970). Parasitoid homogenates were diluted in Tris-buffered saline (TBS; 20 mM Tris–HCl, 500 mM NaCl, pH 7.5) and applied to nitrocellulose at a concentration of 10 μg protein/well. The remainder of the indirect immunodot assay was as described below for parasitized and non-parasitized *H. zea*.

12.2.4 Indirect immunodot assay

Non-parasitized *H. zea* fourth instars and size-matched *H. zea* fourth instars parasitized by second instar *M. croceipes* were prepared for the immunodot assay by homogenization in 500 μl of TBS containing 5 μM phenylthiourea (PTU) to retard melanization, and 1 μM leupeptin, 1 μM pepstatin A, and 1-mM EDTA (Sigma) to inhibit proteases. A 1-μl aliquot of 100 mM phenylmethylsulphonyl fluoride was added to each vial immediately before homogenization of the insect with a plastic pellet pestle. Supernatants were collected by centrifugation for 5 min at 11 000 *g* and the carcasses were discarded. Protein concentrations of the supernatants were determined for five non-parasitized *H. zea* and nine *H. zea* parasitized by *M. croceipes* second instars; the concentrations ranged from 6.5 to 11.3 $\mu g/\mu l$. Whole insects and working aliquots of insect homogenates were kept on ice during handling and were stored at –85°C.

The homogenates were assayed for endogenous enzymes to determine the suitability of peroxidase and phosphatase substrates for the immunodot assay. Aliquots (2 μl) of the homogenates were spotted onto nitrocellulose and incubated in 4-chloro-1-naphthol/3,3′-diaminobenzidine tetrahydrochloride (CN/DAB; Pierce, Rockford, Illinois, USA) to assay for peroxidases, or in nitroblue tetrazolium/5-bromo-4-chloro-3-indolyl phosphate (NBT/BCIP; Bio-Rad) to assay for phosphatases.

All steps of the immunodot assay were performed at room temperature. A nitrocellulose membrane (Bio-Rad) was wetted with TBS and inserted into a Bio-Rad microfiltration dot-blot apparatus. The valve on the apparatus was positioned so that filtration could not occur. The wells were loaded with 100 μl of TBS containing PTU and protease inhibitors. A 2-μl aliquot of insect homogenate was added to the TBS in each well, and the valve was repositioned to initiate filtration. After 30 min, vacuum was applied to the apparatus, the membrane was removed and rinsed in TBS, and the protein binding sites on the membrane were saturated by a 15-min incubation in TBS containing 5% non-fat dry milk. The latter and all subsequent steps were performed with continuous gentle rocking on a rocker table. The membrane was incubated for 30 min in 15 ml of MAb solution diluted to 10 $\mu g/ml$ in TBS containing 0.05% Tween 20 (TTBS) and 1% non-fat dry milk (TTBS–1% blotto). After three 5-min washes in TTBS, the membrane was incubated for 30 min in 15 ml of sheep anti-mouse IgG-horseradish peroxidase conjugate (Sigma) diluted 1:1000 in TTBS–1% blotto. The membrane was washed three times in TTBS, rinsed in TBS to remove the Tween 20, and developed in 10 ml of CN/DAB.

Some blots were developed with a chemiluminescent reagent (Whitehead *et al.*, 1979) or with 3-amino-9-ethylcarbazole (AEC; Moss, Inc., Pasadena, Maryland, USA; Graham *et al.*, 1965). For the chemiluminescence assay, the peroxidase conjugate was diluted 1:3000, and the detection steps were performed according to the instructions supplied by the manufacturer of an ECL Western Blot Kit (Amersham International

PLC, Amersham, UK). X-Ray films were developed in an automatic developer (Eastman Kodak, Rochester, New York, USA). For AEC detection, all procedures were performed exactly as described for CN/DAB, except that the final development was in 10 ml of the AEC reagent. The sensitivity of each detection method was determined against serial two-fold dilutions of dissected *M. croceipes* second instar whole-body homogenates.

A total of 19 non-parasitized *H. zea* fourth instars and 87 *H. zea* fourth instars exposed to adult female *M. croceipes* and allowed to develop long enough for resulting parasitoids to develop to second instars (hereinafter 'parasitized' larvae) were assayed by the indirect immunodot assay. Each individual was assayed three to six times with the CN/DAB detection method, once or twice with AEC, and once with the chemiluminescence assay. Blind assays were performed on insect homogenates randomly selected and applied to wells of the dot-blot apparatus; these blots were developed in CN/DAB, and the wells were scored by eye as positive or negative before the identities of the homogenates were revealed.

12.3 EXPERIMENTAL FINDINGS

12.3.1 ELISA

MAbs from 80 hybridoma lines yielded absorbances at least three-fold higher than controls against *M. croceipes* second instars during initial screening by ELISA. Of these, 43 were further assayed against *M. croceipes* first instars, *C. marginiventris* first and second instars, and non-parasitized *H. zea* fourth instars. Twenty-six reacted exclusively with *M. croceipes* second instars, while 17 reacted with both *M. croceipes* and *C. marginiventris*, or with both parasitoids and *H. zea*. None of the MAbs recognized both first and second instars of *M. croceipes* while maintaining species specificity. We chose to work with MAb 5D1, an IgG1 antibody which is specific for *M. croceipes* that detects the antigen with reproducibly high titres, and is produced by a very robust hybridoma clone.

MAb 5D1 reacted with an antigen that was first detectable 7 days after hatching of the parasitoid. Expression of the antigen began after the parasitoid moulted from first to second instar and continued through the emergence of the third instar from the host (Figure 12.1).

12.3.2 Immunoblot assay

In immunoblots of agarose IEF gels (Figure 12.2), MAb 5D1 recognized components with isoelectric focusing points (pIs) of 6.3 and 6.4 in dissected second instar *M. croceipes*. In blots of native gels (Figure 12.3), the major component detected in dissected second instars was approximately 530 kDa, with minor components recognized at 780 and 310 kDa (lanes 1–5). Antigens recognized in homogenates of *H. zea* parasitized by *M. croceipes* second instars differed among individual animals, but included those found in dissected *M. croceipes*, as well as degradation products ranging in size from 93 to 600 kDa (lanes 6–8). No immunoreactive components were detected in non-parasitized *H. zea* fourth instars (lane 9). In blots of denaturing gels (Figure 12.4), MAb 5D1 recognized major components at 88 and 79 kDa and minor components at 77 and 63 kDa in *M. croceipes* second instars; a 71-kDa antigen was also occasionally seen. First instar *M. croceipes* and larval stages of *C. marginiventris* were not immunoreactive with the MAb.

12.3.3 Immunodot assay

When MAb 5D1 was titrated against whole-body homogenates of dissected second instar *M. croceipes* in the immunodot assay, the minimum amount of protein detected with the CN/DAB reagent was 50 ng (Figure 12.5). Lower limits of detection were 25 ng and 6.25 ng when AEC and chemiluminescence were used, respectively. A positive reaction appeared as a brown dot with CN/DAB, a pink or red dot with AEC, and a black dot or ring on X-ray film with the chemiluminescence assay.

All 19 non-parasitized *H. zea* and 87 *H. zea* parasitized by *M. croceipes* second instars were evaluated by the immunodot assay with each of the three peroxidase detection systems (Figure 12.6). No false positives were observed among the non-parasitized animals regardless of which detection method was used. Of the 87 'parasitized' *H. zea*, 71 (81.6%) were positive in all immunodot assays, nine (10.3%) were consistently negative (animals M31, M33, M69, M81, M85, M90, M98, M102 and M104), and seven (8.0%) gave variable results (Table 12.1). When analysed by dissection, the percentage of larvae found to be parasitized after exposure to ovipositing female parasitoids was typically between 83% and 87% (C.E. Morgan and M.H. Greenstone, unpublished results); results of the immunodot assay were well within this normal range. No *H. zea* parasitized by first instar *M. croceipes* were evaluated in the immunodot assay because dissected first instar *M. croceipes* were not recognized by MAb 5D1 (Figure 12.1). Alkaline phosphatase detection methods were unsuitable for development of nitrocellulose membranes because *H. zea* homogenates possessed endogenous phosphatase activity.

Four blind immunodot assays were performed. The specificity of all assays was 100% (i.e. there were no false positives). The sensitivity of each assay (the percentage of parasitized *H. zea* that were scored as positive) was 94.3%, 88.5%, 90.8% and 84.9%. A representative experi-

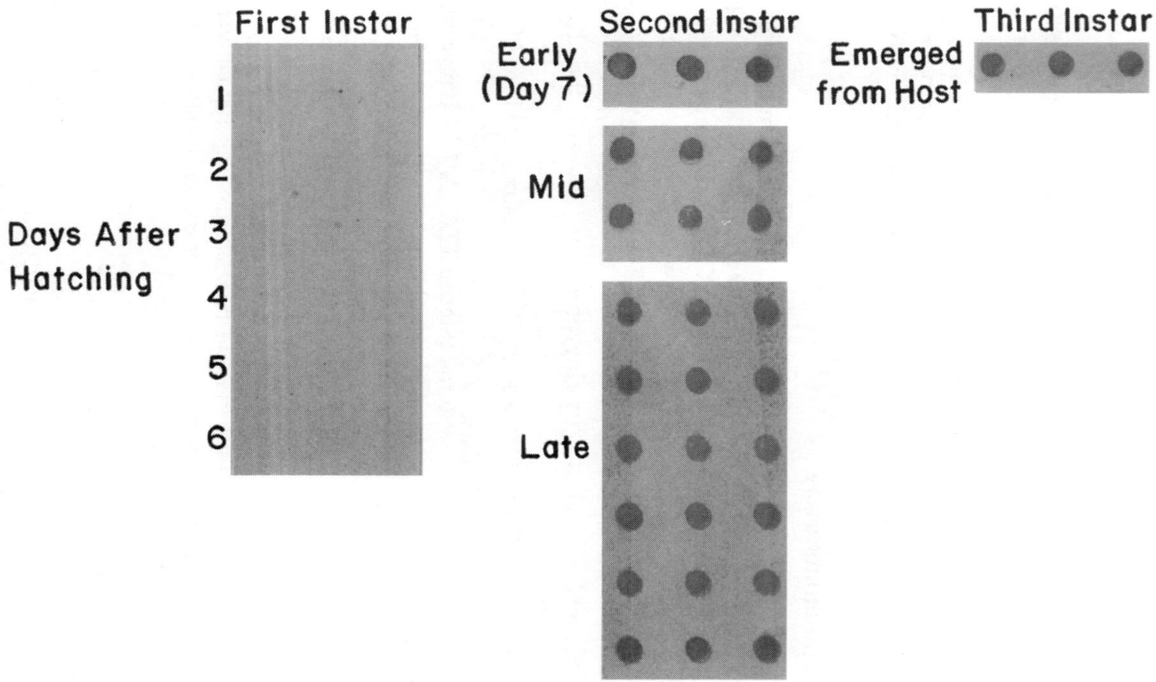
First Instar
Second Instar
Third Instar
Days After Hatching
1
2
3
4
5
6
Early (Day 7)
Mid
Late
Emerged from Host

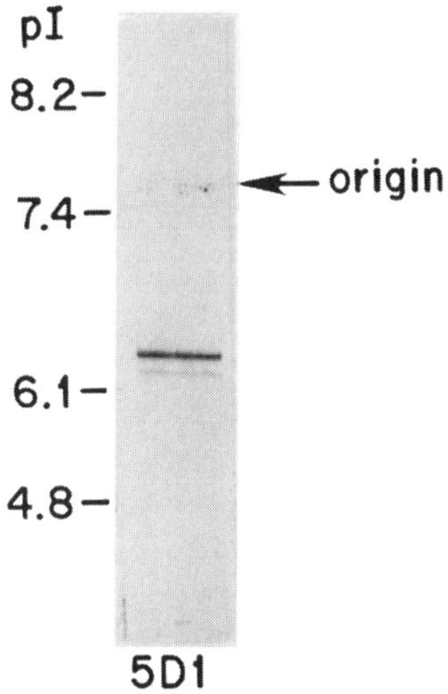

Figure 12.2 Isoelectric focusing gel, pI 3–10, of *M. croceipes* second instar whole-body homogenate (20 μg) immunoblotted with MAb 5D1.

Figure 12.1 (left) Indirect immunodot assay analysis of larval *M. croceipes* dissected from *H. zea*. To determine which stages of *M. croceipes* were reactive with MAb 5D1, parasitoids were dissected free from host material and applied at equivalent protein concentrations (10 μg/well) to a nitrocellulose membrane. First instar *M. croceipes* were collected from *H. zea* on days 1–6 after parasitoid eggs had hatched within their hosts, as determined by dissection. These first instars were applied to the left panel of the figure in rows of triplicate replications. Second instar *M. croceipes* were collected from *H. zea* beginning on day 7 after hatching. These early second instars, as well as mid- and late-second instar *M. croceipes*, were also applied in rows of triplicate replications, to the centre panel in the figure. A third instar *M. croceipes* was collected immediately upon its emergence from the *H. zea* host (right panel, in triplicate). To achieve the desired parasitoid protein concentration of 10 μg/well, it was necessary to pool several parasitoids according to the day they were collected on days 1–7. Two mid-second, six late-second and the single third instar *M. croceipes* were assayed individually. Parasitoid proteins were incubated sequentially with MAb 5D1, anti-mouse IgG-horseradish peroxidase conjugate, and CN/DAB. The blot clearly shows the specificity of MAb 5D1 for second and third instar *M. croceipes*, with no evident reactivity for first instars.

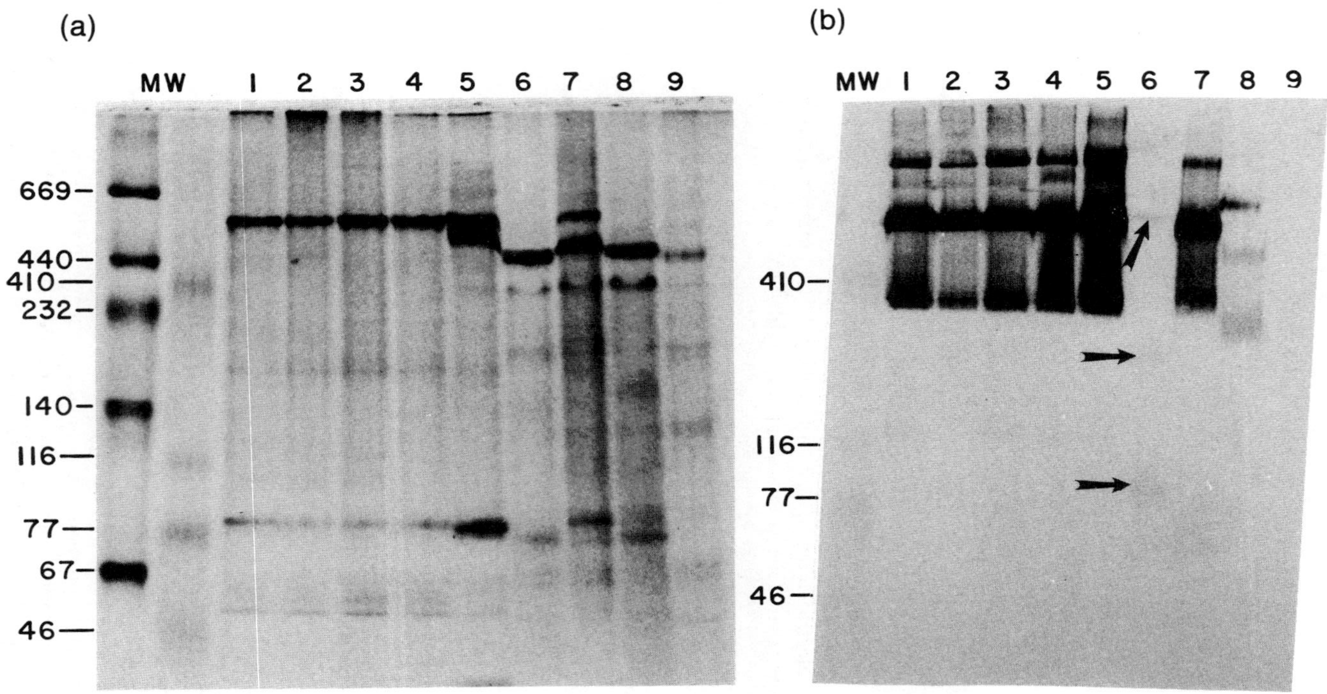

Figure 12.3 Native 5–15% gradient polyacrylamide gel. (a) Stained with Coomassie brilliant blue R-250 or (b) immunoblotted with MAb 5D1. Wells were loaded with homogenized individuals of (1,2) dissected mid-late second instar *M. croceipes*, (3,4) dissected late second instar *M. croceipes*, (5) newly emerged third instar *M. croceipes*, (6–8) *H. zea* larvae parasitized by second instar *M. croceipes* or (9) non-parasitized fourth instar *H. zea*. The amount of protein in each well was 75 μg for the Coomassie-stained gel and 30 μg for the immunoblot. All insects were homogenized in buffer without protease inhibitors. Molecular weights (MW) of the standards are expressed in kDa. Arrows indicate weakly reactive *M. croceipes* antigens of 530, 273 and 93 kDa.

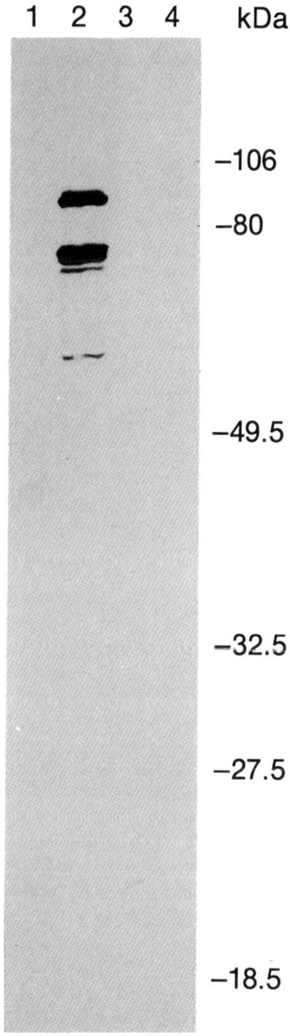

Figure 12.4 A 10% denaturing gel immunoblotted with MAb 5D1. Each well was loaded with 30 μg of dissected *M. croceipes* (1) first or (2) second instars, or *C. marginiventris* (3) first or (4) second instars. Reactive components in *M. croceipes* second instars can be seen at 88, 79, 77 and 63 kDa.

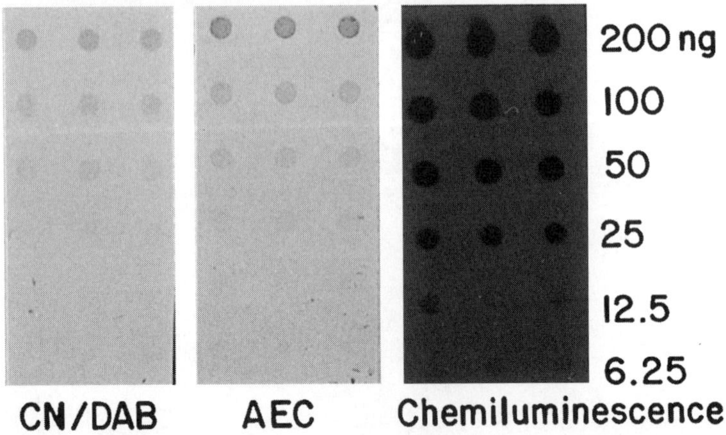

Figure 12.5 Dissected *M. croceipes* second instar whole-body homogenates titrated against MAb 5D1 in the immunodot assay. Each antigen dilution was assayed in triplicate and developed with the reagent indicated. Development in CN/DAB or AEC was allowed to proceed for 5 minutes. X-Ray film was developed after a 10-minute exposure to the nitrocellulose membrane (chemiluminescence assay).

ment (the blind assay displaying a sensitivity of 90.8%) is shown in Figure 12.7.

12.4 DISCUSSION

We have developed a highly specific and sensitive immunodot assay capable of identifying *H. zea* parasitized by second and third instars of *M. croceipes*. Because the MAb used in the immunoassay is species-specific, parasitism by *M. croceipes* can be detected without interference by another braconid parasitoid, *C. marginiventris*, that coexists with *M. croceipes* in much of the United States. The assay is the first of its kind for detecting an insect endoparasitoid within a major field crop pest. A MAb-based, species-specific ELISA was previously developed that identified ectoparasitoids of stored-product insects (Stuart and Burkholder, 1991).

A variety of peroxidase detection kits are commercially available. Of the three examined here, the chemiluminescence kit was the most expensive and cumbersome, requiring a darkroom, an X-ray cassette, and the chemicals necessary for film development. Although it was the most sensitive method for detecting dissected *M. croceipes*, it failed to detect three parasitized *H. zea* (M63, M82 and M83) that were detected by one or both of the other two methods.

CN/DAB and AEC were both easy to use and differed in their sensitivity against whole-body homogenates of dissected *M. croceipes* by only one dilution. Both reagents were supplied as a two-component system that was mixed just before membrane development. Even though CN/DAB was approximately twice the cost of AEC, we preferred CN/DAB because the brown dots deposited by this reagent were more easily visible than the pink dots deposited by AEC. The addition of PTU to insect samples sufficiently inhibited melanization that there was no difficulty in distinguishing the brown dots of a positive immunoreaction from those caused by melanin. Untoward staining of the membrane was further reduced by diluting 2 μl of the insect homogenate in 100 μl/well of TBS rather than spotting 2 μl of the homogenate directly onto the nitrocellulose. The 2-μl volume was optimal because the amount of protein therein approximated the binding capacity of the nitrocellulose (70–120 $\mu g/cm^2$, or about 10 μg/dot; Stott, 1989). Nevertheless, in instances where samples become heavily melanized, AEC may be the superior development reagent because its colour contrasts well with that of melanin (Greenstone, 1995).

Proper preparation and storage of *H. zea* homogenates is critical for successful performance of the immunodot assay. In previous unpublished trials, *H. zea* were homogenized by sonication in phosphate-buffered saline, the larval carcasses were allowed to remain in the vials, PTU and protease inhibitors were omitted from the homogenization buffer, and no attempts were made to keep the homogenates chilled during handling. As a consequence of one or more of these differences, earlier immunodot assays employing MAb 5D1 failed to detect a single parasitized *H. zea*.

Probably the most significant difference in the protocol we currently use is the addition of protease inhibitors to the homogenization buffer. The *H. zea* host, rather than the parasitoid itself, appears to be the source of the most destructive proteases. Degradation of the *M. croceipes* antigen within homogenates of parasitized *H. zea* can be seen in immunoblots of native gels (Figure 12.3), and occurred despite storage of the homogenates at –85°C for less than 1 week before electrophoresis. In contrast, homogenates of dissected *M. croceipes* stored for several weeks at 4°C without protease inhibitors displayed the same native immunoblot pattern as dissected parasitoids that were freshly prepared. This suggests that significant proteolysis of the *M. croceipes* antigen does not occur in the absence of *H. zea*-derived enzymes.

MAb 5D1 makes it possible to screen rapidly for the presence of *M. croceipes* during the final quarter of its life within the larval host (Lewis, 1970), information which gives an indication of impending host mortality. It would also be useful to be able to detect first instar larvae, which make up the bulk of the first half of endoparasitic life. We attempted to

	1	2	3	4	5	6	7	8	9	10	11	12
A	H9	H10	H11	H12	H13	H14	H15	H16	H17	H18	H19	H20
B	H21	H22	H23	H24	H25	H26	H27					
C												
D	M20	M21	M22	M23	M24	M25	M26	M27	M28	M29	M30	M31
E	M32	M33	M34	M35	M36	M37	M38	M39	M40	M41	M42	M43
F	M44	M45	M46	M47	M48	M49	M50	M51	M52	M53	M54	M55
G	M56	M57	M58	M59	M60	M61	M62	M63	M64	M65	M66	M67
H	M68	M69	M70	M71	M72							(+)
FF	M73	M74	M75	M76	M77	M78	M79	M80	M81	M82	M83	M84
GG	M85	M86	M87	M88	M89	M90	M91	M92	M93	M94	M95	M96
HH	M97	M98	M99	M100	M101	M102	M103	M104	M105	M106		

Homogenate Identities

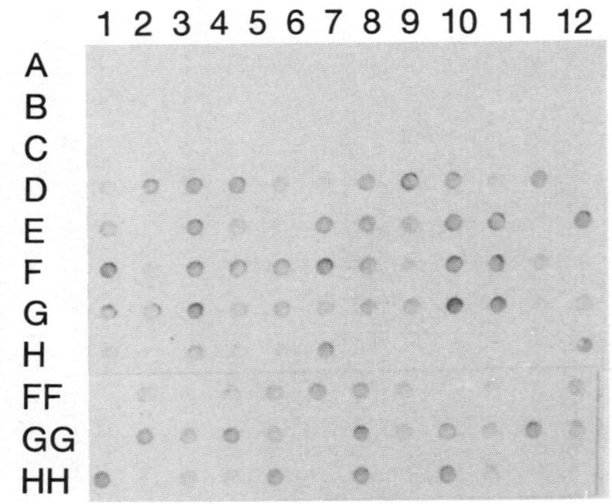

AEC

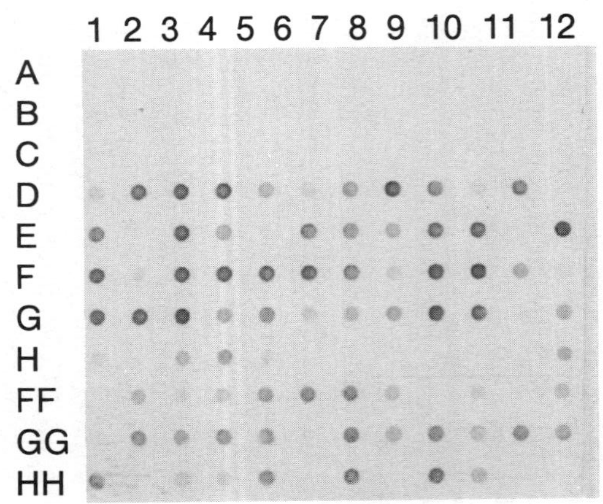

CN/DAB

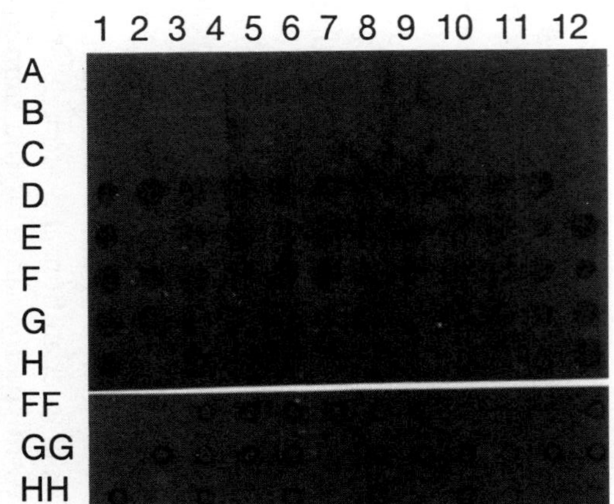

Chemiluminescence

Figure 12.6 Immunodot assay of 106 individual insects blotted with MAb 5D1. Animals designated with the letter 'H' are non-parasitized fourth instar *H. zea*. Animals designated with the letter 'M' are *H. zea* parasitized by second instar *M. croceipes*. Individuals H1–H8 and M1–M19 were used to optimize the assay and are not shown. Replicate blots were developed in the reagent indicated. Positive control wells (+) were loaded with 1 μg of dissected *M. croceipes* second instar whole-body homogenate. In the blot developed with AEC, the membrane was punctured as homogenate M71 was being loaded; M71 was subsequently re-loaded into well H-6.

Table 12.1 A summary of visual scores assigned to animals that yielded variable results in the immunodot assay

	Animal identification						
Assay	*M36*	*M42*	*M50*	*M63*	*M73*	*M82*	*M83*
CN/DAB	+++- -	++- - -	+++++	++++++	- - - -	+++	*NS ++-
AEC	++	- -	+-	++	-	-	-
Chemiluminescence	+	+	+	-	+	-	-

*NS: this animal was not scored in one assay in which it was used (see Figure 12.7). Each (+) or (-) represents a separate assay in which the animal yielded a positive or negative result, respectively.
AEC, 3-amino-9-ethylcarbazole; CN/DAB, 4-chloro-1-naphthol/3,3′-diaminobenzidine tetrahydrochloride.

(a)

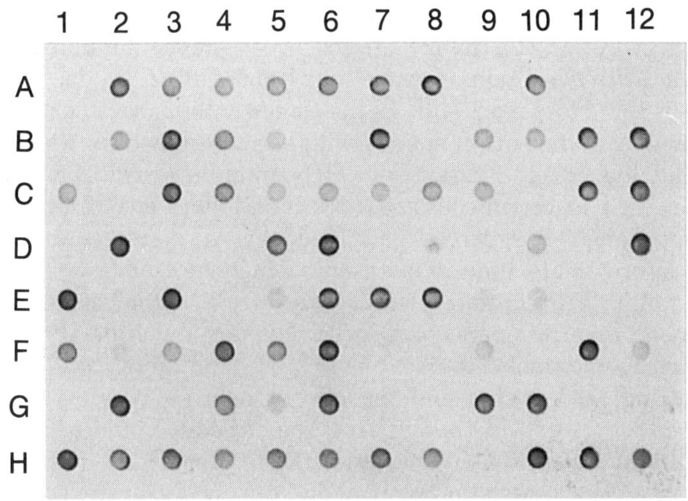

(b)

	1	2	3	4	5	6	7	8	9	10	11	12
A	-	+	+	+	+	+	+	+	-	+	-	-
B	-	+	+	+	+	-	+	-	+	+	+	+
C	+	-	+	+	+	+	+	+	+	-	+	+
D	-	+	-	-	+	+	-	+	+	+	-	+
E	+	-	+	-	+	+	+	+	+	+	-	N.S.*
F	+	+	+	+	+	+	-	-	+	-	+	+
G	-	+	-	+	+	+	-	-	+	+	+	-
H	+	+	+	+	+	+	+	+	-	+	+	+

(c)

	1	2	3	4	5	6	7	8	9	10	11	12
A	H10	M88	M38	M87	M63	M71	M40	M56	H27	M28	H12	H22
B	H18	M92	M43	M26	M20	H16	M52	M33	M62	M84	M89	M95
C	M35	H24	M23	M60	M68	M76	M51	M59	M74	M90	M37	M21
D	H26	M47	M69	H13	M50	M64	H21	M77	M82	M80	H9	M93
E	M44	H25	M22	H20	M75	M41	M86	M34	M36	M100	M73	M83
F	M96	M66	M94	M103	M70	M58	H19	H14	M45	M102	M46	M25
G	M81	M53	H15	M67	M29	M91	M31	H11	M105	M78	M42	H17
H	M65	M32	M30	M61	M45	M39	M101	M24	H23	M49	M57	(+)

Figure 12.7 Immunodot assay of unknown insect homogenates. (a) Membrane developed in CN/DAB; (b) visual scores assigned to the wells; (c) identities of the homogenates. *N.S. indicates a well which could not be scored because it was punctured during loading.

produce a MAb specific for *M. croceipes* first instars that could be used with MAb 5D1 to detect both larval instars of the parasitoid within host homogenates. Electrophoresis was used to identify proteins that appeared unique to *M. croceipes*. These proteins were eluted from gels and used to immunize mice, but the resultant MAbs were not species-specific. As an alternative approach, we have immunized mice with adult *Microplitis* homogenate; this has resulted in the generation of MAbs that cross-react with all life-stages without sacrificing species-specificity, for other insect species with which it has been tried (Stuart and Burkholder, 1991; Stuart *et al.*, 1994). An anti-*Microplitis* antibody of this type, used in conjunction with MAb 5D1, will enable estimates of the proportion of host larvae parasitized by *M. croceipes*, and forecasts of the approximate time of parasitoid emergence and the resultant host mortality. Furthermore, since parasitized larvae cause much less damage than non-parasitized ones (Hopper and King, 1984), such estimates would make it possible to raise economic thresholds, thereby reducing the expense and the loss of natural enemies resulting from unnecessary pesticide application. The antibody could be packaged into a kit which would enable a consultant or grower to determine rates of *Microplitis* parasitism in less than an hour (cf. Greenstone, 1995).

Acknowledgements

We thank William Steiner and Ben Puttler, for providing *Microplitis* and *Cotesia* stocks, Betty Cox, Clyde Morgan and Lauren Shulse for technical assistance, Paula Peters for rearing Lepidoptera, and Davy Jones, Janine Powell, Pamela Marrone, David Ragsdale, Stephen Trowell, and an anonymous reviewer for comments on the manuscript. This work was supported in part by Grant No. 900-306 from the Kirksville College of Osteopathic Medicine to M.K.S., and a USDA-ARS Post Doctoral Fellowship to M.H.G.

This article reports the results of research only. Mention of a proprietary product does not constitute an endorsement or a recommendation for its use by the USDA. All programs and services of the USDA are offered on a nondiscriminatory basis without regard to race, colour, national origin, religion, sex, age, marital status, or handicap.

REFERENCES

Berger, R.S. (1963) *Laboratory Techniques for Rearing* Heliothis *Species on Artificial Medium*, United States Department of Agriculture Agricultural Research Service ARS-33-84, pp. 1–4.

Biever, K.D. and Boldt, P.E. (1970) Utilization of soft X-rays for determining pupal parasitism of *Pieris rapae*. *Annals of the Entomological Society of America*, **63**, 1482–3.

Bottrell, D.G., Young, J.H., Price, R.G. and Adams, R.H. (1968) Parasites reared from *Heliothis* spp. in Oklahoma in 1965 and 1966. *Annals of the Entomological Society of America*, **61**, 1053–5.

Burleigh, J.G. and Farmer, J.H. (1978) Dynamics of *Heliothis* spp. larval parasitism in southeast Arkansas. *Environmental Entomology*, **7**, 692–4.

Campbell, A.M. (1984) Monoclonal antibody technology: the production and characterization of rodent and human hybridomas, in *Laboratory Techniques in Biochemistry and Molecular Biology*, Vol. 13, (eds R.H. Burton and P.H. van Knippenberg), Elsevier Science, New York, pp. 157–60.

Carlsson, M., Hedin, A., Inganas, M. *et al.* (1985) Purification of in vitro produced mouse monoclonal antibodies, a two-step procedure utilizing cation exchange chromatography and gel filtration. *Journal of Immunological Methods*, **79**, 89–98.

Caron, R.E. and Bradley, J.R., Jr (1978) *Heliothis zea* (Boddie): biotic factors affecting egg and larval survival on field corn in North Carolina. *Journal of the Georgia Entomological Society*, **13**, 295–304.

Castañera, P., Loxdale, H.D. and Nowak, K. (1985) Electrophoretic study of enzymes from cereal aphid populations. II. Use of electrophoresis for identifying aphidiid parasitoids (Hymenoptera) of *Sitobion avenae* (F.) (Hemiptera: Aphidiidae). *Bulletin of Entomological Research*, **73**, 659–65.

Eger, J.E., Jr, Sterling, W.L., and Hartstack, A.W., Jr (1982) Population dynamics of *Heliothis* spp. on *Castilleja indivisa*, an unreported host plant, and *Lupinus texanus* in Texas. *Environmental Entomology*, **11**, 327–33.

Graham, R.C., Jr, Lundholm, U. and Karnovsky, M.J. (1965) Cytochemical demonstration of peroxidase activity with 3-amino-9-ethylcarbazole. *Journal of Histochemistry and Cytochemistry*, **13**, 150–2.

Greenstone, M.H. (1995) Bollworm or budworm? Squashblot immunoassay distinguishes eggs of *Heliocoverpa zea* and *Heliothis virescens* (Lepidoptera: Noctuidae). *Journal of Economic Entomology*, **88**, 213–18.

Hamm, J.J., Nordlund, D.A. and Mullinex, B.G., Jr (1983) Interaction of the microsporidium *Vairimorpha* sp. with *Microplitis croceipes* (Cresson) and *Cotesia marginiventris* (Cresson) (Hymenoptera: Braconidae), two parasitoids of *Heliothis zea* (Lepidoptera: Noctuidae). *Environmental Entomology*, **12**, 1547–50.

Hopper, K.R. and King, E.G. (1984) Feeding and movement on cotton of *Heliothis* species (Lepidoptera: Noctuidae) parasitized by *Microplitis croceipes* (Hymenoptera: Braconidae). *Environmental Entomology*, **13**, 1654–60.

Ignoffo, C.M. (1965) The nuclear-polyhedrosis virus of *Heliothis zea* (Boddie) and *Heliothis virescens* (Fabricius). II. Biology and propagation of diet-reared *Heliothis*. *Journal of Invertebrate Pathology*, **7**, 217–26.

Ignoffo, C.M. and Boening, O.P. (1970) Compartmented disposable plastic trays for rearing insects. *Journal of Economic Entomology*, **63**, 1696–7.

Lenz, C.J. and Greenstone, M.H. (1988) Production of a monoclonal antibody to the arylphorin of *Heliothis zea*. *Archives of Insect Biochemistry and Physiology*, **9**, 167–77.

Lewis, W.J. (1970) Life history and anatomy of *Microplitis croceipes* (Hymenoptera: Braconidae), a parasite of *Heliothis* spp. (Lepidoptera: Noctuidae). *Annals of the Entomological Society of America*, **63**, 67–70.

Lewis, W.J. and Brazzel, J.R. (1968) A three-year study of parasites of the bollworm and the tobacco budworm in Mississippi. *Journal of Economic Entomology*, **61**, 673–6.

Lewis, W.J. and Gross, H.R. (1989) Comparative studies on field performance of *Heliothis* larval parasitoids *Microplitis croceipes* and *Cardiochiles nigriceps* at varying densities and under selected host plant conditions. *Florida Entomologist*, **72**, 6–14.

Martin, P.B., Lingren, P.D., Green, G.L. *et al.* (1976) Parasitization of two species of Plusiinae and *Heliothis* spp. after releases of *Trichogramma pretiosum* in seven crops. *Environmental Entomology*, **5**, 991–5.

Mueller, T.F. and Phillips, J.R. (1983) Population dynamics of *Heliothis* spp. in spring weed hosts in southeastern Arkansas: survivorship and stage-specific parasitism. *Environmental Entomology*, **12**, 1846–50.

Nuenzing, H.H. (1969) The biology of the tobacco budworm and the corn earworm in North Carolina. *North Carolina Agricultural Experiment Station Technical Bulletin*, 76 p.

Oatman, E.R. (1966) Parasitization of corn earworm eggs on sweet corn silk in Southern California, with notes on larval infestations and predators. *Journal of Economic Entomology*, **59**, 830–5.

Pair, S.D., Laster, M.L. and Martin, D.F. (1982) Parasitoids of *Heliothis* spp. (Lepidoptera: Noctuidae) larvae in Mississippi associated with sesame interplantings in cotton, 1971–1974: implications of host–habitat interaction. *Environmental Entomology*, **11**, 509–12.

Phillips, J.W. and Barber, G.W. (1933) Egg-laying habits and fate of eggs of the corn earworm moth and factors affecting them. *Virginia Agricultural Experiment Station Technical Bulletin*, **47**, 1–14.

Powell, W. and Walton, M.P. (1989) The use of electrophoresis in the study of hymenopteran parasitoids of agricultural pests, in *Electrophoretic Studies on Agricultural Pests*, (eds H.D. Loxdale and J. Den Hollander), Clarendon Press, Oxford, pp. 443–65.

Puterka, G.J., Slosser, J.E. and Price, J.R. (1985) Parasites of *Heliothis* spp. (Lepidoptera: Noctuidae): parasitism and seasonal occurrence for host crops in the Texas rolling plains. *Environmental Entomology*, **14**, 441–6.

Quaintance, A.L. and Brues, C.T. (1905) The cotton bollworm. *United States Department of Agriculture Bureau of Entomology Bulletin*, **50**, 155 p.

Ridgway, R.L., King, E.G. and Carillo, J.L. (1977) Augmentation of natural enemies for control of plant pests in the Western Hemisphere, in *Biological Control by Augmentation of Natural Enemies*, (eds R.L. Ridgway and S.B. Vinson), Plenum Press, New York, pp. 379–416.

Soteres, K.M., Berberet, R.C. and McNew, R.W. (1984) Parasitic insects associated with lepidopterous herbivores on alfalfa in Oklahoma. *Environmental Entomology*, **13**, 787–93.

Stadelbacher, E.A., Powell, J.E. and King, E.G. (1984) Parasitism of *Heliothis zea* and *H. virescens* (Lepidoptera: Noctuidae) in wild and cultivated host plants in the delta of Mississippi. *Environmental Entomology*, **13**, 1167–72

Stott, D.I. (1989) Immunoblotting and dot blotting. *Journal of Immunological Methods*, **119**, 153–87.

Stuart, M.K. and Burkholder, W.E. (1991) Monoclonal antibodies specific for *Laelius pedatus* (Bethylidae) and *Bracon hebetor* (Braconidae), two hymenopterous parasitoids of stored-product pests. *Biological Control*, **1**, 302–8.

Stuart, M.K., Barak, A.V. and Burkholder, W.E. (1994) Immunological identification of *Trogoderma granarium* Everts (Coleoptera: Dermestidae). *Journal of Stored Products Research*, **30**, 9–16.

Tomiuk, J., Wohrmann, K. and Eggers-Schumacher, H.A. (1979) Enzyme patterns as a characteristic for the identification of aphids. *Zeitschrift für Angewandte Entomologie*, **88**, 440–6.

Van Deusen, R.A. (1983) Making hybridomas, in *Hybridoma Technology in Agricultural and Veterinary Research*, (eds N.J. Stern and H.R. Gamble), Rowman & Allanheld, Beltsville Agricultural Research Center, Agricultural Research Service, US Department of Agriculture, pp. 15–25.

Walton, M.P., Powell, W., Loxdale, H.D. and Allen-Williams, L. (1990) Electrophoresis as a tool for estimating levels of hymenopterous parasitism in field populations for the cereal aphid, *Sitobion avenae. Entomologia Experimentalis et Applicata*, **54**, 271–9.

Whitehead, T.P., Kricka, L.J., Carter, T.J.N. and Thorpe, G.H.G. (1979) Analytical luminescence: its potential in the clinical laboratory. *Clinical Chemistry*, **25**, 1531–46.

Wool, D., van Emden, H.F. and Bunting, S.W. (1978) Electrophoretic detection of the internal parasite *Aphidius matricariae* in *Myzus persicae. Annals of Applied Biology*, **90**, 21–6.

13

Polyclonal, monoclonal and engineered antibodies to investigate the role of predation in slug population dynamics

W.O.C. Symondson and J.E. Liddell

ABSTRACT

The advent of monoclonal antibodies, produced by hybridoma technology, has begun to revolutionize the study of predator–prey interactions. Despite this, however, polyclonal antisera have certain advantages that justify their continued use. In particular they are inexpensive and rapid to produce, and can be highly sensitive. Characterization, however, involves extensive testing to ensure that a degree of cross-reactivity, inevitable with such antisera, does not lead to misidentification of heterologous prey material as target antigen. Controls must be rigorously applied, particularly in quantitative assays, and must allow for reactions between invertebrate remains and both specific and non-specific antibodies in the antisera. Detection limits and antigen decay rates, during digestion in the predator, must be determined using a standard assay protocol, which must be applied throughout the study. Results are presented from a number of studies, including a quantitative examination of the effects of cultivation methods upon predation by *Pterostichus melanarius* Illiger on slugs in arable crops. The data demonstrate that over 80% of beetles, collected over a 3-month period, contained slug antigens. Detailed information was obtained, from enzyme-linked immunosorbent assays, beetle crop weights and population data, on fluctuations in the concentrations and quantities of slug antigens in response to environmental variables. Larger numbers of beetles

The Ecology of Agricultural Pests, Edited by W.O.C. Symondson and J.E. Liddell.
Published in 1996 by Chapman & Hall, London. ISBN 0 412 62190 8

were collected from direct-drilled plots, and these were found to contain more prey material with a higher concentration of slug proteins, than beetles from tilled plots.

A major disadvantage of polygonal antisera is their inability to distinguish between predation upon different species of mollusc. An antiserum raised against one species will also react with other molluscs, with the strength of the reaction related to the taxonomic distance between the species involved, and this presents quantitative difficulties. These two problems can both be resolved with the use of monoclonal antibodies (MAbs). Two species of slug predominate on our collection site – *Deroceras reticulatum* (Müller) and *Arion intermedius* Normand. A genus-specific anti-arionid MAb, that can distinguish predation upon arionids from that upon other genera, was raised against *Arion ater* (L.) haemocyanin. Three species-specific anti-*D. reticulatum* antibodies were produced that can separate this species from the closely related *D. caruanae* (Pollonera). In addition, an anti-slug MAb, that reacts with all slug species, has been raised for initial screening purposes, providing limitless supplies of a uniform reagent to replace polygonal antisera. All of these antibodies have been characterized, and are being applied to field samples. Unfortunately, MAbs take several months to produce using conventional hybridoma technology. We have therefore been developing recombinant phage antibody systems that should allow us to isolate specific antibodies, from an established gene library, in as little as 2 weeks. The advantages that this new class of molecular probe bring to complex ecological studies are discussed.

13.1 INTRODUCTION

13.1.1 Background

Slugs are pests of almost all major arable and horticultural crops in the UK, and frequently cause serious economic damage (Port and Port, 1986; Glen, 1989). Although European slugs are now attacking crops in many parts of the world (South, 1992), following inadvertent introductions, nowhere are they a more severe problem than in Britain. Here, mild wet winters and cool summers allow destructive species such as *Deroceras reticulatum* (Müller) to breed and feed all the year round (Hunter and Symonds, 1971), generating infestation levels within arable crops frequently in excess of 200 per m^2 (Glen *et al.*, 1989). At present, the main means of defence by farmers against slug damage to their crops is the use of molluscicides based upon methiocarb (Port and Port, 1986). Unfortunately this carbamate compound, originally developed as an insecticide, can severely reduce populations of carabid beetles and other non-target predatory arthropods (Purvis and Bannon, 1992). Apart from their potential role as slug predators, carabids are thought to be important natural control agents for many other pest species (Luff, 1987), and may indirectly help to prevent the spread of crop diseases (Kendall *et al.*, 1995).

Interest in an integrated approach to the management of slug populations has grown in response to these problems. In particular, farmers may be encouraged to use metaldehyde-based formulations which are less damaging to arthropods (Büchs *et al.*, 1989), while research has been undertaken to develop new compounds (e.g. Bullock *et al.*, 1992; South, 1992). The recent arrival on the market of the slug parasitic nematode *Phasmarhabditis hermaphrodita* (Schneider) may provide growers in future with a highly specific microbial pesticide (Wilson *et al.*, 1993) that is compatible with farming systems that seek to maximize populations of natural enemies in general, and Carabidae in particular. Such an approach may be combined with appropriate cultural and rotational practices that minimize slug populations and damage (Kendall *et al.*, 1995).

However, before such strategies can be prepared, quantitative information must be obtained on the role of predation in slug population dynamics. Many factors are involved with such an assessment (Sunderland, 1996, this volume), some of which have been investigated using antibodies for gut analysis. In particular, antibodies have been used to identify important slug predators found in the crop environment, to attempt to quantify the biomass of slug ingested, and to monitor the long-term effects of farming practices, such as cultivation systems, and short-term effects of particular events, such as the harvesting of a crop, upon predator–prey interactions (Symondson *et al.*, in press). The paper begins by explaining why we believe antibodies can provide the best means of obtaining detailed information on predation by carabids upon slugs in the field. This is followed by a review of our work over the past few years, and we outline our reasons for progressively using polygonal, monoclonal and now engineered antibodies to answer particular ecological questions.

13.1.2 Why use antibodies?

It is very easy to demonstrate in the laboratory that certain carabids will attack slugs. Starved individuals of large species such as *Pterostichus melanarius* (Illiger), *Abax parallelepipedus* Piller and Mitterpacher or *Carabus violaceous* L. will immediately attack small slugs when offered them after a few days of food deprivation. Furthermore, trials under glass or Polythene have shown that species such as *A. parallelepipedus* are highly effective at controlling slugs, and reducing slug damage, in an immature lettuce crop (Symondson, 1993) and in grass/clover swards (Asteraki, 1993). However such no-choice studies provide limited information about predation in the field, where a wide range of potential prey are available.

Sunderland (1988) provides an excellent review of quantitative methods that have been used to detect and measure predation occurring in the field. However, the methods chosen depend very much upon the habitat and organisms involved. For example, direct observation and the use of video systems may be useful for observing predatory behaviour in some circumstances, and can even provide limited quantitative data. However, both slugs and carabid beetles are principally nocturnal organisms, frequently operating under a closed crop canopy. Although it may be possible to make some observations under red light (Griffiths *et al.*, 1985), it is unlikely that a predation event would be witnessed at natural predator and prey densities, and even if it were, this would provide little quantitative information. Some disturbance of the site would be almost inevitable and may affect the behaviour observed. An alternative approach might be to monitor carabid and slug population densities in the field, and see how these correlate with one another over time, and in response to natural and induced changes in the crop environment (Burn, 1988; Kendall *et al.*, 1995). However, even where positive correlations are found, causal relationships induced by predation cannot be proven by such data alone, particularly where the relevant predator species are unknown. Insect predator–prey relationships can be demonstrated by microscopic analysis of the gut contents of predators, where the predators involved consume recognizable and preferably countable prey remains such as tarsi (Sunderland, 1988; Wallin, 1988). Unfortunately semi-digested slug is not readily identifiable, although occasionally radular teeth or the remains of vestigial shells may be observed (Luff, 1974). These parts may not always be consumed, however, and not all slugs have recognizable shells. Even among Carabidae there are many species that are basically fluid feeders, which may involve pre-oral digestion of their prey and the filtering out of solid remains before ingestion (Forsythe, 1982).

By contrast with the methods outlined above, antibodies can be used in enzyme-linked immunosorbent assays (ELISA) to detect (Symondson and Liddell, 1993a,c), identify to genus, species or stage (Symondson and Liddell, 1993d,e; Symondson *et al.*, 1995) and quantify (Symondson and Liddell, 1993c; Symondson *et al.*, in press) the remains of molluscs extracted from the guts of carabids, even where fluid feeding is involved. The only other detailed study of carabid–slug interactions using antibodies was that of Tod (1973), using the relatively insensitive Ouchterlony agar double-diffusion precipitin test. These are post-mortem techniques which involve minimal disturbance before the capture of the predators to be analysed. In conjunction with population monitoring, meteorological records and information on agricultural variables, such methods can provide valuable ecological information.

13.2 PROGRESS TO DATE

13.2.1 Production and characterization of an anti-slug polyclonal antiserum

Polyclonal antisera have the major advantage that they are relatively quick and simple to produce, requiring no special facilities (Tijssen, 1985; Greenstone, 1996, this volume). The aim must be to inject a rabbit (usually) with a narrow range of proteins, polysaccharides or other antigens that will induce an immunological response to antigenic determinants that are unique, as far as possible, to the prey species involved. For our own study slug haemolymph was chosen for a number of reasons. Approximately 96% of pulmonate haemolymph protein is haemocyanin (Ghiretti, 1966), a protein with a high molecular weight of around 9000 kDa (van Holde and Miller, 1982) that is known to elicit a strong immunological response (Liddell and Cryer, 1991), and is available in a relatively pure form. The structure of this molecule in molluscs is very different to that found in arthropods such as Arachnida, Isopoda, Chilopoda and Diplopoda (advantages of haemocyanin are reviewed in Symondson and Liddell, 1993a).

Full details of the methods used to create and characterize an anti-slug antiserum for *Arion ater* (L.) haemolymph are given in Symondson and Liddell (1993a). This species was chosen for its large size, allowing up to 1 ml of haemolymph to be collected from each slug. After experiments to optimize ELISA conditions, a series of indirect assays was performed in order to establish the sensitivity and specificity of the system. Preliminary tests revealed a cross-reaction, between the antiserum and earthworm mucus, which was effectively eliminated by absorption without major loss of antibody titre. The baseline assay involved diluting invertebrate homogenates to a final dilution of 1:20 000. ELISAs quickly showed that the earthworm-absorbed antiserum reacted strongly with all pulmonate molluscs tested. The assay could clearly detect <1 ng of haemolymph protein, which was a lower detection limit than has been reported in any similar study. Cross-reaction tests against 27 representative species of predator and prey, representing Insecta, Arachnida, Diplopoda, Chilopoda, Isopoda and Annelida demonstrated no major reactions. Initial strong reactions with a few individuals of *Phalangium opilio* L. (Opiliones) indicated slug predation by these harvestmen, with all starved specimens generating minimal responses. The strongest slug-free reaction was with heterologous spider proteins (*Pardosa* sp.). A conservative significance level for the presence of slug haemolymph proteins was therefore set at the mean +2.5 S.D. of the reaction to *Pardosa* sp. All results were quantified with reference to a dilution series of haemolymph included on each ELISA plate. Regression equations derived from these internal standards were used to translate absorbance readings for test samples into equivalent haemolymph concentrations. Background absorbance, and low-level binding with heterologous material, were monitored with parallel tests on all samples with non-immune rabbit sera.

The antiserum was then used to measure the decay rate of the antigens within relevant predators, in this case *A. parallelepipedus* and *Pterostichus madidus* F. (Symondson and Liddell, 1993b). Starved beetles were fed on two species of slug, *Arion hortensis* Férussac and *D. reticulatum*, and then killed after various time periods. Their crop contents were weighed and then ELISA tested. It was found, by covariate analysis, that the rate of decay of slug antigens was very different between the two species. For example, following a 1-hour feed, remains of *D. reticulatum* could be detected for 35 hours in *A. parallelepipedus*, but for 94 hours in *P. madidus*, more than 2.5 times as long. This result appears to contradict previous findings that indicated that larger species within a taxonomic group have longer detection periods (Sunderland and Sutton, 1980; Sopp and Sunderland, 1989). However, the rate of loss of slug mass from the crops of these two species was not significantly different, reaching control values, for starved individuals, after approximately 2.5 days. Clearly, in *A. parallelepipedus*, slug remains decayed beyond antigenic recognition while measurable slug mass could still be detected in the crops of these beetles; however, in *P. madidus*, digestion in the crop must have been minimal, because traces of antibody-recognizable slug could be detected in the crops of this species long after overall crop mass had declined to control values. Consumption rates were not significantly different between the two predators, nor was the mass consumed affected by the species of slug. When *A. parallelepipedus* was fed overnight on *A. hortensis*, the greater quantities of slug consumed, and the stronger immune response to this species of slug, extended the detection limit to a mean of 121 hours, while the control mass was reached after 91 hours. However, one beetle gut sample was still positive for slug antigens after 13 days.

In summary, the characterization experiments demonstrated that a highly sensitive and specific antiserum had been generated to slug haemolymph. The inclusion of an internal dilution series on each ELISA plate allowed haemolymph concentrations to be calculated that were independent of plate-to-plate variability. The w/v method of sample dilution stabilized overall protein levels, and allowed concentrations of haemolymph proteins in beetle crop samples to be measured as a separate variable from haemolymph protein mass. Full quantification of the slug content of gut samples was limited by the fact that the strength of the reaction between the antiserum and different species of slug varied according to the taxonomic distance between the species involved and the antigen species, *A. ater*. In addition, digestion rates varied according to the predator and prey species concerned. These results underline the need for quantitative assays and thorough laboratory characterization experiments before field assessment of predators. If, for example, purely qualitative tests were used to assess natural populations of *A. parallelepipedus* and *P. madidus*, and 2.5 times as many of the latter species reacted positively (which in practice they did not), this might erroneously be interpreted as indicating that *P. madidus* was a more important slug predator, rather than being a function of differential detection periods.

13.2.2 Field data obtained with the polyclonal antiserum

(a) Data on Abax parallelepipedus and Pterostichus madidus

Data were first obtained on slug feeding in the field by *A. parallelepipedus* and *P. madidus* (Symondson and Liddell, 1993c). The former species had been shown to be an effective biological control agent under glass in a no-choice situation, where few alternative prey were available (Symondson, 1993). However, this species could not be considered a useful natural enemy of slugs in gardens and outdoor horticulture without more information on prey choice in the field. Beetles were collected by dry pitfall trapping from a woodland edge site in South Wales from May to September 1990 and tested by ELISA. The only other common species of large carabid trapped was *P. madidus*, which were also collected for inter-species comparisons. Parameters measured included the haemolymph concentration equivalent (HCE) (i.e. the concentration of antibody-recognizable slug haemolymph, at the 1:20 000 dilution used in ELISA tests), crop mass and fresh slug equivalent (i.e. the calculated quantity of undenatured slug protein in the sample, based upon an equation derived from control samples) (Symondson and Liddell, 1993c).

A total of 105 *A. parallelepipedus* were trapped, 90% of which reacted positively with the anti-slug antiserum. Female beetles contained significantly higher concentrations and mass of slug proteins than males, confirming consumption rate data from a previous study (Symondson, 1989). Differences in the concentrations and mass of slug proteins were monitored over time, with peak slug consumption occurring in May and September, possibly in response to higher slug densities at these times of year. Conversely, however, lowest haemolymph concentrations and crop mass in July could have been in part caused by increased digestion rates at higher summer temperatures.

During the same period, 350 *P. madidus* were collected, and 42% of these reacted positively in ELISA tests. There was no significant difference between the sexes in the concentration of haemolymph in the beetle crops, although the mass of material in females' crops was nearly 60% greater than that in males. Although gross biomass of material in the crops of these beetles varied between months, there were no significant differences in slug protein concentrations or mass.

This study therefore revealed that slugs formed a major part of the diet of *A. parallelepipedus*, even where other prey were readily available. Polyphagous predators such as this can be valuable biocontrol agents, in that they can subsist on alternative prey when numbers of the target species decline. By preferentially feeding upon slugs, this carabid can perform a 'lying-in-wait' strategy when mollusc numbers are low. Murdoch *et al.* (1985) have identified such predators as having, in theory, advantages over stenophagous predators and parasitoids, in that their numbers will not decline in response to target prey availability, and local extinction of the pest becomes an achievable goal, rather than reduction of predator and prey densities to a fluctuating equilibrium. Molluscs are of less importance in the diet of *P. madidus*, in that a lower proportion of individuals tested positive, and the mean haemolymph concentration in positive individuals was lower than in *A. parallelepipedus*. Differences between the two species were actually much greater than the data indicate, because the slower digestion rate of *P. madidus* will increase the number and proportion of positive crop samples detected.

Possible sources of error in these results, and those following for *Pterostichus melanarius* (Illiger), might be occasional feeding within pitfall traps upon slugs or predators that contain slug protein in their guts. Dry traps were used, in preference to fluid filled traps, (i) to avoid the danger that moisture would attract slugs to the local trap site, potentially increasing predation artificially, and (ii) because independent measures of gut content mass and prey protein concentration were essential for the purposes of quantification. The traps themselves proved in practice to be unattractive as refugia for slugs, and therefore predation within the traps was considered negligible. It should also be remembered that the proportion of the population of either species of carabid that tests positive, and the quantity of slug protein they contain, are influenced by a great many factors. For example, it may well be the case that well-fed beetles are rarely trapped, because predatory activity is strongly motivated by hunger (Mols, 1993), and active beetles are more likely to fall into pitfall traps.

(b) Data on Pterostichus melanarius

The study above, on *A. parallelepipedus* and *P. madidus*, provided useful information on the dietary preferences of these two carabids on a relatively undisturbed site. However, the role of carabid predators in the dynamic environments found in arable crops are of considerable interest in relation to the planning of integrated systems of pest management, particularly in the context of lower-input production strategies. The most common large carabid beetle within arable crops in many parts of the UK is *P. melanarius*. Experiments were therefore planned that would allow the effects of agricultural and environmental variables on predator–prey interactions to be monitored and quantified. Results from an experiment involving over 2000 beetles, collected over 3 months in 1992, are briefly described below, and serve to illustrate the information-rich nature of the data that can be gathered from such quantitative studies (Symondson *et al.*, in press).

Beetle samples were collected from an on-going field experiment at Long Ashton Research Station in Bristol, and conducted in cooperation with D.M. Glen. The experiment, which has been running for several years, seeks to investigate the effects of different forms of soil cultivation, and straw disposal methods, on, among other things, invertebrate populations and crop yields, and has been described in detail elsewhere (Kendall *et al.*, 1995). The field site consists of 25 experimental plots, 12 m by 30 m, in a randomized block design with five blocks. Beetles were collected twice a week from single pitfall traps in the centre of each plot, with traps set the evening before each early morning sampling session. The five long-term treatments applied to cereal crops over several years were: (i) direct drilled, straw baled; (ii) soil shallow-cultivated, straw baled; (iii) soil shallow-cultivated, straw incorporated; (iv) soil ploughed, straw baled; and (v) soil ploughed, straw incorporated. However, in the year of sampling, there was a break crop of oilseed rape, and just two treatments: (a) baling and removal of rape residues on direct drilled plots (16–18 July); (b) initial shallow-incorporation of rape residues on all tillage treatments (31 July).

The main effects observed applied to current treatments (a and b), but a number of significant effects were detected relating to treatments applied a year previously (i–v). Parameters measured included beetle numbers, gross weights of dissected beetle crops, and the concentration of antibody-recognizable haemolymph protein, while haemolymph quantities were calculated from concentrations and crop weights (Symondson *et al.*, in press). Figure 13.1 shows data for beetle numbers and quantities of slug haemolymph in beetle crop samples, and compares the effects of direct drilling with those of tillage. ANOVAs were performed in which data were pooled for each treatment within

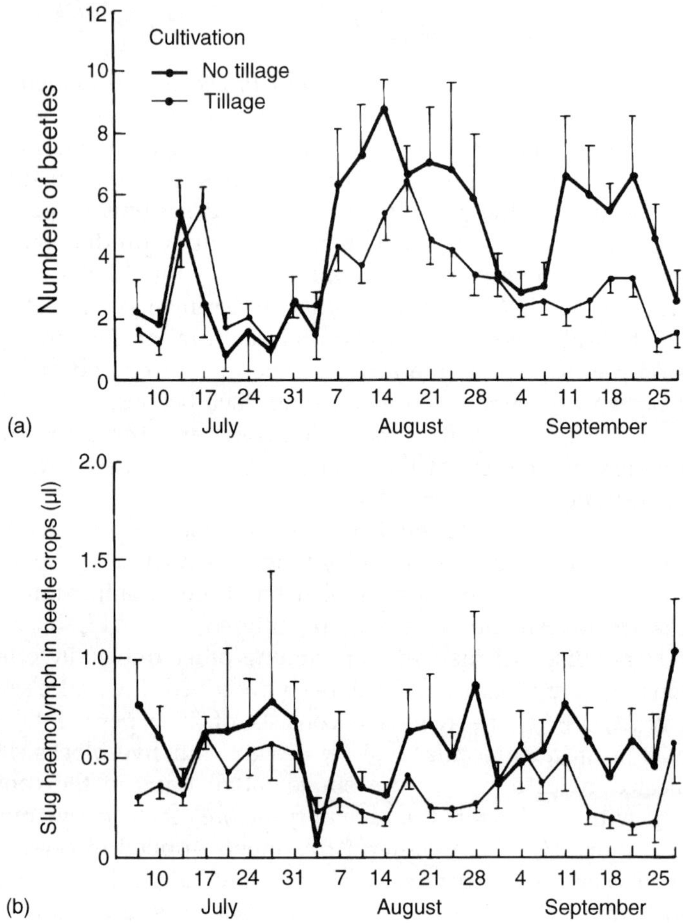

Figure 13.1 *Pterostichus melanarius* collected twice a week over a 3-month period in 1992, comparing no tillage (direct-drilled) with tilled treatments. (a) Mean number of beetles per pitfall trap; (b) mean quantity of slug haemolymph per beetle crop. Bars indicate ± SE, between plots for beetle numbers, between beetles for haemolymph quantities. (Figure reprinted by permission of the editors of the *Journal of Applied Ecology*, published by the British Ecological Society.)

each month, and revealed a great deal of information. They showed, for example, that there were greater numbers of beetles in direct-drilled plots, with greater crop weights, containing higher concentrations of slug proteins and therefore greater quantities of slug haemolymph. Interestingly, although these effects were all significant overall, differences in beetle numbers and crop weights were only shown in August and September, after cultivation of the tilled treatments at the end of July, and therefore primarily reflect current year treatments. However, concentrations and quantities of slug haemolymph were greater even in July in direct-drilled plots, and therefore must relate at this time to cultivation/straw treatment effects persisting from operations the previous autumn. Many other significant differences and interactions were observed. For example, more than twice as many male beetles were trapped as females, with direct drilling apparently favouring male more than female activity–density. However, male crops weighed only about half as much as those of females, and although the concentrations of slug haemolymph proteins were not significantly different between the sexes, females contained approximately twice as much slug haemolymph as males. There were significantly higher concentrations of haemolymph in beetle crop samples in July than in August or September, and many interesting interactions between haemolymph content and the form of tillage, month and beetle sex (Symondson *et al.*, in press). Treatment effects persisting from the previous year included greater numbers of beetles trapped (in August) in the shallow-tilled treatment with straw incorporation, and a greater biomass of both gross crop content and slug haemolymph retrieved in the totals from shallow-tilled plots, compared with those that were ploughed. Overall, 84% of the 2019 *P. melanarius* caught tested positive for the presence of slug antigens, so clearly molluscs form a major part of the diet of these predators.

These experiments were also designed to observe the short-term effects of different agricultural operations and applications. The most obvious of these proved to be the cultivation at the end of July. This had no adverse effects upon beetle populations (in so far as this may be reflected in terms of numbers trapped) (Figure 13.1(a)), and indeed the open ground thus created may in part have been responsible for higher activity–densities recorded thereafter. However, concentrations and quantities of slug haemolymph dropped sharply after cultivation in the tilled treatments, and remained low through most of August (Figure 13.1(b)). A similar drop in the direct-drilled treatment was probably induced by general movement between plots, caused by the disturbance to the tilled treatments and buffer zones, but haemolymph concentrations recovered more rapidly on these plots. If scavenging were important, it might be expected that far from dropping, the proportion of slug in the diet of the beetles would increase following cultivation, which is known to kill large numbers of slugs (Runham and Hunter, 1970).

It was essential to discover whether or not the information obtained above on numbers of beetle trapped and diet related to the effects of the different treatments on slug biomass and/or numbers in the soil. At the same time as information was being gathered on beetle predators, data were collected on slug populations. Soil samples were taken each month from each plot, and slugs extracted by gradual flooding (Glen *et al.*, 1989). Details of the results can be found elsewhere (Symondson *et al.*, in press), and again reflected both contemporary and previous soil treatments. However, a number of regression analyses were performed to examine the relationship between log slug biomass and numbers of beetles trapped, crop weights, concentrations of slug haemolymph and overall haemolymph content. For example, where the regression line was split into three separate lines for the 3 months, with a common slope, beetle numbers were shown to be positively related to log slug biomass in the soil within any month (Figure 13.2(a)). Between months there was an interesting negative relationship, indicating greater beetle activity–density at times of lower prey availability, a behavioural response noted in previous studies where prey were removed by insecticides (Chiverton, 1984; Dixon and McKinlay, 1992). There was no significant relationship between gross crop weight and log slug biomass in the soil, reflecting the fact that these are polyphagous predators known to feed on many non-slug prey items (Sunderland, 1975). The concentration of haemolymph in beetle crops (reflecting the proportion of the diet that was slug) was related to log slug biomass in the soil (Symondson *et al.*, in press), as were haemolymph quantities (Figure 13.2(b)), in a direct relationship that was not improved by separate lines for each month.

All of these results together strongly indicate that *P. melanarius* was aggregating to areas of high slug density, and preferentially feeding in such areas on slugs. Slug biomass, in this and previous experiments, has been shown to be greatest in direct drilled plots, which was where greatest numbers of *P. melanarius* were needed, from a pest control point of view, and where they were found. Although two major pest species of slug were found on the site, principally *D. reticulatum* but also smaller numbers of *Arion intermedius* Normand, analyses with polyclonal antisera could not detect differential predation upon these. Whether or not predation could be responsible, in part, for the relative population densities of these two species could only be investigated using more precise genus- or species-specific probes.

These results demonstrate how quantitative ELISA data can be used in combination with population monitoring, and measurements of the mass of material in predator crops, to obtain valuable comparative information on the ecology of a given species of predator under different conditions. Clearly, absolute information on the total quantity of slug material being taken out of the system by *P. melanarius* is not available, and depends upon many other factors. These include more precise measurement of carabid populations than can be obtained from pitfall-

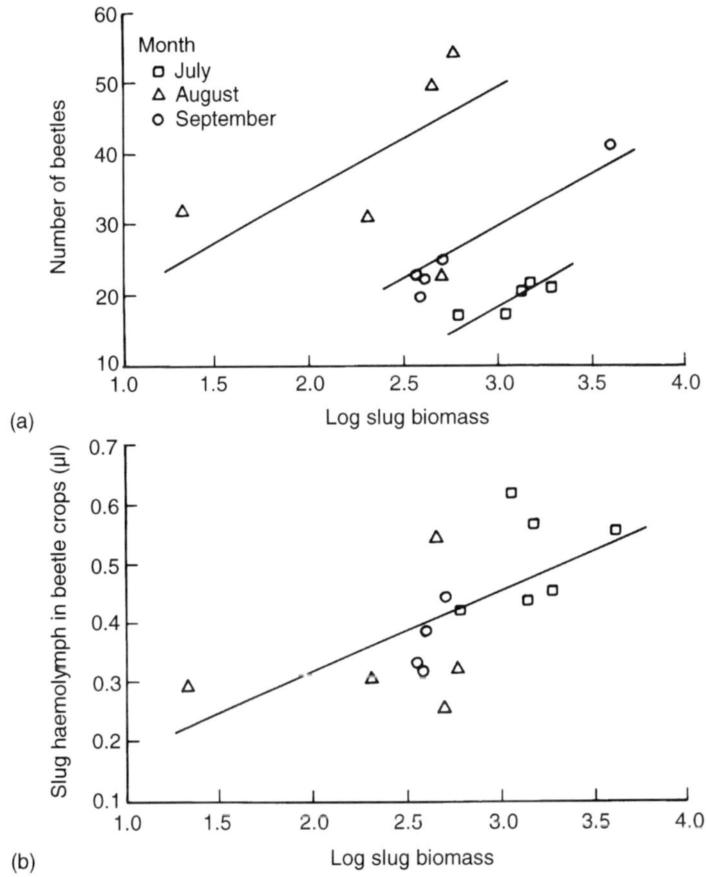

Figure 13.2 Relationship between mean log biomass of slugs extracted each month from soil samples in the five cultivation/crop residue disposal treatments and: (a) beetle activity density, as shown by numbers of *Pterostichus melanarius* caught in pitfall traps; (b) quantities of slug prey material consumed, as shown by the mean quantities of haemolymph in beetle crops. (Figure reprinted by permission of the editors of the *Journal of Applied Ecology*, published by the British Ecological Society.)

traps (which reflect both activity, density and trappability) (Sunderland *et al.*, in press), and calibration of the effects of meal size, subsequent feeding on alternative prey (Symondson and Liddell, 1995), partial consumption of prey and temperature (among other factors) upon estimates of the size of original predator meals based on ELISA data (Greenstone, 1996, this volume; Hagler and Naranjo, 1996, this volume; Sunderland, 1996, this volume). Even when attempts have been made to model predation in a cropping system on the basis of biomass ingested and antigen decay rates within predators (Sopp *et al.*, 1992), this ignores the profound effects that prey size choice may have upon prey population dynamics, particularly of invertebrates such as slugs which are highly variable in size and have relatively long life-cycles [approximately 9 months for *D. reticulatum* (Hunter and Symonds, 1971), 12 months for *A. intermedius* (South, 1989)]. It is difficult to predict how meals of equal mass, comprising, for example, either a number of neonate slugs or half an adult slug, would affect subsequent populations. However, such factors can be largely ignored in a comparative study at one site involving a single predator species, where predator behaviour and the effects of temperature, are given norms, and the differences observed are caused by the direct and indirect effects of treatments.

13.2.3 Monoclonal antibodies to investigate predation

Monoclonal antibodies (MAbs) possess a number of advantages over polyclonal antisera. They are usually highly specific, and have been shown in predation studies to be capable of detecting individual genera, species within genera and even stages, such as eggs or individual instars (Greenstone, 1996, this volume; Hagler and Naranjo, 1996, this volume; Symondson *et al.*, 1995). Monoclonal antibodies, derived as they are from single clones of hybrid cells, are of uniform structure, and can be selected for the required specificity at an early stage. The chances that a single antibody such as this will cross-react with non-target proteins are clearly much less than would be the case for the heterogeneous collection of antibodies in a polyclonal antiserum. Hybridoma clones are potentially immortal. Once a clone has been created, limitless supplies of antibody can be generated from that clone, which can be frozen in liquid nitrogen and revived again at any time (Liddell and Cryer, 1991). This provides therefore a uniform reagent which, unlike the limited supplies of variable polyclonal antisera, can be applied in predation studies over an extended period, allowing results to be compared over time and between laboratories.

A genus-specific MAb was raised against *A. ater* haemocyanin, following ammonium sulphate precipitation of the protein from haemolymph before immunization (Symondson and Liddell, 1993d). Fusion of spleen cells with myeloma cells was effected with polyethylene glycol, followed by growth in a selective medium (HAT), cloning by limiting dilution and ELISA screening for specific antibodies. Chosen hybridoma clones were used to produce MAb-rich ascites at this stage, although currently we are now introducing *in vitro* propagation methods to replace the use of mice at this stage. Details of all protocols may be found in Symondson and Liddell (1993d) and Liddell and Cryer (1991).

This IgG1 MAb, designated AaH-3E5, proved to be genus-specific within the system under study. It reacted strongly with the arionid slugs *Arion subfuscus* (Draparnaud), *A. ater* and *A. intermedius* Normand, but not with major pest species from different genera, such as *Deroceras caruanae* (Pollonera), *D. reticulatum, Tandonia budapestensis* (Hazey) and *T. sowerbyi* (Férussac) (Figure 13.3). The MAb could reliably detect as little as 3.4 ng of haemocyanin, and there were no significant cross-reactions with alternative prey. The detection period for a predation event, using this MAb, was investigated by feeding beetles upon *D. reticulatum* in the laboratory over night and then killing them the next day, and subsequently after a further 24 and 48 hours (L16:D8, 20°C). ELISA tests showed that feeding during the previous night could be detected, but unfortunately not thereafter.

Abbreviated detection periods, compared with those of polyclonal antisera, appear to be a common feature of MAbs and have been found by other workers (e.g. J.R. Hagler, personal communication). In other respects genus-specificity, by this MAb, provided an adequate degree of identity for studies at sites where only one arionid species, *A. intermedius*, was recovered (Symondson *et al.*, in press). Efforts to generate a species-specific MAb to this species are now proceeding using antibody

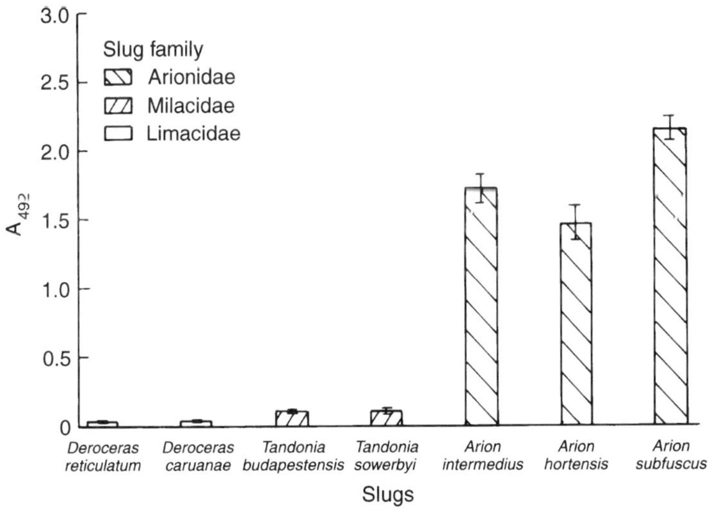

Figure 13.3 Absorbance at A_{492} obtained from ELISA tests of the AaH-3E5 MAb, raised against *Arion ater* haemocyanin, tested against seven species of slug of three genera, after subtraction of background readings. Bars indicate ± SE.

engineering techniques (Liddell and Symondson, 1996, this volume). Attempts were also made to raise MAbs to semi-digested slug remains, taken from the guts of *P. melanarius* that were allowed to feed on *A. intermedius* overnight. The aim was to generate MAbs to breakdown products of slug proteins, or proteins that had resisted digestion. Antibodies with extended detection periods should result from this approach, and early results are looking promising.

The most damaging pest slug species, both at our field sites and world-wide, is *D. reticulatum*, and therefore this species was chosen as the primary target for our studies. Monoclonal antibodies were raised to whole-body homogenates of *D. reticulatum* using similar protocols to those described above (Symondson and Liddell, 1993e). Four hybridoma clones were selected and propagated. Three IgM MAbs proved to be species-specific for *D. reticulatum*, in that they did not react with the closely related *D. caruanae* or any other species tested (Figure 13.4). Detection limit experiments demonstrated high sensitivities: DrW-1G4, 11.6 ng; DrW-1G1, 7.3 ng; DrW-2B12, 10.5 ng.

The fourth MAb selected, DrW-2D11, was an IgG1 isotype, and reacted with all pulmonates tested, not only those in Figure 13.4, but also *Helix aspersa* Müller, *Discus rotundatus* (Müller) and *Lehmannia marginata* (Müller). Tests against 40 alternative prey species demonstrated that only New Zealand Flatworms, *Artioposthia triangulata* (Deny)

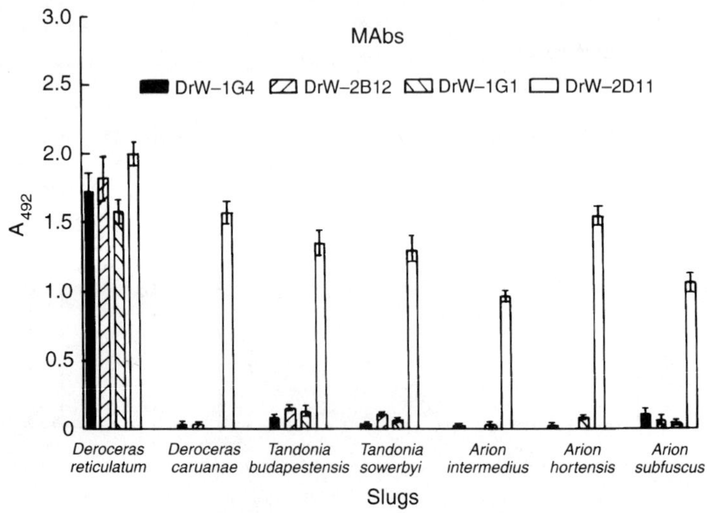

Figure 13.4 Absorbance readings at A_{492} obtained from ELISA tests of four MAbs raised against whole-body extracts of *Deroceras reticulatum*, tested against seven species of slug of three genera, after subtraction of background readings. Bars indicate ± SE.

reacted strongly with this MAb (unpublished data), but as these introduced invertebrates are not found on our study sites in Southern Britain, this presents no problem. This antibody could detect as little as 4.3 ng of slug protein, and should prove to be useful for general screening of predators for mollusc predation. It has largely replaced our polyclonal antiserum for this purpose.

The DrW-2D11 MAb was used in antigen decay rate experiments, to investigate the effect of subsequent feeding on alternative prey upon antigen detection periods (Symondson and Liddell, 1995). This is one of many factors that can affect quantification that has been inadequately investigated in the past, or given equivocal results. *P. melanarius* beetles were starved for a week, and then fed overnight on *D. reticulatum* (L16:D8, 16°C). The beetles were then either starved, or fed *ad libitum* on freshly collected earthworms which were continuously available, and frozen in batches of ten at various time intervals thereafter, before ELISA testing. Figure 13.5 shows decay rates for slug antigens. Analysis of covariance proved that the rate at which detectable antigen declines is greater when no alternative prey is consumed. In practice, this meant, at 16°C, that antigen could be detected for approximately 60 hours when beetles were starved, but 89 hours when allowed to feed subsequently on earthworms. Subsequent feeding, on alternative prey, could therefore extend detection periods by nearly 50%. These data

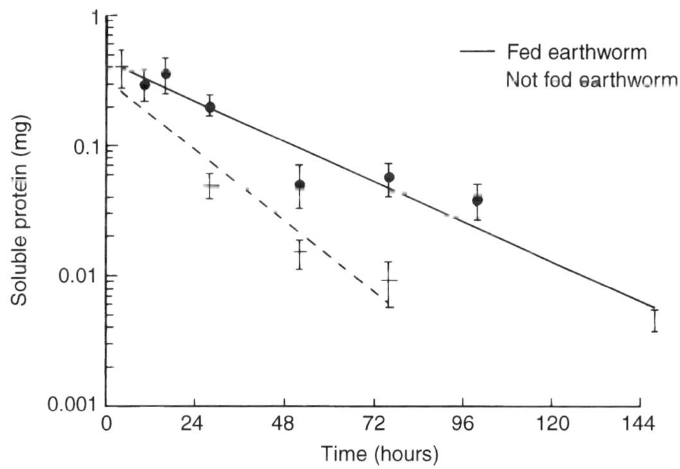

Figure 13.5 Decline in soluble slug protein/beetle crop following overnight feeding by *Pterostichus melanarius* on *Deroceras reticulatum*. Regression lines are for sample means, with (r^2 95.3%) and without (r^2 92.4%) subsequent feeding on alternative prey at 16°C, L16:D8. Bars ± SE for n = 10. (Figure reprinted by permission of Kluwer Academic Publishers.)

contrast with those obtained for spiders, which appear to reduce their rate of digestion in response to starvation (Fichter and Stephen, 1984) as part of a survival strategy involving a general reduction in metabolic rates (Anderson, 1970). Such a strategy appears to be appropriate for predators that ambush their prey, but remain quiescent for extended but variable periods. Carabids, in contrast, have been shown to increase their activity in response to low prey density (e.g. Dixon and McKinlay, 1992; Symondson *et al.*, in press) or to hunger (Mols, 1993), and therefore cannot afford to diminish the rate at which energy is derived from a meal by reducing their rate of digestion (Symondson and Liddell, 1995). Subsequent feeding on alternative prey, or the lack of it, has a profound effect upon detection periods, and therefore on calculations of the time since feeding. It is only by knowing when a prey item was consumed that attempts can be made to calculate the size of the original meal, and therefore estimate the biomass of prey being extracted from a system by predators. In theory, it should be possible to calculate this time interval by raising MAbs to prey proteins that break down at different rates during digestion, if the changing ratios of one protein to another can be calibrated with time (Sunderland, 1987). Current experiments indicate that this might be possible (unpublished data).

13.3 THE WAY AHEAD

Monoclonal antibodies provide a reliable means of obtaining long-term ecological information on predator–prey relationships, using a precise, uniform and highly sensitive form of molecular probe. Once created, the cell lines producing antibodies of the required specificities are immortal, and can be frozen and subsequently revived at any time. The initial investment in time and resources is considerable, but thereafter limitless supplies of antibody can be produced at low cost. For long-term studies therefore, at one or more laboratories, monoclonal antibodies probably provide the best form of detection system in terms of sensitivity and specificity, and the ability to obtain quantitative information, as long as basic cell culturing facilities are available. For short-term studies on a limited budget, ELISAs using polygonal antisera, or electrophoretic techniques (Solomon *et al.*, 1996, this volume), may prove adequate (for a general review, see Powell and Walton, 1995). To date, DNA techniques have not been widely applied in invertebrate predation studies, mainly because DNA appears to be denatured rather more rapidly in predator guts than protein, shortening detection periods (Gokool *et al.*, 1993). Although attempts are being made to devise quantitative gut analyses using DNA techniques (N.D. Martinez, personal communication), there appear to be no published reports of such work to date. Certain DNA techniques, such as the use of PCR, are not normally compatible with quantitative analyses.

Monoclonal antibodies have been generated in the past almost entirely by the cell fusion and hybridoma culturing techniques originally devised by Köhler and Milstein (1975). Recently, however, new methods have been developed that allow antibodies to be created from combinatorial antibody gene libraries, and our progress in applying such methods to the detection of predation upon slugs has been covered in Liddell and Symondson (1996, this volume). The primary advantage of such methods is that they should allow the rapid generation of new specificities within weeks, rather than months, providing a new class of molecular probe suitable for the study of complex ecological interactions. Instead of simply looking at the predators of a given target pest, it should be possible to look at, and quantify, the complete prey range of potentially useful predators. *P. melanarius*, for example, has been shown to feed upon slugs (Symondson *et al.*, in press) and aphids (Chiverton, 1987; Sunderland *et al.*, 1987) in cereal crops, and has been separately assessed as a predator of these organisms in different studies. *P. melanarius* may well be an important predator of many other pests, such as leatherjackets (*Tipula* spp.) (Chapman, 1994), but is also known to feed on potentially useful species including syrphid larvae, spiders, coccinellid beetles, smaller carabids and earthworms (e.g. Sunderland, 1975; Wallin, 1988). Indices of the relative value of different predator species (Sunderland *et al.*, 1987) must ultimately take into consideration the role of key species in the invertebrate community as a whole, and not simply their relationship with a single target species. Recombinant antibodies provide a means of achieving this goal, with parallel analysis of individual gut samples using a wide range of relevant specificities in ELISA tests. An essential component of such analyses is a protocol that uses minimal quantities of antigen per test, and this is achieved by the 1:20 000 dilution used as a standard in our studies. The protocols we have chosen combine the lowest antigen dilution rate with the greatest sensitivities reported in any comparable predation studies.

Acknowledgements

We would like to thank Dr D.M. Glen, Mr C. Langdon and Mr C.W. Wiltshire of Long Ashton Research Station, Bristol, and Prof I.D. Bowen and Prof P.T. Haskell of the School of Pure and Applied Biology, Cardiff, for their valuable help in various ways with the work reviewed here. This work has been funded, over the years, by the Henry Doubleday Research Association, the Environment Foundation and the Biotechnology & Biological Sciences Research Council (formerly AFRC).

REFERENCES

Anderson, J.F. (1970) Metabolic rates of spiders. *Comparative Biochemistry and Physiology*, **33**, 51–72.

Asteraki, E.J. (1993) The potential of carabid beetles to control slugs in grass/clover swards. *Entomophaga*, **38**, 193–8.

Büchs, W., Heimbach, U. and Czarnechi, E. (1989) Effects of snail bait on non-target carabid beetles, in *Slugs and Snails in World Agriculture*, BCPC Monograph **41**, (ed. I. Henderson), British Crop Protection Council, Thornton Heath, pp. 245–52.

Bullock, J.I., Coward, N.P., Dawson, G.W., Henderson, I.F., Larkworthy, L.F., Martin, A.P. and McGrath, S.P. (1992) Contact uptake of metal compounds and their molluscicidal effect on the field slug, *Deroceras reticulatum* (Müller) (Pulmonata, Limacidae). *Crop Protection*, **11**, 329–34.

Burn, A.J. (1988) Assessment of the impact of pesticides on invertebrate predation in cereal crops. *Aspects of Applied Biology*, **17**, 173–9.

Chapman, P.A. (1994) Control of leatherjackets by natural enemies: the potential role of the ground beetle *Pterostichus melanarius. Proceedings of the Brighton Crop Protection Conference, 1994 – Pests and Diseases*, British Crop Protection Council, Thornton Heath, pp. 933–4.

Chiverton, P.A. (1984) Pitfall-trap catches of the carabid beetle *Pterostichus melanarius*, in relation to gut content and prey density, in insecticide treated and untreated spring barley. *Entomologia Experimentalis et Applicata*, **36**, 23–30.

Chiverton, P.A. (1987) Predation of *Rhopalosiphum padi* (Homoptera: Aphididae) by polyphagous predatory arthropods during the aphids' pre-peak period in spring barley. *Annals of Applied Biology*, **111**, 257–69.

Dixon, P.L. and McKinlay, R.G. (1992) Pitfall trap catches of and aphid predation by *Pterostichus melanarius* and *Pterostichus madidus* in insecticide treated and untreated potatoes. *Entomologia Experimentalis et Applicata*, **64**, 63–72.

Fichter, B.L. and Stephen, W.P. (1984) Time-related decay of prey antigens by arboreal spiders as detected by ELISA. *Environmental Entomology*, **13**, 1583–7.

Forsythe, T.G. (1982) Feeding mechanisms of certain ground beetles (Coleoptera: Carabidae). *Coleopterists Bulletin*, **36**, 26–73.

Ghiretti, F. (1966) Molluscan hemocyanins, in *Physiology of Mollusca*, vol. 7, (eds K.M. Wilbur and C.M. Yonge), Academic Press, London, pp. 233–48.

Glen, D.M. (1989) Understanding and predicting slug problems in cereals, in *Slugs and Snails in World Agriculture*, BCPC Monograph **41**, (ed. I. Henderson), British Crop Protection Council, Thornton Heath, pp. 253–62.

Glen, D.M., Milsom, N.F. and Wiltshire, C.W. (1989) Effects of seed-bed conditions on slug numbers and damage to winter wheat in a clay soil. *Annals of Applied Biology*, **115**, 177–90.

Gokool, S., Curtis, C.F. and Smith, D.F. (1993) Analysis of mosquito bloodmeals by DNA profiling. *Medical and Veterinary Entomology*, **7**, 208–15.

Greenstone, M.H. (1996) Serological analysis of arthropod predation: past present and future, in *The Ecology of Agricultural Pests: Biochemical Approaches*, (eds W.O.C. Symondson and J.E. Liddell), Systematics Association Special Volume **53**, Chapman & Hall, London, pp. 265–99.

Griffiths, E., Wratten, S.D. and Vickerman, G.P. (1985) Foraging by the carabid *Agonum dorsale* in the field. *Ecological Entomology*, **10**, 181–9.

Hagler, J.R. and Naranjo, S.E. (1996) Using gut content immunoassays to evaluate predaceous biological control agents: a case study, in *The Ecology of Agricultural Pests: Biochemical Approaches*, (eds W.O.C. Symondson and J.E. Liddell), Systematics Association Special Volume **53**, Chapman & Hall, London, pp. 383–99.

Hunter, P.J. and Symonds, B.V. (1971) The leapfrogging slug. *Nature*, **229**, 349.

Kendall, D.A., Chinn, N.E., Glen, D.M., Wiltshire, C.W., Winstone, L. and Tidboald, C. (1995) Effects of soil management on cereal pests and their natural enemies, in, *Ecology and Integrated Arable Farming Systems*, (eds D.M. Glen, M.P. Greaves and H.M. Anderson), Wiley, London, pp. 83–102.

Köhler, G. and Milstein, C. (1975) Continuous cultures of fused cells secreting antibody of predefined specificity. *Nature*, **256**, 495–7.

Liddell, J.E. and Cryer, A. (1991) *A Practical Guide to Monoclonal Antibodies*, Wiley, Chichester.

Liddell, J.E. and Symondson, W.O.C. (1996) The potential of combinatorial antibody libraries in pest/predator relationship studies, in *The Ecology of Agricultural Pests; Biochemical Approaches*, (eds W.O.C. Symondson and J.E. Liddell), Systematics Association Special Volume **53**, Chapman & Hall, London, pp. 347–66.

Luff, M.L. (1974) Adult and larval feeding habits of *Pterostichus madidus* (F.) (Coleoptera: Carabidae). *Journal of Natural History*, **8**, 403–9.

Luff, M.L. (1987) Biology of polyphagous ground beetles in agriculture. *Agricultural Zoology Reviews*, **2**, 237–78.

Mols, P.J.M. (1993) Walking to survive – searching, feeding and egg production of the carabid beetle *Pterostichus coerulescens* L. (= *Poecilus versicolor* Sturm), Wageningen Agricultural University Papers, The Netherlands.

Murdoch, W.W., Chesson, J. and Chesson, P.L. (1985) Biological control in theory and practice. *American Naturalist*, **125**, 344–66.

Port C.M. and Port, G.R. (1986) The biology and behaviour of slugs in relation to crop damage and control. *Agricultural Zoology Reviews*, **1**, 255–99.

Powell, W. and Walton, M.P. (1995) Populations and communities, in *Insect Natural Enemies*, (eds M. Jervis and N. Kidd), Chapman & Hall, London, pp. 223–92.

Purvis, G. and Bannon, J.W. (1992) Non-target effects of repeated methiocarb slug pellet application on carabid beetle (Coleoptera: Carabidae) activity in winter-sown cereals. *Annals of Applied Biology*, **121**, 401–22.

Runham, N.W. and Hunter, P.J. (1970) *Terrestrial Slugs*, Hutchinson, London.

Solomon, M.G., Fitzgerald, J.D. and Murray, R.A. (1996) Electrophoretic approaches to predator-prey interactions, in *The Ecology of Agricultural Pests: Biochemical Approaches*, (eds W.O.C. Symondson and J.E. Liddell), Systematics Association Special Volume 53, Chapman & Hall, London, pp. 457–68.

Sopp, P.I. and Sunderland, K.D. (1989) Some factors affecting the detection period of aphid remains in predators using ELISA. *Entomologia Experimentalis et Applicata*, **51**, 11–20.

Sopp, P.I., Sunderland, K.D., Fenlon, J.S. and Wratten, S.D. (1992) An improved quantitative method for estimating invertebrate predation in the field using an enzyme-linked immunosorbent assay (ELISA). *Journal of Applied Ecology*, **29**, 295–302.

South, A. (1989) A comparison of the life cycles of the slugs *Deroceras reticulatum* (Müller) and *Arion intermedius* Normand on permanent pasture. *Journal of Molluscan Studies*, **55**, 9–22.

South, A. (1992) *Terrestrial Slugs: Biology, Ecology and Control*, Chapman & Hall, London.

Sunderland, K.D. (1975) The diet of some predatory arthropods in cereal crops. *Journal of Applied Ecology*, **17**, 389–96.

Sunderland, K.D. (1987) A review of methods of quantifying invertebrate predation occurring in the field. *Acta Phytopathologica et Entomologica Hungarica*, **22**, 13–34.

Sunderland, K.D. (1988) Quantitative methods for detecting invertebrate predation occurring in the field. *Annals of Applied Biology*, **112**, 201–24.

Sunderland, K.D. (1996) Progress in quantifying predation using antibody techniques, in *The Ecology of Agricultural Pests: Biochemical Approaches*, (eds W.O.C. Symondson and J.E. Liddell), Systematics Association Special Volume 53, Chapman & Hall, London, pp. 419–55.

Sunderland, K.D. and Sutton, S.L. (1980) A serological study of arthropod predation on woodlice in a dune grassland ecosystem. *Journal of Animal Ecology*, **49**, 987–1004.

Sunderland, K.D., Crook, N.E., Stacey, D.L. and Fuller, B.J. (1987) A study of feeding by polyphagous predators on cereal aphids using ELISA and gut dissection. *Journal of Applied Ecology*, **24**, 907–33.

Sunderland, K.D., de Snoo, G.R., Dinter, A., Hance, T., Helenius, J., Jepson, P., Kromp, B., Samu, F., Sotherton, N.W., Toft, S. and Ulber, B. (in press) Density estimation for invertebrate predators in agroecosystems. *Acta Jutlandica*.

Symondson, W.O.C. (1989) Biological control of slugs by carabids, in, *Slugs and Snails in World Agriculture*, BCPC Monograph **41**, (ed. I. Henderson), British Crop Protection Council, Thornton Heath, pp. 295–300.

Symondson, W.O.C. (1993) The effects of crop development upon slug distribution and control by *Abax parallelepipedus* (Coleoptera: Carabidae). *Annals of Applied Biology*, **123**, 449–57.

Symondson, W.O.C. and Liddell, J.E. (1993a) The development and characterisation of an anti-haemolymph antiserum for the detection of mollusc remains within carabid beetles. *Biocontrol Science and Technology*, **3**, 261–75.

Symondson, W.O.C. and Liddell, J.E. (1993b) Differential antigen decay rates during digestion of molluscan prey by carabid predators. *Entomologia Experimentalis et Applicata*, **69**, 277–87.

Symondson, W.O.C. and Liddell, J.E. (1993c) The detection of predation by *Abax parallelepipedus* and *Pterostichus madidus* (Coleoptera: Carabidae) on Mollusca using a quantitative ELISA. *Bulletin of Entomological Research*, **83**, 641–7.

Symondson, W.O.C. and Liddell, J.E. (1993d) A monoclonal antibody for the detection of arionid slug remains in carabid predators. *Biological Control*, **3**, 207–14.

Symondson, W.O.C. and Liddell, J.E. (1993e) The detection of predator–mollusc interactions using advanced antibody technologies, in, *Annales, A.N.P.P. Third International Conference of Pests in Agriculture*, Montpellier 7–9 Dec., 1993. Association Nationale de Protection des Plantes, Paris, pp. 417–24.

Symondson, W.O.C., and Liddell, J.E. (1995) Decay rates for slug antigens within the carabid predator *Pterostichus melanarius* monitored with a monoclonal antibody. *Entomologia Experimentalis et Applicata*, **75**, 245–50.

Symondson, W.O.C., Mendis, V.H. and Liddell, J.E. (1995) Monoclonal antibodies for the identification of slugs and their eggs. *EPPO Bulletin*, **25**, 377–82.

Symondson, W.O.C., Glen, D.M., Langdon, C., Wiltshire, C.W. and Liddell, J.E. (in press) The effects of cultivation techniques and methods of straw disposal on predation by *Pterostichus melanarius* (Coleoptera: Carabidae) upon slugs (Gastropoda: Pulmonata) in an arable field. *Journal of Applied Ecology*.

Tijssen, P. (1985) *Practice and Theory of Enzyme Immunoassays. Laboratory Techniques in Biochemistry and Molecular Biology*, **15**, Elsevier, Oxford.
Tod, M.E. (1973) Notes on beetle predators of molluscs. *The Entomologist*, **106**, 196–201.
van Holde, K.E. and Miller, K.I. (1982) Haemocyanins. *Quarterly Review of Biophysics*, **15**, 1–129.
Wallin, H. (1988) Mandible wear in the carabid beetle *Pterostichus melanarius* in relation to diet and burrowing behaviour. *Entomologia Experimentalis et Applicata*, **48**, 43–50.
Wilson, M.J., Glen, D.M., George, S.K., Pearce, J.D. and Wiltshire, C.W. (1993) Biological control of slugs in winter wheat using the rhabditid nematode *Phasmarhabditis hermaphrodita. Annals of Applied Biology*, **125**, 377–90.

14

The potential of combinatorial antibody libraries in pest–predator relationship studies

J.E. Liddell and W.O.C. Symondson

ABSTRACT

Immunoassays of predator gut contents can provide valuable information in the study of pest–predator relationships. The ideal antibodies to use in these assays should be highly specific and capable of distinguishing between different prey items at the genus-, species- or even stage-specific level. A wide range of different specificities would be desirable, especially in studies involving polyphagous predators or where the prey is unknown. The antibody should also be capable of detecting its target antigen under test conditions, i.e. semi-digested antigen from insects that may have eaten their prey several hours or days before capture. Monoclonal antibodies produced by hybridoma technology can fulfil these criteria but can take a long time to produce, especially since many pest specific antibodies are not yet available.

This chapter discusses the potential of recombinant antibody technology as an alternative to conventional methods of monoclonal antibody production. The genes coding for the antibody fragments that are responsible for antigen binding can be isolated and cloned into vectors that allow the recombined antibody fragments to be displayed on the surface of phage. Those that bind antigen (together with the genes they contain) can then be isolated relatively easily from the library of millions ($\sim 10^8$) of different specificities. By changing the binding antigen, many different specificities can potentially be isolated from the same gene library. These techniques combine a number of advantages, including improved speed of production,

The Ecology of Agricultural Pests, Edited by W.O.C. Symondson and J.E. Liddell.
Published in 1996 by Chapman & Hall, London. ISBN 0 412 62190 8

increased diversity of specificities obtainable and the facility to fine-tune the affinities and specificities of antigen binding by manipulating the isolated gene sequences. The equipment required, and the techniques employed, are within the scope of any molecular biology laboratory, and offer an improved method for the production of highly specific diagnostic reagents for pest ecologists.

14.1 INTRODUCTION

It is clear from the reviews contained within this volume (Greenstone, 1996; Hagler and Naranjo, 1996; Sigsgaard, 1996; Stuart and Greenstone, 1996; Sunderland, 1996; Symondson and Liddell, 1996) that immunoassays of predator gut contents can provide valuable information in the study of pest–predator relationships. The design of the assay and its components needs careful consideration, however, in order to make the best use of resources and to provide accurate information at the desired level. Sophisticated techniques that are expensive and take a long time to develop, may not always be warranted when all that is required is a simple identification of prey remains. That apparently simple answer can, however, be misleading if the predator samples have not been collected appropriately, if the antibody has not been properly characterized, the assay is insufficiently sensitive, or the statistical analysis is inaccurate. Other chapters in this volume address these issues and their interpretation. This review describes the available choice of antibodies, including new genetic approaches to antibody production that we predict will become the method of choice for pest ecologists.

For pest predation studies the ideal antibodies should be highly specific and capable of distinguishing between different prey items, ideally at the genus-, species- or even stage-specific level. A wide range of different specificities would be desirable, especially in studies involving polyphagous predators or where the prey is unknown. One very important factor is the ability of an antibody to detect its target antigen under test conditions. For instance, even antibodies that are ideal in every other respect would be useless for field studies if they could not detect ingested antigen in insects that may have eaten their prey several hours or days before capture.

The type of assay in which these antibodies are used also needs to be considered. The production of sophisticated monoclonal antibodies is pointless if they are to be used in insensitive precipitation assays, for instance. Indeed, they may not even work at all in this type of assay (see p.350). Sensitive assays are important in this context since very small amounts of antigen need to be detected, but such assays require antibody–antigen binding of high affinity. The development of these antibodies and assays will always require specialist scientists in well-equipped laboratories in industrialized countries and important pest predation analyses are currently being carried out. However, in many situations it may be most useful if, once established, these assays could be used in field locations by non-specialist personnel who do not have access to expensive equipment or even refrigeration facilities.

14.2 ANTIBODY CHOICES

Polyclonal antisera are relatively quick and inexpensive to make but their specificity is dependent on the purity of the immunogen and on the purification of the antiserum. This need not necessarily be a problem. Highly specific antisera have been developed to synthetic peptides, for instance, and when used in the right context, the likelihood of obtaining false positive results is minimal. However, it would be rare for such a source of antigen to be available in predation studies, where the unique target protein may not be known. Careful analysis of the properties of the antiserum is essential, with respect to its cross-reactivity with other potential prey antigens and also the predator gut, while the antigen detection limits of the assay need to be clearly established (Symondson and Liddell, 1993a). The supply of polyclonal antiserum is limited in volume per animal, so all the antibody characterization and assay conditions would need to be repeated with each batch of antiserum, even if the same immunization protocols are followed.

Monoclonal antibodies (MAbs) undoubtedly are the best tools for the precise analysis of prey remains. They have the potential to provide information on species, developmental stage and antigen decay rates that would be extremely difficult to obtain with polyclonal antisera. It is not necessary to immunize with purified immunogen or even to know the identity of the prey protein. Immunization with whole body homogenates has been used successfully to make a range of species-specific anti-slug MAbs (Symondson and Liddell, 1993b; Symondson *et al.*, 1995). Careful characterization is still necessary, but the results are likely to be much less equivocal than those with polyclonal antisera, and limitless supplies of the antibody are available on regeneration of the hybridoma cell line. Producing specific antibody-secreting hybridomas, however, can be a long process (6–12 months) and requires special facilities and expertise (see Figure 11.3 in Greenstone, 1996 this volume). These factors alone deter many pest ecologists from following this route.

The unique specificity of MAbs can sometimes be a disadvantage in that their recognition of a particular epitope is very dependent on the structure of the epitope within the whole protein. This structure can easily be destroyed by different purification and assay procedures, so that the antibody no longer recognizes that protein despite its continued presence. Thus, MAbs that have been selected using native protein or an ELISA-type assay system may not work with different preparations of that protein or in different assays. In particular, the use of MAbs in Western blotting detection systems often leads to disappointing results because proteins are denatured in the process. MAbs often do not work in precipitation assays since these are dependent on formation of a matrix of antibody and antigen (although adding an anti-mouse polyclonal antibody to the reaction mixture often solves this problem).

This level of characterization may not be possible in the early stages of production without a great deal of extra work. So, one may find that at the end of a year, despite having a range of antibodies, the ideal MAb for your application has still not been isolated. One is then faced with beginning all over again. However, these drawbacks to the production of useful MAbs can now be overcome by applying recently developed techniques of genetic manipulation which are described below. The genetic approach has gained momentum due to the failure of the hybridoma method to reliably produce human monoclonal antibodies for therapeutic purposes. These techniques combine a number of advantages, including improved speed of production, increased diversity of specificities obtainable and the facility to fine-tune the affinities and specificities of antigen binding by manipulating the isolated genes. It is also now possible to bypass the immune system altogether and generate specific antibodies without the need to immunize animals (see section 14.5).

14.3 ANTIBODY STRUCTURE AND GENETICS

The basic structure of an antibody molecule consists of two identical 'heavy' chains paired with two identical shorter 'light' chains, forming an approximate Y-shape (Figure 14.1). The chains are sub-divided into domains and are linked by a variable number of disulphide bonds. IgG antibodies consist of one such unit, whereas IgM is made up of five. The CH and CL domains are relatively constant in sequence and are responsible for the effector functions of the antibody, i.e. its binding to complement and cells involved in the immune response. The VH and VL domains, however, are more variable in sequence. Within the variable regions of each chain there are three areas of hypervariable sequence, known as complementarity determining regions or CDRs, each forming a loop structure supported by a framework of β sheets (Figure 14.1). The six CDRs on each arm of the antibody Y together form the antigen binding site. Replacement of the gene segments coding for each of these CDRs in a human antibody with the corresponding CDRs from a specific mouse monoclonal antibody, a technique known as CDR grafting, results in a chimaeric antibody which retains the specificity of the original mouse antibody. Indeed, it has been shown (see section 14.5) that changing the sequence and, therefore, structure, of only the third CDR loop of the heavy chain can radically alter antibody specificity.

The genetic manipulation of antibodies is greatly facilitated by the domain structure of the protein, each domain being encoded by different gene exons in the immunoglobulin gene complex. The genes encoding the variable region of the heavy chain consist of V, D and J segments. In germ line DNA, one of several hundred V segments will combine with one of twelve D segments and one of four J segments, to produce a functional VDJ gene. Similarly, light chain variable genes are made up from a choice of several hundred V segments and four *kappa* J segments or two *lambda* J segments. The constant region genes are arranged downstream from the VDJ or VJ regions in such a way that any of the C genes can recombine with the VDJ or VJ genes, giving rise to antibodies of different isotypes (see Figure 14.1). Additional diversity is conferred on the final antibody product by somatic mutation and combination of different light and heavy chains (Roitt *et al.*, 1993).

This natural generation of antibody diversity *in vivo*, arising out of the recombination of many different gene fragments, can be exploited in the laboratory, in order to produce large gene libraries from which many different antibodies can be isolated and redesigned, if necessary, for specific requirements. The techniques are described in the following section and also in recent reviews such as Winter *et al.* (1994).

14.4 TECHNIQUES OF ANTIBODY GENE LIBRARY CONSTRUCTION AND SELECTION

The ability to construct gene libraries of antibody fragments has only recently been possible due to several technological developments in molecular biology. The full DNA sequences of many antibodies are now recorded, and the polymerase chain reaction (PCR) technique of amplifying small pieces of DNA has greatly facilitated the isolation of the appropriate genes. The manipulation of antibody genes was hampered by the fact that whole antibody molecules cannot be expressed in *E. coli* since they require glycosylation mechanisms that bacteria do not possess. In 1988, however, two groups demonstrated that antigen-binding fragments (Fv and Fab) of antibody could be expressed in *E. coli* (Better *et al.*, 1988; Skerra and Plückthun, 1988) which enabled the cloned genes to be quickly and easily amplified and analysed. Even more recently, the technique of expressing the antibody fragments on the surface of phage (phage display; first demonstrated for peptide fragments by Smith, 1985) has made the isolation of the

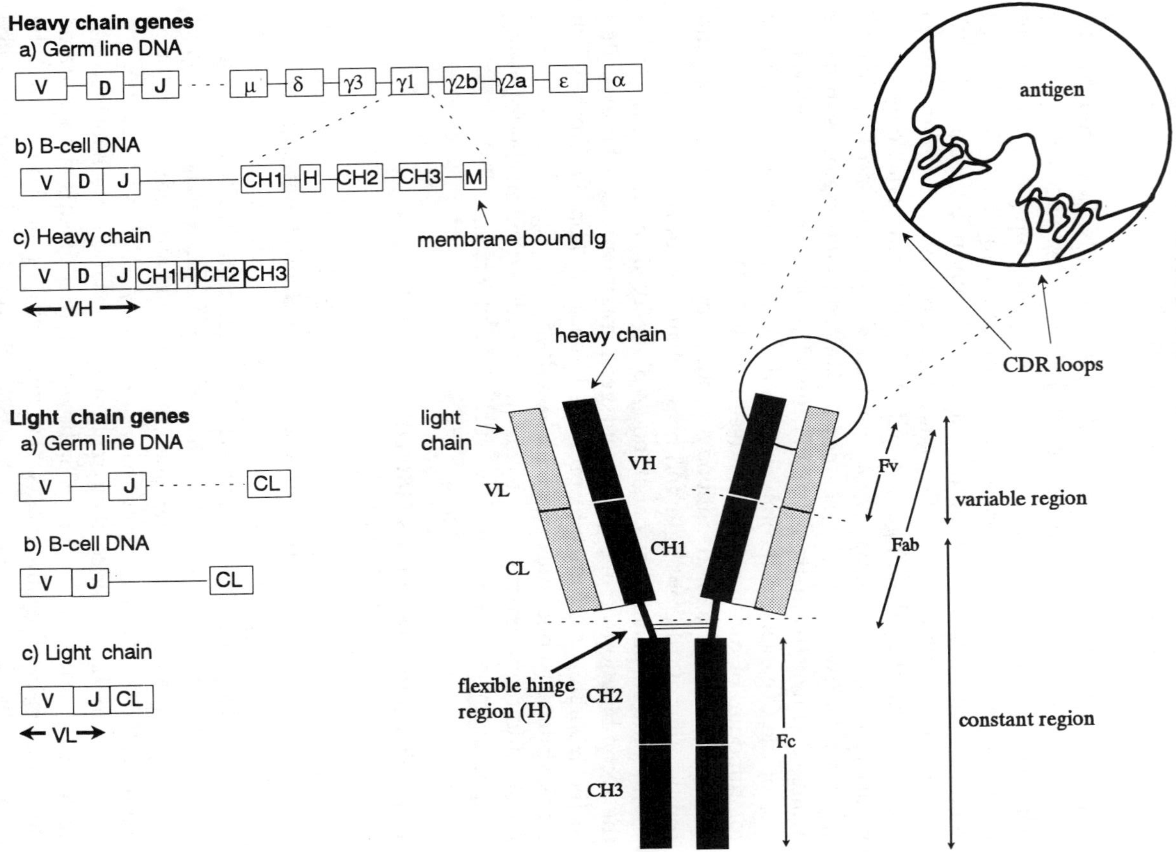

Heavy chain genes
a) Germ line DNA
V
D
J
μ
δ
γ3
γ1
γ2b
γ2a
ε
α
b) B-cell DNA
V
D
J
CH1
H
CH2
CH3
M
membrane bound Ig
c) Heavy chain
V
D
J
CH1
H
CH2
CH3
VH
Light chain genes
a) Germ line DNA
V
J
CL
b) B-cell DNA
V
J
CL
c) Light chain
V
J
CL
VL
antigen
CDR loops
heavy chain
light chain
VH
VL
CH1
CL
Fv
Fab
variable region
flexible hinge region (H)
CH2
CH3
Fc
constant region

Figure 14.1 Antibody structure and gene construction. The structure of a human IgG1 molecule is shown. Other species and classes of antibody vary in the number and position of the disulphide bonds binding the chains and in the number and composition of the CH domains. Antibody diversity is based on the selection of many different variable gene fragments (V,D,J for heavy chains and V,J (κ and λ) for light chains) combined with different constant region genes, one for each chain being selected for expression by a B cell. Additional diversity is conferred by the combination of different heavy and light chains and by somatic mutation. See also section 14.3.

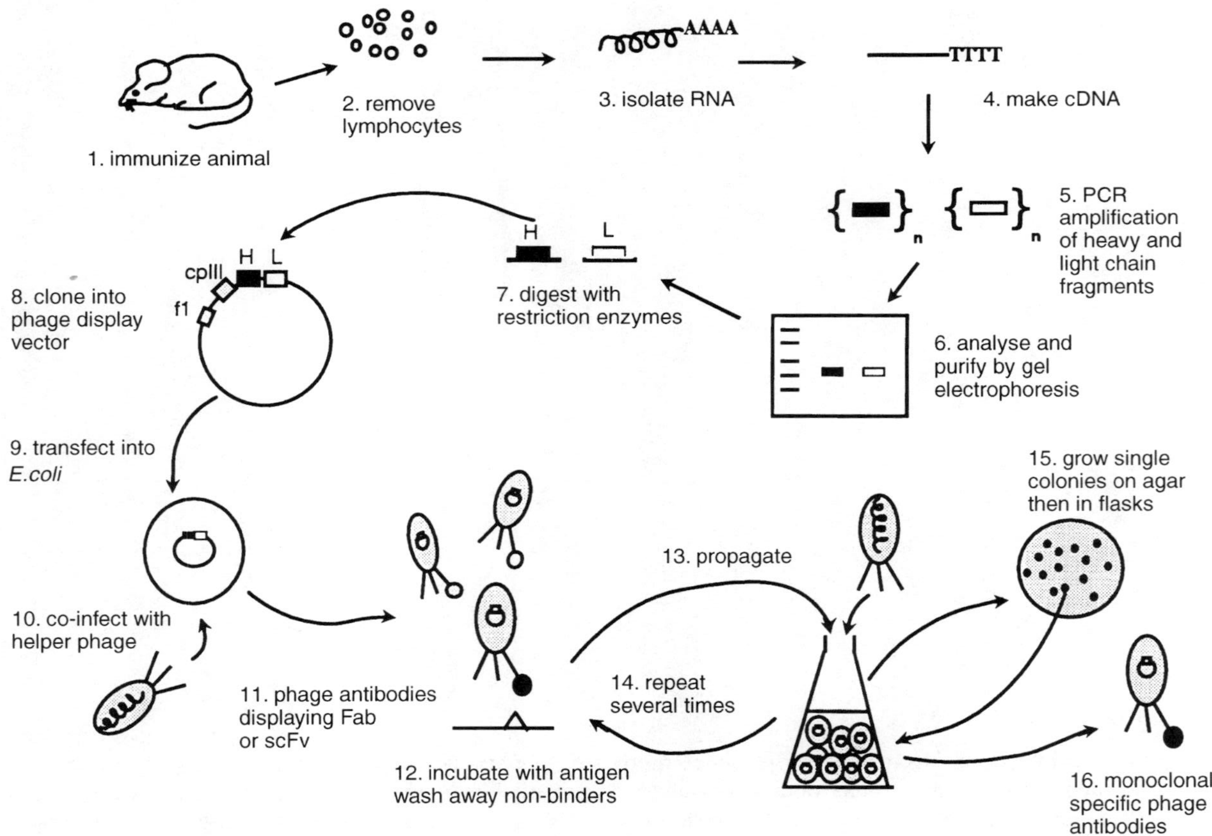

Figure 14.2 General strategy for recombinant antibody production. For full explanation, see section 14.4.1.

desired antibody specificities from libraries of 10^8 or more both easier and faster. All these technical advances, when combined, have resulted in the isolation of the genes coding for the variable regions of heavy and light antibody chains, their cloning into expression vectors, and their expression as specific antigen-binding fragments (Huse *et al.*, 1989; McCafferty *et al.*, 1990; Barbas *et al.*, 1991; Hoogenboom *et al.*, 1991).

14.4.1 Library construction

The general strategy of library construction is shown in Figure 14.2. First, RNA is isolated from immune lymphocytes such as the spleen, peripheral blood or bone marrow. From this template, cDNA is made using oligo-dT or antibody-specific primers, and amplified by PCR using pairs of specific primers complementary to either end of the desired antibody gene fragment. The heavy and light chain genes are amplified separately, and the size of the fragment will depend on whether whole Fab fragments or single chain variable fragments (scFv) are to be made (see section 14.4.3 and Figure 14.4). The heavy and light chain gene fragments are isolated by gel electrophoresis, purified and their ends digested by restriction enzymes compatible with the appropriate expression vector. They are then ligated into the vector which is transfected into *E. coli* by electroporation (Figure 14.2, 1–9). Antibody fragments are assembled in the bacterial periplasm anchored by the coat protein (cpIII). They cannot be packaged into free phage particles without co-infection with helper phage (Figure 14.2, 10). This results in secreted phage with monovalent display of the antibody fragment (Figure 14.2, 11), which can then be selected against specific antigen by affinity enrichment or panning (Figure 14.2, 12–16 and Figure 14.3).

The original combinations of heavy and light chains forming the specific antibodies in the original repertoire are, of course, destroyed in the process of library construction and the separate chains recombined at random. Typically, 10^8 different recombinations and therefore potential specificities and binding affinities can be obtained. The size of the library, and the range of specificities, are directly dependent on the transformation efficiency of the bacteria with the vector, which is steadily improving as the technology improves. In order to isolate those colonies that express recombined antibody fragments that bind specifically to the desired antigen, alternative screening methods to the conventional time-consuming filter lift plaque techniques must be used. For this purpose, phage display vectors, in which the gene product is displayed on the surface of phage, are ideal. The desired phage containing the right gene combination can then be selected by affinity purification on the basis of specific binding of the surface protein to antigen. Once a

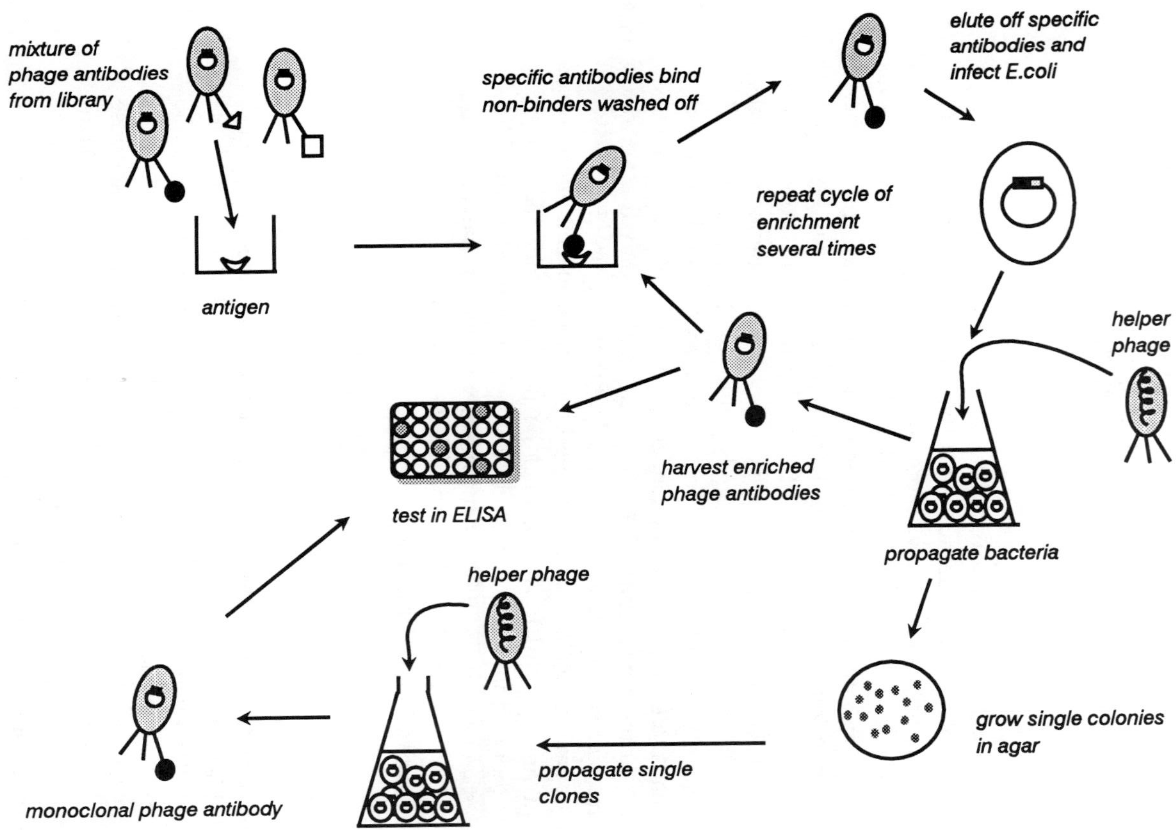

Figure 14.3 Selection of antibodies obtained from a combinatorial antibody gene library. Each round of panning can enrich the specific phage antibody population by several thousand-fold enabling very rare binders to be isolated. A library can be panned against many different antigens potentially yielding many different antibody specificities.

library has been constructed, of course, it can be screened for antibody binding specificities against many other antigens.

14.4.2 Phage display vectors

In order to display a protein on the outer surface of a bacteriophage, the gene coding for that protein has to be inserted into the vector in conjunction with the gene coding for one of the phage coat proteins. The foreign gene will then be processed with the coat protein gene such that the foreign protein will be displayed on the phage surface attached to the coat protein. The preferred coat protein for antibody display is known as gene III, which has approximately three copies on the tip of the phage particle.

Phage vectors tend to be less efficient at transforming DNA than plasmid vectors. The greater the transformation efficiency, the larger will be the library size and therefore the breadth of gene diversity and the likelihood of obtaining high-affinity antibody binders. Plasmid vectors are, therefore, preferred but would result in all the gene III coat proteins being expressed with displayed antibody fragments. This is a disadvantage for two reasons: first, the function of the gene III coat protein is to initiate infection of bacteria, but if all have antibody fragments attached, this function is impaired; second, multivalent expression of antibody is not desirable for the phage selection process because several copies of poorer affinity binders will appear equivalent to high-affinity binders and it will be difficult to select for the higher-affinity binders.

The expression vectors currently most in favour are phagemids, which combine the advantages of high transformation efficiency and control over the number of antibody fragments displayed. Phagemids are plasmids with an origin of replication for filamentous bacteriophage; thus, they are incapable of packaging the DNA to produce free phage particles expressing the protein on the surface, without co infection of the bacteria with helper phage. The helper phage has a defective origin of replication so the secretion of the phagemid package is favoured. The experimental conditions can be adjusted so that only one of the gene III coat proteins carries the antibody fragment, thus the display is monovalent.

Phage expressing the desired specificities can now be selected by affinity purification against specific antigens. The process is normally known as panning (Figure 14.3), whereby antigen is attached to a solid phase, the mixture of phage display antibodies from the library is allowed to bind, non-binders are washed off, binders are eluted with acid or excess antigen, and the phage is allowed to re-infect *E. coli* which are allowed to grow and therefore amplify the selected phage. Secondary infection with helper phage releases more phage particles which are re-exposed to antigen and the process repeated several times. With each round of panning the proportion of specific phage in the population can increase several thousand-fold. Ultimately, the bacterial colonies are plated out, and single colonies are selected and propagated for monoclonal production of phage antibodies. At this stage the antibody fragment is still attached to the phage and can be used in assays with a labelled antibody directed against the phage surface. Alternatively, free antibody fragments can be made by minor adjustments to the vector which depend upon the particular vector used.

14.4.3 Fab versus scFv

The type of antibody fragment that is obtained depends on the portion of the antibody gene that is amplified at the PCR stage (Figure 14.4). There are two choices. Either the whole Fab arm is obtained or only the variable regions of each chain. In order to obtain Fab fragments the whole of the light chain is amplified and the VH and CH1 domains of the heavy chain (Barbas *et al.*, 1991). The 3′ PCR primer of the heavy chain hybridizes to a hinge region sequence which includes a Cys involved in the disulphide bond that naturally holds the two chains together. These primers (one for each isotype) are paired with several different primers for the 5′ end of the gene, therefore representing a large proportion of the different antibody variations in the original repertoire. The light chain is normally cloned into the vector first and amplified before insertion of the heavy chain into the same vector. Only the heavy chain is linked to the cpIII gene so that, on expression, the cpIII coat protein becomes anchored in the bacterial periplasm with the heavy chain attached. The light chain is secreted into the periplasm separately, but is able to associate with the heavy chain via the disulphide bond. The Fab fragment structure, therefore, closely resembles the native form with its natural flexibility which is so important for binding to antigen. Once selection of a specific phage antibody has been made, soluble Fab fragments can easily be produced by cutting out the cpIII gene with restriction enzymes and re-ligating. Detection of soluble Fabs simply requires the application of commonly available, inexpensive anti-Fab label conjugates similar to those that might be used in immunoassays. Yields of antibody obtainable from bacterial cultures can be expected to be several mg per litre.

The alternative antibody fragment obtainable from a combinatorial library (single chain Fv or scFv) is smaller and represents only the variable regions (VH and VL) of both antibody chains (Figure 14.4). More primers are needed for the 3′ end to account for the greater diversity in this region and there is no natural disulphide bond to hold the two chains together. Consequently, the two gene fragments are linked

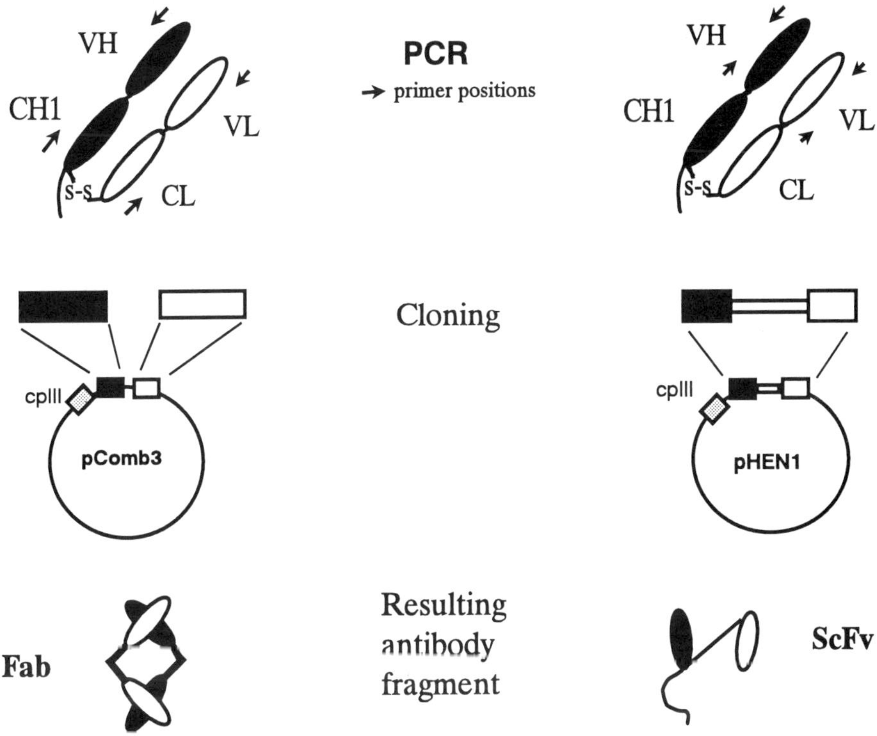

Figure 14.4 Essential differences between Fab and ScFv construction.

together by a small piece of DNA before cloning into the vector. There is, therefore, only one cloning stage compared with that for Fab construction and a smaller fragment is produced which could facilitate intracellular penetration should it be required. In order to use free scFv fragments, a *c-myc* gene is included in the vector so that the scFv is expressed with a *c-myc* tag that can be detected with anti-*c-myc* antibody labels.

There is currently considerable debate as to whether scFvs are less stable and flexible than Fabs. ScFvs, because of their smaller size (25 kDa), are likely to be more useful where penetration of cells is required, for instance in intracellular delivery of toxins. For diagnostic purposes, such as detection of pest predation, however, production of Fabs is currently more practical since there are more anti-Fab label reagents available commercially. It is likely that as more experience is gained of the different approaches and the technology is simplified, one or other or the fragment types will be found to be most appropriate for different applications.

14.4.4 Altering antibody characteristics

One important advantage of the combinatorial library approach over the hybridoma method is the ease and speed with which the initial antibody can be changed and improved upon without having to begin again. By genetic manipulation, antibody–antigen binding affinities can be improved, undesirable cross-reactivities can be eliminated, specificities can even be totally changed, bispecific antibodies can be constructed, and structural analysis of the antibody combining site can assist in defining the antigen structure. In addition, once a library has been established, many other antibodies with different binding characteristics can be selected from it within days. The selection process can be biased towards eliminating undesirable cross-reactivities, broadening cross-reactivity in certain cases (for instance, to improve recognition of rapidly evolving viral antigens) and improving affinity.

The antibody–antigen binding affinities of recombinant antibodies obtained from immune sources tend to be in the range of 10^7–10^9 M^{-1}, which are similar to those expected from an *in vivo* secondary immune response. These can be further improved upon *in vitro* by several different strategies, including chain shuffling (Barbas, 1993) and mutagenesis (Myers *et al.*, 1985; Schaaper, 1988; Leung *et al.*, 1989). Chain shuffling involves retaining the original heavy chain and recombining it with the original light chain library, thus producing a family of specific antigen binders some of which may have improved binding characteristics (300-fold increases in affinity have been reported). The converse, where the light chain is retained and the heavy chain is changed, results in loss of antigen binding. Random or site-directed mutagenesis of the variable region genes can be induced by standard chemical means, by PCR or by *in vivo* mutagenesis using mutator bacterial strains. Our knowledge of the most important targets to change or retain is accumulating as more antibody combining sites are being analysed (Sharon *et al.*, 1993; Wilson and Stanfield, 1993).

14.5 CURRENT STATUS OF RECOMBINANT ANTIBODY TECHNOLOGY IN MEDICAL THERAPEUTICS

One of the aims of producing human recombinant antibodies was to bypass immunization altogether, and this has been shown to be possible even to the extent that semi-synthetic antibodies have been made (Marks *et al.*, 1991; Barbas *et al.*, 1992). Antibody gene libraries derived from naive or non-immunized lymphocytes can be screened for virtually any antigen binding. The larger the library, the greater will be the chance of selecting a desired specificity. Antibodies have been obtained with binding affinities comparable with those found after primary immunisation (10^6–10^7 M^{-1}). Antibodies derived from naive libraries are likely to have a different spectrum of specificities than those from immunized sources, where immunodominant epitopes are favoured at the expense of the less immunogenic, including self antigens that are very difficult to achieve by conventional methods. A similar approach has even led to the production of bacterial cell surface peptides that recognize small inorganic molecules such as Fe_3O_4 (Brown, 1992) that would never be possible using *in vivo* methods.

Semi-synthetic libraries have been constructed from pooled gene libraries, or single clones which have been subjected to recombination with randomized nucleotide sequences to replace the CDR3 loop of the heavy chain. Selection of alternative specificities is made by panning against different antigens. In this way, it has been possible to convert an anti-tetanus toxoid antibody to one specific to fluorescein (Barbas *et al.*, 1992). Several reports have demonstrated the wide variety of different antibody specificities that can be obtained from a single gene library (Lerner *et al.*, 1992; Williamson *et al.*, 1993; Nissim *et al.*, 1994).

The breadth of antibody diversity that can be obtained from a library is dependent on the size of the library, which has been restricted to a maximum of around 10^9 because of the limits of bacterial transformation efficiency. This limit has recently been overcome by applying the method of combinatorial infection (Waterhouse *et al.*, 1993). Antibody libraries of 6.5×10^{10} have been constructed (Griffiths *et al.*, 1994) by combining very diverse heavy chain plasmid libraries containing random CDR3 sequences, with a light chain phage library in the same bacteria. Lox P sites are contained within each of these vectors which can combine in the presence of Cre recombinase (introduced by co-infection with a phage containing the enzyme); thus, the two vectors – and therefore the heavy and light chain genes – are randomly combined. Library sizes are potentially equivalent to the number of bacteria infected.

14.6 PARTICULAR ADVANTAGES FOR APPLICATIONS IN PEST–PREDATOR RELATIONSHIP STUDIES

Producing antibodies for analysis of pest remains does not have the same constraints on immunization as there are for human production, so it is not necessary to forego immunization of the source animal. This improves the chances of obtaining reasonable antibody affinities without extra gene manipulation. The main advantages are the speed of library production once the methods are established (a few weeks) and the diversity of specificities that can be obtained from that library (particularly when the precise antigen target is unknown). Screening of the library for different specificities and affinities, and for reactivities with digestion-resistant epitopes, for instance, can be done in a week. The equipment required is no more than that which is standard in any molecular biology facility (i.e. electrophoresis systems, PCR machine, electroporator, ultracentrifuge, shaking incubators, etc.) The initial cost of reagents, such as PCR primers and restriction enzymes, can be expensive, but this has to be balanced against the saving in labour and the range of products that can be obtained. One then has all the advantages of having a range of different monoclonal antibodies plus the ability to manipulate the genes to eliminate any undesired cross-reactivities that are revealed. Propagation of the antibodies is also facilitated by the ability to grow in bacteria, which is both quicker and cheaper than *in vitro* methods of hybridoma propagation, and more ethically acceptable than ascitic fluid production in mice.

14.7 PROGRESS IN PRODUCING RECOMBINANT ANTIBODIES FOR PEST PREDATION STUDIES

We have constructed the first recombinant antibody gene library for use in pest predation studies, full details of which will be published elsewhere (J.E. Liddell, N. Walkley and W.O.C. Symondson, in preparation). The source of RNA was the spleen of a mouse immunized with whole-body homogenates of the slug *Arion intermedius* Normand. Antibody gene fragments representing heavy chain (VH and CH1) of IgG1, IgG2a and IgM, and kappa light chain, were amplified (for mouse primer sequences, see Kang *et al.*, 1991). The light chain and then the heavy chain were ligated into the vector pComb3H (Barbas *et al.*, 1991). A library size of approximately 10^7 was obtained. The phage display Fab fragments were panned initially against whole-body homogenates of *A. intermedius*. The resulting antibodies when fully characterized will be compared with our earlier range of monoclonal antibodies obtained by the hybridoma method, for use in ELISAs to study predation of slugs by carabid beetles (Symondson and Liddell, 1996, this volume, and references contained therein).

14.8 CONCLUSIONS

Monoclonal antibodies, however they are produced, can provide the fine specificity necessary to distinguish between different prey proteins – an especially important point when analysing the diets of polyphagous predators. However, conventional hybridoma production methods cannot compete with the speed of production and diversity of specificity offered by genetic engineering.

Recombinant antibody phage display technology is developing rapidly, fuelled by the race to produce novel and lucrative therapeutic reagents. The equipment required, and the techniques employed to produce useful diagnostic reagents, are within the scope of any molecular biology laboratory. It has already been shown that antibody gene libraries can be generated from immune or non-immune and semi-synthetic sources, from which potentially any antigen specificity can be selected. We predict that antibody gene libraries, perhaps for a single pest genus or family, will become the source of choice for pest ecologists, from which the desired antibody specificities can be quickly and cheaply isolated.

Alternative methods of analysis, based on DNA rather than protein detection, have been used successfully to identify insect vectors of disease (Hill and Crampton, 1994) and to analyse the bloodmeals of haematophagous insects. However, selecting suitable oligonucleotide probes to well-characterized human DNA is a relatively easy task compared with the work and time involved in developing new probes to the huge variety of different prey invertebrates that might form the diet of a given target predator. In addition, DNA is invariant during the life-cycle of the prey so it would not be possible to distinguish different stages within that cycle.

Developing both antibody and DNA probes is a specialist activity, but once developed, antibodies – unlike DNA probes – can easily be applied in simple, cheap and sensitive assays, that can be used anywhere by any untrained fieldworker. For instance, qualitative, colorimetric ELISAs in a dipstick format, are already used by the public in home pregnancy test kits, without the need for sophisticated equipment or training. No DNA-based assay can yet match these requirements.

Precise biochemical analysis of predator gut contents is only one of the factors in the equation that will ultimately provide a profile of a predator's role in a given system. Detailed information on climate conditions, farming practices and behavioural characteristics are also required to interpret the impact of the predator on pest populations. These interpretations, however, are limited without accurate, quantitative analysis of predator diets for which immunoassays are ideal.

Acknowledgements

We are grateful for the gift of the pComb3 vector from Professor Dennis Burton and Dr Carlos Barbas III of the Scripps Research Institute, La Jolla, California and to the Biotechnology and Biological Sciences Research Council, UK for partly funding this work.

REFERENCES

Barbas, C.F., III (1993) Recent advances in phage display. *Current Opinion in Biotechnology*, **4**, 526–30.

Barbas, C.F., III, Kang, A.S., Lerner, R.A. and Benkovic, S.J. (1991) Assembly of combinatorial antibody libraries on phage surfaces: the gene III site. *Proceedings of the National Academy of Sciences, USA*, **88**, 7978–82.

Barbas, C.F., III, Bain, J.D., Hoekstra, D.M. and Lerner, R.A. (1992) Semi-synthetic combinatorial antibody libraries: a chemical solution to the diversity problem. *Proceedings of the National Academy of Sciences, USA*, **89**, 4457–61.

Better, M., Chang, C.P., Robinson, R.R. and Horwitz, A.H. (1988) *Escherischia coli* secretion of an active chimeric antibody fragment. *Science*, **240**, 1041–3.

Brown, S. (1992) Engineered iron oxide-adhesion mutants of the *Escherichia coli* phage *lambda* receptor. *Proceedings of the National Academy of Science, USA*, **89**, 8651–5.

Greenstone, M.H. (1996) Serological analysis of arthropod predation: past, present and future, in *The Ecology of Agricultural Pests: Biochemical Approaches*, (eds W.O.C. Symondson and J.E. Liddell), Systematics Association, Special Volume No. 53, Chapman & Hall, London, pp. 265–300.

Griffiths, A.D., Williams, S.C. Hartley, O., *et al.* (1994) Isolation of high affinity human antibodies directly from large synthetic repertoires. *EMBO Journal*, **13**, 3245–60.

Hagler, J.R. and Naranjo, S.E. (1996) Using gut content immunoassays to evaluate predaceous biological control agents: a case study, in *The Ecology of Agricultural Pests: Biochemical Approaches*, (eds W.O.C. Symondson and J.E. Liddell), Systematics Association, Special Volume No. 53, Chapman & Hall, London, pp. 385–99.

Hill, S.M. and Crampton, J.M. (1994) DNA-based methods for the identification of insect vectors. *Annals of Tropical Medicine and Parasitology*, **88**, 227–50.

Hoogenboom, H.E., Griffiths, A.D., Johnson, K.S., Chiswell, D.J., Hudson, P. and Winter, G. (1991) Multi-subunit proteins on the surface of filamentous phage: methodologies for displaying antibody (Fab) heavy and light chains. *Nucleic Acids Research*, **19**, 4133–7.

Huse, W.D., Sastry, L., Iverson, S.A., Kang, A.S., AltingMees, M., Burton, D.R., Benkovic, S.J. and Lerner, R.A. (1989) Generation of a large combinatorial library of the immunoglobulin repertoire in phage lambda. *Science*, **246**, 1275–81.

Kang, A.S., Barbas, C.F., Janda, K.D., Benkovic, S.J. and Lerner, R.A. (1991) Linkage of recognition and replication functions by assembling combinatorial antibody Fab libraries along phage surfaces. *Proceedings of the National Academy of Sciences, USA*, **88**, 4363–6.

Lerner, R.A., Kang, A.S., Bain, J.D., Burton, D.R. and Barbas, C.F., III (1992) Antibodies without immunisation. *Science*, **258**, 1313–14.

Leung, D.W., Chen, E. and Goeddel, D.V. (1989) A method for random mutagenesis of a defined DNA segment using a modified polymerase chain reaction. *Journal of Methods in Cellular and Molecular Biology*, **1**, 11–15.

Marks, J.D., Hoogenboom, H.R., Bonnert, T.P., McCafferty, J., Griffiths, A.D. and Winter, G. (1991) By-passing immunization: human antibodies from V-gene libraries displayed on phage. *Journal of Molecular Biology*, **222**, 581–97.

McCafferty, J., Griffiths, A.D., Winter, G. and Chiswell, D.J. (1990) Phage antibodies: filamentous phage displaying antibody variable domains. *Nature*, **348**, 552–4.

Myers, R.M., Lerman, L.S. and Maniatis, T. (1985) A general method for saturation mutagenesis of cloned DNA fragments. *Science*, **229**, 242–7.

Nissim, A., Hoogenboom, H.R., Tomlinson, I.M., Flynn, G., Midgley, C., Lane, D. and Winter, G. (1994) Antibody fragments from a 'single pot' phage display library as immunochemical reagents. *EMBO Journal*, **13**, 692–8.

Roitt, I., Brostoff, J. and Male, D. (1993) *Immunology*, 3rd ed, Gower Medical Publishing, London.

Schaaper, R.M. (1988) Mechanisms of mutagenesis in the *Escherichia coli* mutator *mutD5*: role of DNA mismatch repair. *Proceedings of the National Academy of Sciences, USA*, **85**, 8126–30.

Sharon, J., Kao, C.-Y. Y. and Sompuram, S.R. (1993) Oligonucleotide-directed mutagenesis of antibody combining sites. *International Review of Immunology*, **10**, 113–27.

Skerra, A., and Plückthun, A. (1988) Assembly of a functional immunoglobulin Fv fragment in *Escherichia coli*, *Science*, **240**, 1038–41.

Smith, G.P. (1985) Filamentous fusion phage: novel expression vectors that display cloned antigens on the surface of the virion. *Science*, **228**, 1315–17.

Sigsgaard, L. (1996) Serological analysis of predators of *Helicoverpa armigera* Hübner (Lepidoptera: Noctuidae) eggs in sorghum–pigeonpea intercropping at ICRISAT, India: a preliminary field study, in *The Ecology of Agricultural Pests: Biochemical Approaches*, (eds W.O.C. Symondson and J.E. Liddell), Systematics Association, Special Volume No. 53, Chapman & Hall, London, pp. 367–81.

Stuart, M.K. and Greenstone, M.H. (1996) Serological diagnosis of parasitism: a monoclonal antibody-based immunodot assay for *Microplitis croceipes*, in *The Ecology of Agricultural Pests: Biochemical Approaches*, (eds W.O.C. Symondson and J.E. Liddell), Systematics Association, Special Volume No. 53, Chapman & Hall, London, pp. 301–21.

Sunderland, K.D. (1995) Progress in quantifying predation using antibody techniques, in *The Ecology of Agricultural Pests: Biochemical Approaches*, (eds W.O.C. Symondson and J.E. Liddell), Systematics Association, Special Volume No. 53, Chapman & Hall, London, pp. 419–55.

Symondson, W.O.C. and Liddell, J.E. (1993a) The development and characterisation of an anti-haemolymph antiserum for the detection of mollusc remains within carabid beetles. *Biocontrol Science and Technology*, **3**, 261–75.

Symondson, W.O.C. and Liddell, J.E. (1993b) The detection of predator–mollusc interactions using advanced antibody technologies, in *Annales, A.N.P.P. Third International Conference of Pests in Agriculture*, Montpellier 7–9 December, 1993, pp. 417–24.

Symondson, W.O.C. and Liddell, J.E. (1996) Polyclonal, monoclonal and engineered antibodies to investigate the role of predation in slug population dynamics, in *The Ecology of Agricultural Pests: Biochemical Approaches*, (eds W.O.C. Symondson and J.E. Liddell), Systematics Association, Special Volume No. 53, Chapman & Hall, London, pp. 323–45.

Symondson, W.O.C., Mendis, V.W. and Liddell, J.E. (1995) Monoclonal antibodies for the identification of slugs and their eggs. *E.P.P.O. Bulletin*, **25**, 377–82.

Waterhouse, P., Griffiths, A.D., Johnson, K.S. and Winter, G. (1993) Combinatorial infection and *in vivo* recombination: a strategy for making large phage antibody repertoires. *Nucleic Acids Research*, **21**, 2265–6.

Williamson, R.A., Burioni, R., Sanna, P.P. and Partridge, L.J. (1993) Human monoclonal antibodies against a plethora of viral pathogens from single combinatorial libraries. *Proceedings of the National Academy of Sciences, USA*, **90**, 4141–5.

Wilson, I.A. and Stanfield, R.L. (1993) Antibody–antigen interactions. *Current Opinion in Structural Biology*, **3**, 113–18.

Winter, G., Griffiths, A.D., Hawkins, R.E. and Hoogenboom, H.R. (1994) Making antibodies by phage display technology. *Annual Review of Immunology*, **12**, 433–55.

15

Serological analysis of predators of *Helicoverpa armigera* Hübner (Lepidoptera: Noctuidae) eggs in sorghum–pigeonpea intercropping at ICRISAT, India: a preliminary field study

Lene Sigsgaard

ABSTRACT

Helicoverpa armigera (Hübner) is a serious pest in many regions of the Old World, including South India. Heavy reliance on chemical insecticides has led to high levels of resistance as well as environmental hazards. This has raised interest in alternative pest management approaches, especially control by natural enemies. While the role of parasitism has been relatively easy to document, evidence of predation has been primarily circumstantial or indirect, based largely on correlations between pest and predator population estimates. An objective of the present study has been to identify and begin to evaluate predators as potential biological control agents of *H. armigera*. Eggs and potential predators were censused in the field, and predators collected and frozen during the censuses were assayed by enzyme-linked immunosorbent assay (ELISA) using a monoclonal antibody for helioithine eggs. The ELISA data identified the following egg predators: *Cheiracanthium inornatum* O.P.-Cambridge (Araneae: Clubionidae), *Menochilus sexmaculatus* F. (Coleoptera: Coccinellidae), *Orius tantillus* (Motschulsky) (Heteroptera: Anthochoridae), *Formicomus* sp. (Coleoptera: Anthicidae), and *H.*

The Ecology of Agricultural Pests, Edited by W.O.C. Symondson and J.E. Liddell.
Published in 1996 by Chapman & Hall, London. ISBN 0 412 62190 8

armigera larvae. The pattern of predation over the cropping period was established by associating data from the field sampling of eggs and predators with data from ELISA of the same predators. The present study illustrates the use of immunoassay of field-collected predators in conjunction with pest and predator population estimates in evaluating the role of predators as biological control agents.

15.1 INTRODUCTION

Helicoverpa armigera (Hübner) (Lepidoptera: Noctuidae) is a pest throughout Africa, the Middle East, southern Europe, India, central and northern Australia, New Zealand, and many eastern Pacific Islands. It is highly polyphagous and mobile (Farrow and Daly, 1987; Pedgley *et al.*, 1987; McCaffery *et al.*, 1989; Fitt, 1989; King *et al.*, 1990). Pigeonpea is among the crops sustaining the highest losses due to *H. armigera* in the areas where it is grown (Fitt, 1989). It is an important smallholder crop in India, where it is normally intercropped with sorghum. Although oviposition by *H. armigera* tends to be high in sorghum, losses in sorghum tend to be low in South India.

Chemical insecticide spraying has been the standard management tool for *H. armigera*. This had led to chronically high levels of pesticide resistance, and hence reduced pesticide efficacy (Armes *et al.*, 1992, 1993). Furthermore, since naturally occurring levels of biological control of this pest can be substantial, intensive pesticide spraying contributes to subsequent outbreaks through destruction of natural enemy populations (King *et al.*, 1982; Zalucki *et al.*, 1986; Abdelrahman and Munir, 1989; van den Berg and Cock, 1993).

Quantitative information on the abundance and impact of predators is required before we can develop alternative management strategies to reduce reliance on chemical insecticides. A variety of methods have been used to gather data on predation (for an extensive review see Sunderland, 1988). Of these, serological gut analysis is the most efficient and least disruptive to ecosystem interactions (Greenstone, 1989). The primary objective of the present study was to identify predators of *H. armigera* and then to begin an evaluation of their impact on the population. As egg disappearance rates are high (Duffield, 1993), predation on eggs was given first priority. The predators chosen for the study were: *Orius tantillus* (Motschulsky) (Heteroptera: Anthochoridae)[1], *Menochilus sexmaculatus* F. (Coleoptera: Coccinellidae), *Formicomus* sp. (Coleoptera: Anthicidae), larval *H. armigera*, and the spider *Cheiracanthium inornatum* O.P.-Cambridge (Araneae: Clubionidae)[2]. The first two were chosen on the basis of earlier records of egg predation (Pawar *et al.*, 1986; Duffield, 1993), the anthicid because it was seen to be very active in the pigeonpea canopy, and the spider because of its abundance during *H. armigera* oviposition (L. Sigsgaard, unpublished results). Larvae of *H. armigera* were chosen because egg and larval cannibalism has been recorded previously in this species (Twine, 1971; Twine *et al.*, 1983; Zalucki *et al.*, 1986).

15.2 EXPERIMENTAL APPROACH

15.2.1 Field sampling

Oviposition in pigeonpea began by the onset of budding and flowering. Sampling began at that point and continued until 16 December when there were 8.2 eggs per plant (Table 15.2). Sampling took place once or twice weekly between 08:30 and 11:30, in a 0.72-ha experimental field of intercropped sorghum and pigeonpea at the International Crops Research Institute for the Semi-Arid Tropics (ICRISAT), in Andhra Pradesh, India. The field was a split-plot experiment with pesticide treatment as whole-plot factor (eight whole plots, one spray) and cowpea interplanting as split-plot factor (16 split-plots). During this period there was no rain, and windspeeds were moderate. Plant density was 4.5 pigeonpea plants per m^2. Plant size was recorded three times during the sampling period.

Eggs and predators were sampled on 320 pigeonpea plants. Within each of the 16 subplots four locations were selected at random and five plants sampled in each location. Most oviposition occurs in the upper third of the pigeonpea plant (Venugopal Rao *et al.*, 1990). It is difficult to sample in the large pigeonpea plants (1–2 m tall and 40–150 cm in diameter) with many main and side branches; sampling was therefore restricted to the upper third of plants. Eggs were counted on the terminal 10–15 cm of one upper shoot, where buds, flowers and shoots, and also eggs are concentrated. Arthropods were collected by beating three upper shoots from each plant into a plastic bag. Arthropods from one location were pooled, giving a total of 64 bags.

In order to retard digestion of ingested prey during transport from the field to the laboratory, the bags were placed in iceboxes containing cooling elements capable of maintaining temperatures of 10–12°C. Upon arrival at the laboratory they were transferred to a 2°C refrigerator in order to both retard digestion and reduce arthropod mobility to facilitate handling. Predators were placed in individual 1.5-ml Eppendorf tubes labelled with date, split-plot number, sample number, species and instar. They were kept in the refrigerator for no longer than

[1,2] Identified by Identification Services, International Institute of Entomology.

2 hours before being transferred to a –80°C freezer to await assay. Other arthropods present in the bags were recorded.

Later in the season, *O. tantillus* and *M. sexmaculatus* were collected once from a post-rain sorghum field where *H. armigera* egg densities were high (an average of 2.7 eggs per head recovered by brushing eggs off the sorghum head by hand). Fields were also sampled three times in a village in Warangal district of Andhra Pradesh, once at high egg density (4.04 eggs per shoot). With the exception of the *H. armigera* larvae, all of the collected insect predators were adults; the spiders ranged from small immatures to adults (2–7 mm).

15.2.2 ELISA

Predators were thawed and homogenized in 0.25 ml PBS Tween–BSA (Greenstone and Morgan, 1989) and then centrifuged for 3 min at 14 000 *g* at 4°C. The homogenates were kept on ice and the supernatants used as the first reagent in an indirect ELISA, essentially as described by Lenz and Greenstone (1988). The only differences were that the plates were rotated only briefly at the beginning and end of each incubation because no mechanical rotator was available, and substrate development was allowed to proceed for 90 min. The first antibody used was a monoclonal antibody specific for heliothine eggs (Greenstone and Trowell, 1994), produced in the laboratory of M.H. Greenstone. Goat anti-mouse alkaline phosphatase conjugate and *p*-nitrophenyl phosphate were purchased from Sigma Chemical Company, St. Louis, USA. Optimal antigen and antibody concentrations were determined by box titration.

Assays were performed in flat-bottom plates (Griener, Frickenhausen, Germany) and read with a type 314 Multiscan Plus plate reader (Labsystems, Helsinki, Finland). The method of Sopp and Sunderland (1989) was used to determine the negative threshold for scoring ELISA results. In addition to field samples, each plate contained negative and positive controls. Negatives comprised PBS (Greenstone and Morgan, 1989) (two wells) and predators that had been fed alternative prey (1–4 wells); positives were *H. armigera* egg homogenate (two wells each with 0.16 egg/0.1 ml) and, on some plates, predators fed *H. armigera* egg.

(a) Detectability interval of an egg in the gut

To estimate the detection period of an *H. armigera* egg after ingestion, predators were each offered a single egg, and then frozen at intervals at –80°C before ELISA. The animals were confined for these experiments in 1.5-ml Eppendorf tubes as larger cages resulted in extended searching periods before feeding. Except for *H. armigera*, which were tested as neonates within 12 hours of hatching, all predators were starved for 24 hours and only offered water on a cotton swab to ensure no traces of eggs ingested in the field remained in the gut. *Cheiracanthium inornatum* were offered eggs on small pieces of pigeonpea. The insects were offered free eggs, which adhered to the walls of the tube by static electricity. The animals were observed at hourly intervals to record the time of feeding. Alternative prey were offered during the 24–40-hour interval, thrips to *O. tantillus*, aphids to *M. sexmaculatus*, jassids to *C. inornatum*, and pigeonpea buds to *H. armigera*. The numbers tested depended upon the availability of predator individuals. The time allowed before individuals began to feed was based on experience of how robust each species was.

As digestion rates are temperature-dependent (Sopp and Sunderland, 1989; Greenstone and Hunt, 1993), the experiment was carried out in a laboratory with netted windows, affording temperatures close to those recorded outdoors (day 25–28°C, night 10–15°C) and natural light conditions (L : D, approximately 11 : 13).

15.3 RESULTS

15.3.1 Field sampling and ELISA

Numbers collected and percentage of positive results by ELISA for the most abundant predator species in the ICRISAT samples are presented in Table 15.1. The number of spiders was high during this period; the single most abundant spider (68.8%) – and also the most abundant predator – was *C. inornatum*. Some 11% overall of those assayed were positive for *H. armigera* egg antigen. Numbers of the next most abundant species, *H. armigera*, were low at the beginning of the study (the start of oviposition) but then rose; 7% of all individuals assayed were positive, confirming prior mention of cannibalism in this species (Twine *et al.*, 1983; Zalucki *et al.*, 1986) and explaining correlations between egg loss and first instar larval densities in cage studies (L. Sigsgaard, unpublished results). *Formicomus* sp. was the least abundant predator, with 6% of individuals assayed being positive.

Numbers of early instar *H. armigera* larvae collected were less than expected. This was probably due to difficulties in dislodging feeding larvae from within flowers and buds but may also reflect to some extent the difficulty of recovering very small and semi-transparent larvae from samples in the laboratory.

Only seven *O. tantillus* and 34 *M. sexmaculatus* were collected at ICRISAT, with 14% and 12% respectively, of assayed individuals positive. These numbers are too small to draw any conclusions about

Table 15.1 ELISA assay on field-collected predators, *Cheiracanthium inornatum* O.P.-Cambridge, *Formicomus* sp. and *Helicoverpa armigera* (Hübner) larval instars 1–2, 3–4, and 5–6. (The method of Sopp and Sunderland (1989) was used to determine the negative threshold.)

Date	*C. inornatum*			*Formicomus* sp.			*H. armigera* instars 1–2			*H. armigera* instars 3–4			*H. armigera* instars 5–6		
	No. collected	*No. tested*	*% positive*	*No. collected*	*No. tested*	*% positive*	*No. collected*	*No. tested*	*% positive*	*No. collected*	*No. tested*	*% positive*	*No. collected*	*No. tested*	*% positive*
4/11	76	73	15	50	46	0	0	0	0	0	0	0	1	0	0
9/11	63	39	21	31	21	5	3	3	0	16	13	23	2	2	0
11/11	23	10	30	13	10	30	1	0	0	6	0	0	0	0	0
15/11	110	89	13	25	20	10	3	3	0	25	22	32	4	3	33
18/11	138	78	10	38	35	0	24	22	0	63	42	0	9	3	0
22/11	80	56	9	33	27	14	18	17	17	189	60	10	26	5	0
24/11	57			13			71			218			29		
26/11	58	32	0	25	12	0	55	21	0	202	21	0	18	1	0
3/12	23	16	0	22	3	33	12	6	0	226	39	3	28	4	0
9/12	30	25	3	4	1	0	18	16	0	144	36	0	42	3	0
16/12	50			9			31			210			43		
Total	718	418	11	262	175	6	236	88	5	1299	233	7	203	21	5

their impact on *H. armigera* populations. However, 14% of 118 *O. tantillus* collected in a post-rain sorghum field on 15 December, 1993 – when *H. armigera* egg densities were high (an average of 2.7 eggs per head recovered by brushing eggs off by hand) – were positive by ELISA. Some 25% of 12 *M. sexmaculatus* collected from the same field were positive.

Egg densities were high (4.04 eggs per shoot) on 11 November, 1993 at the Warangal village site. In total, 4% of 23 *C. inornatum*, 10% of 82 *H. armigera*, and 13% of 24 *M. sexmaculatus* were positive in the ELISA; no *O. tantillus* or *Formicomus* sp. were found. These results may be conservative, since the village samples were kept in iceboxes for more than 10 hours after sampling, and antigens may have been degraded by the prolonged period of digestion. Village fields were also smaller and more diverse than those at ICRISAT, and had many more ants. Nevertheless, the results from the Warangal village site are in general agreement with those from ICRISAT, showing moderate levels of predation by the most abundant species.

15.3.2 Detectability interval of eggs in predator guts

Although no predators died prematurely during the experiment, attempts to determine egg detection intervals were frustrated by several unexpected problems:

1. Many predators were unwilling to consume eggs at the start of the experiment. It was therefore necessary to begin with excess individuals and select for assay only those which ate within a reasonable length of time. While 49% of *H. armigera* fed during the first 4 hours, it took 6 hours for 36% of *O. tantillus* and 43% of *M. sexmaculatus* to feed, and 24 hours for 68% of *C. inornatum* to feed. *Cheiracanthium inornatum* fed only in darkness. Reluctance to feed might be due to problems with egg quality such as infertility or unknown handling effects, or factors intrinsic to individual predators such as preparation for moulting (Greenstone, 1979).
2. Portions of the egg were often not consumed, and the exact amount eaten was not known. Variations in the size of a meal can bias or increase the variance of estimates of the detectability interval (Sopp and Sunderland, 1989).
3. Because of delays in the onset of feeding it was only possible to observe predators at intervals of maximum 1 hour to determine individual times of feeding. Since the exact time of feeding could not always be determined by observation, feeding sometimes had to be confirmed by microscopic examination of the eggs. However, evidence of feeding can be difficult to detect. For example what

appeared to be probing or feeding by *O. tantillus* for periods of 3–4 minutes left no visible trace on the egg.

4. None of the 18 *Formicomus* sp. observed to have fed was positive in the ELISA.

These results do not permit a reliable estimate of the detectability interval, and a conservative interval was used of one day (where 50% positives were expected) for the analysis which follows. Larger sample sizes would probably reduce the impact of some of the problems encountered (Greenstone and Hunt, 1993). Maximum detectability intervals (beyond which no positives were found) were 24 hours in *C. inornatum* and *H. armigera,* and 48 hours in *O. tantillus* and *M. sexmaculatus.*

15.3.3 Impact of predation on the *H. armigera* egg population

Only *C. inornatum* and *H. armigera* larvae had high enough densities to warrant analysis of their impact on the *H. armigera* egg populations for the dates studied. A Poisson model of predation was used to convert percent positives in ELISA to an estimate of mean egg numbers per gut (Nakamura and Nakamura, 1977; Greenstone, 1979; Lister *et al.*, 1987). This model assumes that predation is a rare event and that the presence of an item in the gut does not affect the probability of consumption of another item of the same type. The model has been used twice for spiders (Nakamura and Nakamura, 1977; Greenstone, 1979) whose tendency, at least in some circumstances, is to take prey more or less as encountered (Riechert and Lockley, 1984) and supports the model's use in the case of these species. Since *H. armigera* egg cannibalism is probably incidental to feeding on plant structures it is also an appropriate model for that species.

Using this model, the mean number of items per gut is given by $-\ln(1 - p)$, where p is the proportion of positives in the ELISA. Then the number of eggs killed by predators of a given species is given by:

$$P[-\ln(1 - p)]/D,$$

where P is predator density and D is the detection period. This number was then divided by the egg population to estimate the proportion of eggs predated. Since eggs require an average of 3 days to hatch under the temperatures obtaining in Andhra Pradesh during this study (Twine, 1978), this estimate was multiplied by three to convert it to an estimate of the proportion predated on one night. Estimates of the average number of upper shoots (based on plant growth data from 10 November, 30 November and 20 December, 1993, of 100, 112 and 80 pigeonpea plants respectively) were used to convert the estimates to a per-plant basis.

The predation rate estimates, as percent of the egg population consumed, are presented in Table 15.2. The maximum rate for *C. inornatum* was 3.6% on 4 November, 1993. This high rate may be explained by the lack of alternative prey on that date. The mean predation rate for *C. inornatum* was 1.4%. Maximum rates for *H. armigera* larvae occurred around the time of peak oviposition (Figure 15.1), with 4.0% and 19.7% on 15 November and 22 November, 1993, respectively. The mean rate of consumption by *H. armigera* was 2.6%.

15.4 DISCUSSION

15.4.1 Predators of *H. armigera* eggs

This study has identified *O. tantillus, M. sexmaculatus, C. inornatum, Formicomus* sp. and *H. armigera* larvae as predators of *H. armigera* eggs. Various other anthocorids, coccinellids, spiders and conspecifics, as well as ants, chrysopids, *Geocoris* spp., *Nabis* spp. and others have been demonstrated or suspected to be predators of members of the bollworm/budworm complex (*Helicoverpa* and *Heliothis* spp.) in North America, Africa, Australia and Asia (Bell and Whitcomb, 1965; Lingren *et al.*, 1968; van den Bosch *et al.*, 1969; McDaniel and Sterling, 1979; Hogg and Nordheim, 1983; Zalucki *et al.*, 1986; Kogan *et al.*, 1989; Manjunath *et al.*, 1989; Nyambo, 1990; van den Berg and Cock, 1993). *Orius (Dimorphella) maxidentex* Ghauri, and *M. sexmaculatus* have been recorded as egg predators of *H. armigera* (Pawar *et al.*, 1986). The earlier record indicates that there could be more than one *Orius* species feeding on *H. armigera* in this area. *Orius tantillus* is recorded as preying upon *H. armigera* in Africa (van den Berg and Cock, 1993). An anthicid, *Notoxus calceratus* Horn, has been found to feed on eggs of *Heliothis virescens* (F.) (Butler, 1966).

Spiders, including Clubionids are known predators of *Helicoverpa* and *Heliothis* spp. from all continents (McDaniel and Sterling, 1979; Zalucki *et al.*, 1986; Patel and Pillai, 1988; Manjunath *et al.*, 1989). In Texas, *Cheiracanthium inclusum* (Hentz) was found feeding on eggs of *H. virescens* (McDaniel and Sterling, 1982) and eggs of *Helicoverpa zea* (Boddie) (Nuessly and Sterling, 1994). Earlier lists of *H. armigera* predators include species not discussed here, and additional predators will probably be discovered. For example, ground-living carabids might enter the plants at night to feed. Based on present observations in villages, the effect of ants may be very important in small fields. Finally, animals not normally considered to be predators may turn out to feed on *H. armigera* eggs, as demonstrated here for conspecific larvae, mirroring the situation in *H. virescens* (McDaniel and Sterling, 1979).

Table 15.2 Predation on *Helicoverpa armigera* (Hübner) eggs by *Cheiracanthium inornatum* O.P.-Cambridge and *Helicoverpa armigera* larvae in the upper third of pigeonpea plants using the Poisson model of predation

Date	*Ratio positives in ELISA*	*Mean no. of eggs/gut**	*Shoots*†	*Predators/ plant*‡	*Expected no. of eggs eaten*§	*Eggs/ plant***	*% Eaten*††
C. inornatum							
4/11	0.15	0.16	9.3	0.74	0.12	10.08	3.6
9/11	0.21	0.23	9.4	0.62	0.14	26.00	1.6
11/11	0.30	0.36	9.5	0.22	0.08	23.57	1.0
15/11	0.13	0.14	9.6	1.10	0.15	27.33	1.6
18/11	0.10	0.11	9.7	1.39	0.15	27.95	1.6
22/11	0.09	0.09	9.8	0.82	0.07	12.65	1.7
26/11	0.00	0.00	9.9	0.60	0.00	7.33	0.0
3/12	0.00	0.00	10.2	0.24	0.00	10.14	0.0
9/12	0.03	0.03	10.5	0.33	0.01	10.24	0.0
H. armigera larvae, all instars cumulated							
4/11	0.00	0.00	9.3	0.01	0.00	10.08	0.0
9/11	0.17	0.19	9.4	0.21	0.13	26.00	1.5
11/11	0.00	0.00	9.5	0.07	0.00	23.57	0.0
15/11	0.29	0.34	9.6	0.32	0.36	27.33	4.0
18/11	0.00	0.00	9.7	0.97	0.00	27.95	0.0
22/11	0.10	0.11	9.8	2.38	0.83	12.65	19.7
26/11	0.00	0.00	9.9	2.83	0.00	7.33	0.0
3/12	0.02	0.02	10.2	2.83	0.06	10.14	0.7
9/12	0.00	0.00	10.5	2.23	0.00	10.24	0.0

*Mean number of eggs per gut estimated as $-\ln(1 - p)$
†Average number of shoots in the upper third of a pigeonpea plant based on plant growth data from 10 November, 30 November and 20 December 1993 of $n = 100$, 112, and 80 plants respectively
‡Number of predators found/320) × (shoots/3), as three shoots were sampled per plant
§Expected number of eggs eaten (predators/plant × mean no. of eggs/gut/(D) [D = days detection interval, set nominally at 1])
**Average number of eggs per plant (total egg count/320 × average number of shoots)
††Per cent of one night's oviposition of eggs eaten. Since eggs require an average of 3 days to hatch, a multiplication by three gives an estimate of the proportion predated on 1 night (eaten × 3/(eggs/plant)/100).

15.4.2 The impact of predation

A life-table study conducted at ICRISAT in 1992–1993 (Duffield, 1993) found that most mortality in both sorghum and pigeonpea was in the egg stage. In different pigeonpea cultivars it was from 83.7% to 94.2%, with disappearance accounting for 74.3% to 88.3%, respectively, of egg mortality. Similar results were found in sorghum, maize and sunflower in a life-table study in Kenya (van den Berg and Cock, 1993). The mechanical effects of wind, rain and 'roll off' of eggs due to desiccation can be substantial (Kyi *et al.*, 1991). An aim of the present study was to begin to estimate the contribution of predators to egg disappearance.

The two most important predators of *H. armigera* eggs in this study were *C. inornatum* and conspecific larvae. *C. inornatum* were abundant at the onset of oviposition, and they had their highest per-capita impact on the egg population at that time. Whether this was due to the lack of alternate prey or to an innate feeding preference remains to be determined, but in any event it demonstrated once again the value of spiders in attacking pest populations at low densities (Riechert and Lockley, 1984). The highest estimated contribution to egg predation was due to cannibalism by *H. armigera* larvae at peak egg density. However, since fewer young larvae were collected than anticipated, the present estimates are probably low. Remedies might include destructive whole plant sampling or fumigation. Such modifications might also raise the estimated contributions of other predator species.

Although the Poisson model has been used several times, its assumptions have never been tested. For example, probabilities of attacking a prey item may not be independent of prior feeding behaviour (see Sunderland, 1988). Extensive laboratory and cage tests will be required to determine the validity of the model (Greenstone, 1996, this volume).

Since cannibalism rates appear to be high, feeding on *H. armigera* larvae may lead to some false positives for egg predation (cf. Sunderland *et al.*, 1987). For example *C. inornatum* was observed feeding on small larvae, and feeding on larvae by *O. tantillus* and *M. sexmaculatus* was demonstrated in cage studies (L. Sigsgaard, unpublished observations).

The monoclonal antibody used in this research is egg- and subfamily-specific (Greenstone and Trowell, 1994). The only other eggs encountered in significant numbers in pigeonpea during this study were those of blue butterflies, comprising the species *Lampides boeticus* (L.) and *Catochrysops strabo* (Fabricius) (Lepidoptera: Lycaenidae). When tested by ELISA they were negative (L. Sigsgaard, unpublished results). In Andhra Pradesh, *H. armigera* is by far the most dominant heliothine species in pigeonpea and sorghum. *Helicoverpa peltigera* Schiff is found at low densities in safflower and the weed *Acanthospermum hispidum*, and *Helicoverpa assulta* Guenee is present at low densities primarily on the weed *Datura metel* (Jadhav et al., 1985). The dominance of *H. armigera* allows one to be confident that ELISA positives in this research were primarily due to this species.

This study has shown that stage-specific ELISA is a valuable tool in identifying *H. armigera* egg predators, and has demonstrated its potential for estimating rates of predation in the field. More extensive studies of the stage-specific impact of predators are needed to provide a foundation for the development of biological control programmes against this pest. Pesticide resistance problems add to the urgency of this task.

Acknowledgements

I am grateful to Dr M.H. Greenstone for assistance with the final preparation of the manuscript, to P. Geetanjali for laboratory and field assistance, to Dr M.H. Greenstone and C. Morgan for providing antibody for the experiment, and to Drs M.H. Greenstone, K.D. Sunderland and Professor P. Esbjerg for valuable comments and suggestions on earlier drafts of this paper. I would also like to thank ICRISAT for providing field and laboratory facilities and ICRISAT staff, Drs K.F. Nwanze, T.G. Shanower, S.J. Duffield, M.M. Anders and others for inspiration and assistance.

REFERENCES

Abdelrahman, A.A. and Munir, B. (1989) Sudanese experience in integrated pest management of cotton. *Insect Science and its Application*, **10**, 787–94.

Armes, N.J., Jadhav, D.R., Bond, G.S. and King, A.B.S. (1992) Insecticide resistance in *Helicoverpa armigera* in South India. *Pesticide Science*, **34**, 355–64.

Armes, N.J., Jadhav, D.R., Lonergqan, P. and Forrester, N. (1993) Pyrethroid and endosulfan resistance in *Helicoverpa armigera* in India. *Resistant Pest Management*, **5**, 34–9.

Bell, K.O. and Whitcomb, W.H. (1965) Field studies on egg predators of the bollworm, *Heliothis zea* (Boddie). *The Florida Entomologist*, **47**, 171–80.

Butler, G.D. (1966) Insect predators of bollworm eggs. *Progress in the Agriculture of Arizona*, **18**, 26–7.

Duffield, S.J. (1993) *The role of natural enemies in sorghum short-duration pigeonpea intercrops. Report of work.* International Crops Research Institute for the Semi-Arid Tropics: ODA/ICRISAT.

Farrow, R.A. and Daly, J.C. (1987) Long-range movements as an adaptive strategy of the genus *Heliothis* (Lepidoptera: Noctuidae): a review of its occurrence and detection in four pest species. *Australian Journal of Zoology*, **35**, 1–24.

Fitt, G.P. (1989) The ecology of *Heliothis* species in relation to agroecosystems. *Annual Review of Entomology*, **34**, 17–52.

Greenstone, M.H. (1979) Spider feeding behaviour optimises dietary essential amino acid composition. *Nature*, **282**, 501–3.

Greenstone, M.H. (1989) Foreign exploration for predators: a proposed new methodology. *Environmental Entomology*, **18**, 195–200.

Greenstone, M.H. (1996) Serological analysis of arthropod predation: past, present and future, in *The Ecology of Agricultural Pests: Biochemical Approaches*, (eds W.O.C. Symondson and J.E. Liddell), Chapman & Hall, London, pp. 265–300.

Greenstone, M.H. and Hunt, J.H. (1993) Determination of prey antigen half-life in *Polistes metricus* using a monoclonal antibody-based immunodot assay. *Entomologia Experimentalis et Applicata*, **68**, 1–7.

Greenstone, M.H. and Morgan, C.E. (1989) Predation on *Heliothis zea* (Lepidoptera: Noctuidae); an instar-specific ELISA for stomach analysis. *Annals of the Entomological Society of America*, **83**, 45–9.

Greenstone, M.H. and Trowell, S.C. (1994) Arthropod predation: a simplified immunodot format for predator gut analysis. *Annals of the Entomological Society of America*, **87**, 214–17.

Hogg, D.B. and Nordheim, E.V. (1983) Age-specific survivorship analysis of *Heliothis* spp. populations on cotton. *Research in Population Ecology*, **25**, 280–97.

Jadhav, D.R., Bhatnagar, V.S. and Pawar, C.S. (1985) The species status of *Heliothis armigera* (Hub.) (Lepidoptera: Noctuidae) in Andhra Pradesh, India on the basis of aedegal cornutal spines. *Current Science*, **54**, 239–40.

King, A.B.S., Armes, N.J. and Pedgley, D.E. (1990) A mark-capture study of *Helicoverpa amigera* dispersal from pigeonpea in southern India. *Entomologia Experimentalis et Applicata*, **55**, 257–66.

King, E.G., Powell, J.E. and Smith, J.W. (1982) Prospects for Utilization of Parasites and Predators for Management of *Heliothis* spp., in *Proceedings of the International Workshop on Heliothis Management*, (eds W. Reed and V. Kumble) ICRISAT, 15–20 November 1981, ICRISAT Center, Patancheru, A.P., India, pp. 103–22.

Kogan, M., Helm, C.G., Kogan, J. *et al.* (1989) Distribution and economic importance of *Heliothis virescens* and *Heliothis zea* in North, Central, and South America, and of their natural enemies and host plants, in *Proceedings of the Workshop on Biological Control of Heliothis: Increasing the Effectiveness of Natural Enemies*, (eds E.G. King, R.D. Jackson and V. Kumble), Far Eastern Regional Research Office, Office of International Cooperation and Development (OICD) USDA, 11–15 Nov. 1985, Chanakya Puri, New Delhi, India, pp. 241–98.

Kyi, A., Zalucki, M.P. and Titmarsh, I.J. (1991) An experimental study of early stage survival of *Helicoverpa armigera* (Lepidoptera: Noctuidae) on cotton. *Bulletin of Entomological Research*, **81**, 263–72.

Lenz, C.J. and Greenstone, M.H. (1988) Production of a monoclonal antibody to the arylphorin of *Heliothis zea*. *Archives of Insect Biochemistry and Physiology*, **9**, 167–77.

Lingren, P.D., Ridgway, R.I. and Jones, S.L. (1968) Consumption by several arthropod predators of eggs and larvae of two *Heliothis* species that attack cotton. *Annals of the Entomological Society of America*, **61**, 613–18.

Lister, A., Usher, M.B. and Block, W. (1987) Description and quantification of field attack rates by predatory mites: an example using an electrophoresis method with a species of Antarctic mite. *Oecologia*, **72**, 185–91.

Manjunath, T.M., Bhatnagar, V.S., Pawar, C.S. *et al.* (1989) Economic importance of *Heliothis* spp. in India and an assessment of their natural enemies and host plants, in *Proceedings of the Workshop on Biological Control of Heliothis: Increasing the Effectiveness of Natural Enemies*, (eds E.G. King, R.D. Jackson and V. Kumble) Far Eastern Regional Research Office, Office of International Cooperation and Development (OICD) USDA, 11–15 November 1985, Chanakya Puri, New Delhi, India, pp. 197–228.

McCaffery, A.R., King, A.B.S., Walker, A.J. *et al.* (1989) A mark-capture study of *Helicoverpa armigera* dispersal from pigeonpea in southern India. *Pesticide Science*, **27**, 65–76.

McDaniel, S.G. and Sterling, W.L. (1979) Predator determination and efficiency on *Heliothis virescens* eggs in cotton using ^{32}P. *Environmental Entomology*, **8**, 1083–7.

McDaniel, S.G. and Sterling, W.L. (1982) Predation of *Heliothis virescens* (F.) eggs on cotton in East Texas. *Environmental Entomology*, **11**, 60–6.

Nakamura, N. and Nakamura, K. (1977) Population dynamics of the chestnut gall wasp, *Dryocosmys kuryphilus* Yasumatsu (Hymenoptera: Cynipidae) V. Estimation of the effect of predation by spiders on the mortality of imaginal wasps based in the precipitin test. *Oecologia*, **27**, 97–116.

Nuessly, G.S. and Sterling, W.L. (1994) Mortality of *Helicoverpa zea* (Lepidoptera: Noctuidae) eggs in cotton as a function of oviposition sites, predator species, and desiccation. *Environmental Entomology*, **23**, 1189–202.

Nyambo, B.T. (1990) Effect of natural enemies on the cotton bollworm, *Heliothis armigera* Hübner (Lepidoptera: Noctuidae) in Western Tanzania. *Tropical Pest Management*, **36**, 50–8.

Patel, B.H. and Pillai, G.K. (1988) Studies on the spider fauna of groundnut fields in Gujarat, India. *Journal of Biological Control*, **2**, 83–8.

Pawar, C.S., Bhatnagar, V.S. and Jadhav, D.R. (1986) *Heliothis* species and their natural enemies, with their potential for biological control. *Proceedings of the Indian Academy of Science (Animal Science)*, **95**, 695–703.

Pedgley, D.E., Tucker, M.R. and Pawar, C.S. (1987) Windborne migration of *Heliothis armigera* (Hübner) (Lepidoptera: Noctuidae) in India. *Insect Science and its Application*, **8**, 599–604.

Riechert, S.E. and Lockley, T. (1984) Spiders as biological control agents. *Annual Review of Entomology*, **29**, 299–320.

Sopp, P.I. and Sunderland, K.D. (1989) Some factors affecting the detection period of aphid remains in predators using ELISA. *Entomologia Experimentalis et Applicata*, **51**, 11–20.

Sunderland, K.D. (1988) Quantitative methods for detecting invertebrate predation occurring in the field. *Annals of Applied Biology*, **112**, 201–24.

Sunderland, K.D., Crook, N.E., Stacy, D.L. *et al.* (1987) A study of feeding by polyphagous predators on cereal aphids using ELISA and gut dissection. *Journal of Applied Ecology*, **24**, 907–33.

Twine, P.H. (1971) Cannibalistic behaviour of *Heliothis armigera* (Hübner) (Lepidoptera: Noctuidae). *Queensland Journal of Agriculture and Animal Science*, **28**, 153–7.

Twine, P.H. (1978) Effect of temperature on the development of larvae and pupae of the corn earworm, *Heliothis armigera* (Hübner) (Lepidoptera: Noctuidae). *Queensland Journal of Agriculture and Animal Science*, **35**, 23–8.

Twine, P.H., Kay, I.R. and Lloyd, R.J. (1983) *Heliothis* in sorghum. *Queensland Agricultural Journal*, **4**, 185–8.

Van den Berg, H. and Cock, M.J.W. (1993) Stage-specific mortality of *Helicoverpa armigera* in three smallholder crops in Kenya. *Journal of Applied Ecology*, **30**, 640–53.

van den Bosch, R., Leigh, T.F., González, D. *et al.* (1969) Cage studies on predators of the bollworm in cotton. *Journal of Economic Entomology*, **62**, 1486–9.

Venugopal Rao, N., Tirumala Rao, K. and Subba Rao, A. (1990) Present status of '*Heliothis*' on pulses and strategies for its management in Andhra Pradesh, in *First National Workshop on Heliothis Management: Current Status and Future*

Strategies, (ed. J.N. Sachan), Directorate of Pulses Research, 30–31 August 1990, Kanpur, Uttar Pradesh, India, pp. 100–8.

Zalucki, M.P., Daglish, G., Firempong, S. *et al.* (1986) The Biology and Ecology of *Heliothis armigera* (Hübner) and *H. punctigera* Wallengren (Lepidoptera: Noctuidae) in Australia: what do we know? *Australian Journal of Zoology*, **34**, 779–814.

16

Using gut content immunoassays to evaluate predaceous biological control agents: a case study

J.R. Hagler and S.E. Naranjo

ABSTRACT

Pest-specific monoclonal antibodies (MAbs) can be used to identify key predaceous biological control agents. A MAb library was used to (i) identify key indigenous predators of the sweetpotato whitefly, *Bemisia tabaci* (Gennadius) and pink bollworm, *Pectinophora gossypiella* (Saunders) and (ii) monitor the efficacy of an augmentative predator release targeted for these two pests.

The gut contents of over 10 000 individual predators indigenous to Arizona cotton fields were examined for the presence of whitefly and pink bollworm egg antigens by enzyme-linked immunosorbent assay (ELISA). Five predator species with frequent positive assays for whitefly proteins were (% positive): *Collops vittatus* (Say) (55%), *Geocoris* spp. nymphs (39%), *Orius tristicolor* (White) (38%) and *Hippodamia convergens* Guerin–Menceville (33%). The most frequent pink bollworm egg predator was *Lygus hesperus* Knight (adults 30%, nymphs 20%). *L. hesperus* was also the predator species with the highest percentage of individuals scoring positive for both pest egg antigens (12%).

In a separate experiment, we compared the efficacy of commercially reared *Orius insidiosus* (Say) against indigenous populations of *O. tristicolor*. The commercially reared predators were released into field cages, recaptured 36 hours later and examined by ELISA for the presence of whitefly and pink bollworm egg antigens. Some 4% of the released *O. insidiosus* and 26% of the indigenous *O. tristicolor* had whitefly egg antigen in their guts. Pink bollworm egg antigen

The Ecology of Agricultural Pests, Edited by W.O.C. Symondson and J.E. Liddell.
Published in 1996 by Chapman & Hall, London. ISBN 0 412 62190 8

was found in the guts of 8% of the *O. insidiosus* and 5% of the *O. tristicolor*.

The immunologically based techniques described in this review can expedite the search for key indigenous predator species and also serve as a tool for monitoring the efficacy of augmentative biological control agents. The advantages and limitations of our techniques are discussed.

16.1 INTRODUCTION

Despite the heavy use of pesticides over the past half century, crop losses have almost doubled due to damage by insects (Wilson, 1990). With increasing concerns about the over-use of pesticides, growers are seeking more environmentally benign methods for controlling insect pests. One such method receiving renewed attention includes using predaceous natural enemies in an integrated pest management programme (Stern *et al.*, 1959). While it is accepted that predators are generally important in controlling pest populations, it is difficult to assess the ecological impact of specific predator species in cropping systems. The accurate identification of key entomophagous insects in any given system has been virtually impossible for several reasons: observers of predator–prey interactions disrupt the normal foraging process; predators, unlike parasites, rarely leave evidence of attack; and both predators and prey are often small, elusive and nocturnal.

Historically, many experimental methods have been employed for measuring and analysing predator–prey interactions. Laboratory and glasshouse experiments have been used to evaluate predator–prey interactions (e.g. Orphanides *et al.*, 1971; Henneberry and Clayton, 1985; Hagler and Cohen, 1991); however, these types of studies seldom translate to actual field situations. The microscopic examination of predator gut contents (James, 1961) is labour-intensive and unreliable, and unusable for predator species that liquefy prey contents before consumption (Hengeveld, 1980). Indirect gut analysis, including the use of radioactive markers (Baldwin *et al.*, 1955; Jenkins, 1963; McDaniel and Sterling, 1979; McCarty *et al.*, 1980) for tagging potential prey and electrophoresis (Murray and Solomon, 1978) have also been used, but such techniques either pose potential dangers to users and the environment, are too time-consuming, or do not possess the necessary specificity and sensitivity (Lister *et al.*, 1987; Sunderland, 1987; Luck *et al.*, 1988). These difficulties have resulted in a dearth of information on the impact that predators have on suppressing key insect pest populations.

Perhaps the most promising techniques for measuring predation are immunologically based tests employing pest-specific monoclonal antibodies (MAbs) (Whitten and Oakshott, 1990). Immunological methods of studying predation avoid many of the pitfalls inherent with the more traditional methods of study (Brooke and Proske, 1946; West, 1950; Boreham and Ohiagu, 1978; Miller, 1979). Insect pest antigens can stimulate an immunological response in a vertebrate which culminates in the production of serum polyclonal antibodies to the insect antigen. Once the antibodies have been produced in the vertebrate, they can be harvested and used serologically to probe predator gut contents. Unfortunately, most polyclonal antibodies tend to cross-react with other insect species (Davies, 1969; Lund and Turpin, 1977; Miller, 1981; Gardner *et al.*, 1981; Doane *et al.*, 1985). Cross-reactivity is unacceptable when working with an insect complex as diverse as that found in cotton. However, with advances in hybridoma technology, investigators now can isolate individual antibody-producing cells grown *in vitro* and harvest antibodies of single specificity (Köhler and Milstein, 1975). The results are monoclonal antibodies (MAbs) that offer specificity unachievable with conventional polyclonal antisera (Lenz and Greenstone, 1988). Furthermore, once the MAbs have been developed, the immunological assays employed are simple, rapid and economical.

Given all of the positive attributes of using MAb-based gut content immunoassays to investigate predator–prey interactions, it is somewhat surprising that very few investigators have actually adapted this technology to field research. This is due, in part, to the fact that the development of pest-specific MAbs is costly, time-consuming, and requires expertise in hybridoma technology. However, as more researchers from a wide variety of disciplines incorporate MAbs into their research, the techniques will probably become more refined and less expensive (Liddell and Symondson, 1996, this volume). Private industry has begun to develop MAbs on a contract basis. As the competition stiffens, the cost of developing pest-specific MAbs will decrease. Researchers interested in developing gut content immunoassays should be fully aware of the benefits and limitations of this qualitative technology (Sunderland, 1996, this volume).

The purpose of this chapter is to provide an example of how we have incorporated MAb technology into applied research aimed at identifying potential key predator species of the egg stage of the sweetpotato whitefly, *Bemisia tabaci* (Gennadius) (strain B) and pink bollworm, *Pectinophora gossypiella* (Saunders) in a cotton agroecosystem. The first part of the chapter summarizes the techniques that we have used to mass-screen indigenous predator species for the presence of whitefly and pink bollworm prey remains in the gut. Here, we summarize our research findings and discuss how these pest-specific MAbs have aided our efforts to identify key predators. The second part of the chapter presents preliminary data on how we have used our MAbs to assess the effectiveness of augmentative releases of the minute pirate bug, *Orius insidiosus* (Say) on these two pests.

16.2 SURVEY OF INDIGENOUS PREDATORS OF PINK BOLLWORM AND SWEETPOTATO WHITEFLY

16.2.1 Introduction

Dozens of insect species are common to cotton fields in the southwestern United States. Of these, many are potential biological control candidates. Due to the difficulties mentioned above, it is not feasible to investigate all of them by traditional methods (i.e. bioassays and behavioural experiments). However, with our pest-specific MAbs, we can quickly screen indigenous predators for the presence of prey remains in their guts. Our major emphasis here is to demonstrate how we utilized pest-specific MAbs to characterize the native predator complex on two economically important cotton pests, the pink bollworm and sweetpotato whitefly.

16.2.2 Methods

(a) Predator sampling

Indigenous predators were collected from 7 June to 6 September, 1992 from experimental cotton fields located on two, 2-ha sites at The University of Arizona's Maricopa Agricultural Research Center, Maricopa, Arizona, and one, 0.5-ha site at The Western Cotton Research Laboratory, Phoenix, Arizona.

Predators were collected at weekly and bi-weekly intervals using a modified vacuum sampler described by Ellington *et al.* (1984). A randomly selected, continuous 30-m row of cotton was sampled in each of four quadrants per site. The contents from each vacuum sample were immediately placed on dry ice and transported to the laboratory where they were frozen at −80°C. Many of the predators collected in the vacuum samples were potentially contaminated with whitefly antigen (i.e. adult females) during the sampling process. To eliminate any possibility of external contamination each predator was cleaned by removing externally attached whiteflies and then irrigated with PBS. This procedure was effective in cleansing any extraneous whitefly debris (unpublished data).

On several sampling dates, additional predators were collected and kept alive for use as negative controls. The predators were fed cabbage looper, *Tricoplusia ni* (Hübner) larvae and water *ad libitum* for several days to ensure that any potential whitefly or pink bollworm antigens were eliminated from their guts (Hagler and Cohen, 1990). The negative controls were macerated in 250 μl phosphate-buffered saline (PBS) (pH 7.4) and assayed against pink bollworm and whitefly MAbs by the ELISAs described below. The mean (±SD) absorbance values were calculated for each species. The negative controls permit evaluation of any predator constituents that may react with our antibodies and provide estimates of any inherent background noise associated with the ELISAs.

(b) ELISAs

The indirect ELISA procedure described originally by Voller *et al.* (1976) was modified by Hagler *et al.* (1992) and Hagler and Naranjo (1994a,b). Whole bodies of individual field-collected predators were ground in 250 μl of PBS. A 50 μl aliquot of each of the macerated insects was pipetted into individual wells of two uncoated 96-well ELISA plates (Figure 16.1). Each plate was incubated at 4°C overnight. Following incubation, the insect samples were discarded from each well and a 330 μl aliquot of 1% non-fat dry milk in distilled water was added to each well for 30 min at 37°C. The non-fat dry milk was discarded and a 50 μl aliquot of anti-*B. tabaci* MAb (Hagler *et al.*, 1993) was added to the first ELISA plate and a 50 μl aliquot of anti-*P. gossypiella* MAb (Hagler *et al.*, 1994) was added to the second plate (Figure 16.1). Plates were incubated for 1 hour at 37°C, after which the MAbs were discarded and the wells were briefly rinsed three times with PBS–Tween 20 (0.05%) and twice with PBS. Aliquots (50 μl) of goat anti-mouse IgG/IgM conjugated to alkaline phosphatase (TAGO Inc., Burlingame, CA) diluted to 1:500 in 1% non-fat milk were added to each well of both plates. Plates were incubated for 1 hour after which the conjugated antibody was discarded, and the wells were rinsed as noted above. A 50-μl aliquot of *p*-nitrophenylphosphate (pNPP) substrate (Sigma, St Louis, MO) was added to each well for 1 hour and the absorbance at 405 nm of each well was measured with a Cambridge Technology Model 750 (Waterton, MA) microplate reader. Field-collected predators were scored positive for whitefly egg or pink bollworm egg antigen if the absorbance values exceeded the mean negative predator control reading by three standard deviations (Schoof *et al.*, 1986; Sutula *et al.*, 1986).

16.2.3 Results of the indigenous predator survey

The gut contents of over 10 000 individual predators representing nine of the most common predator species found in the cotton agroecosys-

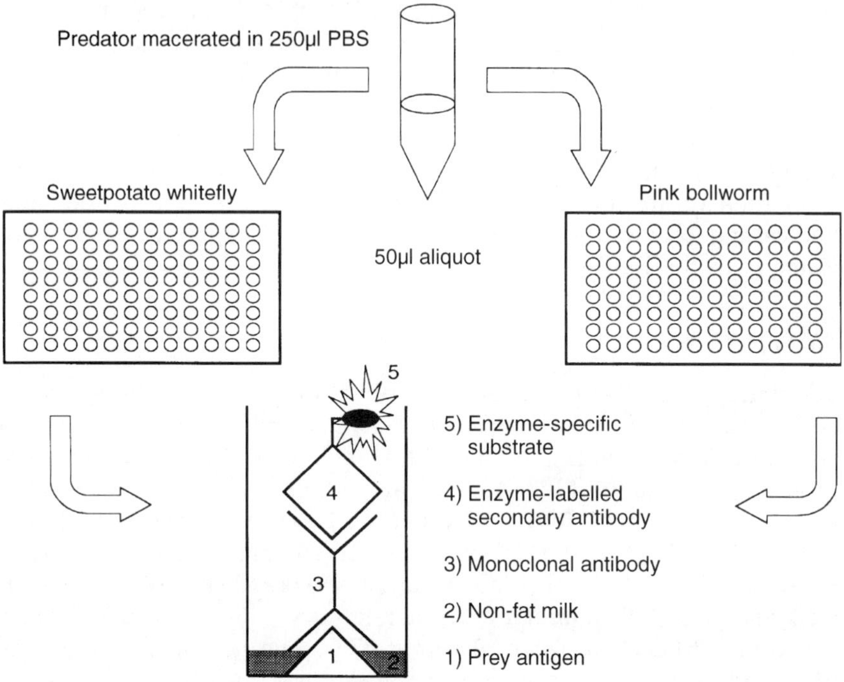

Figure 16.1 A diagrammatic representation of the steps in an enzyme-linked immunosorbent assay (ELISA): 1, Wells are coated with insect antigen; 2, wells are blocked with non-fat dry milk; 3, sweetpotato whitefly or pink bollworm egg-specific monoclonal antibody is added; 4, wells are washed and an enzyme-labelled secondary antibody is added; 5, wells are washed and an enzyme-specific substrate is added; and 6, absorbance is measured with an ELISA microplate reader. PBS, phosphate-buffered saline.

tem for the presence of whitefly and pink bollworm egg antigen were screened (Table 16.1). Almost one-third (28.9%) of the individuals contained whitefly egg antigen, a high proportion which was not surprising. Whitefly egg densities ranged from <1 egg per leaf early in the season to >5000 eggs per leaf late in the season (Naranjo and Hagler, in press). The population of individual species of predators scoring positive for whitefly egg antigen ranged from 4.0% for *Nabis* spp. (primarily *N. alternatus* Parshley) to 55.0% for *Collops vittatus* (Say) (Hagler and Naranjo, 1994a,b).

Pink bollworm egg antigen was not detected in as many predators as whitefly egg antigen, with only 12.8% of the individuals screened scoring positive (Table 16.1). This relatively low proportion of individuals is probably due to low pink bollworm egg densities in our fields,

Table 16.1 Frequency of predators scoring positive for the presence of sweetpotato whitefly, pink bollworm, and both pest egg antigens in their gut. Predators were collected from 7 June to 6 September, 1992

Insect	Stage	Number assayed	Percentage of predators scoring positive for		
			Sweetpotato whitefly	*Pink bollworm*	*Both*
Geocoris spp.	Nymphal	315	39.4	4.1	3.8
Orius tristicolor	Adult/nymphal	1402	38.4	8.0	3.3
G. pallens	Adult	2349	33.7	10.2	8.0
G. punctipes	Adult	615	31.7	3.6	2.4
Lygus hesperus	Nymphal	1342	27.7	19.8	12.4
L. hesperus	Adult	1709	20.1	30.1	12.5
Sinea confusa	Adult/nymphal	42	14.3	14.3	7.1
Zelus spp.	Adult/nymphal	303	12.2	11.9	4.3
Nabis spp.	Adult/nymphal	1101	4.0	0.7	0.3
Collops vittatus	Adult	680	55.0	14.0	11.0
Hippodamia convergens	Adult	614	33.0	4.0	2.0
Grand total and mean percentages		10 472	28.9	12.8	7.1

values ranging from 0.3 eggs per plant early in the season to only 1.8 eggs per plant late in the season (Naranjo and Hagler, in press). The proportion of individual species of predators scoring positive for pink bollworm egg antigen ranged from 0.7% for *Nabis* spp. (primarily *N. alternatus*) to 30.1% for adult *Lygus hesperus* Knight. Interestingly, although *L. hesperus* is considered a major pest, over one-quarter of the *L. hesperus* population surveyed scored positive for pink bollworm egg antigen. The correlation between *Lygus'* phytophagous feeding behaviour and their high predatory activity on pink bollworm eggs might explain why so many individuals scored positive for pink bollworm eggs, even when pest densities were low. *Lygus* feed on meristematic and reproductive tissues of the cotton plant, the same areas of the plant where pink bollworms oviposit (Brazzel and Martin, 1957; Henneberry and Clayton, 1982). These sites may increase the probability of discovery of pink bollworm eggs by foraging *Lygus* (Hagler and Naranjo, 1994b).

Overall, a relatively small proportion (7.1%) of the predators examined had preyed on both pest species. The percentage of predators scoring positive for the presence of both whitefly and pink bollworm prey ranged from 0.3% for *Nabis* spp. to 12.5% for adult *L. hesperus* (Table 16.1).

16.3 EXAMINING THE EFFICACY OF AUGMENTATIVE RELEASES OF *Orius insidiosus*

16.3.1 Introduction

Despite the lack of predator efficacy data in the scientific literature, many producers of biological control agents promote predaceous insects as 'an environmental-friendly alternative to chemical control'. Currently, there are over 132 companies in North America alone selling beneficial organisms (i.e. predators and parasites) for augmentative biological control use (Hunter, 1994). Although probably environmentally safe (see Harris, 1990 for a review), these biological agents might be serving only as a placebo to the end-user. Hoy *et al.* (1991) suggests that action must be taken soon to evaluate these organisms before they are sold to consumers on a wide scale. Studies are needed to evaluate both the efficacy of augmentative releases and the quality of these 'domesticated' predators (Roush, 1990). If predators are improperly used or their effectiveness is misrepresented to the public, the biological control industry could suffer a loss of consumer confidence. Efficacy and quality issues need to be examined under realistic field conditions.

The effectiveness of two pirate bug species was assessed, indigenous *Orius tristicolor* (White) and mass-released *O. insidiosus*, using our pest-specific MAbs and multiple ELISAs. *O. insidiosus* was selected for this study because previous studies had indicated that a member of this genus frequently preyed on whitefly and pink bollworm (Table 16.1) and is readily available from producers of beneficials. Although the data set presented here is preliminary, it should give the reader an idea of how pest-specific MAbs can be incorporated into a programme designed to monitor the efficacy and quality of mass-produced biological control agents.

16.3.2 Methods

Predators

Adult *O. insidiosus* were purchased from Arbico (Oracle, Arizona) and released into 1.83 × 1.83 m field cages located at The University of Arizona's Maricopa Research Station; 50, 100 or 250 *O. insidiosus* were released per cage. A fourth cage contained a no-release control (i.e. indigenous *O. tristicolor* only). These releases were made throughout the season on 28 June, 19 July and 26 July, 1993. Predators were released at approximately 8:00 p.m. on each date and allowed to forage within each cage for 36 hours, after which each cage was vacuumed as described above. Within minutes after collection the predators were frozen in dry ice and transported to the laboratory where they were assayed by ELISA for the presence of whitefly and pink bollworm egg antigens (Figure 16.1).

16.3.3 Results from the augmentative release

Very few minute pirate bugs were collected in the vacuum sampler on each of the sampling dates. This was due, in part, to the relatively low release rates and the difficulty in finding these small predators. Because the recovery rates were so low for both the released and indigenous populations it was difficult to measure any treatment effects over time. To obtain an estimate of the proportion of individuals scoring positive for whitefly and pink bollworm throughout the entire season all of the vacuum samples were pooled. ELISA results from this experiment showed that 4.0% of the released *O. insidiosus* (n = 215) and 26.0% of the indigenous *O. tristicolor* (n = 133) had whitefly egg antigen in their guts. Pink bollworm egg antigen was found in the guts of 8.0% of the *O. insidiosus* (n = 215) and 5.0% of the *O. tristicolor* (n = 133). Overall, a relatively small proportion of the *Orius* examined had preyed on both pest species. Only 4.0% of the mass-released *O. insidiosus* (n = 215) and none (n = 133) of the indigenous *O. tristicolor* scored positive for the presence of both whitefly and pink bollworm egg antigens.

16.4 DISCUSSION

Pest-specific MAbs can be used to identify key indigenous predator species and to monitor the efficacy of introduced (augmented) biological control agents. *Collops vittatus, Hippodamia convergens* Guerin–Meneville, *O. tristicolor* and *Geocoris* spp. were the most active native predator species, particularly on whitefly. An interesting result from the indigenous predator survey was that *L. hesperus*, considered a major cotton pest, was also a significant predator of pink bollworm and whitefly (Hagler and Naranjo, 1994b). While the predatory nature of *L. hesperus* has been reported previously (Dunbar and Bacon, 1972), it was surprising that such a high percentage of these predators scored positive for both pests.

Once the key predators were identified, the research was focused on evaluating the efficacy of mass releases of *O. insidiosus*. Although *O. tristicolor* was ordered from the distributor of the beneficials, *O. insidiosus* was actually received. Consequently, the experiment was conducted using *O. insidiosus*.

Caged experiments revealed that the indigenous *O. tristicolor* populations were more frequent predators on whitefly eggs than the commercially reared *O. insidiosus*. However, the augmented predators had a slightly greater percentage of individuals scoring positive for pink bollworm than the indigenous predators. Unfortunately, because of the misidentification of these predators by the distributor it was impossible to separate differences due to a species effect (i.e. *O. insidiosus* versus *O. tristicolor*) from that due to an origin effect (i.e. mass-produced versus indigenous predators). The experiment still provides an example of how a pest-specific MAb library can be used to monitor (i.e. quality control) the efficacy of augmentative biological control agents. Clearly, a more thorough investigation is needed before any conclusions are made on the feasibility of using *Orius* spp. as augmentative biological control agents.

16.5 LIMITATIONS OF PREDATOR GUT CONTENT IMMUNOASSAYS

While gut content immunoassays are useful for studying predator–prey interactions it would be remiss if their limitations were not discussed. Potential pitfalls common with gut content immunoassays and other indirect methods of assessing predation (i.e. radiolabelling and electrophoresis) include the possibility of obtaining false positive reactions due to third trophic level interactions, scavenger feeding, or predators that ingest enough prey antigens to score positive in the immunoassay but fail to kill their prey. Any of these interactions would lead to an overestimation of predation (Breene and Sterling, 1988). An additional limitation that especially hinders the widespread use of MAbs is the high degree of technical knowledge and specialized facilities needed to produce useful antibodies. Moreover, MAbs are targeted for pest egg antigens; therefore, a positive response can also occur if a gravid adult female has been consumed (Hagler *et al.*, 1993, 1994). Adult pink bollworms are probably too large and elusive for most small predators to capture. However, adult whiteflies can be preyed on by all of the species surveyed. This uncontrollable variable must be considered when interpreting results.

The most important limitation of using gut content immunoassays (and other indirect methods) to study predation is that the immunoassays are not quantitative. Antigen detection methods, such as the ELISA, dot-blot or radial immunodiffusion provide quantitative results in that strength of response is directly proportional to the concentration of antigen. However, translating these quantitative readings into an estimate of prey consumed is complicated by several interacting and uncontrollable factors (McIver, 1981; Naranjo and Hagler, in press; Greenstone, 1996, this volume; Sunderland, 1996, this volume). The interaction of the size of a predator's meal and the subsequent digestive rate are key factors to the strength of an immunoassay response. Thus, without additional knowledge of either the size of the meal or the elapsed time since ingestion, a given immunoassay response could translate to a small meal eaten recently or a large meal eaten several hours earlier. Moreover, if a predator continually consumes additional prey items after an initial meal the same assay response could have many interpretations depending on when the predator was assayed relative to its first meal. An additional issue is the effect of temperature on the rate of digestion (Fichter and Stephen, 1981, 1984; Sopp and Sunderland, 1989; Hagler and Cohen, 1990). Again, a single assay response could have many interpretations depending on the ambient temperature preceding the assay. Prey age (i.e. prey phenology) can also affect the quantitative outcome of gut content immunoassays. ELISAs are five times more responsive to freshly oviposited *L. hesperus* eggs than they are to 7-day-old eggs (Hagler *et al.*, 1992).

A factor that can also affect the quantitative and qualitative response of gut content immunoassays is the variable digestive rates exhibited by different predator species. Symondson and Liddell (1993) have shown that two closely related carabid beetles have vastly different detection intervals after feeding on the same prey species in the laboratory. We have similar data suggesting that pink bollworm egg antigen can on average be detected for only 1 hour in *H. convergens* and for 48 hours in *O. insidiosus* (J.R. Hagler *et al.*, in preparation).

Given all of the drawbacks and limitations of gut content immunoassays, they remain valuable tools for conducting field research. The results from such assays when combined with field (predator and prey population densities) and laboratory (feeding experiments) studies can yield more accurate qualitative estimates of predation. Sopp *et al.* (1992) provide an excellent summary of some of the predator efficiency indices (PEIs) developed over the past 35 years. All of these PEIs are based on key assumptions that might be useful in certain situations. For instance, many such indices assume that a positive ELISA response is the result of a single prey consumed (Dempster, 1960; Boreham and Ohiagu, 1978; Ragsdale *et al.*, 1981). These indices are more applicable if food resources are limited. Rothschild (1966) developed a PEI based on prey consumption under laboratory conditions, though this index is more applicable when prey populations are abundant and feeding is continuous. Another index has been suggested by Nakamura and Nakamura (1977) and Greenstone (1979) in which prey consumption is estimated as the zero term of a poisson distribution. Sopp *et al.* (1992) provide the only PEI that attempts to use the quantitative relationship between an ELISA response and the number of prey consumed in the laboratory. We have proposed a new PEI that is a modification of previously published PEIs, the major difference being that it incorporates prey consumption rates measured under more realistic conditions (i.e. greenhouse plant arenas rather than Petri dishes in the laboratory), takes into account the detection interval as a function of fluctuating field temperatures, and incorporates predator functional response behaviour (predator feeding as a function of prey density) (Naranjo and Hagler, in press). Surprisingly, none of the previously published PEIs has accounted for a predator's functional response.

16.6 SUMMARY

In summary, gut content immunoassays combined with field, laboratory and greenhouse studies can lead to better qualitative assessments of predation. However, uncontrollable variables inherent to gut content immunoassays of field-collected predators make precise quantitative assessments impossible. Biotic factors such as predator–prey phenology, number of prey eaten and prior metabolic status can all affect the outcome of an ELISA (McIver, 1981; Fichter and Stephen, 1981, 1984; Lovei *et al.*, 1985, Sopp and Sunderland, 1989; Hagler *et al.*, 1992). Furthermore, differential prey digestive rates of each predator species must be examined thoroughly in the laboratory before reliable qualitative assessments of predation can be made (Lovei *et al.*, 1990; Symondson and Liddell, 1993; J.R. Hagler *et al.*, in preparation).

In this review, an attempt has been made to present a realistic view of the benefits and drawbacks of using pest-specific MAbs to study predator feeding behaviour. Whether the advantages of using pest-specific MAbs to study insect predator–prey interactions outweigh the disadvantages is ultimately dependent on the researchers' needs. Data gathered using pest-specific MAbs can lead to a more effective utilization of predators in a biological control programme because it tests a predator's effect under natural and undisturbed field conditions. Pest-specific MAbs provide quick, efficient and cost-effective tools to identify qualitatively the effects that key predators have on key insect pests. Additionally, this technology can facilitate the stewardship of mass-produced predators into viable and trustworthy biological control programmes.

Acknowledgements

We thank Christie Durand, Melissa Erickson, Scott Machtley and Sally Wright for their excellent technical assistance. Special thanks to Debbie Hagler, Gloria DeGrandi-Hoffman, Justin Schmidt and two anonymous reviewers for providing helpful suggestions to earlier versions of this manuscript. We especially thank Eric Erickson and the staff at the Carl Hayden Honey Bee Laboratory for providing the laboratory facilities used in these experiments.

This article reports the results of research only. Mention of a proprietary product does not constitute an endorsement or a recommendation for its use by the USDA.

REFERENCES

Baldwin, W.F., James, H.G. and Welch, H.E. (1955) A study of predators of mosquito larvae and pupae with a radio-active tracer. *Canadian Entomologist*, **87**, 350–6.

Boreham, P.F.L. and Ohiagu, C.E. (1978) The use of serology in evaluating invertebrate predator–prey relationships: a review. *Bulletin of Entomological Research*, **68**, 171–94.

Brazzel, J.R. and Martin, D.F. (1957) Oviposition sites of the pink bollworm on the cotton plant. *Journal of Economic Entomology*, **50**, 122–4.

Breene, R.G. and Sterling, W.L. (1988) Quantitative phosphorus-32 labeling method for analysis of predators of the cotton fleahopper (Hemiptera: Miridae). *Journal of Economic Entomology*, **81**, 1494–8.

Brooke, M.M. and Proske, H.O. (1946) Precipitin test for determining the natural predators of immature mosquitoes. *Journal of the National Malaria Society*, **5**, 45–56.

Davies, R.W. (1969) The production of antisera for detecting specific triclad antigens in the gut contents of predators. *Oikos*, **20**, 248–60.

Dempster, J.P. (1960) A quantitative study of the predators on the eggs and larvae of the broom beetle, *Phytodecta olivacea* Forster, using the precipitin test. *Journal of Animal Ecology*, **29**, 148–67.

Doane, J.F., Scotti, P.D., Sutherland, O.R. and Pottinger, R.P. (1985) Serological identification of wireworm and staphylinid predators of the Australian soldier fly and wireworm feeding on plant and animal food. *Entomologia Experimentalis et Applicata*, **38**, 65–72.

Dunbar, D.M. and Bacon, O.G. (1972) Feeding, development, and reproduction of *Geocoris punctipes* (Heteroptera: Lygaeidae) on eight diets. *Annals of the Entomological Society of America*, **65**, 892–5.

Ellington, J., Kiser, K., Cardenas, M., Duttle, J. and Lopez, Y. (1984) The insectavac: a high-clearance, high-volume arthropod vacuuming platform for agricultural ecosystems. *Environmental Entomology*, **13**, 259–65.

Fichter, B.L. and Stephen, W.P. (1981) Time related decay in prey antigens ingested by the predator *Podisus maculiventris* as detected by ELISA. *Oecologia*, **51**, 404–7.

Fichter, B.L. and Stephen, W.P. (1984) Time related decay of prey antigens ingested by arboreal spiders as detected by ELISA. *Environmental Entomology*, **13**, 1583–7.

Gardner, W.A., Shepard, M. and Noblet, R. (1981) Precipitin test for examining predator–prey interactions in soybean fields. *Canadian Entomologist*, **113**, 365–9.

Greenstone, M.H. (1979) Spider feeding behaviour optimises dietary essential amino acid composition. *Nature*, **181**, 501–3.

Greenstone, M.H. (1996) Serological analysis of predation: past, present and future, in *The Ecology of Agricultural Pests: Biochemical Approaches*, (eds W.O.C. Symondson and J.E. Liddell), Chapman & Hall, London, pp. 265–300.

Hagler, J.R. and Cohen, A.C. (1990) Effects of time and temperature on digestion of purified antigen by *Geocoris punctipes* (Hemiptera: Lygaeidae) reared on artificial diet. *Annals of the Entomological Society of America*, **83**, 1177–80.

Hagler, J.R. and Cohen, A.C. (1991) Prey selection by *in vitro*-reared and field-reared *Geocoris punctipes. Entomologia Experimentalis et Applicata*, **59**, 201–5.

Hagler, J.R. and Naranjo, S.E. (1994a) Qualitative survey of two coleopteran predators of *Bemisia tabaci* (Homoptera: Aleyrodidae) and *Pectinophora gossypiella* (Lepidoptera: Gelechiidae) using a multiple prey gut content ELISA. *Annals of the Entomological Society of America*, **23**, 193–7.

Hagler, J.R. and Naranjo, S.E. (1994b) Determining the frequency of heteropteran predation on sweetpotato whitefly and pink bollworm using multiple ELISAs. *Entomologia Experimentalis et Applicata*, **72**, 63–70.

Hagler, J.R., Cohen, A.C., Bradley-Dunlop, D. and Enriquez, F.J. (1992) Field evaluation of predation on *Lygus hesperus* using a species- and stage-specific monoclonal antibody. *Environmental Entomology*, **21**, 896–900.

Hagler, J.R., Brower, A.G., Tu, Z., Byrne, D.N., Bradley-Dunlop, D. and Enriquez, F.J. (1993) Development of a monoclonal antibody to detect predation of the sweetpotato whitefly, *Bemisia tabaci. Entomologia Experimentalis et Applicata*, **68**, 231–6.

Hagler, J.R., Naranjo, S.E., Bradley-Dunlop, D., Enriquez, F.J. and Henneberry, T.J. (1994) A monoclonal antibody to pink bollworm (Lepidoptera: Gelechiidae) egg antigen: a tool for predator gut analysis. *Annals of the Entomological Society of America*, **87**, 85–90.

Harris, P. (1990) Environmental impact of introduced biological control agents, in *Critical Issues in Biological Control*, (eds M. Mackauer, L.E. Ehler and J. Roland). VCH Publishers, New York, NY, pp. 289–300.

Hengeveld, R. (1980) Qualitative and quantitative aspects of the food of ground beetles. A review. *Netherlands Journal of Zoology*, **30**, 555–63.

Henneberry, T.J. and Clayton, T.E. (1982) Pink bollworm: seasonal oviposition, egg predation, and square and boll infestation in relation to cotton plant development. *Environmental Entomology*, **11**, 663–6.

Henneberry, T.J. and Clayton, T.E. (1985) Consumption of pink bollworm and tobacco budworm eggs by some predators commonly found in cotton fields. *Environmental Entomology*, **14**, 416–19.

Hoy, M.A., Nowerski, R.M., Johnson, M.W. and Flexner, J.L. (1991) Issues and ethics in commercial releases of arthropod natural enemies. *American Entomologist*, Summer, 74–5.

Hunter, C.D. (1994) *Suppliers of Beneficial Organisms in North America*, California Environmental Protection Agency, Sacramento, California.

James, H.G. (1961) Some predators of *Aedes stimulans* and *Aedes trichurus* in woodland pools. *Canadian Journal of Zoology*, **39**, 533–40.

Jenkins, D.W. (1963) Use of radionuclides in ecological studies of insects, in *Radioecology*, (eds V. Schultz and A.W. Klement), Reinhold, New York. pp. 431–40.

Köhler, G. and Milstein, C. (1975) Continuous cultures of fused cells secreting antibody of predefined specificity. *Nature*, **256**, 495–7.

Lenz, C.J. and Greenstone, M.H. (1988) Production of a monoclonal antibody to the arylphorin of *Heliothis zea*. *Archives of Insect Biochemistry & Physiology*, **9**, 167–77.

Liddell, J.E. and Symondson, W.O.C. (1996) The potential of combinatorial antibody libraries in pest–predator relationship studies, in *The Ecology of Agricultural Pests: Biochemical Approaches*, (eds W.O.C. Symondson and J.E. Liddell), Chapman & Hall, London, pp. 347–66.

Lister, A., Usher, M.B. and Block, W. (1987) Description and quantification of field attack rates by predatory mites: an example using an electrophoresis method with a species of Antarctic mite. *Oecologia*, **72**, 185–91.

Lovei, G.L., Monostori, E. and Ando, I. (1985) Digestion rate in relation to starvation in the larva of a carabid predator, *Poecilus cupreus*. *Entomologia Experimentalis et Applicata*, **37**, 123–7.

Lovei, G.L., Sopp, P.I. and Sunderland, K.D. (1990) Digestion rate in relation to alternative feeding in three species of polyphagous predators. *Ecological Entomology*, **15**, 293–300.

Luck, R.F., Shepard, B.M., Kenmore, P.E. (1988) Experimental methods for evaluating arthropod natural enemies. *Annual Review of Entomology*, **33**, 367–91.

Lund, R.D. and Turpin, F.T. (1977) Serological investigations of black cutworm larval consumption by ground beetles. *Annals of the Entomological Society of America*, **70**, 322–4.

McCarty, M.T., Shepard, M. and Turnipseed, S.G. (1980) Identification of predaceous arthropods in soybeans by using autoradiography. *Environmental Entomology*, **9**, 199–203.

McDaniel, S.G. and Sterling, W.L. (1979) Predator determination and efficacy on *Heliothis* eggs in east Texas cotton using P32. *Environmental Entomology*, **8**, 1083–7.

McIver, J. (1981) An examination of the utility of the precipitin test for evaluation of arthropod predator–prey relationships. *Canadian Entomologist*, **113**, 213–22.

Miller, M.C. (1979) Serology in insect predator–prey studies. *Entomological Society of America Miscellaneous Publication*, **11**, 1–84.

Miller, M.C. (1981) Evaluation of enzyme-linked immunosorbent assay of narrow- and broad-spectrum anti-adult southern pine serum. *Annals of the Entomological Society of America*, **74**, 279–82.

Murray, R.A. and Solomon, M.G. (1978) A rapid technique for analyzing diets of invertebrate predators by electrophoresis. *Annals of Applied Biology*, **90**, 7–10.

Nakamura, M. and Nakamura, K. (1977) Population dynamics of the chestnut gall wasp, *Dryocosmus kuriphilus* Yasumatsu (Hymenoptera: Cynipidae). V. Estimation of the effect of predation by spiders on the mortality of imaginal wasps based on the precipitin test. *Oecologia*, **27**, 97–116.

Naranjo, S.E. and Hagler, J.R. (in press) Characterizing and estimating the effect of heteropteran predation, in *Predatory Heteroptera in Agroecosystems: Their Ecology and Use in Biological Control*, (eds M. Coll and J. Ruberson), Thomas Say Symposium Proceedings, Entomological Society of America.

Orphanides, G.M., Gonzales, D. and Bartlett, B.R. (1971) Identification and evaluation of pink bollworm predators in southern California. *Journal of Economic Entomology*, **64**, 421–4.

Ragsdale, D.W., Larson, A.P. and Newsome, L.D. (1981) Quantitative assessment of the predators of *Nezara viridula* eggs and nymphs within a soybean agroecosystem using an ELISA. *Environmental Entomology*, **10**, 402–5.

Rothschild, G.H.L. (1966) A study of a natural population of *Conomelus anceps* including observations on predation using the precipitin test. *Journal of Animal Ecology*, **53**, 413–34.

Roush, R.T. (1990) Genetic variation in natural enemies: critical issues for colonization in biological control, in *Critical Issues in Biological Control*, (eds M. Mackauer, L.E. Ehler and J. Roland), VCH Publishers, New York, NY, pp. 263–88.

Schoof, D.D., Palchick, S. and Tempelis, C.H. (1986) Evaluation of predator–prey relationships using an enzyme immunoassay. *Annals of the Entomological Society of America*, **79**, 91–5.

Sopp, P.I. and Sunderland, K.D. (1989) Some factors affecting the detection period of aphid remains in predators using ELISA. *Entomologia Experimentalis et Applicata*, **51**, 11–20.

Sopp, P.I., Sunderland, K.D., Fenlon, J.S. and Wratten, S.D. (1992) An improved quantitative method for estimating invertebrate predation in the field using an enzyme-linked immunosorbent assay (ELISA). *Journal of Applied Ecology*, **29**, 295–302.

Stern, V.M., Smith, R.F., van den Bosch, R. and Hagen, K.S. (1959) The integrated control concept. *Hilgardia*, **29**, 81–101.

Sunderland, K.D. (1987) A review of methods of quantifying invertebrate predation occurring in the field. *Acta Phytopathologia et Entomologia Hungarica*, **22**, 13–34.

Sunderland, K.D. (1996) Progress in quantifying predation using antibody techniques, in *The Ecology of Agricultural Pests: Biochemical Approaches*, (eds W.O.C. Symondson and J.E. Liddell), Chapman & Hall, London, pp. 419–55.

Sutula, C.L., Gillett, J.M., Morrissey, S.M. and Ramsdell, D.C. (1986) Interpreting ELISA data and establishing the positive–negative threshold. *Plant Disease*, **70**, 722–6.

Symondson, W.O.C. and Liddell, J.E. (1993) Differential antigen decay rates during digestion of molluscan prey by carabid predators. *Entomologia Experimentalis et Applicata*, **69**, 277–87.

Voller, A., Bidwell, D. and Bartlett, A. (1976) Microplate enzyme immunoassays for the immunodiagnosis of virus infections, in *Manual of Clinical Immunology*, (eds N.R. Rose and H. Friedman), American Society of Microbiology, Washington, pp. 506–12.

West, A.S. (1950) The precipitin test as an entomological tool. *Canadian Entomologist*, **82**, 241–4.
Whitten, M.J. and Oakshott, J.G. (1990) Biocontrol of insects and weeds. *Agricultural Biotechnology*, **2**, 123–42.
Wilson, E.O. (1990) First word. *Omni*, **12**, 6.

17

An environmental risk assessment for release of an exotic microsporidium for European corn borer control in North America

D.W. Ragsdale and C.T. Oien

ABSTRACT

The European corn borer, *Ostrinia nubilalis* (Hübner) (Lepidoptera: Pyralidae), is the most destructive pest of field and sweet corn in the mid-western US. In Minnesota alone, the average annual yield loss is $113 million. In an effort to establish biological control of the European corn borer, 24 exotic parasitoids have been released over the past 70 years. Only a few species have become widely established and parasitism rates are often low. In Minnesota average cumulative parasitism by the four known established exotic species was only 6.1% in a 4 year study. A microsporidium, *Nosema pyrausta* (Paillot) (Microspora: Nosematidae), is the most frequently encountered natural enemy of the European corn borer and it plays a significant role in regulating European corn borer populations in North America. An exotic microsporidium, *N. furnacalis* Wen and Sun, whose normal host is the Asian corn borer, *O. furnacalis* (Guenée), is a candidate biological control agent for the European corn borer. Using *N. furnacalis* as a classical biological control agent offers great potential for consistent, reliable control of the European corn borer. However, *N. pyrausta* and *N. furnacalis* cannot be distinguished morphologically using light microscopy. The inability to determine which microsporidian is infecting the European corn

The Ecology of Agricultural Pests, Edited by W.O.C. Symondson and J.E. Liddell.
Published in 1996 by Chapman & Hall, London. ISBN 0 412 62190 8

borer has curtailed interest in employing *N. furnacalis* as a classical biological control agent. We developed a species-specific immunoassay that can discriminate between *N. pyrausta* and *N. furnacalis*. Using this immunoassay has enabled the laboratory host range of *N. furnacalis* to be studied. Our data indicate that no generalist predator is a host for this exotic microsporidium. However, the European corn borer parasitoid, *Macrocentrus grandii* Goidanich, can be infected with *N. furnacalis*. The fact that *N. furnacalis* infects *M. grandii* is not surprising. Many lepidopteran-infecting *Nosema* spp. are known to infect parasitoids. Further study is needed to determine the extent to which *M. grandii* and other corn borer parasitoids are affected by *N. furnacalis*. Once these data are available a decision whether or not to release *N. furnacalis* over a wide geographic area can be made.

17.1 INTRODUCTION

Microsporidia are frequently encountered obligate intracellular protozoan pathogens of insects. Pathogenic effects range from rapid death to sublethal effects manifested as reduced fitness, longevity and fecundity. All microsporidia form a resistant spore that is capable of surviving outside of host tissue. Much of the past applied research on microsporidia has focused on using the spore stage as the active component of a biopesticide. Unfortunately, microsporidia-based biopesticides have not given consistent reliable control of the target pests (Henry, 1990). Reasons for such poor performance are not fully understood, but may include the need to produce microsporidia in live insects with inherent risks in selecting for strains which produce high spore numbers but have diminished virulence. Spores produced in live insects also require at least partial purification, which may adversely affect pathogenicity. In general, until propagation procedures improve so that spores of consistent virulence are produced, microsporidia-based biopesticides do not appear to be a viable approach to insect pest population suppression.

Microsporidia can play a major role in pest population regulation, a phenomenon which has been particularly well documented for the microsporidium *Nosema pyrausta* (Paillot) (Microspora: Nosematidae)–European corn borer *Ostrinia nubilalis* (Hübner) (Lepidoptera: Pyralidae) pathosystem (Maddox *et al.*, 1992). This chapter focuses on characterizing a new pathosystem consisting of the European corn borer and the exotic protozoan *Nosema furnacalis* Wen and Sun. The native host of *N. furnacalis* is the Asian corn borer, *O. furnacalis* (Guenée) (Lepidoptera: Pyralidae). *N. furnacalis* is a candidate biological control agent for the European corn borer. If *N. furnacalis* can be established in North America it may augment existing biological control of the European corn borer.

17.2 EUROPEAN CORN BORER – PEST STATUS AND NATURAL ENEMY COMPLEX

The European corn borer is a major pest of corn throughout North America (Mason *et al.*, 1994; Orr *et al.*, 1994) causing hundreds of millions of dollars in losses each year. In Minnesota there are approximately two million hectares of field corn planted each year. During European corn borer outbreak years (which occur every 6–8 years), 13% of the crop is lost. The average annual loss to European corn borer in Minnesota alone is estimated at $113 million (Noetzel *et al.*, 1985). Injury by the European corn borer is cryptic since larvae bore into the stalk and evidence of feeding is not readily apparent. Thus, many economically damaging populations remain untreated, even in outbreak years. Most of the estimated $113 million annual loss is direct yield loss with chemical and application costs representing only a small percent of this total. In addition to corn, the European corn borer is a minor pest in several other crops, potato, snap bean, peppers, sorghum and onions (Showers *et al.*, 1989).

Given the pest status and damage potential of the European corn borer, establishing biological control of this insect has been a priority for agricultural scientists for decades. Corn pest management has focused on two principal strategies for European corn borer control, namely the development of resistant corn hybrids (host plant resistance) and the release of exotic parasitoids (classical biological control). With the latter approach, 24 exotic parasitoid species were released between 1920 and 1940 (Mason *et al.*, 1994). A few of these parasitoids became established and spread throughout the range of the European corn borer (Mason *et al.*, 1994). In general, European corn borer parasitism rates are low; for example, the two most common parasitoids in Minnesota, *Macrocentrus grandii* Goidanich and *Eriborus terebrans* Gravenhorst, were found in less than half the counties surveyed (Winnie *et al.*, 1981). Subsequently, Sreenivasam *et al.* (1993) showed the cumulative percent parasitism of the European corn borer by the four established parasitoid species in Minnesota to be 6.1% over a 4-year period. Although higher parasitism rates for the European corn borer have been reported elsewhere (Landis and Haas, 1992), biological control of this species using exotic parasitoids has not successfully reduced its pest status in North America. Reasons for the low parasitism levels are not fully understood. Landis and Haas (1992) showed that *E. terebrans* built up to effective levels when landscape structure benefited parasitoid survival. *Lydella thompsoni* Herting, a once widely established and abundant tachinid parasitoid, has since disappeared for unknown reasons (Mason *et al.*, 1994). Recent efforts to re-establish *L. thompsoni* (Mason *et al*, 1994) represent an attempt to match the climate of the European collection site with that of the initial release sites in the US.

17.2.1 Prevalence of *Nosema pyrausta*

Of all the natural enemies of the European corn borer, *N. pyrausta* is the most abundant (Lewis and Lynch, 1978; Hill and Gary, 1979). For example, in a 3-year study in Illinois, average *N. pyrausta* infection in first-brood adults was 29.7% (Siegel *et al.*, 1988). In Minnesota, diapausing European corn borer larvae were collected in late autumn 1992 and 1993 and held at 4°C for 3 months, after which time larvae were placed into 30-ml diet cups containing 2% agarose and incubated at a constant 27°C. Larvae were held until parasitoids emerged, adult corn borers eclosed, or larvae died. All non-parasitized European corn borer larvae were examined microscopically for presence of *N. pyrausta*. Some 42% (n = 248) and 44% (n = 251) of the larvae dissected in 1992 and 1993, respectively, were infected with *N. pyrausta*. In Minnesota, a European corn borer larvae is seven times more likely to be infected with *N. pyrausta* than to be parasitized. Interestingly, *N. pyrausta* was not deliberately introduced into North America, Maddox (1986) suggesting that the species was introduced accidentally via release of *N. pyrausta*-contaminated parasitoids. From these initial unintentional releases, *N. pyrausta* has spread throughout North America and is now considered the major mortality factor of *O. nubilalis* (Zimmack *et al.*, 1954; Onstad and Maddox, 1989).

The importance of horizontal transmission (*per os*) in the development of a *N. pyrausta* epizootic has been well documented (Hill and Gary, 1979; Andreadis, 1986, 1987; Maddox, 1986; Onstad and Ruesink, 1986; Siegel *et al.*, 1988; Onstad and Maddox, 1989). Generally, European corn borer larval density in non-outbreak years is too low in the first generation for horizontal transmission to play a significant role in the spread of microsporidia (Siegel *et al.*, 1988). European corn borer population levels required for an epizootic to occur far exceed recommended economic thresholds and thus reliance upon epizootics to control European corn borer is unrealistic (Maddox, 1986; Onstad and Maddox, 1989; Henry, 1990; MacVean and Capinera, 1992). Baculoviruses suppress pest populations by causing epizootics (Fuxa, 1990). In contrast, microsporidia suppress pest populations by causing sublethal effects such as reduced adult fitness, fecundity and longevity.

In non-outbreak years, vertical transmission (transovarial) is critically important to maintain *N. pyrausta* in the population. Infected European corn borer larvae that survive the winter give rise to adults capable of transovarial transmission of *N. pyrausta*. It is known that although extracorporeal survival of microsporidian spores varies from species to species the following generalizations are valid: (i) spores of most species do not persist in the environment for more than 1 year (Lewis and Cossentine, 1986), (ii) temperatures above 35°C greatly reduce the viability of most spores; (iii) drying is harmful; and (iv) naked spores cannot withstand exposure to direct sunlight for more than a few days (Maddox, 1973). For most species of microsporidia, the principal means of persistence in the environment is in infected living tissue.

Recognition that *N. pyrausta* plays such an important role in population suppression of the European corn borer has fostered an interest in using microsporidia as classical biological control agents (McManus *et al.*, 1989; Bechnel, 1992; Lipa and Hokkanen, 1992; Bauer and Pankratz, 1993). Employing microsporidia in such a way offers the greatest potential for consistent, reliable control (Maddox *et al.*, 1992) as opposed to using microsporidia as bioinsecticides.

17.3 *NOSEMA FURNACALIS*, A CANDIDATE BIOLOGICAL CONTROL AGENT FOR THE EUROPEAN CORN BORER

Nosema furnacalis is a microsporidian whose native host is the Asian corn borer, *O. furnacalis*, a close relative of the European corn borer (Wen and Sun, 1988). *N. furnacalis* can also infect the European corn borer and infects many of the same organs and tissues as *N. pyrausta* (Wen and Sun, 1988). The ultimate goal is to release and establish *N. furnacalis* in the North American European corn borer population. Before this microsporidian can be considered for release over a wide geographic area, an assessment of the risks associated with release of an exotic insect pathogen must be completed. To fully understand the benefits and risks associated with releasing *N. furnacalis* requires that we understand the pathogenicity of this exotic organism with respect to its host range (other Lepidoptera), transmission characteristics within the European corn borer population (vertical and horizontal transmission), and interactions with existing natural enemies (predators, parasitoids and other microsporidia).

Unfortunately, *N. furnacalis* and *N. pyrausta* spores are indistinguishable from each other using light microscopy. Simple observation of spores cannot determine if *N. furnacalis* infects a particular host. Microsporidian species are usually distinguished using a combination of ultrastructure, host range, tissue specificity and morphology of vegetative stages. Before *N. furnacalis* could be evaluated as a potential biological control agent, a means of distinguishing between these two microsporidia was needed. Although techniques such as protein electrophoresis (Street and Briggs, 1982), pulsed-field electrophoresis (Munderloh *et al.*, 1990; Malone and McIvor, 1993), and rDNA sequencing of polymerase chain reaction (PCR)-amplified genes (Vossbrinck *et al.*,

1993; Baker *et al.*, 1994) can distinguish between closely related microspora, these assays require a relatively large amount of purified spores or extensive sample preparation. The necessity of processing a large number of samples limits the utility of these methods as monitoring tools. Moreover, the inability reliably to distinguish *N. furnacalis* from *N. pyrausta* has diminished interest in developing a biological control programme using *N. furnacalis*.

17.3.1 Electrophoresis as a diagnostic tool

Fowler and Reeves (1974a,b) and Knell (1975) were the first to attempt to use electrophoresis to characterize microsporidia. However, the electrophoretic patterns of hydrophilic spore protein extracts (Fowler and Reeves, 1974b) were variable under different conditions and consequently found to be unsuitable for characterization. Hydrophobic spore protein extracts (Fowler and Reeves, 1974a) can presumably be used to distinguish between species. However, difficulties with the extraction medium (Knell, 1975) showed this method to be unreliable. Subsequently, Street and Briggs (1982) investigated sodium dodecyl sulphate (SDS)–polyacrylamide gel electrophoresis (PAGE) using homogenized spore extracts. These authors found that SDS–PAGE can be used to identify microsporidian isolates, provided that electrophoretic profiles of the individual species are available for reference.

Munderloh *et al.* (1990) used pulsed-field electrophoresis to separate chromosomes isolated from germinated sporoplasms from *N. pyrausta* and *N. furnacalis*. They found that separation based on size alone may not be sufficiently sensitive to distinguish between these two closely related species. Malone and McIvor (1993) also used pulsed-field electrophoresis and demonstrated that a different number of bands were obtained when a single microsporidian species was extracted from two different hosts. These results provide further evidence that pulsed-field electrophoresis is not a reliable method for discriminating between two closely related microsporidian species or two strains of the same species.

Baker *et al.* (1994) and Vossbrinck *et al.* (1993) amplified and sequenced two segments of the rDNA gene that read into highly variable regions of that gene. Vossbrinck *et al.* (1993) demonstrated that restriction digestion of amplified rDNA genes yields different patterns on agarose gels and that such differences among distantly related species can be used to separate species. However, with closely related microsporidia, the amplified rDNA fragment needed to be sequenced for reliable species identification. In fact, Baker *et al.* (1994) showed that between two closely related microsporidia, sequence data from the rDNA region they chose to amplify was either identical or varied by as little as a single nucleotide. It appears that using this portion of the rDNA gene may not discriminate reliably between two closely related species, and further work is needed to identify more variable portions of the genome. PCR has a distinct advantage over other detection methods because in theory only a single copy of the target gene is needed, thus yielding an assay that is orders of magnitude more sensitive.

17.3.2 Immunology as a diagnostic tool

Serological tests have long played an important role in detection and identification of insect pathogens. A few serological assays have been developed to detect microsporidia. Niederkorn *et al.* (1980) found that sera obtained from individual rabbits, each immunized with one of six species of microsporidia, produced antisera that cross-reacted across genus and species boundaries using immunofluorescence. These results suggest there are antigenic similarities in the structure of the spore wall among these species. Hayasaka and Ayuzawa (1987) found that latex beads coated with antibody directed against *Nosema bombycis* Naegeli showed cross-reactivity to another species of *Nosema*. Again, this seems to suggest that antigenic determinants on spore coats are highly conserved.

Greenstone (1983) described the first enzyme-linked immunosorbent assay (ELISA) developed to detect a microsporidium. The antigen was prepared by homogenizing *Amblyospora* sp. with glass beads until spores were disrupted, followed by low-speed centrifugation to remove particulate matter. Greenstone (1986) later investigated the specificity of this assay and found differences in the regression slopes comparing absorbance and homogenate protein mass for the five species tested. His data support a taxonomic distinction for the family Amblyosporidae, although his ELISA cross-reacted with three species of *Nosema* and one of *Vairimorpha*.

Widespread cross-reactivity is commonly reported when the antigen consisted of whole spores, spore coat proteins or spore homogenate (Niederkorn *et al.*, 1980; Greenstone, 1983, 1986; Hayasaka and Ayuzawa, 1987; Kawarabata and Hayasaka, 1987; Ke *et al.*, 1990). These data support our hypothesis that the spore coat (a mixture of chitin and proteins) is a highly conserved structure and immunoassays based on these antigens cross-react broadly.

(a) Immunoassays to detect Nosema furnacalis

In previous work we developed a species-specific immunoassay to detect *N. furnacalis* within the European corn borer (Oien and Ragsdale,

1992). The antigen was obtained by purifying *N. furnacalis* spores following procedures in Oien and Ragsdale (1992, 1993). Briefly, sporoplasm from *N. furnacalis* was obtained by germinating spores (1×10^9) *in vitro* in 500 μl of germination buffer as described by Munderloh *et al.* (1990). Germinated spores were gently sonicated to disrupt the membrane-bound sporoplasm, followed by centrifugation at 15 600*g* for 1 min to pellet spore walls, immature (non-refractile) spores and polar filaments. Soluble sporoplasm proteins contained in the supernatant were precipitated using high salt concentrations to partially purify the low-molecular weight proteins (<20 kDa) used as immunogens. Either mice (Robertsonian RBF/dnj or BALB/c) or New Zealand White rabbits were given subcutaneous injections of 1.0 or 10.0 mg soluble protein, respectively, in phosphate-buffered saline (PBS), pH 7.4, emulsified with an equal volume of Freund's complete adjuvant for the first injection and with Freund's incomplete adjuvant for all remaining booster injections given at 3-week intervals. Once a sufficient titre was obtained, animals were exsanguinated and immunoglobulin G (IgG) purified by passing dilute sera through a Protein A column (Bio-Rad). IgG was concentrated to 1 mg/ml and frozen (–20°C) in 1-mg aliquots.

Using the *N. furnacalis* antisera, an ELISA was developed that can determine if any insect has an active microsporidian infection caused by this microsporidium. Given that partially purified proteins derived from germinated spores (sporoplasm) were used (Oien and Ragsdale, 1992) rather than using homogenized spores or spore coat proteins, the immunoassay did not react with outer spore coat proteins of either *N. furnacalis* or *N. pyrausta* or with six other microsporidian species (Table 17.1; Oien and Ragsdale, 1992). Westem blot analysis confirmed specificity of the ELISA, and showed that we could discriminate among *N. furnacalis* infected, *N. pyrausta* infected, and uninfected *O. nubilalis* (Oien and Ragsdale, 1992). This ELISA was the first species-specific immunoassay ever developed for microsporidia.

An added benefit of this ELISA is that a positive reaction is obtained only when *N. furnacalis* spores have germinated and sporoplasm is liberated. Because there is no reaction in the ELISA with proteins associated with spore coats, mere accumulation of spores in the gut of an insect is not detected. The latter has been a major limitation of using light microscopy to determine if predators or parasitoids are infected. Predators, parasitoids, and non-target Lepidoptera are known to accumulate spores within their midgut (Thomson, 1958). By using microscopic examination alone, if a predator or parasitoid had a high spore number, it might be concluded that the natural enemy has an active microsporidian infection. Using the *N. furnacalis*-specific immunoassay, it was considered that an active infection was occurring only when spore counts increased in concert with absorption values in the ELISA

Table 17.1 Specificity of the ELISA directed at sporoplasm proteins of *Nosema furnacalis*. (From Oien and Ragsdale, 1992.)

Species tested in ELISA	*n*	OD_{410} (*mean* ± *SE*)
N. furnacalis purified spores		
Intact spores	76	0.08 ± 0.07
Germinated spores (10 μg protein)	20	1.99 ± 0.01
Homogenized spores	4	>2.00*
N. furnacalis infecting:		
Ostrinia nubilalis	38	1.87 ± 0.15
Helicoverpa zea	23	1.25 ± 0.25
H. zea cell culture	7	1.25 ± 0.11
Other microsporidia from infected host insects:		
Nosema pyrausta	66	0.01 ± 0.01
Nosema locustae	12	0.00 ± 0.00
Nosema apis	14	0.02 ± 0.01
Nosema bombycis	49	0.01 ± 0.01
Nosema sp.	11	0.01 ± 0.01
Vairimorpha necatrix	7	0.08 ± 0.01
Vairimorpha sp. 1	11	0.00 ± 0.00
Vairimorpha sp. 2	11	0.00 ± 0.00

*Maximum absorption in ELISA plate reader is 2.00

(Figure 17.1). The added advantage of the ELISA over electrophoresis is that the ELISA is quantitative, so that absorption values can be converted to spore density (Figure 17.2). ELISA is easier to perform, requiring little or no sample preparation, whereas electrophoresis and PCR require extensive sample preparation and yield results that are not easily quantified.

17.4 NON-TARGET EFFECTS

Host specificity is of great concern to biological control specialists. Unfortunately, with entomopathogens host range is not easily defined. Although laboratory infection is not a good predictor of ecological host range (Maddox *et al.*, 1992), it does implicate possible negative interactions. If a microsporidium causes a slight infection in a non-target host, but produces a low number of infective forms that cannot be transmitted to its offspring, such an infection may be detrimental to an individual non-target host but it is unlikely to be biologically significant. The native hosts for *N. furnacalis* include the Asian corn borer and two

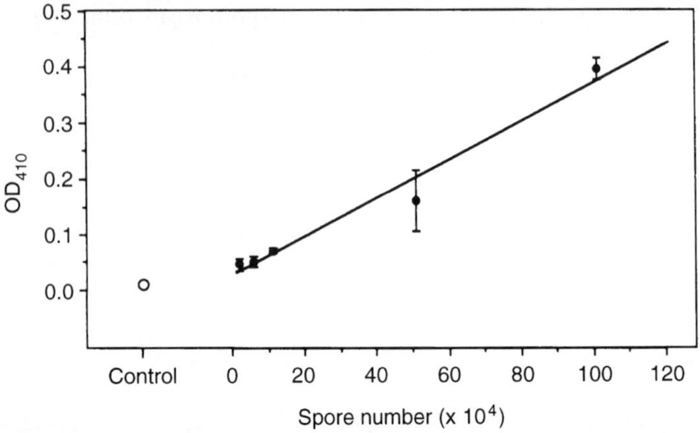

Figure 17.1 Relationship between *Nosema furnacalis* spore number ($\times 10^4$) and mean $OD_{410}(\pm SE)$ of four replicated experiments. The equation $y = 0.00345x + 0.0345$ explains 97% of the variation; the line was fitted by simple linear regression. (From Oien and Ragsdale, 1992, reprinted with permission.)

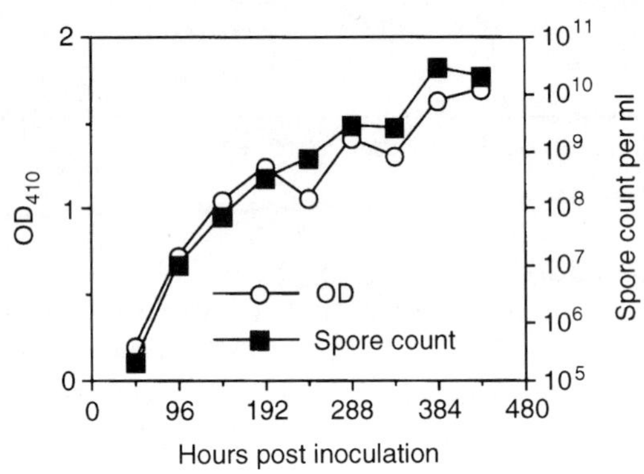

Figure 17.2 Relationship between time post-inoculation of *Ostrinia nubilalis* spore number per millilitre ($\log_{10}$), and mean OD_{410} of four replicated experiments when neonate larvae were infected with *Nosema furnacalis*. (From Oien and Ragsdale, 1992, reprinted with permission.)

of its parasitoids (Wen and Sun, 1988; Nafus and Schreiner, 1991). Experimentally, *N. furnacalis* can infect several Lepidoptera (Wen and Sun, 1988). Having a broad experimental host range is a characteristic of many microsporidia, among them the common lepidopteran microsporidium, *Vairimorpha necatrix* (Kramer), and the grasshopper pathogen, *Nosema locustae* Canning. The ability to infect several hosts may allow a pathogen to survive periods of low host density. Although host range appears to be broad, especially in the genus *Nosema*, analysis of rDNA sequence data (Baker *et al.*, 1994) suggests that the Lepidoptera-infecting *Nosema* spp. form a natural assemblage and that the genus is likely composed of many unrelated species.

It is not uncommon for Lepidoptera-infecting *Nosema* to also infect the host's parasitoids. In general, there is strong evidence to suggest that a microsporidian-infected insect is a less fit host for parasitoids (McLaughlin and Adams, 1966; Brooks and Cranford, 1972; Brooks, 1973; Andreadis 1980, 1982; Huger, 1984; Own and Brooks, 1986; Cossentine and Lewis, 1987). Adverse effects of parasitizing an infected host include failure of the parasitoid to emerge, and reduced fitness of any parasitoids that do emerge in terms of having a lower fecundity and diminished longevity. Such detrimental effects may occur even if the parasitoid itself is not infected because an infected host is often smaller and perhaps nutritionally compromised when compared with an uninfected host (Laigo and Tamashiro, 1967). Spores can also accumulate in a parasitoid's blind gut (Thomson, 1958) and interfere with nutrient uptake.

Nosema pyrausta is known to infect at least three European corn borer parasitoids: *M. grandii* (Andreadis, 1980, 1982; Siegel *et al.*, 1986, Cossentine and Lewis, 1987); *Trichogramma nubilale* Ertle and Davis (Sajap and Lewis, 1988); and *T. evanescens* Westwood (Huger, 1984). However, Cossentine and Lewis (1988) showed that *L. thompsoni* was not a host for *N. pyrausta*. This is important because after *L. thompsoni* became widely established in North America, its population declined rapidly so that by 1977 no established field populations remained (Mason *et al.*, 1994). It was often speculated that *N. pyrausta* was at least partially responsible for the decline of *L. thompsoni* (Lewis, 1982). Although factors responsible for the decline of *L. thompsoni* are not fully understood, the work by Cossentine and Lewis (1988) supports the conclusion that the microsporidium was not involved in the disappearance of this parasitoid.

There have been only a few studies on the fate of microsporidia in predaceous insects (Bechnel, 1992). Van Essen and Anthony (1976) tested nine predator species for susceptibility to microsporidia. Only one, *Notonecta undulata* Say, was susceptible to *Nosema algerae* Vavra and Undeen, after feeding on diseased mosquito larvae. Sajap and Lewis (1989) studied one European corn borer predator, *Chrysoperla carnea* (Stephens), and found this predator was not infected by *N. pyrausta*. Lipa and Hokkanen (1992) tested five insect species including *Apis mellifera* L., *Coccinella septempunctata* L., *Galleria mellonella* L., *Tenebrio molitor* L. and *Tribolium castaneum* Herbst for susceptibility to

Nosema meligethi Issi and Raditscheva. None of the five species showed any sign of infection when examined up to 30 days after exposure to *N. meligethi*.

17.4.1 Host range studies of *N. furnacalis*

Some important potential alternate hosts of *N. furnacalis* include the predators and parasitoids of the European corn borer. Insect predators may play an important role in suppressing some European corn borer populations (Sparks *et al.*, 1966; Andow, 1990).

Studies to determine host ranges of microsporidia often require lengthy trials to measure subtle effects that an infection might have on developmental rate and reproductive capability (Andreadis, 1982; Sajap and Lewis, 1988, 1989). The immunoassay we developed for *N. furnacalis* has provided a novel means to carefully examine the host range of *N. furnacalis*. Because our ELISA detects only germinated spores and not intact spores, mere accumulation of spores in a predator's or parasitoid's gut is not detected. Although an assessment of host range in the laboratory cannot be considered equivalent to its ecological host range (Maddox *et al.*, 1992), such analyses can indicate if active infections are possible thus demonstrating where more detailed studies are needed.

The ELISA has been employed to assess the impact of *N. furnacalis* on a broad array of natural enemies of *O. nubilalis* (Oien and Ragsdale, 1993). Several hundred specimens of eight species of generalist predators were fed either suspensions of purified *N. furnacalis* spores or an infected caterpillar. Although a few predators tested weakly positive in the ELISA for *N. furnacalis* sporoplasm (Oien and Ragsdale, 1993), there was no evidence an active infection was occurring in any of the predators. However, the most common European corn borer parasitoid found in Minnesota, *M. grandii*, tested strongly positive each time in the ELISA (average OD_{410} = 1.36), indicating that large amounts of sporoplasm proteins were present. Mean spore number in *M. grandii* was relatively high (1.4×10^7), supporting the conclusion that an active infection was occurring. Vegetative forms were found in whole-body smears of this parasitoid when stained with Wright's stain, demonstrating that *M. grandii* could be infected with *N. furnacalis*. Subsequently, a chronically infected *M. grandii* colony was established, confirming that *N. furnacalis* can infect *M. grandii*. Five consecutive generations of infected *M. grandii* were maintained on uninfected European corn borers in the laboratory. All parasitoid larvae from a chronically infected female who oviposited in an uninfected European corn borer larva were infected with *N. furnacalis*. Chronically infected *M. grandii* larvae appeared to give rise to less fecund females and as adults these parasitoids were shorter-lived than uninfected *M. grandii*, but these data need to be confined in additional experiments.

Not all interactions between a microsporidian infected host and its natural enemy complex are negative. Interactions can be neutral and result in no measurable effect on the natural enemy (Allen, 1954; Lipa and Hokkanen, 1992). However, predators and parasites can frequently act as mechanical vectors of microsporidia (Maddox *et al.*, 1992), spreading the pathogen to larvae that otherwise would have remained uninfected. In experiments with mechanical transmission of *N. furnacalis* by *M. grandii*, uninfected *M. grandii* were allowed access to *N. furnacalis*-infected European corn borers for 24 hours. These *M. grandii* females were removed and given access to uninfected European corn borers for 24 hours and 30% of these larvae became infected with *N. furnacalis*. Since it was demonstrated that *M. grandii* can mechanically vector and transovarially transmit *N. furnacalis*, this exotic microsporidium may spread rapidly through the North American European corn borer population.

17.5 CONCLUSIONS

The ELISA detects only active *N. furnacalis* infection because it reacts to sporoplasm rather than outer spore coat proteins. Using the ELISA, we demonstrated that the parasitoid *M. grandii* was likely infected with *N. furnacalis*, but that no generalist predator was actively infected (Oien and Ragsdale, 1993). Subsequent studies with *M. grandii* confirmed that this parasitoid can indeed be infected with *N. furnacalis*. Although recent experimental evidence indicates that *N. furnacalis* does adversely affect *M. grandii*, the extent to which the protozoan reduces fitness of *M. grandii* has not yet been determined. Given that we were able to maintain a chronically infected *M. grandii* colony for five generations suggests that detrimental effects of *N. furnacalis* are not devastating to *M. grandii*. Moreover, since the latter can transovarially and mechanically transmit *N. furnacalis*, the parasitoid may accelerate the spread of this exotic pathogen. We know that parasitism in Minnesota by *M. grandii* is generally under 4%, while incidence of *N. pyrausta* exceeds 40%. These data suggest that microsporidia are far more important than parasitoids as natural control agents of the European corn borer in the mid-western US. Thus, even if *N. furnacalis* caused a decrease in *M. grandii* abundance, we would not expect to see a significant increase in European corn borer density.

Altogether, *N. furnacalis* appears to be a promising candidate biological control agent for the European corn borer. Ultimately, the question we need to address is whether *N. furnacalis* should be released. The ELISA developed provides the necessary diagnostic tool to follow the environmental fate of *N. furnacalis* in both the European corn borer and non-target hosts. Clearly, further work on effects of *N. furnacalis* on non-target hosts and characterization of *N. furnacalis* biology in the European corn borer is needed to guide a decision to release the microsporidium in North America.

REFERENCES

Allen, H.W. (1954) Nosema disease of *Gnorimoschema operculella* (Zeller) and *Macrocentrus ancylivorus* Rohwer. *Annals of the Entomological Society of America*, **47**, 407–24.

Andow, D.A. (1990) Characterization of predation on egg masses of *Ostrinia nubilalis* (Lepidoptera: Pyralidae). *Annals of the Entomological Society of America*, **83**, 482–6.

Andreadis, T.G. (1980) *Nosema pyrausta* infection in *Macrocentrus grandii*, a braconid parasite of the European corn borer, *Ostrinia nubilalis*. *Journal of Invertebrate Pathology*, **35**, 229–33.

Andreadis, T.G. (1982) Impact of *Nosema pyrausta* on field populations of *Macrocentrus grandii*, an introduced parasite of the European corn borer, *Ostrinia nubilalis*. *Journal of Invertebrate Pathology*, **39**, 298–302.

Andreadis, T.G. (1986). Dissemination of *Nosema pyrausta* in feral populations of the European corn borer, *Ostrinia nubilalis*. *Journal of Invertebrate Pathology*, **48**, 335–43.

Andreadis, T.G. (1987) Horizontal transmission of *Nosema pyrausta* (Microsporida Nosematidae) in the European corn borer, *Ostrinia nubilalis* (Lepidoptera: Pyralidae). *Environmental Entomology*, **16**, 1124–9.

Baker, M.D., Vossbrinck, C.R., Maddox, J.V. and Undeen, L.H. (1994) Phylogenetic relationships among *Vairimorpha* and *Nosema* species (Microspora) based on ribosomal RNA sequence data. *Journal of Invertebrate Pathology*, **64**, 100–6.

Bauer, L.H. and Pankratz, H.S. (1993) *Nosema scripta* N.Sp. (Microsporida: Nosematidae), a microsporidian parasite of the cottonwood leaf beetle, *Chrysomela scripta* (Coleoptera: Chrysomelidae). *Journal of Eukariotic Microbiology*, **40**, 135–41.

Bechnel, J.C. (1992) Safety of *Edhazardia aedis* (Microspora: Amblyosporidae) for nontarget aquatic organisms. *Journal of the American Mosquito Control Association*, **8**, 256–60.

Brooks, W.M. (1973) Protozoa: host–parasite–pathogen interrelationships. *Miscellaneous Publications of the Entomological Society of America*, **9**, 105–11.

Brooks, W.M. and Cranford, J.D. (1972) Microsporidoses of the hymenopterous parasites, *Campoletis sonorensis* and *Cardiochiles nigriceps*, larval parasites of *Heliothis* species. *Journal of Invertebrate Pathology*, **20**, 77–94.

Cossentine, J.E. and Lewis, L.C. (1987) Development of *Macrocentrus grandii* Goidanich within microsporidian-infected *Ostrinia nubilalis* (Hübner) host larvae. *Canadian Journal of Zoology*, **65**, 2532–5.

Cossentine, J.E. and Lewis, L.C. (1988) Impact of *Nosema pyrausta, Nosema* sp., and a nuclear polyhedrosis virus on *Lydella thompsoni* within infected *Ostrinia nubilalis* hosts. *Journal of Invertebrate Pathology*, **51**, 126–32.

Fowler, J.L. and Reeves, E.L. (1974a) Detection of relationships among microsporidian isolates by electrophoretic analysis: hydrophobic extracts. *Journal of Invertebrate Pathology*, **23**, 3–12.

Fowler, J.L. and Reeves, E.L. (1974b) Detection of relationships among microsporidian isolates by electrophoretic analysis: hydrophilic extracts. *Journal of Invertebrate Pathology*, **23**, 63–9.

Fuxa, J.R. (1990) New directions for insect control with baculoviruses, in *New Directions in Biological Control: Alternatives for Suppressing Agricultural Pests and Diseases*, (eds R.R. Baker and P.E. Dunn), Alan R. Liss, Inc., New York, pp. 97-113.

Greenstone, M.H. (1983) An enzyme-linked immunosorbent assay for the *Amblyospora* sp. of *Culex salinarius* (Microspora: Amblyosporidae). *Journal of Invertebrate Pathology*, **41**, 250–5.

Greenstone, M.H. (1986) The ELISA for *Amblyospora* sp.: reproducibility, sensitivity, and cross-reactivity with other microsporidian species. *Journal of the Kansas Entomological Society*, **59**, 658–65.

Hayasaka, S. and Ayuzawa, C. (1987) Diagnosis of microsporidians, *Nosema bombycis* and closely related species by antibody-sensitized latex. *Journal of Sericultural Science of Japan*, **56**, 169–70.

Henry, J.E. (1990) Control of insects by protozoa, in *New Directions in Biological Control: Alternatives for Suppressing Agricultural Pests and Diseases*, (eds R.R. Baker and P.E. Dunn), Alan R. Liss, Inc., New York, pp. 161–76.

Hill, R.E., and Gary, W.J. (1979) Effects of the microsporidium, *Nosema pyrausta*, on field populations of European corn borers in Nebraska. *Environmental Entomology*, **8**, 91–5.

Huger, A.M. (1984) Susceptibility of the egg parasitoid *Trichogramma evanescens* to the microsporidium *Nosema pyrausta* and its impact on fecundity. *Journal of Invertebrate Pathology*, **44**, 228–9.

Kawarabata, T. and Hayasaka, S. (1987) An enzyme-linked immunosorbent assay to detect alkali-soluble spore surface antigens of strains of *Nosema bombycis* (Microspora: Nosematidae). *Journal of Invertebrate Pathology*, **50**, 118–23.

Ke, Z., Xie, W., Wang, X. *et al.* (1990) A monoclonal antibody to *Nosema bombycis* and its use for identification of microsporidian spores. *Journal of Invertebrate Pathology*, **56**, 395–400.

Knell, J.D. (1975) *The Separation of Species of Microsporidia Using Serology, Microelectrophoresis and Isoelectric Focusing*, PhD Dissertation, University of Florida, Gainesville.

Laigo, F.M. and Tamashiro, M. (1967) Interactions between a microsporidian pathogen of the lawn-armyworm and the hymenopterous parasite *Apanteles marginiventris*. *Journal of Invertebrate Pathology*, **9**, 46–54.

Landis, D.A. and Haas, M.J. (1992) Influence of landscape structure on abundance and within field distribution of European corn borer larval parasitoids in Michigan. *Environmental Entomology*, **21**, 409–16.

Lewis, L.C. (1982) Present status of introduced parasitoids of the European corn borer, *Ostrinia nubilalis* (Hübner), in Iowa. *Iowa State Journal of Research*, **56**, 429–36.

Lewis, L.C. and Cossentine, J.E. (1986) Season long intraplant epizootics of entomopathogens, *Beauveria bassiana* and *Nosema pyrausta*, in a corn agroecosystem. *Entomophaga*, **31**, 363–9.

Lewis, L.C., and Lynch, R.E. (1978) Foliar application of *Nosema pyrausta* for suppression of populations of European corn borer. *Entomophaga*, **23**, 83–8.

Lipa, J.J., and Hokkanen, H.M.T. (1992) Safety of *Nosema meligethi* I. and R. (Microsporida) to *Apis mellifera* L. and *Coccinella septempunctata* L. *Journal of Invertebrate Pathology*, **60**, 310–11.

McLaughlin, R.E. and Adams, C.H. (1966) Infection of *Bracon mellitor* (Hymenoptera: Braconidae) by *Mattesia grandis* (Protozoa: Neogregarinida). *Annals of the Entomological Society of America*, **59**, 800–2.

McManus, M.L., Maddox, J.V., Jeffords, M.R. and Webb, R.E. (1989) Evaluation and selection of candidate European microsporidia for introduction into U.S. gypsy moth populations, in *The Lymantriidae: A Comparison of Features of New and Old World Tussock Moths*, US, Department of Agriculture Forest Service General Technical Report NE-125, Broomall, PA., pp. 455–68.

MacVean, C.M. and Capinera, J.L. (1992) Field evaluation of two microsporidian pathogens, an entomophagous nematode, and carbaryl suppression of the Mormon cricket, *Anabrus simplex* Hald. (Orthoptera: Tettigoniidae). *Biological Control*, **2**, 59–65.

Maddox, J.V. (1973) The persistence of the microsporidia in the environment. *Miscellaneous Publications of the Entomological Society of America*, **9**, 99–104.

Maddox, J.V. (1986) Possibilities for manipulating epizootics caused by Protozoa: a representative case-history of *Nosema pyrausta*, in *Fundamental and Applied Aspects of Invertebrate Pathology*, (eds R.A. Samson, J.M. Vlak and D. Peters), Wageningen, The Netherlands, pp. 563–5.

Maddox, J.V., McManus, M.L., Jeffords, M.R. and Webb, R.E. (1992) Exotic insect pathogens as classical biological control agents with an emphasis on regulatory considerations, in *Selection Criteria and Ecological Consequences of Importing Natural Enemies*, (eds W.C. Kauffman and J.R. Nechols), Thomas Say Publications in Entomology, Entomological Society of America, Lanham, Maryland, pp. 27–39.

Malone, L.M. and McIvor, C.A. (1993) Pulsed-field gel electrophoresis of DNA from four microsporidian isolates. *Journal of Invertebrate Pathology*, **61**, 203–5.

Mason, C.E., Romig, R.R., Wendel, L.E. and Wood, L.A. (1994) Distribution and abundance of larval parasitoids of European corn borer (Lepidoptera: Pyralidae) in the East Central United States. *Environmental Entomology*, **23**, 521–31.

Munderloh, U.G., Kurtti, T.J. and Ross, S.E. (1990) Electrophoretic characterization of chromosomal DNA from two microsporidia. *Journal of Invertebrate Pathology*, **56**, 243–8.

Nafus, D.M. and Schreiner, I.H. (1991) Review of the biology and control of the Asian corn borer, *Ostrinia furnacalis* (Lep: Pyralidae). *Tropical Pest Management*, **37**, 41–56.

Niederkorn, J.Y., Shadduck, J.A. and Weidner, E. (1980) Antigenic cross-reactivity among different microsporidian spores as determined by immunofluorescence. *Journal of Parasitology*, **66**, 675–7.

Noetzel, D.M., Cutkomp, L.K. and Harein, P.K. (1985) Estimated annual losses due to insects in Minnesota 1981–1983. University of Minnesota Agricultural Bulletin, **254**.

Oien, C.T. and Ragsdale, D.W. (1992) A species-specific enzyme-linked immunosorbent assay for *Nosema furnacalis* (Microspora: Nosematidae). *Journal of Invertebrate Pathology*, **60**, 84–8.

Oien, C.T. and Ragsdale, D.W. (1993) Susceptibility of nontarget hosts to *Nosema furnacalis* (Microsporida: Nosematidae), a potential biological control agent of the European corn borer, *Ostrinia nubilalis* (Lepidoptera: Pyralidae). *Biological Control*, **3**, 323–8.

Onstad, D.W. and Maddox, J.V. (1989) Modeling the effects of the microsporidium, *Nosema pyrausta*, on the population dynamics of the insect, *Ostrinia nubilalis*. *Journal of Invertebrate Pathology*, **53**, 410–21.

Onstad, D.W. and Ruesink, W.G. (1986) Simulation model of the epizootiology of a microsporidium infecting an insect, in *Proceedings, 4th International Colloquium of Invertebrate Pathology*, Veldhoven, The Netherlands, pp. 576–9.

Orr, D.B., Lewis, L.C. and Obrycki, J.J. (1994) Behavior and survival in corn plants of *Ostrinia nubilalis* (Lepidoptera: Pyralidae) larvae when infected with

Nosema pyrausta (Microspora: Nosematidae) and parasitized by *Macrocentrus grandii* (Hymenoptera: Braconidae). *Environmental Entomology*, **23**, 1020-4.

Own, O.S. and Brooks, W.M. (1986) Interactions of the parasite *Pediobius foveolatus* (Hymenoptera: Eulophidae) with two *Nosema* spp. (Microsporida: Nosematidae) of the Mexican bean beetle (Coleoptera: Coccinellidae). *Environmental Entomology*, **15**, 32–9.

Sajap, A.S. and Lewis, L.C. (1988) Effects of the microsporidium *Nosema pyrausta* (Microsporida: Nosematidae) on the egg parasitoid, *Trichogramma nubilale* (Hymenoptera: Trichogrammatidae). *Journal of Invertebrate Pathology*, **52**, 294–300.

Sajap, A.S. and Lewis, L.C. (1989) Impact of *Nosema pyrausta* (Microsporida: Nosematidae) on a predator, *Chrysoperla carnea* (Neuroptera: Chrysopidae). *Environmental Entomology*, **18**, 172–6.

Showers, W.B., Witkowski, J.F., Mason, C.E. *et al.* (1989) European corn borer development and management. *North Central Regional Publication No. 327*, Iowa State University, Ames.

Siegel, J.P., Maddox, J.V. and Ruesink, W.G. (1986) Impact of *Nosema pyrausta* on a braconid, *Macrocentrus grandii*, in central Illinois. *Journal of Invertebrate Pathology*, **47**, 271–6.

Siegel, J.P., Maddox, J.V. and Ruesink, W.G. (1988) Seasonal progress of *Nosema pyrausta* in the European corn borer, *Ostrinia nubilalis*. *Journal of Invertebrate Pathology*, **52**, 130–6.

Sparks, A.N., Chiang, H.C., Burkhardt, C.C. *et al.* (1966) Evaluation of the influence of predation on corn borer populations. *Journal of Economic Entomology*, **59**, 104–7.

Sreenivasam, D., Luhman, J., Wilson, N., and Gagne, R. (1993) *Plant Pest Survey, Detection and Biological Control Program 1993 Activity Summary*, Minnesota Department of Agriculture, Plant Protection Division.

Street, D.A. and Briggs, J.D. (1982). An evaluation of sodium dodecyl sulfate–polyacrylamide gel electrophoresis for the identification of microsporidia. *Journal of Invertebrate Pathology*, **40**, 159–65.

Thomson, H.M. (1958) The effect of a microsporidian parasite of the spruce budworm *Chorisoneura fumiferana* (Clem.), on two internal hymenopterous parasites. *Canadian Entomologist*, **90**, 694–5.

Van Essen, F.W. and Anthony, D.W. (1976) Susceptibility of nontarget organisms to *Nosema algerae* (Microsporida: Nosematidae), a parasite of mosquitoes. *Journal of Invertebrate Pathology*, **28**, 77–85.

Vossbrinck, C.R., Baker M.D., Didier, E.S. *et al.* (1993) Ribosomal DNA sequences of *Encephalitozoon hellem* and *Encephalitozoon cuniculi*: species identification and phylogenetic construction. *Journal of Eukariotic Microbiology*, **10**, 354–62.

Wen, J. and Sun, C. (1988) Two new species of *Nosema* (Microspora: Nosematidae). *Acta Zootaxonomica Sinica*, **13**, 105–11.

Winnie, W.V., Sreenivasam, D.D. and Chiang, H.C. (1981) European corn borer parasitoids; distribution in southern Minnesota. *Journal of the Minnesota Academy of Science*, **47**, 3–5.

Zimmack, H.L., Arbuthnot, K.D. and Brindley, T.A. (1954) Distribution of the European corn borer parasite *Perezia pyraustae*, and its effect on the host. *Journal of Economic Entomology*, **47**, 641–5.

18

Progress in quantifying predation using antibody techniques

K. D. Sunderland

ABSTRACT

The majority of early research, employing biochemical approaches for estimating predation of agricultural pests, simply used the proportion of positive results in the precipitin test as an index of predation. This approach was gradually improved by incorporating estimates of predator density into predation indices and by allowing for the temperature-dependent rates of antigen digestion characteristic of each species of predator. Further advances came with the use of ELISA, which enabled estimation of the biomass of prey in the gut of an individual predator, and with the development of monoclonal antibodies, which greatly increased the specificity of prey detection in predator guts.

The above approach, in common with other indirect or post-mortem techniques (such as electrophoresis), estimates the **consumption rate** of prey by predators, rather than the **predation rate**. It is necessary to make this distinction because prey which have not been killed by the predator can, nevertheless, be consumed by it. Conversely, the predator may kill prey but fail to ingest any prey material. Thus consumption and predation are not always synonymous. A knowledge of consumption rates is useful in energetics studies, but information on predation rates is required in investigations of pest control by predators. To convert consumption rates to predation rates (or simply to obtain a crude estimate of the extent to which the measured consumption rates are reliable indices of predation rates) requires, potentially, a large amount of additional information about the specific predator–prey system under investigation. Examples of the extra information required include: (i) the propor-

The Ecology of Agricultural Pests, Edited by W.O.C. Symondson and J.E. Liddell.
Published in 1996 by Chapman & Hall, London. ISBN 0 412 62190 8

tion of the prey biomass detected per predator that is due to ingestion of dead prey (scavenging), and ingestion of moribund prey (in the early stages of parasitism and pathogenesis); (ii) the incidence of secondary predation, which results in mis-identification of the primary consumer of the prey (food chain errors); (iii) the extent to which there is pre-oral digestion, partial prey consumption, killing without ingestion (wasteful killing), and wounding of prey that eventually results in death; and (iv) prey size availability in relation to the prey size preference of the predator. Each of these factors may vary according to predator species, hunger level of the predator, prey species, season, crop type and location within the habitat.

While it is not feasible to collect all of the above information for each predator–prey system under study, it would be extremely valuable to begin to estimate the order of magnitude of the main categories of potential error in predation rate estimates, at least for a few representative predator–prey systems. This would separate factors of negligible significance from potentially important factors, which might then have to be measured routinely in predation studies. Without such investigations, the possibility will remain that many published predation rate estimates are grossly in error, and that predators have been incorrectly ranked in terms of their probable contribution to the control of specific pests.

18.1 INTRODUCTION

There is often a need to measure predation, free from major errors, either to estimate its impact on a pest population or to rank the component species of a predator fauna in terms of their relative value for pest control. Although great technical progress has been made in the quantification of consumption rates (reviewed briefly below), these rates are not necessarily synonymous with predation rates. The extent to which the two rates differ will depend on the relative importance, in any particular predator–prey system, of a number of factors including scavenging, feeding on moribund prey, food chain errors, pre-oral digestion, wounding, wasteful killing, partial consumption and prey size preference and availability. These factors are discussed below and, since many of them are affected by predator–prey distributions and encounter rates, these too are briefly discussed. There is very little quantitative information in the literature on the extent to which the above factors can cause error in predation rate estimates derived from antibody techniques (and the methodology to do this is largely undeveloped), and so the emphasis here is on an exploration of the biology and ecology of predators and pests in an attempt to determine in what contexts major errors might be occurring.

Although the comments here are directed towards antibody techniques, they apply equally to other indirect or post-mortem methods (Sunderland, 1987) such as electrophoresis, gut dissection and radiotracers. The quantification of prey consumption is much easier to achieve than the quantification of predation. Consumption rates need to be measured in energetics studies and can also be relevant to longer-term predator–prey interactions, such as in perennial crops where a numerical response to prey abundance is more likely. Usually, however, the need is to quantify predation. A functional definition of predation is most relevant to the applied ecologist and relates to 'irreplaceable mortality' (Jones, 1982) of the pest caused by the predator. This would include cases of wounding which will be fatal, but exclude predation of moribund prey (i.e. living prey that have already been attacked by other natural enemies, such as parasitoids and pathogens, and consequently will soon die even if not attacked by the predator). It is nevertheless important to quantify the amount of consumption of moribund prey for three reasons: (i) to remove this error from predation rate estimates; (ii) because moribund prey can still cause plant damage and even reproduce; and (iii) because it determines what analysis of the effect of predation on pest population dynamics is most appropriate (Morris, 1965; Carey, 1989; Barclay, 1992). It can also be of relevance for pest control outside the context of mortality due to predation, e.g. predators feeding on diseased prey can disseminate the pathogen (Abbas and Boucias, 1984; Young and Hamm, 1985; Ruberson *et al.*, 1991; Young and Kring, 1991).

Finally – although not strictly within the scope of this review – it is important to note that the pest control value of predators is not confined to predation. Predators – and even parasitoids – (Van Driesche, 1983) can cause indirect mortality by stimulating the dispersal of pests which may fail to relocate a feeding site. For example, various ladybirds (Coccinellidae), spiders (Araneae), wasps (Hymenoptera) and heteropteran bugs can cause the dispersal of aphids (Homoptera), caterpillars (Lepidoptera) and pentatomid bugs (Heteroptera) (MacLellan, 1973; Roitberg and Myers, 1979; Mansour *et al.*, 1981; Yeargan and Braman, 1986; Riechert, 1992) and the mortality rate of the aphid *Schizaphis graminum* (Theobald) resulting from dislodgement by the ladybird *Coccinella 7-punctata* L, can actually be greater than the effect of direct predation (Obrycki, 1992).

18.2 TECHNICAL PROGRESS IN QUANTIFYING CONSUMPTION

18.2.1 Precipitin test and ELISA

Serological techniques were used in predator–prey studies as long ago as 1946 (Frank, 1979), and to identify blood meals of mosquitoes as early as 1921 (Greenstone, 1996 this volume). Greenstone (1996, this volume) gives a comprehensive account of the evolution of antibody techniques for predation studies. Initially, the percentage of predators positive in the precipitin test was used in an index of predation, but this was found to be biased seriously by variation in rate of antigen deactivation between different taxonomic groups of predator (Sunderland *et al.*, 1987). The prey antigen detection period is now usually incorporated into equations for estimating consumption rate (Sopp *et al.*, 1992). It varies with temperature (West and Eligh, 1952; McIver, 1981; Chiverton, 1987; Sopp and Sunderland, 1989; Hagler and Cohen, 1990; Sopp *et al.*, 1992) and is sometimes affected by meal size (Titova, 1970; McIver, 1981) but not always (Sopp and Sunderland, 1989). Alternative food given during the detection period sometimes affects the duration of the detection period (Scheller, 1984; Symondson and Liddell, 1996, this volume) but it is often unaffected (Titova, 1970; Fichter and Stephen, 1984; Lövei *et al.*, 1987, 1990). It is also unaffected by moulting (Fichter and Stephen, 1981). For many species of predator the detectability of prey antigen undergoes an exponential decay during digestion (Sopp and Sunderland, 1989) and in these cases detectability half-life is considered a more appropriate measure than the maximum detection period (Greenstone and Hunt, 1993; Symondson and Liddell, 1993c). Nearly one-third of predation studies using antibody techniques have incorporated measures of predator density into the predation rate estimates. The density estimates will, however, have been rather inaccurate in some cases because of methodological difficulties in estimating the density of some species of predator in agroecosystems (Sunderland *et al.*, 1995). Since the advent of the precipitin test many other serological techniques, suitable for predation studies, have become available (Sunderland, 1987), but enzyme-linked immunosorbent assay (ELISA) is probably now the most established method (Stuart and Greenstone, 1990). ELISA is a very sensitive test suitable for screening large numbers of predators, but the immunodot assay is an inexpensive alternative that does not require a plate reader (Stuart and Greenstone, 1990).

18.2.2 Monoclonal antibodies (MAbs)

Monoclonal antibodies (MAbs) represent an advance over the use of polyclonal antisera in that they confer much greater specificity and can be produced in almost limitless quantities, enabling reproducibility of assays over time. MAbs do, however, tend to have lower sensitivity and shorter detection periods than polyclonal antisera (Symondson and Liddell, 1996, this volume). Occasionally, a small number of cross-reactions have been experienced (Hagler *et al.*, 1993; Symondson and

Liddell, 1993b), but MAbs have also proved capable of species- and stage-specificity [e.g. to eggs of the bug *Lygus hesperus* Knight (Heteroptera) (Hagler *et al.*, 1991, 1992a), to eggs of *Pectinophora gossypiella* (Saunders) (Lepidoptera) (Hagler and Naranjo, 1992) and to the fifth instar of *Helicoverpa zea* Boddie (Lepidoptera) (Greenstone and Morgan, 1989)]. Three species-specific MAbs have been produced to the pest slug *Deroceras reticulatum* (Müller), together with a 'general mollusc' MAb, which is superior to an equivalent polyclonal because it is non-variable (Symondson and Liddell, 1993d). Antibody engineering is beginning to permit the production of designed antibodies with required characteristics. The process is quicker and less labour-intensive than hybridoma technology and the affinity and specificity can be manipulated (Symondson and Liddell, 1993d; Liddell and Symondson, 1996, this volume).

18.2.3 Quantification of consumption

Winder *et al.* (1994), studying predation of the cereal aphid *Sitobion avenae* (F.), used ELISA data to estimate the predator–prey encounter rate, then obtained an upper estimate of consumption rate by assuming that each predator positive in ELISA had consumed aphids at its 'maximum' (laboratory-measured) consumption rate (the method of Rothschild, 1966). A lower estimate was obtained by assuming that each positive predator had consumed one aphid during the detection period (the method of Dempster, 1960). The true consumption rate was assumed to lie somewhere between the upper and lower estimates. The same approach was used by Sunderland and Sutton (1980) to estimate consumption rates of woodlice in dune grassland. ELISA can be used to quantify the biomass of prey in individual predators. When this is combined with data on predator density and temperature-specific detection periods, it is possible to estimate the rate at which prey biomass is consumed by predators in the field (Sopp *et al.*, 1992).

18.2.4 Size of meal and time since feeding

Quantitative antibody techniques currently do not distinguish between a large meal eaten some time ago and a small meal eaten recently. It would be valuable to be able to separate meal size from time since feeding as this would allow greater accuracy of consumption rate estimation and provide useful additional information on the degree of feeding synchrony between individuals in a population. In theory, MAbs against prey antigens that are labile and refractory during digestion could, because of their different decline rate characteristics, enable separation of meal size from time since feeding (Sunderland, 1987;

Symondson and Liddell, 1993c). Similarly, since the rate of loss of reactivity of prey antigen in predators can be different for monoclonal and polyclonal systems, use of these rates for individual predators could determine the time since feeding and the size of the original meal (Symondson and Liddell, 1993b; Symondson and Liddell, 1996, this volume). However, results may still be difficult to interpret if more than one meal was eaten within the detection period. An alternative tactic would be to use a labile-only system so that the time since feeding could be narrowed down to a few hours. In this case a large sample size would be needed on each occasion, together with a high sampling frequency and great care to operate an unbiased sampling regime, so as to provide an accurate measure of consumption. Care would also be needed in the choice of labile protein to avoid selection of one that disappears during pre-oral digestion (see later). Prey consumption rates by some groups of predator will not be quantified easily by this approach, e.g. web-spiders store surplus semi-digested prey, swathed in silk, for later use (Vollrath, 1979, 1984), so bouts of feeding on a single prey are separated in time. Similarly, predatory mites feeding in prey aggregations sometimes return to the dead prey for repeated feedings (Yao and Chant, 1990).

18.3 SCAVENGING

The problem of serological tests not distinguishing between active predation and scavenging (feeding on dead prey) has long been acknowledged (Fox and MacLellan, 1956). Consumption of carrion has no direct pest control value but might be of some indirect value by (i) contributing to a numerical response (because most scavengers are also predators), and (ii) arresting predators in a prey-rich patch and signalling that there are living prey in the vicinity (e.g. dead prey on the ground might cause predators to climb the vegetation above; dead aphids are more available than living aphids to ground-based predators (Griffiths *et al.*, 1985; Winder *et al.*, 1994) and some predators will search the plant above which aphids have been found (Chiverton, 1988)).

Lövei (1986) suggested that since isoenzyme electrophoresis relies on active enzymes which may alter after death, predation could be separated from carrion feeding by using a combination of electrophoretic and serological tests, i.e. if serological methods detect prey but electrophoresis does not, then carrion feeding may be suspected. This merits further investigation, but each specific system needs testing because some enzymes (e.g. some esterases) are probably too stable for this application (R.A. Murray, personal communication). It is possible that labile proteins disappear quickly by decomposition, in which case the use of MAbs to labile and refractory proteins would enable separation of predation from scavenging, except where very fresh carrion had been consumed.

18.3.1 Predators as facultative scavengers

Various Miridae (Wheeler, 1974), Chrysopidae, Coccinellidae (Wheeler *et al.*, 1968; Wheeler, 1977), Anthocoridae (Smith, 1966) and Carabidae (Powell *et al.*, 1987) are known to eat dead, mummified aphids. Predation of mummies may not give a positive reaction in serological tests if little aphid protein remains, but this should be established at the specificity testing stage of the study.

Sunderland *et al.* (1987) recorded 48 species of predator that would readily feed on dead (unmummified) aphids; some predators even preferred dead to living prey (Sunderland and Sutton, 1980). The carabid beetle *Calosoma sayi* DeJean survived for long periods in the laboratory on an exclusive diet of dead caterpillars (Young, 1984b) suggesting that dietary requirements are met equally from dead or living prey. Although staphylinid beetles are known to scavenge (Evans, 1964), they will not feed on cadavers that have dried out (Evans, 1967). Small carabids will only feed on dead prey with hard exoskeletons if the cadavers have been damaged to expose the soft internal tissues (Mitchell, 1963). There are many records of predators behaving as scavengers, including Carabidae, Diptera, Heteroptera, Formicidae and Dermaptera consuming aphids, weevils, caterpillars and a wide range of other invertebrates (Young, 1984a; Greenstone and Morgan, 1989; Barnes, 1990; Cameron and Reeves, 1990; Symondson and Liddell, 1993a,c). Tod (1973) suspected small species of Carabidae to be scavengers of slugs. Some species of spiders in the families Lycosidae, Salticidae and Thomisidae have been seen to eat dead food (Foelix, 1982; Aitchison, 1984; Wolff, 1986) in spite of the absence of prey movement which might be thought necessary to stimulate feeding (Knost and Rovner, 1975). Predators and non-predators can both be kleptoparasitic in spiders' webs, stealing the spiders' stored food. Examples include scorpion flies (Mecoptera) (which also occasionally kill slugs), harvestmen (Opiliones), scavenger beetles (Thornhill, 1975), wasps, damselflies (Odonata), milichiid flies (Diptera), spiders (Vollrath, 1979), Noctuidae, Miridae, Nabidae and *Arachnocoris* (Heteroptera) (Vollrath, 1984). The potential scale on which scavenging could occur should not be underestimated; Wheeler (1971) recorded 11 species of predator, on alfalfa plants, feeding on flies (*Hylemya* spp.) that had been killed by the entomogenous fungus *Entomophthora muscae* (Cohn). This appeared to be happening on a large scale because 40 mirids (*Plagiognathus politus* Uhler) were seen feeding on these cadavers in just 30 minutes of searching time.

As a first stage in the interpretation of prey consumption rate data, laboratory screening can be used to ascertain which species of predator, positive in antibody tests, are 'obligate scavengers' with respect to a particular prey type (Sunderland and Sutton, 1980; Lesiewicz *et al.*, 1982).

18.3.2 Non-predators feeding on dead prey

Some predominantly herbivorous Miridae frequently eat insect cadavers and aphid mummies on plants (Wheeler, 1974) so they could be mislabelled as predators by antibody techniques. Other suspected non-predators, that registered positive as caterpillar consumers, included Isopoda and Diplopoda (Tod, 1973; Leslie and Boreham, 1981; Dudevoir and Reeves, 1990).

18.3.3 Availability of prey corpses

Winder *et al.* (1994) estimated the number of dead aphids (*Sitobion avenae*) falling to the ground in winter wheat using estimates of the proportion of the aphid population dying due to old age. In Winder's UK study the relative availability of dead aphids was greater than that of living aphids on the ground, because the residence time of live aphids on the ground (estimated from laboratory observations) was short. However, the relative availability of dead aphids on the ground in countries with high daytime summer soil surface temperatures is likely to be far greater because live aphids that fall from the plant die very quickly on the soil surface (e.g. within 10 seconds in Oklahoma, USA; Ruth *et al.*, 1975). The availability of prey corpses will usually increase after a pesticide application. For example, more dead aphids (*Macrosiphum euphorbiae* (Thomas)) fell to ground in potatoes treated with the insecticide demeton-*S*-methyl than in control areas and gut dissection showed that 78% of the carabid *Pterostichus madidus* (F.) had eaten aphids in sprayed areas compared with 23% in unsprayed areas, 1 day after spray application (Dixon and McKinlay, 1992). This effect is not necessarily confined to insecticides, as some fungicides also have insecticidal properties. For example, Heimbach (1990) recovered large numbers of dead insects on the soil surface in a field of winter wheat after application of the fungicide pyrazophos. Other factors augmenting the supply of cadavers include lepidopteran pupation failures (Dempster, 1971) and caterpillars drowned in rainstorms (Dempster, 1967). There is likely to be virtually no carrion available to predators in ant-dominated crops, because ants are supremely efficient scavengers, often operating by day and night (Fellers and Fellers, 1982). Rates of removal of carrion by ants (and to a much lesser extent by earwigs

(Dermaptera), vespid wasps and silphid beetles) are typically more than 80% within 24 hours, and can be up to 100% in less than 1 hour (Sunderland and Sutton, 1980; Seastedt *et al.*, 1981; Fellers and Fellers, 1982; Young, 1984a, 1985; Retana *et al.*, 1991). Very localised scavenging by skylarks and mice (U. Heimbach, personal communication) can introduce a stochastic element into carrion availability.

An index of carrion availability in the field might be obtained by testing obligate scavengers; low values in this trophic group would suggest that prey consumption rates by predators are not seriously inflated by scavenging.

18.3.4 Marking living and dead prey to study predation relative to scavenging

If living and dead prey are marked with different labels, their relative rates of consumption by predators can be studied in laboratory and field. Hagler *et al.* (1992b) marked prey externally, for up to 8 days, by dipping them into rabbit IgG; the label can then be detected using antibody techniques. This would not be an effective method for detecting consumption by fluid-feeding predators, but insects can also be marked internally using rabbit IgG added to an artificial diet (Hagler and Cohen, 1990; Hagler and Naranjo, 1996, this volume). Prey can also be labelled by treating host plants with solutions of rare elements or by adding these to artificial diets. This is reported to be a very sensitive system (Jackson, 1991).

18.4 FEEDING ON MORIBUND PREY

The significance, in pest control terms, of predation on moribund pests (i.e. in the early stages of parasitism and pathogenesis), lies somewhere between that of scavenging and predation on healthy pests. This is because a proportion of moribund pests retain the capacity to damage plants, and even to reproduce (see later). Predators may kill and eat prey that are infected by microsporidia (Protozoa) (Oien and Ragsdale, 1993; Ragsdale and Oien, 1996, this volume), viral pathogens (Young and Kring, 1991), the bacterium *Bacillus thuringiensis* Berl. (Salama *et al.*, 1982, 1991), entomogenous fungi (Wheeler, 1971) and entomopathogenic nematodes (Service, 1973b; Desender and Pollet, 1985), or that are parasitized (Obrycki, 1992). The availability of moribund prey to predators will probably be greater in IPM programmes where parasitoids have been released or microbial pesticides have been applied (Soares *et al.*, 1994). It is probably not uncommon for predators to consume multiply-moribund prey; individual pests can be attacked simultaneously by parasitoids and entomogenous fungi (Powell *et al.*, 1986) and some nuclear polyhedrosis viruses (Vinson and Iwantsch, 1980a), microsporidia (Ragsdale and Oien, 1996, this volume) and *Bacillus thuringiensis* (Salama *et al.*, 1982; Soares *et al.*, 1994) are transmitted via the ovipositor of parasitoids. Consumption of moribund prey in the egg stage may also occur (Rothschild, 1966; Jones, 1982). Predators may consume a higher proportion of moribund prey than the incidence of moribund prey in the habitat would suggest if: (i) their distribution, (in relation to the distribution of predators) is different from that of healthy prey; (ii) their size, development rate and phenology is altered; and (iii) their activity, defences or palatability are modified. These possibilities are discussed later. A corollary of predation on moribund pests is that it will cause underestimation of the potential of parasitoids and pathogens to cause pest mortality.

It would be very difficult to solve this problem with antibody techniques. A battery of MAbs to a wide range of parasitoid and pathogen species would be needed and they would have to be sensitive enough to detect as little as one parasitoid egg or a tiny quantity of pathogen. Previous attempts, using electrophoretic methods (Castanera *et al.*, 1983; Walton *et al.*, 1990) and MAbs (Stuart and Greenstone, 1996, this volume) were confined to well-developed parasitoid larvae. More than one MAb per parasitoid species would be required if the egg, larva and pre-adult of the parasitoid were antigenically different. However, electrophoretic techniques are capable of detecting parasitoid eggs and it might be possible to detect prey, pathogens and parasitoids in individual predators using these techniques (M.G. Solomon, personal communication). The problem of distinguishing consumption of a parasitized prey from consumption of unparasitized prey together with an adult parasitoid (in separate meals) would remain.

Virus-infected sawfly larvae and caterpillars can continue to live for 3 weeks or more (Smirnoff, 1960; Tatchell, 1981). Aphids live for about 9 days after attack by parasitoids (except in cases of incomplete parasitization, where premature death of the parasitoid also kills the host; Stary, 1989) and 4 days after attack by entomogenous fungi (Powell *et al.*, 1986), so continued plant damage is possible. In some host–parasitoid relationships the developing parasitoid has evolved to avoid the host's vital organs for as long as possible and some caterpillars continue feeding until only 24 hours before emergence of the adult parasitoid (Yeargan and Braman, 1986). Parasitism or disease can increase the feeding rate of caterpillars (Hunter and Stoner, 1975), but feeding rate and plant damage are usually reduced (Hopper and King, 1984), sometimes by more than 90% (Tatchell, 1981; Coop and Berry, 1986), and this can be, economically, the most important aspect of pest control (Hill, 1988).

Some parasitoids sterilize the host (Vitzthum, 1933; Service, 1973b;

Smith Trail, 1980), but when adults or higher instars of aphids are parasitized they can usually achieve, before death, about 10–20% of the reproductive output of unparasitized aphids (Stary, 1970; Cate *et al.*, 1977).

18.4.1 Distribution

Some parasitoids tend to concentrate their activity in a limited range of vertical strata (Gardner and Dixon, 1985), so predators searching in these areas are more likely to encounter moribund prey [but parasitoid-searching stratum can also be affected by sublethal pesticide residues (Jiu and Waage, 1990)]. Parasitism can cause caterpillars to leave the vegetation and descend to the ground (Ryan, 1985; Yeargan and Braman, 1986; Nealis and Regniere, 1987), to remain exposed on the vegetation for long periods (Hopper and King, 1984), or to seek higher strata of vegetation (Stamp, 1981). Similarly, virus-infected caterpillars may move down (Smirnoff, 1960), stay longer on the plant (Ohbayashi and Iwabuchi, 1991), or seek the tops of vegetation (Shapiro, 1976), while those attacked by entomogenous fungi may spend more time wandering on the soil surface (Ohbayashi and Iwabuchi, 1991) or climb to the top of vegetation (Shapiro, 1976; Ohbayashi and Iwabuchi, 1991). Caterpillars infected with *B. thuringiensis* are more likely to fall off plants (H.D. Burges, personal communication), as are aphids infected with some species of entomophthoraceous fungi (Hagen and Van den Bosch, 1968). Aphids attacked by certain species of parasitoid typically seek the ground or lower strata of vegetation to mummify (Powell, 1980; Brodeur and McNeil, 1989). All these alterations in vertical distribution could modify the encounter rate with predators, the exact outcome depending on the species and crop concerned. In UK cereals, for example, about two-thirds of 279 species of predator appear to be confined to the ground or lower plant (Sunderland *et al.*, 1986, 1988), so if there is a large-scale movement of moribund prey into this zone, predation rate estimates for ground predators could be significantly affected.

Analogous horizontal effects have also been recorded. Parasitization is greater at the periphery than in the centre of colonies of diprionid sawfly larvae (Hymenoptera) on pine; heteropteran predators also attack on the periphery with the result that 69% of fifth instar larvae attacked by these predators were already parasitized (Tostowaryk, 1971).

18.4.2 Size, development rate and phenology

Parasitoids can accelerate or reduce the development rates of their hosts (Vinson and Iwantsch, 1980b); this is sometimes mediated via viruses or non-paralysing venoms injected by the parasitoid (Shaw, 1981). Reduced development rates (Hunter and Stoner, 1975; Shapiro, 1976; Ryan, 1985) increase exposure time to predators and may also result in the moribund prey attaining a greater maximum weight than healthy prey (Hunter and Stoner, 1975), thus potentially altering its acceptability to some predators. Some caterpillars infected with *B. thuringiensis* are available to predators and parasitoids for 3 days longer than healthy caterpillars (Soares *et al.*, 1994). Effects on phenology, such as early or late emergence of a particular growth stage of a moribund prey (Nealis and Regniere, 1987), may result in predators encountering a greatly altered ratio of moribund to healthy prey, for short periods. Some hymenopteran parasitoids in the genus *Aphidius* can cause an aphid nymph to develop into an aptera rather than an alata (Hagen and Van den Bosch, 1968; Smith Trail, 1980); this suggests that late in the season a high proportion of aphids available to predators could be parasitized because of alate emigration in response to poor plant quality.

18.4.3 Activity, defences and palatibility

If parasitism, disease or other factors affect the activity, defensive behaviour or palatability of the prey, this could result in an altered predation rate. Parasitism is known to reduce the activity level of some caterpillars (Hopper and King, 1984) and infection by Protozoa, entomogenous fungi and entomopathogenic nematodes can have the same effect on mosquito larvae (Service, 1973b, 1975). Caterpillars infected with *B. thuringiensis* are less active than healthy caterpillars, which could affect attack rates by predators (Soares *et al.*, 1994). Parasitized (Morris, 1963) and virus-infected (Young and Kring, 1991) caterpillars are less well able to defend themselves from some predators, but other predators attack healthy and moribund caterpillars with equal success (Abbas and Boucias, 1984). There can be complex interactions between prey size, activity and defence and vulnerability to predation (Iwao and Wellington, 1970). Invertebrates suffering sublethal pesticide poisoning can be in semi-paralysis for up to 4 days, and then make a full recovery (i.e. they are temporarily moribund), if protected from predation and desiccation in the laboratory. Such poisoned invertebrates, however, have little or no defence against predators and suffer increased mortality from predation (Everts *et al.*, 1991). Prey that have become injured (e.g. after a non-fatal attack by a predator or parasitoid – see 'Wounding', p. 434) can be killed and eaten by other species of predator, such as some staphylinids (Evans, 1964, 1967) and carabids (Forsythe, 1982) that are unable to kill the healthy prey.

Host–parasitoid co-evolution involves host-driven and parasitoid-driven modifications of host behaviour (including availability to predators). In cases where parasitoid-driven modifications are dominant, it is likely that exposure to predation will be reduced to increase the probability of parasitoid survival (Smith Trail, 1980; Fritz, 1982). In cases where host-driven modifications are dominant 'host suicide' may occur. McAllister *et al.* (1990) provide an example of host suicide in pea aphids attacked by the hymenopteran parasitoid *Aphidius ervi* Haliday. If second instars are attacked they will be unable to reproduce before death and their behaviour (e.g. readiness to fall off the plant in response to alarm pheromone) results in a greater risk of predation compared with the case of parasitized fourth instars, which are capable of some reproduction before death, and which exhibit behaviour that minimizes predation risk. The higher mortality rate of non-reproducing second instars reduces the chances of kin being attacked by parasitoids. There are theoretical reasons for considering that host suicide could also be common in Lepidoptera (Smith Trail, 1980), and Shapiro (1976) claims an example in the Nymphalidae (Lepidoptera).

Fritz (1982) quotes various examples of moribund invertebrate prey being preferred or rejected by vertebrate predators, but little information seems to be available for invertebrate predators.

18.5 HOST-FEEDING BY PARASITOIDS

Antibody techniques could register consumption of pest antigens by parasitoids, because many adult parasitoids consume some of the haemolymph or tissues of the host, sometimes killing them in the process (i.e. they can act as predators). Antibody techniques do not, however, seem to have been used, so far, in this context (Jervis *et al.*, 1992). Care would be needed in devising sampling protocols for collecting material for testing with antibody techniques because some species of parasitoid exhibit distinct diel cycles of host-feeding (Jervis and Kidd, 1986). Also, antibody techniques would not distinguish between destructive and non-destructive host-feeding.

Host-feeding can be non-destructive (associated with parasitoid oviposition), or destructive (with no oviposition and immediate or delayed host death) (Jervis and Kidd, 1986). It is estimated that about one-third of the world's parasitoid fauna (*c.* 100 000 species) are host-feeders (Kidd and Jervis, 1989) and a significant proportion of these are destructive host-feeders (M. Jervis, personal communication); this involves 20 families of Hymenoptera (Bartlett, 1964) including most species in the Aphelinidae, which are widely used in biocontrol (Hagen and Van den Bosch, 1968; Jervis and Kidd, 1991), and some Tachinidae

(Diptera) (Nettles, 1987). The Lepidoptera, Homoptera and Diptera are among the most frequently attacked (Bartlett, 1964; M. Jervis, personal communication). It can be a significant cause of pest mortality; for example, the hymenopteran parasitoid *Aphelinus asychis* (Walker) killed 1.5 aphids (*Schizaphis graminum*) per parasitoid per day by destructive host-feeding (Cate *et al.*, 1973). In general, host-feeding is responsible for 11–50% of parasitoid-induced mortality (Van Driesche, 1983), with the early instars of hosts being the main target (Flanders, 1953) but with non-hosts also being attacked (Jervis and Kidd, 1986). The behaviour of parasitoids as predators (by destructive host-feeding) also renders them more effective as parasitoids, because it increases their fecundity (Sahragard *et al.*, 1991; Sunderland *et al.*, 1996).

18.6 FOOD CHAIN ERRORS

Initial laboratory screening of an antiserum should establish that there are no cross-reactions with what the pest has been feeding on (e.g. plant extracts) or with pest by-products, such as honeydew or cast exuviae (Crook and Sunderland, 1984; Chiverton, 1987), that might otherwise subsequently cause misinterpretation of field results.

The 'secondary predation' problem (i.e. that the true predator may be mis-identified when a predator consumes another predator that has eaten the prey; Sopp *et al.*, 1992) has been recognized for many years (Davies, 1953; Fox and MacLellan, 1956). Instances of predators killing and consuming other predators ('hyperpredation'; Sabelis, 1992) are extremely common in the complex food matrices of agroecosytems (see later). Many other trophic interactions, that would result in misinterpretation of test results, are possible. Some examples are: (i) scavenging on dead predators that have eaten the prey: (ii) predation on scavengers that have eaten dead prey, (iii) scavenging on scavengers, and so on. It is difficult to estimate the scale on which such interactions might be occurring. A suspected case of predation (by beetles) on scavengers (ants) that had eaten the target prey (woodlice) was reported by Sunderland and Sutton (1980).

A possible approach to avoiding food chain errors would be to adjust aspects of the test system (e.g. sensitivity, dilution) to reduce greatly the probability of detecting consumption of target prey in distant trophic levels. The attendant penalty would be loss of the capacity to detect some cases of true predation (e.g. for very small predators or predators that have fast digestion rates). There are few examples in the literature of antisera being assessed for sensitivity in this context, but the antiserum of Tod (1973) did appear to be capable of detecting hyperpredation, while that of Dennison and Hodkinson (1983) did not.

18.6.1 Hyperpredation

The literature is replete with records of hyperpredation, which suggests that food chain errors might be common in predation studies which utilize antibody techniques (or other indirect, post-mortem methods). Spiders have been observed to kill other spiders, ants, ladybirds, lacewings (Neuroptera) and nabid, mirid and anthocorid bugs (Wheeler, 1973, 1974; Foelix, 1982; Vollrath, 1984; Hagley and Allen, 1989; Young, 1989; Nyffeler *et al.*, 1992). Carabid beetles will kill hoverfly larvea (Diptera), ladybird larvae and spiders (Chiverton, 1984; Pollet and Desender, 1987; Everts *et al.*, 1991; Sunderland *et al.*, 1994), while staphylinid beetles kill carabids, spiders, earwigs and other staphylinids (Good and Giller, 1988; Sunderland *et al.*, 1994). Nabid bugs can prey on spiders and mirid bugs (Wheeler, 1974; Sunderland *et al.*, 1994), dragonflies (Odonata) and harvestmen kill spiders (Adams, 1984; Young and Lockley, 1988), stigmaeid mites kill phytoseiid mites (Croft and MacRae, 1992), centipedes (Chilopoda) kill adult and larval staphylinids (Poser, 1988) and there are species of ant that specialize in eating spider eggs and centipedes (Holldobler and Wilson, 1990). There are even records of adult parasitoids killing predators by predation rather than by parasitism (Bartlett, 1964). Studies of the pest control implications of hyperpredation are in their infancy (Moran and Hurd, 1994; Sunderland *et al.*, 1994), but Rosenheim *et al.* (1993) demonstrated that hyperpredation on chrysopid larvae by heteropteran bugs in cotton caused pest problems by releasing aphids (*Aphis gossypii* Glover) from lacewing control.

18.7 PRE-ORAL DIGESTION, WOUNDING, WASTEFUL KILLING AND PARTIAL CONSUMPTION

The use of antibody techniques to provide indices or quantification of predation relies on the assumption that some pest antigens are ingested at the time of predation. However, for various physiological or behavioural reasons, significant levels of predation can sometimes be accompanied by little or no consumption of detectable antigen; some cases are discussed below.

18.7.1 Pre-oral digestion

Predators that exhibit pre-oral digestion (also called extra-oral or extra-intestinal digestion), in its extreme form, vomit extra-intestinal midgut fluid, containing proteases, onto the prey to achieve partial or complete digestion of food before final ingestion. This behaviour, or attenuated versions of it, has been recorded in all carabid larvae, some cantharid larvae (Coleoptera), some staphylinid larvae (where there can be associated structural modifications of the mouthparts; Good and Giller, 1988), many adult staphylinids, and some adult carabids (Evans, 1964, 1967; Hengeveld, 1980; Forsythe, 1982; Vaje and Mossakowski, 1986; Good and Giller, 1988). Spiders often alternately suck and then relax the sucking muscles when feeding, so that fluid oscillates between the prey and the spider's gut (Pollard, 1990); the protein carried in these oscillating digestive fluids can be as much as 3.5% of the spider's total body protein (Riechert and Harp, 1987). Adult Asilidae (Diptera) secrete a fluid from the labial glands which dissolves prey tissues (Hobby, 1931).

Sometimes negative results are recorded in serological and electrophoretic tests even when the predator is tested immediately after having been observed to consume the prey. This occurs with beetle larvae, staphylinid adults and elaterid and silphid beetle predators, and pre-oral digestion is a possible explanation (Tod, 1973; Doane *et al.*, 1985; Sunderland *et al.*, 1987; Good and Giller, 1988).

Most cases of pre-oral digestion are likely to be discovered and screened out during the measurement of detection periods in the laboratory. It should be noted, however, that the experimental protocol adopted for measuring detection periods (e.g. whether the predators are starved or well fed before the main detection period feed) can influence the intensity of pre-oral digestion, and the direction of influence will vary according to the species of predator under test. Enzyme production increases with feeding rate in fluid-feeders (Lucas and Grafen, 1985), but the reverse is true for some adult carabids (Vaje and Mossakowski, 1986).

18.7.2 Wounding

Attacks by predators and parasitoids that cause fatal wounding of pests are not accompanied by ingestion of pest antigens, but they do, nevertheless, make a legitimate contribution to pest control. Larvae of predatory Muscidae (Diptera) are known to wound leatherjackets (Griffiths *et al.*, 1984), and the integuments of soldier fly larvae (Diptera) can be punctured by the mandibles of predatory wireworm larvae (Coleoptera); 6% found injured in this way were expected to die due to haemolymph loss and infection (Doane *et al.*, 1985). The majority of rice planthoppers (Hemiptera) wounded by dryinid parasitoids (Hymenoptera) die within 3 days (Kitamura, 1982), but most species of host wounded by parasitoids die within a few hours (Jervis *et al.*, 1992). Stung aphids (*Schizaphis graminum*) that moved before the parasitoid *Aphelinus asychis* could host-feed on them became paralysed, swelled and were dead within a few hours (Cate *et al.*, 1977). There are many records of pests being killed by parasitoids without feeding or successful parasitism (Jervis and Kidd, 1986). This 'host mutilation' (Bartlett, 1964), which sometimes involves injection of venom (Hagen and Van den Bosch, 1968; Kidd and Jervis, 1989) can also result from repeated probing by the ovipositor of the parasitoid, either in an attempt to oviposit in an instar too small for successful parasitism (Rahman, 1970; Jervis *et al.*, 1992), or in an unsuccessful attempt to induce exudation of host haemolymph for host-feeding (Bartlett, 1964).

18.7.3 Wasteful killing

Sometimes a predator will kill more prey than it consumes. This is usually, but not always, associated with a satiated predator in an aggregation of prey. One hypothesis to explain this behaviour is that the foregut of the predator becomes full before the midgut, but the midgut is the seat of hunger and until food enters the midgut to suppress hunger the predator continues killing, but is physically unable to ingest more food because the foregut is full (Johnson *et al.*, 1975). Wasteful killing has been observed, or inferred, for spiders attacking aphids and caterpillars (Fichter and Stephen, 1984; Mansour and Heimbach, 1993; Samu and Bíró, 1993), carabids attacking dipteran larvae (Allen and Hagley, 1990), heteropteran bugs attacking caterpillars (Fichter and Stephen, 1981) and phytoseiid mites attacking tetranychid mites (Sandness and McMurtry, 1972). When attacking egg masses of chrysomelid beetles, earwigs and staphylinids ate some eggs and damaged the rest (Sotherton *et al.*, 1984).

18.7.4 Partial consumption

Partial consumption is defined by Ernsting and Van der Werf (1988) as occurring when 'a prey is abandoned before all its ingestible biomass is consumed'. This presents no difficulties in cases where antibody techniques are used to provide an index of predation, but, if quantification of predation is being attempted, a variable degree of partial consumption will render interpretation of results problematical.

Partial consumption often occurs as the predator is approaching satiation, and this has been recorded in spiders (Haynes and Sisojevic, 1966; Samu, 1993) and scathophagid flies (Diptera) (Failes *et al.*, 1992). It has also been associated with large prey size, for adult carabids (Ernsting and Van der Werf, 1988; Weseloh, 1988; Wallin, 1991) and staphylinid larvae (Vickerman *et al.*, 1988), with physical diffulties in extracting the last 30% of a prey, in hoverfly larvae (Barlow and Whittingham, 1986) and with mutual interference during conspecific prey sharing by phytoseiid mites (Yao and Chant, 1990). It is considered to be a typical property of predators foraging in areas of high prey density (Sabelis, 1992).

18.7.5 Assessing hunger level

Since the hunger level of predators can influence the degree of pre-oral digestion, wasteful killing and partial consumption (and some of the other error-generating factors mentioned in this review), it would be useful to have a simple assay of hunger level in the field, in order to assess the probability of error in predation rate estimates arising from these factors. Possible approaches to assaying hunger level include measurement of biomass gain by individual predators brought in from the field and given food under standard conditions in the laboratory (Makowski and Szyszko, 1986), and weight of faeces produced under standard laboratory conditions during the first 24 hours ex-field (Honek, 1986). The nutritional status of field populations of spiders can be inferred from mean intermoult periods and growth ratios (from length measurements of exuviae), but this method is only applicable to habitats with fairly stable environmental conditions and large web-building species with high web site tenacities (Vollrath, 1988).

18.8 PREY SIZE PREFERENCE

Where antibody techniques are used to provide indices of predation (e.g. for ranking the relative value of predator species), measurement of the size distribution of pests killed is not crucial. It does, however, become necessary if estimation of the predation rate and the impact of predators on the pest population is the goal. Models of predator–pest interactions suggest that pest age distribution can be a very important variable (Baumgaertner *et al.*, 1981). Occasionally, a predator shows no size preference in relation to a specific prey species (Loughridge and Luff, 1983; Raworth, 1984), but usually a limited size range is preferred. The size distribution of pests killed will be influenced by: (i) the relative distribution, density and behaviour of predators in relation to that of the various pest instars (see next section); and (ii) prey size preference by the predator, which has both structural and behavioural aspects (see later).

A general knowledge of the factors affecting prey size selection could be valuable for devising appropriate experimental protocols. Nevertheless, prey size preference data for use in serological studies of predation must be obtained for each predator–prey combination in any specific study (and such observations are usually made in the laboratory; e.g. MacLellan, 1973; Ragsdale *et al.*, 1981; Young, 1984b; Weseloh, 1988; Cherrill and Begon, 1989; Hagler and Naranjo, 1996, this volume).

It is often convenient to express prey size preference in terms of a frequency distribution of instar size consumed, but it should be noted that biomass within an instar can be very variable (e.g. four-fold in some aphids; Raworth, 1984). The majority of predators eat prey of a size similar to or smaller than that of their own body. Maximum prey size increases with size of predator but the minimum size increases very little (Sabelis, 1992). In the case of spiders, Nentwig and Wissel (1986) found that prey 50–80% of the spider's size had the highest acceptance rates, and, not surprisingly, web-builders were superior to non-web-builders in catching large prey.

18.8.1 Prey size preference: structural aspects

There are positive correlations between size of prey taken and mandible gape in the carabid tribe Pterostichini (Wheater, 1988), size of chelicerae in mesostigmatid mites (Buryn and Brandl, 1992) and leg geometry in some other mites (Usher and Bowring, 1984), but these correlations tend to be weak. Long legs and the habit of wrapping prey in silk enable some spiders to prey on large, dangerous prey (Riechert, 1992). For some carabids, ladybird larvae and spiders, capture efficiency declines with increase in prey size, within any one species of prey (Baumgaertner *et al.*, 1981; Sunderland *et al.*, 1986; Young and Lockley, 1986; Ernsting and Van der Werf, 1988).

18.8.2 Prey size preference: behavioural aspects

Some predators seem unable to detect very small prey (e.g. large ladybird larvae in the presence of first instar aphids; Frazer *et al.*, 1981). In cases where small instars are preferred, there can nevertheless be a switch to larger instars at low prey density, in agreement with optimal foraging theory (Ernsting and Mulder, 1981). Sociality in spiders and ants can enable prey to be taken that are larger than the predators themselves (Rypstra and Tirey, 1991; Traniello and Beshers, 1991). Late instars of some colonial caterpillars are more vulnerable to predation because in that stage the effectiveness of colonial defence is weakened by movement on the tree in the search for food.

18.9 FACTORS AFFECTING THE DISTRIBUTION OF PREDATOR AND PREY AND THEIR ENCOUNTER RATES

The distribution of predators and prey and their encounter rates have an important influence on most of those error-generating factors (scavenging, feeding on moribund prey, food chain errors, pre-oral digestion, wounding, wasteful killing, partial consumption, prey size preference), discussed earlier, which are likely to be encountered in predation studies using antibody techniques. For this reason, some major factors affecting distribution and encounter are discussed briefly below. Relevant aspects of distribution include vertical components (crop, weeds, ground surface, fissures in ground), horizontal components (e.g. prey patchiness, predator aggregative responses) and temporal components (e.g. diel cycles of vertical migration). Other factors affecting encounter rates include temperature, activity level of prey, predator searching behaviour and the size of the predator searching universe.

18.9.1 Vertical distribution of predator and prey

Vertical stratification of predators in crops is well documented (Pasquet, 1984; Sunderland *el al.*, 1986, 1988; Provencher *et al.*, 1988). Prey species are also often stratified vertically, but the stratum occupied can be affected by factors such as plant disease (Bohnke and Zebitz, 1990) or pesticide residues (Mack and Smilowitz, 1981). Seasonal changes in prey stratification may bring prey into contact with stratified predator populations (Rothschild, 1966; Ferran *et al.*, 1991). Herbivores often leave the food plant voluntarily (Sunderland *et al.*, 1986; Sopp *et al.*, 1987; Holmes, 1988) and the rate of leaving is increased on plant species exhibiting antixenotic resistance (Sunderland *et al.*, 1988; Bhuiyan and Wratten, 1994). Some caterpillars are only available to ground predators when pre-pupae leave the plants to enter the soil for pupation (Lesiewicz *et al.*, 1982). Prey can be flushed off the plant by searching predators (MacLellan, 1973; Roitberg and Myers, 1979; Frazer *et al.*, 1981; Mansour *et al.*, 1981; Yeargan and Braman, 1986; Chiverton, 1988; Obrycki, 1992; Riechert, 1992) and parasitoids (Tamaki *et al.*, 1970; Ruth *et al.*, 1975). Many will subsequently succeed in regaining the plant, if they are not killed by ground-active predators (Winder, 1990; Winder *et al.*, 1994). Climbing carabid and staphylinid beetles (Price and Shepard, 1978; Griffiths *et al.*, 1985; Kennedy *et al.*, 1986) can pursue them on the plant. Climbing frequency of carabids is affected by temperature (Loughridge and Luff 1983), degree of hunger (Wallin, 1991) and previous success at finding prey on the ground below the plant (Chiverton, 1988).

Many predators and prey in agroecosystems are known to make diel vertical migrations (Archer and Musick, 1976; Holmes, 1984; Griswold and Trumble, 1985; Kennedy *et al.*, 1986; Dudevoir and Reeves, 1990). Some marine copepods descend at night to escape from predators that are ascending; the disadvantage of having to make a daily migration is overcome if predation is reduced by as little as 16% (Ohman *et al.*, 1983). It is possible that a similar reverse diel migration phenomenon is occurring in agroecosystems, since many predators ascend crop plants at night and there are more hemipterous herbivores on the plants during the daytime (Vickerman and Sunderland, 1975; Leathwick and Winterbourn, 1984).

18.9.2 Horizontal distribution

The horizontal distribution of predators in fields is often aggregated (Kennedy *et al.*, 1986; Nyffeler and Breene, 1992; Topping and Sunderland, 1994). Some spiders (Riechert, 1992), carabids and staphylinids (Bryan and Wratten, 1984) show aggregative numerical responses to prey density, while spiders (Sunderland and Topping, 1995) and carabids (Speight and Lawton, 1976; Coombes and Sotherton, 1986) may aggregate in weedy patches. Some carabids can locate Collembola (Bauer, 1986), slugs (Symondson and Liddell, 1996, this volume) and aphids (A. Evans, personal communication) by means of chemical cues. Neonate caterpillars are often intercepted by spiders as they move from egg masses to the apical terminals of branches in orchards (MacLellan, 1973). Encounter rates are likely to be high when heteropteran predators (Morris, 1963) or earwigs (Danthanarayana, 1983) enter the nests of colonial caterpillars and remain there for long periods, or where hoverfly larvae (Hagen and Van den Bosch, 1968; Chambers, 1991) or cecidomyiid larvae (Raworth, 1984) emerge among a superabundance of aphid prey because the adult fly is adapted to search out, and oviposit in, aphid colonies.

18.9.3 Other factors affecting predator–prey encounter rates

For pests such as aphids and caterpillars that drop off the plant onto the ground and then climb back again, the residence time on the ground affects the probability of encounter with ground-based predators (Weseloh, 1990; Winder *et al.*, 1994) and residence times are likely to be temperature-dependent. Prey that are inactive at low temperatures may become available to predators with low-temperature activity thresholds; some species of carabid remain active below 5°C while other species become comatose (Chiverton, 1988). Some species of spider can catch prey at temperatures below 0°C (Aitchison, 1984). The effect of temperature on prey capture can be non-linear. For example, springtails were active at 2°C but their carabid predator, *Agonum dorsale* (Pont.) was not; above 10°C springtails jumped very efficiently and escaped from *A. dorsale*, but between 5 and 10°C the predator had an activity advantage and predation was maximal (Griffiths, 1983). During unseasonally hot springs, forest caterpillars suffer greatly inflated rates of predation from ants whose foraging activity has been boosted by the high temperatures (Olofsson, 1992).

Immobile prey, or prey with a reduced level of activity (such as eggs, pupae, or prey that are emerging from pupae, moulting or are paralysed by sublethal doses of pesticides) do not have the usual capacity for escape and so could be very vulnerable to predation, including by animals not normally considered to be predators (e.g. springtails, woodlice and pseudoscorpions; Dempster, 1971; Leslie, 1988). Predator screening should therefore be wide enough to admit these possibilities. Some spiders attack moulting prey, which they probably could not overcome between moults (Titova and Yegorova, 1978; Vollrath, 1984). Some 14% of emerging mosquitoes were eaten at the water surface by predatory Diptera (Service, 1973a) and 84% of gall midges emerging from soil were eaten by polyphagous predators (Basedow and Schutte, 1982).

Some plants manipulate predators by attracting them (e.g. ants to nectaries) when there are high herbivore loads on the plant (Price *et al.*, 1980). Specialist predators may be attracted by plant odours, or respond to plant textures (Sheehan, 1986). Herbivores on semi-resistant varieties of plant often develop at slower rates and so are exposed to predation for longer periods (Price *et al.*, 1980).

Predator foraging behaviour can have a strong influence on encounter rate. Social predators, such as ants, can communicate a prey source to others in the colony (MacLellan, 1973). Predators with fixed search paths, such as anthocorids that search along leaf margins and midribs (Price *et al.*, 1980), will not encounter prey outside the search path. In contrast, other predators, such as carabids, chrysopids and phytoseiids, often switch from random search to area-restricted search after encountering a prey (Mitchell, 1963; Nordlund and Morrison, 1990; Hanna and Wilson, 1991) or prey kairomone (McEwen *et al.*, 1993), which increases the probability of encountering more prey, since many prey species have aggregated distributions. Some carabids return repeatedly to search patches of habitat where aphids were found previously (Wratten *et al.*, 1984; Chiverton, 1988). Although this may be a kairomone-mediated response, the possibility of it resulting from true learning cannot be dismissed, as carabids have been demonstrated to learn and remember for several hours (Plotkin, 1979). Learning has also been demonstrated in spiders (Vollrath, 1984) and ants (Holldobler and Wilson, 1990). Predators may develop a search image for a particular species of prey, which will affect their foraging behaviour and the subsequent probability of encountering prey. Pre-conditioning with a single prey type had no effect on the subsequent foraging behaviour of ants (Agnew and Sterling, 1982) and scathophagid flies (Failes *et al.*, 1992), but it did foster a search image in coccinellids (Houck, 1986) and dragonfly larvae (Bergelson, 1985).

Finally, it should not be forgotten that the searching universe of predators is constantly changing as plants grow and senesce and this, in effect, changes the availability of prey, even when prey density (per unit area of ground) does not change. In winter wheat crops the carabid *Agonum dorsale* spends most of its time on the ground and Griffiths *et al.* (1985) attributed this to the effective density of prey being much greater in the searching universe of the ground than that of a wheat plant. Predators were able to maintain predation rates on caterpillars in soybean in spite of a four-fold increase in leaf area (O'Neil and Stimac, 1988). Visually or tactilely searching predators will have to search more vegetation area to find the same number of prey in diverse agroecosystems, which usually have greater plant density (Sheehan, 1986).

18.10 CONCLUSIONS

While there has been great technical progress in the development of antibody techniques for quantifying **consumption rates** of pests there has been much less progress in developing protocols for converting consumption rates to **predation rates**.

The potential for various aspects of the biology and ecology of predator and prey (scavenging, feeding on moribund prey, food chain errors, pre-oral digestion, wounding, wasteful killing, partial consumption, and prey size preference) to generate serious errors, not only in predation rate estimates, but also in crude rankings of predator importance, does not seem to be generally appreciated. In a survey of 72 publications (1956–1994) that reported the use of antibody techniques to measure predation, the percentage mentioning the following topics as being relevant was; detection periods 57%, predator density 28%, prey size preference 19%, scavenging 15%, food chain errors 7%, other topics combined <3%. There was no clear indication of a change in the situation during the period 1990–1994 (18 publications).

Preliminary testing of an antiserum will detect food chain errors arising from the food of the pest or from pest by-products, and preliminary screening of predators is likely to indicate which species exhibit extreme pre-oral digestion. If the predator-testing programme is drawn sufficiently wide, detection of pest consumption by 'unconventional' predators, such as Isopoda, Collembola and parasitoids, becomes possible. Quantification of all the other factors with potential to generate errors in predation estimates will rarely, if ever, be possible in a single study, and in many cases the methods still remain to be developed. It would, nevertheless, be very valuable to be able to estimate the order of magnitude of the main categories of potential error for some representative predator–prey systems. This would alert researchers as to which factors are likely to be of negligible significance and which are potentially important, requiring routine measurement in predation studies.

Acknowledgement

Constructive comments on the manuscript were made by Mark Tatchell, Gabor Lövei, Mark Jervis and Linton Winder. I am very grateful for the assistance of the HRI Library Staff. The work was funded by the UK Ministry of Agriculture, Fisheries and Food.

REFERENCES

Abbas, M.S.T. and Boucias, D.G. (1984) Interaction between nuclear polyhedrosis virus-infected *Anticarsia gemmatalis* [Lepidoptera: Noctuidae] larvae and predator *Podisus maculiventris* (Say)[Hemiptera: Pentatomidae]. *Environmental Entomology*, **13**, 599–602.

Adams, J. (1984) The habitat and feeding ecology of woodland harvestmen (Opiliones) in England. *Oikos*, **42**, 361–70.

Agnew, C.W. and Sterling, W.L. (1982) Predation rates of the red imported fire ant on eggs of the tobacco budworm. *Protection Ecology*, **4**, 151–8.

Aitchison, C.W. (1984) Low temperature feeding by winter-active spiders. *Journal of Arachnology*, **12**, 297–305.

Allen, W.R. and Hagley, E.A.C. (1990) Epigeal arthropods as predators of mature larvae and pupae of the apple maggot (Diptera: Tephritidae). *Environmental Entomology*, **19**, 309–12.

Archer, T. C. and Musick, G. H. (1976) Responses of black cutworm larvae to light at several intensities. *Annals of the Entomological Society of America*, **69**, 476–8.

Barclay, H.J. (1992) Combining methods of insect pest control: partitioning mortality and predicting complementarity. *Researches on Population Ecology*, **34**, 91–107.

Barlow, C.A. and Whittingham, J.A. (1986) Feeding economy of larvae of a flower fly: *Metasyrphus corollae* (Dip.: Syrphidae): partial consumption of prey. *Entomophaga*, **31**, 49–57.

Barnes, J.K. (1990) Life history of *Dohrniphora cornuta* (Bigot)(Diptera: Phoridae), a filth-inhabiting humpbacked fly. *Journal of the New York Entomological Society*, **98**, 474–83.

Bartlett, B.R. (1964) Patterns in the host-feeding habit of adult parasitic Hymenoptera. *Annals of the Entomological Society of America*, **57**, 344–50.

Basedow, T. and Schutte, F. (1982) Die Populationsdynamik der Weizengallmucken *Contarinia tritici* (Kirby) und *Sitodiplosis mosellana* (Gehin) (Dipt., Cecidomyiidae) in zwei norddeutschen Weizenanbaugebieten von 1969 bis 1976. *Zoologischer Jahrbucher, Abteilung fur Systematik Okologie und Geographie der Tiere*, **109**, 33–82.

Bauer, T. (1986) How to capture springtails on the soil surface: the method of *Loricera pilicornis* F., in *Feeding Behaviour and Accessibility of Food for Carabid Beetles*, (eds P.J. den Boer, L. Grum and J. Szyszko), Warsaw Agricultural University Press, Warsaw, pp. 43–8.

Baumgaertner, J.U., Frazer, B.D., Gilbert, N., Gill, B., Gutierrez, A.P., Ives, P.M., Nealis, V., Raworth, D.A. and Summers, C.G. (1981) Coccinellids (Coleoptera) and aphids (Homoptera): the overall relationship. *Canadian Entomologst,* **113**, 975–80.

Bergelson, J.M. (1985) A mechanistic interpretation of prey selection by *Anax junius* larvae (Odonata: Aeschnidae). *Ecology,* **66**, 1699–705.

Bhuiyan, M.S.I. and Wratten, S.D. (1994) Grain aphid populations and their fall-off rate on different cultivars of wheat. *IOBC wprs Bulletin,* **17**, 27–35.

Bohnke, B. and Zebitz, C.P.W. (1990) Einfluss der Pflanzengesundheit auf das Wirtswahlverhalten von *Aphis fabae. Mitteilungen der Deutschen Gesellschaft fur Allgemeine und Angewandte Entomologie,* **7**, 439–42.

Brodeur, J. and McNeil, J.N. (1989) Seasonal microhabitat selection by an endoparasitoid through adaptive modification of host behaviour. *Science,* **244**, 226–8.

Bryan, K. M. and Wratten, S.D. (1984) The responses of polyphagous predators to prey spatial heterogeneity: aggregation by carabid and staphylinid beetles to their cereal aphid prey. *Ecological Entomology,* **9**, 251–9.

Buryn, R. and Brandl, R. (1992) Are the morphometrics of chelicerae correlated with diet in mesostigmatid mites (Acari)? *Experimental and Applied Acarology,* **14**, 67–82.

Cameron, E.A. and Reeves, R.M. (1990) Carabidae (Coleoptera) associated with Gypsy Moth (*Lymantria dispar* (L.)(Lepidoptera: Lymantriidae)) populations subjected to *Bacillus thuringiensis* Berliner treatments in Pennsylvania. *Canadian Entomologist,* **122**, 123–9.

Carey, J.R. (1989) The multiple decrement life table: a unifying framework for cause-of-death analysis in ecology. *Oecologia,* **78**, 131–7.

Castanera, P., Loxdale, H.D. and Nowak, K. (1983) Electrophoretic study of enzymes from cereal aphid populations. II. Use of electrophoresis for identifying aphidiid parasitoids (Hymenoptera) of *Sitobion avenae* (F.)(Hemiptera: Aphididae). *Bulletin of Entomological Research,* **73**, 659–65.

Cate, R.H., Archer, T.L., Eikenbary, R.D., Starks, K.J. and Morrison, R.D. (1973) Parasitization of the greenbug by *Aphelinus asychis* and the effect of feeding by the parasitoid on aphid mortality. *Environmental Entomology,* **2**, 549–53.

Cate, R.H., Eikenbary, R.D. and Morrison, R.D. (1977) Preference for and effect of greenbug parasitism and feeding by *Aphelinus asychis. Environmental Entomology,* **6**, 547–50.

Chambers, R.J. (1991) Oviposition by aphidophagous hoverflies (Diptera: Syrphidae) in relation to aphid density and distribution in winter wheat, in *Behaviour and Impact of Aphidophaga,* (eds L. Polgar, R.J. Chambers, A.F.G. Dixon and I. Hodek), SPB Academic Publishing bv, The Hague, The Netherlands, pp. 115–21.

Cherrill, A.J. and Begon, M. (1989) Predation on grasshoppers by spiders in sand dune grasslands. *Entomologia Experimentalis et Applicata,* **50**, 225–31.

Chiverton, P.A. (1984) Pitfall-trap catches of the carabid beetle *Pterostichus melanarius,* in relation to gut contents and prey densities, in insecticide treated and untreated spring barley. *Entomologia Experimentalis et Applicata,* **36**, 23–30.

Chiverton, P.A. (1987) Predation of *Rhopalosiphum padi* (Homoptera: Aphididae) by polyphagous predatory arthropods during the aphids' pre-peak period in spring barley. *Annals of Applied Biology,* **111**, 257–69.

Chiverton, P.A. (1988) Searching behaviour and cereal aphid consumption by *Bembidion lampros* and *Pterostichus cupreus,* in relation to temperature and prey density. *Entomologia Experimentalis et Applicata,* **47**, 173–82.

Coombes, D.S. and Sotherton, N.W. (1986) The dispersal and distribution of polyphagous predatory Coleoptera in cereals. *Annals of Applied Biology*, **108**, 461–74.

Coop, L.B. and Berry, R.E. (1986) Reduction in variegated cutworm (Lepidoptera: Noctuidae) injury to peppermint by larval parasitoids. *Journal of Economic Entomology*, **79**, 1244–8.

Croft, B.A. and MacRae, I.V. (1992) Persistence of *Typhlodromus pyri* and *Metaseiulus occidentalis* (Acari: Phytoseiidae) on apple after inoculative release and competition with *Zetzellia mali* (Acari: Stigmaeidae). *Environmental Entomology*, **21**, 1168–77.

Crook, N.E. and Sunderland, K.D. (1984) Detection of aphid remains in predatory insects and spiders by ELISA. *Annals of Applied Biology*, **105**, 413–22.

Danthanarayana, W. (1983) Population ecology of the light brown apple moth, *Epiphyas postvittana* (Lepidoptera: Tortricidae). *Journal of Animal Ecology*, **52**, 1–33.

Davies, M.J. (1953) The contents of the crops of some British carabid beetles. *Entomologist's Monthly Magazine*, **89**, 18–23.

Dempster, J.P. (1960) A quantitative study of the predators of the eggs and larvae of the Broom Beetle, *Phytodecta olivacea* Forster, using the precipitin test. *Journal of Animal Ecology*, **29**, 149–67.

Dempster, J.P. (1967) The control of *Pieris rapae* with DDT. I. The natural mortality of the young stages. *Journal of Applied Ecology*, **4**, 485–500.

Dempster, J.P. (1971) The population ecology of the cinnabar moth, *Tyria jacobaeae* (Lepidoptera, Arctiidae). *Oecologia*, **7**, 26–67.

Dennison, D.F. and Hodkinson, I.D. (1983) Structure of the predatory beetle community in a woodland soil ecosystem. I. Prey selection. *Pedobiologia*, **25**, 109–15.

Desender, K. and Pollet, M. (1985) Ecological data on *Clivina fossor* (Coleoptera, Carabidae) from a pasture ecosystem. II. Reproduction, biometry, biomass, wing polymorphism and feeding ecology. *Revue d'Ecologie et de Biologie du Sol*, **22**, 233–46.

Dixon, P.L. and McKinlay, R.G. (1992) Pitfall trap catches of and aphid predation by *Pterostichus melanarius* and *Pterostichus madidus* in insecticide treated and untreated potatoes. *Entomologia Experimentalis et Applicata*, **64**, 63–72.

Doane, J.F., Scotti, P.D., Sutherland, O.R.W. and Pottinger, R.P. (1985) Serological identification of wireworm and staphylinid predators of the Australian soldier fly (*Inopus rubriceps*) and wireworm feeding on plant and animal food. *Entomologia Experimentalis et Applicata*, **38**, 65–72.

Dudevoir, D.S. and Reeves, R.M. (1990) Feeding activity of carabid beetles and spiders on gypsy moth larvae (Lepidoptera: Lymantriidae) at high-density prey populations. *Journal of Entomological Science*, **25**, 341–56.

Ernsting, G. and Mulder, A.J. (1981) Components of predatory behaviour underlying density-dependent prey-size selection by *Notiophilus biguttatus* F. (Carabidae, Coleoptera). *Oecologia*, **51**, 169–74.

Ernsting, G. and Van der Werf, D.C. (1988) Hunger, partial consumption of prey and prey size preference in a carabid beetle. *Ecological Entomology*, **13**, 155–64.

Evans, M.E.G. (1964) A comparative account of the feeding methods of the beetles *Nebria brevicollis* (F.) (Carabidae) and *Philonthus decorus* (Grav.) (Staphylinidae). *Transactions of the Royal Society of Edinburgh*, **66**, 91–109.

Evans, M.E.G. (1967) Notes on feeding in some predaceous beetles of the woodland floor. *The Entomologist*, **100**, 300–3.

Everts, J.W., Willemsen, I., Stulp, M., Simons, L., Aukema, B. and Kammenga, J. (1991) The toxic effect of deltamethrin on linyphiid and erigonid spiders in connection with ambient temperature, humidity and predation. *Archives of Environmental Contamination and Toxicology*, **20**, 20–4.

Failes, E.S., Whistlecraft, J.W. and Tomlin, A.D. (1992). Predatory behaviour of *Scatophaga stercoraria* under laboratory conditions. *Entomophaga*, **37**, 205–13.

Fellers, G.M. and Fellers, J.H. (1982) Scavenging rates of invertebrates in an eastern deciduous forest. *American Midland Naturalist*, **107**, 389–92.

Ferran, A., Iperti, G., Lapchin, L. and Rabasse, J.M. (1991) La localisation, le comportement et les relations 'proie-predateur' chez *Coccinella septempunctata* dans un champ de ble. *Entomophaga*, **36**, 213–25.

Fichter, B.L. and Stephen, W.P. (1981) Time-related decay in prey antigens ingested by the predator *Podisus maculiventris* (Hemiptera: Pentatomidae) as detected by ELISA. *Oecologia*, **51**, 404–7.

Fichter, B.L. and Stephen, W.P. (1984) Time-related decay of prey antigens ingested by arboreal spiders as detected by ELISA. *Environmental Entomology*, **13**, 1583–7.

Flanders, S.E. (1953) Predatism by the adult hymenopterous parasite and its role in biological control. *Journal of Economic Entomology*, **46**, 541–4.

Foelix, R.F. (1982) *Biology of Spiders*, Harvard University Press, USA.

Forsythe, T.G. (1982) Feeding mechanisms of certain ground beetles (Coleoptera: Carabidae). *The Coleopterists Bulletin*, **36**, 26–73.

Fox, C.J.S. and MacLellan, C.R. (1956) Some Carrabidae and Staphylinidae shown to feed on a wireworm, *Agriotes sputator* (L.), by the precipitin test. *Canadian Entomologist*, **88**, 228–31.

Frank, J.H. (1979) The use of precipitin technique in predator–prey studies to 1975, in *Serology in Insect Predator–Prey Studies*, (ed. M.C. Miller), Miscellaneous Publication 11, Entomological Society of America, pp. 1–15.

Frazer, B.D., Gilbert, N., Ives, P.M. and Raworth, D.A. (1981) Predation of aphids by coccinellid larvae. *Canadian Entomologist*, **113**, 1043–6.

Fritz, R.S. (1982) Selection for host modification by insect parasitoids. *Evolution*, **36**, 283–8.

Gardner, S.M. and Dixon, A.F.G. (1985) Plant structure and the foraging success of *Aphidius rhopalosiphi* (Hymenoptera: Aphidiidae). *Ecological Entomology*, **10**, 171–9.

Good, J.A. and Giller, P.S. (1988) A contribution to a check-list of Staphylinidae (Coleoptera) of potential importance in the integrated protection of cereal and grass crops, in *Integrated Crop Protection in Cereals*, (eds R. Cavalloro and K.D. Sunderland), A.A. Balkema, Rotterdam, pp. 81–98.

Greenstone, M.H. (1996) Serological analysis of predation: past, present and future, in *The Ecology of Agricultural Pests: Biochemical Approaches*, (eds W.O.C. Symondson and J.E. Liddell), Chapman & Hall, London, pp. 265–300.

Greenstone, M.H. and Hunt, J.H. (1993) Determination of prey antigen half-life in *Polistes metricus* using a monoclonal antibody-based immunodot assay. *Entomologia Experimentalis et Applicata*, **68**, 1–7.

Greenstone, M.H. and Morgan, C.E. (1989) Predation by *Heliothis zea* (Lepidoptera: Noctuidae): an instar-specific ELISA assay for stomach analysis. *Annals of the Entomological Society of America*, **82**, 45–9.

Griffiths, C., Carter, J.B. and Overend, J. (1984) *Phaonia signata* (Meigen)(Diptera: Muscidae) larvae predatory upon leatherjackets, *Tipula paludosa* (Meigen) (Diptera: Tipulidae) larvae. *Entomologists Gazette*, **35**, 53–5.

Griffiths, E. (1983) *The feeding ecology of the carabid beetle* Agonum dorsale *in cereal crops*, PhD Thesis, University of Southampton.

Griffiths, E., Wratten, S.D. and Vickerman, G.P. (1985) Foraging by the carabid *Agonum dorsale* in the field. *Ecological Entomology*, **10**, 181–9.

Griswold, M.J. and Trumble, J.T. (1985) Responses of *Spodoptera exigua* (Lepidoptera: Noctuidae) larvae to light. *Environmental Entomology*, **14**, 650–3.

Hagen, K.S. and Van den Bosch, R. (1968) Impact of pathogens, parasites and predators of aphids. *Annual Review of Entomology*, **13**, 325–84.

Hagler, J.R. and Cohen, A.C. (1990) Effects of time and temperature on digestion of purified antigen by *Geocoris punctipes* (Hemiptera: Lygaeidae) reared on artificial diet. *Annals of the Entomological Society of America*, **83**, 1177–80.

Hagler, J.R. and Naranjo, S.E. (1992) Use of serological techniques for identifying predators of major cotton pests. *Proceedings of the 1992 Beltwide Cotton Conference*, National Cotton Council of America, **2**, 909–10.

Hagler, J.R. and Naranjo, S.E. (1996) Using gut content immunoassays to evaluate predaceous biological control agents: a case study, in *The Ecology of Agricultural Pests: Biochemical Approaches*, (eds W.O.C. Symondson and J.E. Liddell), Chapman & Hall, London, pp. 383–99.

Hagler, J.R., Cohen, A.C., Enriquez, F.J. and Bradley-Dunlop, D. (1991) An egg-specific monoclonal antibody to *Lygus hesperus*. *Biological Control*, **1**, 75–80.

Hagler, J.R., Cohen, A.C., Bradley-Dunlop, D. and Enriquez, F.J. (1992a) Field evaluation of predation on *Lygus hesperus* (Hemiptera: Miridae) using a species- and stage specific monoclonal antibody. *Environmental Entomology*, **21**, 896–900.

Hagler, J.R., Cohen, A.C., Bradley-Dunlop, D. and Enriquez, F.J. (1992b) New approach to mark insects for feeding and dispersal studies. *Environmental Entomology*, **21**, 20–5.

Hagler, J.R., Brower, A.G., Tu, Z., Byrne, D.N., Bradley-Dunlop, D. and Enriquez, F.J. (1993) Development of a monoclonal antibody to detect predation of the sweetpotato whitefly, *Bemisia tabaci*. *Entomologia Experimentalis et Applicata*, **68**, 231–6.

Hagley, E.A.C. and Allen, W.R. (1989) Prey of the cribellate spider, *Dictyna annulipes* (Araneae, Dictynidae), on apple tree foliage. *The Journal of Arachnology*, **17**, 366–7.

Hanna, R. and Wilson, L.T. (1991) Prey preference by *Metaseiulus occidentalis* (Acari: Phytoseiidae) and the role of prey aggregation. *Biological Control*, **1**, 51–8.

Haynes, D.L. and Sisojevic, P. (1966) Predatory behaviour of *Philodromus rufus* Walckenaer (Araneae: Thomisidae). *Canadian Entomologist*, **98**, 113–33.

Heimbach, U. (1990) Side effect of the fungicide pyrazophos on the aphid *Sitobion avenae*, in *'Euraphid' Network: Trapping and Prognosis*, (ed. R. Cavalloro), Commission of the European Communities, pp. 285–92.

Hengeveld, R. (1980) Food specialization in ground beetles: an ecological or a phylogenetic process? (Coleoptera, Carabidae). *Netherlands Journal of Zoology*, **30**, 585–94.

Hill, M.G. (1988) Analysis of the biological control of *Mythimna separata* (Lepidoptera: Noctuidae) by *Apanteles ruficrus* (Braconidae: Hymenoptera) in New Zealand. *Journal of Applied Ecology*, **25**, 197–208.

Hobby, B.M. (1931) The British species of Asilidae (Diptera) and their prey. *Transactions of the Entomological Society of Southern England*, **6**, 1–42.

Holldobler, B. and Wilson, E.O. (1990) *The Ants*, Springer-Verlag, Berlin.

Holmes, P.R. (1984) A field study of the predators of the grain aphid, *Sitobion avenae* (F.) (Hemiptera: Aphididae), in winter wheat in Britain. *Bulletin of Entomological Research*, **74**, 623–31.

Holmes, P.R. (1988) Mobility of apterous grain aphids *Sitobion avenae* within wheat fields. *Entomologia Experimentalis et Applicata*, **46**, 275-9.

Honek, A. (1986) Production of faeces in natural populations of aphidophagous coccinellids (Col.) and estimation of predation rates. *Zeitschrift für angewandte Entomologie*, **5**, 467–76.

Hopper, K.R. and King, E.G. (1984) Feeding and movement on cotton of *Heliothis* species (Lepidoptera: Noctuidae) parasitized by *Microplitis croceipes* (Hymenoptera: Braconidae). *Environmental Entomology*, **13**, 1654–60.

Houck, M.A. (1986) Prey preference in *Stethorus punctum* (Coleoptera: Coccinellidae). *Environmental Entomology*, **15**, 967–70.

Hunter, K.W. and Stoner, A. (1975) *Copidosoma tuncatellum*: effect of parasitization on food consumption of larval *Trichoplusia ni*. *Environmental Entomology*, **4**, 381–2.

Iwao, S. and Wellington, W.G. (1970) The influence of behavioural differences among tent-caterpillar larvae on predation by pentatomid bug. *Canadian Journal of Zoology*, **48**, 896-8.

Jackson, C.G. (1991) Elemental markers for entomophagous insects. *Southwestern Entomologist*, **14**, 65–70.

Jervis, M.A. and Kidd, N.A.C. (1986) Host-feeding strategies in hymenopteran parasitoids. *Biological Reviews of the Cambridge Philosophical Society*, **61**, 395–434.

Jervis, M.A. and Kidd, N.A.C. (1991) The dynamic significance of host-feeding by insect parasitoids – what modelers ought to consider. *Oikos*, **62**, 97–9.

Jervis, M.A., Kidd, N.A.C. and Walton, M. (1992) A review of methods for determining dietary range in adult parasitoids. *Entomophaga*, **37**, 565–74.

Jiu, G.D. and Waage, J.K. (1990) The effect of insecticides on the distribution of foraging parasitoids, *Diaeretiella rapae* (Hym.: Braconidae) on plants. *Entomophaga*, **35**, 49–56.

Johnson, D.M., Akre, B.G. and Crowley, P.H. (1975) Modeling arthropod predation: wasteful killing by damselfly naiads. *Ecology*, **56**, 1081–93.

Jones, D. (1982) Predators and parasites of temporary row crops: agents of irreplaceable mortality or scavengers acting prior to other mortality factors? *Entomophaga*, **27**, 245–66.

Kennedy, T.F., Evans, G.O. and Feeney, A.M. (1986) Studies on the biology of *Tachyporus hypnorum* F. (Col. Staphylinidae), associated with cereal fields in Ireland. *Irish Journal of Agricultural Research*, **25**, 81–95.

Kidd, N.A.C. and Jervis, M.A. (1989) The effects of host-feeding behaviour on the dynamics of parasitoid–host interactions, and the implications for biological control. *Researches on Population Ecology*, **31**, 235–74.

Kitamura, K. (1982) Comparative studies on the biology of dryinid wasps in Japan. I. Preliminary report on the predacious and parasitic efficiency of *Haplogonatopus atratus* Esaki et Hashimoto (Hymenoptera: Dryinidae). *Bulletin of the Faculty of Agriculture of Shimane University*, **16**, 172–6.

Knost, S.J. and Rovner, J.S. (1975) Scavenging by Wolf Spiders (Araneae: Lycosidae). *The American Midland Naturalist*, **93**, 239–44.

Leathwick, D.M. and Winterbourn, M.J. (1984) Arthropod predation on aphids in a lucerne crop. *New Zealand Entomologist*, **8**, 75–80.

Lesiewicz, D.S., Lesiewicz, J.L., Bradley, J.R. and Van Duyn, J.W. (1982) Serological determination of carabid (Coleoptera, Adephaga) predation on corn earworm (Lepidoptera, Noctuidae) in field corn. *Environmental Entomology*, **11**, 1183–6.

Leslie, G.W. (1988) The identification and importance of predators of *Eldana saccharina* (Lepidoptera: Pyralidae). *Proceedings of the 62nd Annual Congress of the South African Sugar Technologists' Association*, 124–8.

Leslie, G.W. and Boreham, P.F.L. (1981) Identification of arthropod predators of *Eldana saccharina* Walker (Lepidoptera: Pyralidae) by cross-over electrophoresis. *Journal of the Entomological Society of South Africa*, **44**, 381-8.

Liddell, J.E. and Symondson, W.O.C. (1996) The potential of combinatorial antibody libraries in pest/predator relationship studies, in *The Ecology of Agricultural Pests: Biochemical Approaches*, (eds W.O.C. Symondson and J.E. Liddell), Chapman & Hall, London, pp. 347–66.

Loughridge, A.H. and Luff, M.L. (1983) Aphid predation by *Harpalus rufipes* (DeGeer) (Coleoptera: Carabidae) in the laboratory and field. *Journal of Applied Ecology*, **20**, 451–62.

Lövei, G.L. (1986) The use of biochemical methods in the study of carabid feeding: the potential of isoenzyme analysis and ELISA, in *Feeding Behaviour and Accessibility of Food for Carabid Beetles*, (eds P.J. den Boer, L. Grum and J. Szyszko), Warsaw Agricultural University Press, Warsaw, pp. 21–7.

Lövei, G.L., Sopp, P.I. and Sunderland, K.D. (1987) The effect of mixed feeding on the digestion of the carabid *Bembidion lampros*. *Acta Phytopathologica et Entomologica Hungarica*, **22**, 403–7.

Lövei, G.L., Sopp, P.I. and Sunderland, K.D. (1990) Digestion rate in relation to alternative feeding in three species of polyphagous predators. *Ecological Entomology*, **15**, 293–300.

Lucas, J.R. and Grafen, A. (1985) Partial prey consumption by ambush predators. *Journal of Theoretical Biology*, **113**, 455-73.

Mack, T.P. and Smilowitz, Z. (1981) The vertical distribution of green peach aphids and its effect on the relationship between green peach aphids and a predator. *American Potato Journal*, **58**, 345–53.

MacLellan, C.R. (1973) Natural enemies of the light brown apple moth, *Epiphyas postvittana*, in the Australian Capital Territory. *Canadian Entomologist*, **105**, 681–700.

Makowski, M. and Szyszko, J. (1986) On the possibility to use individual biomass to investigate the accessibility of food for Carabidae (Coleoptera), in *Feeding Behaviour and Accessibility of Food for Carabid Beetles*, (eds P.J. den Boer, L. Grum and J. Szyszko), Warsaw Agricultural University Press, Warsaw, pp. 123–30.

Mansour, F. and Heimbach, U. (1993) Evaluation of lycosid, micryphantid and linyphiid spiders as predators of *Rhopalosiphum padi* (Hom.: Aphididae) and their functional response to prey density – laboratory experiments. *Entomophaga*, **38**, 79–87.

Mansour, F., Rosen, D. and Shulov, A. (1981) Disturbing effect of a spider on larval aggregations of *Spodoptera littoralis*. *Entomologia Experimentalis et Applicata*, **29**, 234–7.

McAllister, M.K., Roitberg, B.D. and Weldon, L. (1990) Adaptive suicide in pea aphids: decisions are cost sensitive. *Animal Behaviour*, **40**, 167–75.

McEwen, P.K., Clow, S., Jervis, M.A. and Kidd, N.A.C. (1993) Alteration in searching behaviour of adult female green lacewings *Chrysoperla carnea* (Neur.: Chrysopidae) following contact with honeydew of the black scale *Saissetia oleae* (Hom.: Coccidae) and solutions containing acid hydrolysed L-tryptophan. *Entomophaga*, **38**, 347–54.

McIver, J.D. (1981) An examination of the utility of the precipitin test for evaluation of arthropod predator–prey relationships. *Canadian Entomologist*, **113**, 213–22.

Mitchell, B. (1963) Ecology of two carabid beetles, *Bembidion lampros* (Herbst) and *Trechus quadristriatus* (Schrank). I. Life cycles and feeding behaviour. *Journal of Animal Ecology*, **32**, 289–99.

Moran, M.D. and Hurd, L.E. (1994) Short-term responses to elevated predator densities: noncompetitive intraguild interactions and behaviour. *Oecologia*, **98**, 269–73.

Morris, R.F. (1963) The effect of predator age and prey defence on the functional response of *Podisus maculiventris* Say to the density of *Hyphantria cunea* Drury. *Canadian Entomologist*, **95**, 1009–20.

Morris, R.F. (1965) Contemporaneous mortality factors in population dynamics. *Canadian Entomologist*, **97**, 1173–84.

Nealis, V. and Regniere, J. (1987) The influence of parasitism by *Apanteles fumiferanae* (Hymenoptera: Braconidae) on spring dispersal and changes in the distribution of larvae of the spruce budworm (Lepidoptera: Tortricidae). *Canadian Entomologist*, **119**, 141–6.

Nentwig, W. and Wissel, C. (1986) A comparison of prey lengths among spiders. *Oecologia*, **68**, 595–600.

Nettles, W.C. (1987) *Eucelatoria bryani* (Diptera: Tachinidae): effect on fecundity of feeding on hosts. *Environmental Entomology*, **16**, 437–40.

Nordlund, D.A. and Morrison, R.K. (1990) Handling time, prey preference and functional response for *Chrysoperla rufilabris* in the laboratory. *Entomologia Experimentalis et Applicata*, **57**, 237–42.

Nyffeler, M. and Breene, R.G. (1992) Dominant insectivorous polyphagous predators in winter wheat: high colonization power, spatial dispersion patterns and probable importance of the soil surface spiders (Araneae). *Deutsche Entomologische Zeitschrift*, **39**, 177–88.

Nyffeler, M., Dean, D.A. and Sterling, W.L. (1992) Diets, feeding specialization and predatory role of two lynx spiders, *Oxyopes salticus* and *Peucetia viridans* (Araneae: Oxyopidae), in a Texas cotton agroecosystem. *Environmental Entomology*, **21**, 1457–65.

Obrycki, J.J. (1992) Techniques for evaluation of predators of Homoptera. *Florida Entomologist*, **75**, 472–6.

Ohbayashi, T. and Iwabuchi, K. (1991) Abnormal behaviour of the common armyworm *Pseudaletia separata* (Walker) (Lepidoptera: Noctuidae) larvae infected with an entomogenous fungus, *Entomophaga aulicae*, and a nuclear polyhedrosis virus. *Applied Entomology and Zoology*, **26**, 579–85.

Ohman, M.D., Frost, B.W. and Cohen, E.B. (1983) Reverse diel migration: an escape from invertebrate predators. *Science, Washington*, **220**, 1404–6.

Oien, C.T. and Ragsdale, D.W. (1993) Susceptibility of nontarget hosts to *Nosema furnacalis* (Microsporidia: Nosematidae), a potential biological control agent of the European Corn Borer, *Ostrinia nubilalis* (Lepidoptera: Pyralidae). *Biological Control*, **3**, 323–8

Olofsson, E. (1992) Predation by *Formica polyctena* Foerster (Hym., Formicidae) on newly emerged larvae of *Neodiprion sertifer* (Geoffroy)(Hym., Diprionidae). *Journal of Applied Entomology*, **114**, 315–19.

O'Neil, R.J. and Stimac, J.L. (1988) Measurement and analysis of arthropod predation on velvetbean caterpillar, *Anticarsia gemmatalis* (Lepidoptera: Noctuidae) in soybeans. *Environmental Entomology*, **17**, 821–6.

Pasquet, A. (1984) Proies capturées et strategies prédatrices chez deux espèces d'araignées orbitèles: *Argiope bruennichi* et *Araneus marmoreus*. *Entomologia Experimentalis et Applicata*, **36**, 177–84.

Plotkin, H.C. (1979) Learning in a carabid beetle (*Pterostichus melanarius*). *Animal Behaviour*, **27**, 567–75.

Pollard, S.D. (1990) A methodological constraint influencing measurement of food intake rates in sucking predators. *Oecologia*, **82**, 569–71.

Pollet, M. and Desender, K. (1987) The consequences of different life histories in ground beetles for their feeding ecology and impact on other pasture arthropods. *Mededelingen van de Fakulteit Landbouwwetenschappen Rijksuniversiteit Gent*, **52**, 179–90.

Poser, T. (1988) Chilopoden als Pradatoren in einem Laubwald. *Pedobiologia*, **31**, 261–81.
Powell, W. (1980) *Toxares deltiger* (Haliday)(Hymenoptera: Aphidiidae) parasitising the cereal aphid, *Metopolophium dirhodum* (Walker)(Hemiptera: Aphididae), in Southern England: a new host–parasitoid record. *Bulletin of Entomological Research*, **70**, 407–9.
Powell, W., Wilding, N., Brobyn, P.J. and Clark, S.J. (1986) Interference between parasitoids [Hym.: Aphidiidae] and fungi [Entomophthorales] attacking cereal aphids. *Entomophaga*, **31**, 293–302.
Powell, W., Ashby, J. and Wright, A.F. (1987) Encouraging polyphagous predators, Rothamsted Experimental Station, Report for 1986, p. 90.
Price, J.F. and Shepard, M. (1978) *Calosoma sayi* and *Labidura riparia* predation on noctuid prey in soybeans and locomotor activity. *Environmental Entomology*, **7**, 653–6.
Price, P.W., Bouton, C.E., Gross, P., McPheron, B.A., Thompson, J. N. and Weis, A.E. (1980) Interactions among three trophic levels: influence of plants on interactions between insect herbivores and natural enemies. *Annual Review of Ecology and Systematics*, **11**, 41–65.
Provencher, L., Coderre, D. and Dondale, C.D. (1988) Spiders (Araneae) in corn fields in Quebec. *Canadian Entomologist*, **120**, 97–100.
Ragsdale, D.W. and Oien, C.T. (1996) An environmental risk assessment for release of an exotic microsporidium for European corn borer control in North America, in *The Ecology of Agricultural Pests: Biochemical Approaches*, (eds W.O.C. Symondson and J.E. Liddell), Chapman & Hall, London, pp. 401–77.
Ragsdale, D.W., Larson, A.D. and Newsom, L.D. (1981) Quantitative assessment of the predators of *Nezara viridula* eggs and nymphs within a soybean agroecosystem using an ELISA. *Environmental Entomology*, **10**, 402–5.
Rahman, M. (1970) Mutilation of the imported cabbageworm by the parasite *Apanteles rubecula*. *Journal of Economic Entomology*, **63**, 1114-16.
Raworth, D.A. (1984) Population dynamics of the cabbage aphid, *Brevicoryne brassicae* (Homoptera: Aphididae) at Vancouver, British Columbia. IV Predation by *Aphidoletes aphidimyza* (Diptera: Cecidomyiidae). *Canadian Entomologist*, **116**, 889–93.
Retana, J., Cerda, X. and Espadaler, X. (1991) Arthropod corpses in a temperate grassland: a limited supply? *Holarctic Ecology*, **14**, 63–7.
Riechert, S.E. (1992) Spiders as representative 'sit-and-wait' predators, in *Natural Enemies: the Population Biology of Predators, Parasites and Diseases*, (ed. M.J. Crawley), Blackwell Scientific Publications, Oxford, pp. 313–28.
Riechert, S.E. and Harp, J.M. (1987) Nutritional ecology of spiders, in *Nutritional Ecology of Insects, Mites, Spiders and Related Invertebrates*, (eds F. Slansky and J.G. Rodriguez), Wiley & Sons, New York, pp. 645–72.
Roitberg, B.D. and Myers, J.H. (1979) Behavioural and physiological adaptations of pea aphids (Homoptera: Aphididae) to high ground temperatures and predator disturbance. *Canadian Entomologist*, **111**, 515–19.
Rosenheim, J.A., Wilhoit, L.R. and Armer, A. (1993) Influence of intraguild predation among generalist insect predators on the suppression of an herbivore population. *Oecologia*, **96**, 439–49.
Rothschild, G. (1966) A study of a natural population of *Conomelus anceps* Germar (Homoptera: Delphacidae) including observations on predation using the precipitin test. *Journal of Animal Ecology*, **35**, 413–34.
Ruberson, J.R., Young, S.Y. and Kring, T.J. (1991) Suitability of prey infected by nuclear polyhedrosis virus for development, survival, and reproduction of the

predator *Nabis roseipennis* (Heteroptera: Nabidae). *Environmental Entomology,* **20**, 1475–9.

Ruth, W.E., McNew, R.W., Caves, D.W. and Eikenbary, R.D. (1975) Greenbugs (Hom: Aphididae) forced from host plants by *Lysiphlebus testaceipes* (Hym.: Braconidae). *Entomophaga,* **20**, 65–71.

Ryan, R.B. (1985) A hypothesis for decreasing parasitization by larch casebearer (Lepidoptera: Coleophoridae) on larch foliage by *Agathis pumila. Canadian Entomologist,* **117**, 1573–4.

Rypstra, A.L. and Tirey, R.S. (1991). Prey size, prey perishability and group foraging in a social spider. *Oecologia,* **86**, 25–30.

Sabelis, M.W. (1992) Predatory arthropods, in *Natural Enemies: the Population Biology of Predators, Parasites and Diseases,* (ed. M.J. Crawley), Blackwell Scientific Publications, Oxford, pp. 225–64.

Sahragard, A., Jervis, M.A. and Kidd, N.A.C. (1991) Influence of host availability on rates of oviposition and host-feeding and on longevity in *Dicondylus indianus* Olmi (Hym., Dryinidae), a parasitoid of the Rice Brown Planthopper, *Nilaparvata lugens* Stal (Hem., Delphacidae). *Journal of Applied Entomology,* **112**, 153–62.

Salama, H.S., Zaki, Z.N. and Sharaby, A. (1982) Effect of *Bacillus thuringiensis* Berliner on parasites and predators of the cotton leafworm *Spodopera littoralis. Zeitschrift fur angewandte Entomologie,* **94**, 498–504.

Salama, H.S., El-Moursy, A., Zaki, F.N., Aboul-Ela, R. and Abdel-Razek, A. (1991) Parasites and predators of the meal moth *Plodia interpunctella* Hbn. as affected by *Bacillus thuringiensis* Berl. *Journal of Applied Entomology,* **112**, 244–53.

Samu, F. (1993) Wolf spider feeding strategies: optimality of prey consumption in *Pardosa hortensis. Oecologia,* **94**, 139–45.

Samu, F. and Bíró, Z. (1993) Functional response, multiple feeding and wasteful killing in a wolf spider (Araneae: Lycosidae). *European Journal of Entomology,* **90**, 471–6.

Sandness, J.N. and McMurtry, J.A. (1972) Prey consumption behaviour of *Amblyseius largoensis* in relation to hunger. *Canadian Entomologist,* **104**, 461–70.

Scheller, H.V. (1984) The role of ground beetles (Carabidae) as predators on early populations of cereal aphids in spring barley. *Zeitschtift für angewandte Entomologie,* **97**, 451–63.

Seastedt, T.R., Mameli, L. and Gridley, K. (1981) Arthropod use of invertebrate carrion. *American Midland Naturalist,* **105**, 124–9.

Service, M.W. (1973a) Study of the natural predators of *Aedes cantans* (Meigen) using the precipitin test. *Journal of Medical Entomology,* **10**, 503–10.

Service, M.W. (1973b) Mortalities of the larvae of the *Anopheles gambiae* Giles complex and detection of predators by the precipitin test. *Bulletin of Entomological Research,* **62**, 359–69.

Service, M.W. (1975). *Coelomomyces* infections in larvae of *Anopheles gambiae* and *Culex univittatus* collected from Kenya. *Transactions of the Royal Society of Tropical Medicine and Hygiene,* **69**, 181.

Shapiro, A. (1976) Beau geste? *American Naturalist,* **110**, 900–2.

Shaw, M.R. (1981) Delayed inhibition of host development by the nonparalyzing venoms of parasitic wasps. *Journal of Invertebrate Pathology,* **37**, 215–21.

Sheehan, W. (1986) Response by specialist and generalist natural enemies to agroecosystem diversification: a selective review. *Environmental Entomology,* **15**, 456–61.

Smirnoff, W.A. (1960) Observations on the migration of larvae of *Neodiprion swainei* Midd. (Hymenoptera: Tenthredinidae). *Canadian Entomologist,* **92**, 957–8.

Smith, B.D. (1966) Effects of parasites and predators on a natural population of the aphid *Acyrthosiphon spartii* (Koch) on broom (*Sarothamnus scoparius* L.). *Journal of Animal Ecology*, **35**, 255–67.

Smith Trail, D.R. (1980) Behavioural interactions between parasites and hosts: host suicide and the evolution of complex lifecycles. *American Naturalist*, **116**, 77–92.

Soares, G.G., Lewis, W.J., Strong-Gunderson, J.M., Waters, D.J. and Hamm, J.J. (1994) Integrating the use of MVP^R Bioinsecticide, a unique B.t-based product, with natural enemies of noctuid pests: a novel concept in cotton IPM, in *Proceedings of the 2nd Canberra Meeting on Bacillus thuringiensis*, (ed. R.J. Akhurst), CSIRO, Australia, pp. 133–45.

Sopp, P.I. and Sunderland, K.D. (1989) Some factors affecting the detection period of aphid remains in predators using ELISA. *Entomologia Experimentatis et Applicata*, **51**, 11–20.

Sopp, P.I., Sunderland, K.D. and Coombes, D.S. (1987) Observations on the number of cereal aphids on the soil in relation to aphid density in winter wheat. *Annals of Applied Biology*, **111**, 53–7.

Sopp, P.I., Sunderland, K.D., Fenlon, J.S. and Wratten, S.D. (1992) An improved quantitative method for estimating invertebrate predation in the field using an enzyme-linked immunosorbent assay (ELISA). *Journal of Applied Ecology*, **29**, 295–302.

Sotherton, N.W., Wratten, S.D. and Vickerman, G.P. (1984) The role of egg predation in the population dynamics of *Gastrophysa polygoni* (Coleoptera) in cereal fields. *Oikos*, **43**, 301–8.

Speight, M.R. and Lawton, J.H. (1976) The influence of weed-cover on the mortality imposed on artificial prey by predatory ground beetles in cereal fields. *Oecologia*, **23**, 211–23.

Stamp, N.E. (1981) Behaviour of parasitized aposematic caterpillars advantageous to the parasitoid or the host? *American Naturalist*, **118**, 715–25.

Stary, P. (1970) *Biology of Aphid Parasites (Hymenoptera: Aphidiidae) with respect to Integrated Control*. Dr W. Junk, The Hague, Netherlands.

Stary, P. (1989) Incomplete parasitization in aphids and its role in pest management (Hymenoptera: Aphidiidae). *Acta Entomologica Bohemoslovaka*, **86**, 356–67.

Stuart, M.K. and Greenstone, M.H. (1990). Beyond ELISA: a rapid, sensitive, specific immunodot assay for identification of predator stomach contents. *Annals of the Entomological Society of America*, **83**, 1101–7.

Stuart, M.K. and Greenstone, M.H. (1996) Serological diagnosis of parasitism: a monoclonal antibody-based immunodot assay for *Microplitis croceipes* (Hymenoptera: Braconidae), in *The Ecology of Agricultural Pests: Biochemical Approaches*, (eds W.O.C. Symondson and J.E. Liddell), Chapman & Hall, London, pp. 301–21.

Sunderland, K.D. (1987) A review of methods of quantifying invertebrate predation occurring in the field. *Acta Phytopathologica et Entomologica Hungarica*, **22**, 13–34.

Sunderland, K.D. and Sutton, S.L. (1980) A serological study of arthropod predation on woodlice in a dune grassland ecosystem. *Journal of Animal Ecology*, **49**, 987–1004.

Sunderland, K.D. and Topping, C.J. (1995) Estimating population densities of spiders in cereals, in *Arthropod Natural Enemies in Arable Land. Density, Spatial Heterogeneity and Dispersal*, (eds S. Toft and W. Riedel), *Acta Jutlandica*, **70**, Aarhus University Press, Denmark, pp. 13–22.

Sunderland, K.D., Fraser A.M. and Dixon, A.F.G. (1986) Field and laboratory studies on money spiders (Linyphiidae) as predators of cereal aphids. *Journal of Applied Ecology*, **23**, 433–47.

Sunderland, K.D., Crook, N.E., Stacey, D.L. and Fuller, B.J. (1987) A study of feeding by polyphagous predators on cereal aphids using ELISA and gut dissection. *Journal of Applied Ecology*, **24**, 907–33.

Sunderland, K.D., Chambers, R.J. and Carter, O.C.R. (1988) Potential interactions between varietal resistance and natural enemies in the control of cereal aphids in *Integrated Crop Protection in Cereals*, (eds R. Cavalloro and K.D. Sunderland), A.A. Balkema, Rotterdam pp. 41–56.

Sunderland, K.D., Ellis, S.J., Weiss, A., Topping, C.J. and Long, S.J. (1994) The effects of polyphagous predators on spiders and mites in cereal fields, *Proceedings BCPC Conference, Pests and Diseases, 1994*, pp. 1151–6.

Sunderland, K.D., De Snoo, G.R., Dinter, A., Hance, T., Helenius, J., Jepson, P., Kromp, B., Samu, F., Sotherton, N.W., Toft, S. and Ulber, B. (1995) Density estimation for invertebrate predators in agroecosystems, in *Arthropod Natural Enemies in Arable Land. Density, Spatial Heterogeneity and Dispersal*, (eds S. Toft and W. Riedel), *Acta Jutlandica*, **70**, Aarhus University Press, Denmark, pp. 133–64.

Sunderland, K.D., Bilde, T., Den Nijs, L.J.F.M., Dinter, A., Heimbach, U., Lys, J.A., Powell, W. and Toft, S. (1996) Reproduction of beneficial predators and parasitoids in agroecosystems in relation to habitat quality and food availability. *Acta Jutlandica*, in press.

Symondson, W.O.C. and Liddell, J.E. (1993a) The detection of predation by *Abax parallelepipedus* and *Pterostichus madidus* (Coleoptera: Carabidae) on Mollusca using a quantitative ELISA. *Bulletin of Entomological Research*, **83**, 641–7.

Symondson, W.O.C. and Liddell, J.E. (1993b) A monoclonal antibody for the detection of arionid slug remains in carabid predators. *Biological Control*, **3**, 207-14.

Symondson, W.O.C. and Liddell, J.E. (1993c) Differential antigen decay rates during digestion of molluscan prey by carabid predators. *Entomologia Experimentalis et Applicata*, **69**, 277–97.

Symondson, W.O.C. and Liddell, J.E. (1993d) The detection of predator–mollusc interactions using advanced antibody technologies. *Annales, A.N.P.P. Third NJ International Conference of Pests in Agriculture, Montpellier 7–9 December, 1993*, Association Nationale de Protection des Plantes, Paris, Tome 1, pp. 417–24.

Symondson, W.O.C. and Liddell, J.E. (1996) Polyclonal, monoclonal and engineered antibodies to investigate the role of predation in slug population dynamics, in *The Ecology of Agricultural Pests: Biochemical Approaches*, (eds W.O.C. Symondson and J.E. Liddell), Chapman & Hall, London, pp. 323–45.

Tamaki, G. Halfhill, J.E. and Hathaway, D.O. (1970) Dispersal and reduction of colonies of pea aphids by *Aphidius smithi* (Hymenoptera: Aphidiidae). *Annals of the Entomological Society of America*, **63**, 973–80.

Tatchell, G.M. (1981) The effects of a granulosis virus infection and temperature on the food consumption of *Pieris rapae* [Lep.: Pieridae]. *Entomophaga*, **26**, 291–9.

Thornhill, R. (1975) Scorpionflies as kleptoparasites of web-building spiders. *Nature*, **258**, 709.

Titova, E.V. (1970) Use of the precipitin test in a study of inter-relationships between *Eurygaster integriceps* Put. (Heteroptera, Scutelleridae) and predatory arthropods. *Entomological Reviews*, **49**, 155–62.

Titova, E.V. and Yegorova, N.S. (1978) Serological evaluation of the trophic connection between spiders and *Eurygaster integriceps* (Heteroptera, Scutelleridae). *Entomological Reviews*, **57**, 197–201.

Tod, M.E. (1973) Notes on beetle predators of molluscs. *The Entomologist*, **106**, 196–201.

Topping, C.J. and Sunderland, K.D. (1994) Methods for quantifying spider density and migration in cereal crops. *Bulletin of the British Arachnological Society*, **9**, 209–13.

Tostowaryk, W. (1971) Relationship between parasitism and predation of diprionid sawflies. *Annals of the Entomological Society of America*, **64**, 1424–7.

Traniello, J.F.A. and Beshers, S.N. (1991) Maximization of foraging efficiency and resource defence by group retrieval in the ant *Formica schaufussi*. *Behavioural Ecology and Sociobiology*, **29**, 283–9.

Usher, M.B. and Bowring, M.F.B. (1984) Laboratory studies of predation by the Antarctic mite *Gamasellus racovitzai* (Acarina: Mesostigmata). *Oecologia*, **62**, 245–9.

Vaje, S. and Mossakowski, D. (1986) Occurrence of proteolytic emzymes in relation to accessibility of food in carabid beetles, in *Feeding Behaviour and Accessibility of Food for Carabid Beetles*, (eds P.J. den Boer, L. Grum and J. Szyszko), Warsaw Agricultural University Press, Warsaw, pp. 29–34.

Van Driesche, R.G. (1983) Meaning of 'percent parasitism' in studies of insect parasitoids. *Environmental Entomology*, **12**, 1611–22.

Vickerman, G.P. and Sunderland, K.D. (1975) Arthropods in cereal crops: nocturnal activity, vertical distribution and aphid predation. *Journal of Applied Ecology*, **12**, 755–66.

Vickerman, G.P., Dowie, W. and Playle, K.E. (1988) The potential of *Tachyporus* spp. (Coleoptera: Staphylinidae) as predators of cereal aphids, in *Integrated Crop Protection in Cereals*, (eds R. Cavalloro and K.D. Sunderland), A.A. Balkema, Rotterdam, pp. 69–80.

Vinson, S.B. and Iwantsch, G.F. (1980a) Host suitability for insect parasitoids. *Annual Review of Entomology*, **25**, 397–419.

Vinson, S.B. and Iwantsch, G.F. (1980b) Host regulation by insect parasitoids. *Quarterly Review of Biology*, **55**, 143–65.

Vitzthum, H. (1933) Einiges über *Microtrombidium dermeijerei* und sein Atmungssystem. *Zoologischer Anzeiger*, **8**, 217–20.

Vollrath, F. (1979) Behaviour of the kleptoparasitic spider *Argyrodes elevatus* (Araneae, Theridiidae). *Animal Behaviour*, **27**, 515–21.

Vollrath, F. (1984) Kleptobiotic interactions in invertebrates, in *Producers and Scroungers*, (ed. C. Barnard), Croom Helm, London, pp. 61–94.

Vollrath, F. (1988) Spider growth as an indicator of habitat quality. *Bulletin of the British Arachnological Society*, **7**, 217–19.

Wallin, H. (1991) Movement patterns and foraging tactics of a caterpillar hunter inhabiting alfalfa fields. *Functional Ecology*, **5**, 740-9.

Walton, M.P., Powell, W., Loxdale, H.D. and Allen-Williams, L. (1990) Electrophoresis as a tool for estimating levels of hymenopterous parasitism in field populations of the cereal aphid, *Sitobion avenae*. *Entomologia Experimentalis et Applicata*, **54**, 271–9.

Weseloh, R.M. (1988) Prey preferences of *Calosoma sycophanta* L. (Coleoptera: Carabidae) larvae and relationship of prey consumption to predator size. *Canadian Entomologist*, **120**, 873–80.

Weseloh, R.M. (1990) Simulation of litter residence times of young gypsy moth larvae and implications for predation by ants. *Entomologia Experimentalis et Applicata*, **57**, 215–21.

West, A.S. and Eligh, G.S. (1952) The rate of digestion of blood in mosquitoes: precipitin test studies. *Canadian Journal of Zoology*, **30**, 267–72.

Wheater, C.P. (1988) Predator–prey size relationships in some Pterostichini (Coleoptera: Carabidae). *Coleopterists Bulletin*, **42**, 237–40.

Wheeler, A.G. (1971) Studies on the arthropod fauna of alfalfa. Insect feeding on *Hylemya* flies (Diptera: Anthomyiidae) killed by a phycomycosis. *Journal of the New York Entomological Society*, **79**, 225–7.

Wheeler, A.G. (1973) Studies on the arthropod fauna of alfalfa. V. Spiders (Araneida). *Canadian Entomologist*, **105**, 425–32.

Wheeler, A.G. (1974) Studies on the arthropod fauna of alfalfa. VI. Plant bugs (Miridae). *Canadian Entomologist*, **106**, 1267–75.

Wheeler, A.G. (1977) Studies on the arthropod fauna of alfalfa. VII. Predaceous insects. *Canadian Entomologist*, **109**, 423–7.

Wheeler, A.G., Hayes, J.T. and Stephens, J.L. (1968) Insect predators of mummified pea aphids. *Canadian Entomologist*, **100**, 221–2.

Winder, L. (1990) Predation of the cereal aphid *Sitobion avenae* by polyphagous predators on the ground. *Ecological Entomology*, **15**, 105–10.

Winder, L., Hirst, D.J., Carter, N., Wratten, S.D. and Sopp, P.I. (1994) Estimating predation of the grain aphid *Sitobion avenae* by polyphagous predators. *Journal of Applied Ecology*, **31**, 1–12.

Wolff, R.J. (1986) Scavenging by jumping spiders (Araneae: Salticidae). *The Great Lakes Entomologist*, **19**, 121–2.

Wratten, S.D., Bryan, K., Coombes, D. and Sopp, P. (1984) Evaluation of polyphagous predators of aphids in arable crops. *Proceedings of the 1984 British Crop Protection Conference: Pests and Diseases*, pp. 271–6.

Yao, D.S. and Chant, D.A. (1990) Changes in body weight of two species of predatory mites (Acarina: Phytoseiidae) as a consequence of feeding in an interactive system. *Experimental and Applied Acarology*, **8**, 195–220.

Yeargan, K.V. and Braman, S.K. (1986) Life history of the parasite *Diologaster facetosa* (Hymenoptera: Braconidae) and its behavioural adaptation to the defensive response of a lepidopteran host. *Annals of the Entomological Society of America*, **79**, 1029–33.

Young, O.P. (1984a) The utilization of dead insects on the soil surface in row crop situations. *Environmental Entomology*, **13**, 1346–51.

Young, O.P. (1984b) Prey of adult *Calosoma sayi* (Coleoptera: Carabidae). *Journal of the Georgia Entomological Society*, **19**, 503–7.

Young, O.P. (1985) Survival of a carrion beetle, *Necrodes surinamensis* (Coleoptera, Silphidae) on a diet of dead fall armyworm (Lepidoptera, Noctuidae). *Journal of Entomological Science*, **20**, 359–66.

Young, O.P. (1989) Interactions between the predators *Phidippus audax* (Araneae, Salticidae) and *Hippodamia convergens* (Coleoptera, Coccinellidae) in cotton and in the laboratory. *Entomological News*, **100**, 43–7.

Young, O.P. and Hamm, J.J. (1985) Compatibility of 2 fall armyworm pathogens with the predaceous beetle, *Calosoma sayi* (Coleoptera, Carabidae). *Journal of Entomological Science*, **20**, 212-18.

Young, O.P. and Lockley, T.C. (1986) Predation of striped lynx spider, *Oxyopes salticus* (Araneae: Oxyopidae), on tarnished plant bug, *Lygus lineolaris* (Heteroptera: Miridae): a laboratory evaluation. *Annals of the Entomological Society of America*, **79**, 879–83.

Young, O.P. and Lockley, T.C. (1988) Dragonfly predation upon *Phidippus audax* (Araneae, Salticidae). *Journal of Arachnology*, **16**, 121–2.

Young, S.Y. and Kring, T.J. (1991) Selection of healthy and nuclear polyhedrosis virus infected *Anticarsia gemmatalis* Hubner (Lep.: Noctuidae) as prey by nymphal *Nabis roseipennis* Reuter (Hemiptera: Nabidae) in laboratory and on soybean. *Entomophaga*, **36**, 265–73.

19

Electrophoretic approaches to predator–prey interactions

M.G. Solomon, J.D. Fitzgerald and R.A. Murray

ABSTRACT

Isoenzyme analysis by electrophoresis offers a convenient approach to the examination of predator–prey relationships by detecting prey material in the gut of a predator. In particular, this technique avoids the potential problem associated with all serological methods, i.e. the need to prepare specific antisera.

The use of concentration gradient polyacrylamide gel electrophoresis has provided a successful technique for detecting prey proteins in the predator gut in a range of ecological situations. As a preliminary to using the technique for a field investigation, it is necessary to run samples of the potential prey species, and of starved predators and predators that have been fed on the prey in the laboratory. For the technique to be applicable, it must be demonstrated that, following staining (typically for esterase activity) at least one enzyme of prey origin should be detectable in the fed predator sample, and absent from the starved predator sample. If no suitable band is apparent, then other enzyme-specific stains can be tried. It is necessary to check the band pattern of other potential prey species in the field situation to be examined; it may be possible to detect a range of different prey in the gut of a predator if they all have sufficiently different characteristic band patterns.

The intensity of the stain in a band is, within limits, proportional to the quantity of enzyme present in the sample. Densitometry can quantify this band intensity and provide an index of the concentration of the enzyme in the predator gut. This is a function of the initial quantity ingested and the extent to which it has been broken

The Ecology of Agricultural Pests, Edited by W.O.C. Symondson and J.E. Liddell.
Published in 1996 by Chapman & Hall, London. ISBN 0 412 62190 8

down in the gut; this in turn is a function of the time since ingestion. Thus, the technique can be used as the basis for determining the consumption rates of field-collected predators.

19.1 INTRODUCTION

Field ecologists investigating arthropod predation have a range of choices to make about the experimental techniques to be used. For some species observation of predatory activity may be possible in an artificial laboratory situation, or even in the field. More usually, however, this approach is not practical, and it is necessary to examine the predator gut-content to seek material of prey origin. Predators that chew their prey may ingest fragments of cuticle that can be identified, but if the prey has no durable cuticle, or the predator does not ingest such fragments, then it is necessary to examine the liquid content of the gut by biochemical means. The possible biochemical techniques may be categorized as:

1. **Serological techniques**. ELISA (enzyme-linked immunosorbent assay) is often used as a sensitive assay method (Sunderland, 1988; Symondson and Liddell, 1993,a,b; Greenstone, 1996, this volume) and monoclonal antibody techniques are allowing the preparation of extremely specific antisera (Symondson and Liddell, 1993c; Hagler and Naranjo, 1996, this volume). All such methods, however, depend on the production of an antiserum for each prey species to be investigated, and as such are most suited to studies involving one or a few prey species rather than many prey species.
2. **Molecular techniques**. Various DNA techniques such as RAPD (random amplified polymorphic DNA) and RFLP (restriction fragment length polymorphism) analyses are now being exploited by entomologists (Loxdale *et al.*, 1996, this volume). At present these techniques are being used primarily in genetic studies, but it may be possible to use them in future in predator–prey investigations.
3. **Isoenzyme analysis by electrophoresis**. Electrophoretic separation of isoenzymes, followed by staining to detect particular enzyme groups, had been widely exploited by biologists investigating phylogenetic relationships, when it was shown to be applicable to the detection of prey isoenzymes in the gut of a predator (Murray and Solomon, 1978). Once a basic set of procedures has been established, very little further preliminary work is required to apply it to predator–prey interactions. It has advantages, therefore, when examining feeding patterns involving a wide range of possible prey species.

19.2 ELECTROPHORETIC TECHNIQUES

19.2.1 Electrophoresis systems

Electrophoresis on polyacrylamide concentration gradient slab gels, with a crosslinker gradient, gives a good separation of enzymes, with sharply focused bands after staining. The distances travelled by the migrating enzymes are very reproducible, provided they are run to pore limit. Ready-made gradient gels can be purchased, or they can be prepared in the laboratory; this offers the possibility of adjusting the slope of the concentration gradient if it is expedient to increase the separation of critical marker enzyme bands from different prey species. Concentration gradient polyacrylamide gels can also be prepared in microcapillary tubes (Neuhoff, 1973). This may provide an increased sensitivity, by confining the migrating enzymes within the narrow (0.45-mm diameter) gel, and preventing lateral diffusion. Such gels are simple to prepare, but the gradient is less reproducible than in slab gels. Microcapillary gels can be run in batches (Solomon and Murray, 1986).

Homogenous polyacrylamide gels may give a satisfactory separation of enzymes (Giller, 1982, 1984), but bands are not sharply focused and their position depends on running conditions. In this situation it is probably advantageous to include standard marker enzymes to calibrate the gels.

Isoelectric focusing can also be used to separate enzymes, providing sharp focusing of bands in a position dependent on the isoelectric point of the molecule. Höller and Braune (1988) used the technique successfully to assess parasitism in aphid populations. However, some enzymes lose activity at their isoelectric point (Loxdale, 1994).

As an alternative to polyacrylamide gels, cellulose acetate sheets can be used as the medium for electrophoresis. Bands are not focused, but the thinness of the gel should increase sensitivity (Easteal and Boussy, 1987). Cellulose acetate gels have been used to distinguish spider mite species (Grafton-Cardwell *et al.*, 1988), but have had limited use in predation studies (van der Geest and Overmeer, 1985). A possible disadvantage is that some carboxylesterases appear to be inactivated on cellulose acetate gels (Loxdale, 1994).

19.2.2 Sample preparation

In order to prepare a sample containing as little predator material as possible, it would be ideal to take the gut content of the predator and macerate it in buffer. This is seldom likely to be possible, though with some species of predatory insect the entire gut can be excised without any loss of content, and with small predatory insects it is often possible to remove the abdomen together with the foregut. However, with very small predatory insects and mites the only practical approach is to include the entire organism in the sample. Murray and Solomon (1985) describe a microhomogenizer suitable for macerating such small insects and mites in volumes of buffer as small as 5 μl. Brookes and Loxdale (1985) describe a convenient device for simultaneously homogenizing multiple samples of around 50 μl each.

19.2.3 Staining

Enzyme staining is inherently more sensitive than total protein staining because, theoretically at least, stain continues to be generated until the substrate is depleted. Thus, if very low concentrations of enzyme are present, the gel can be left in the staining solution for a long period, to increase the sensitivity of detection. Esterases have usually been chosen as the target enzymes (Murray and Solomon, 1978; Giller, 1984; Lister, 1984; van der Geest and Overmeer, 1985); esterase staining produces a dye that is visible at low concentrations and the procedure is a simple one-step operation. If esterase staining does not give clear marker bands for the target species, it is possible to stain for other enzyme groups (see Murray *et al.*, 1989).

19.2.4 Procedure

As a preliminary to an investigation of predator diet it is necessary to select a marker band or bands characteristic of each prey species, and absent from the predator. Predator individuals are maintained in the laboratory and starved for several days. Some individuals are then fed with particular prey species, and observed to confirm that feeding has taken place. Samples are then prepared from starved and fed predators, and from individuals of each of the prey species. After electrophoresis and staining, band patterns are compared, and marker bands selected. If no suitable marker bands are apparent, then the procedure can be repeated for a different group of enzymes.

Different life-stages of a prey species may not produce the same banding pattern; if different stages are likely to be subject to predation, then the occurrence of the marker band in all stages must be checked. Some polyphagous pest species may produce different banding patterns depending on the plant species on which they have fed (Osakabe, 1984); this possibility should be considered when planning preliminary tests.

19.2.5 Predation studies using electrophoresis

Electrophoretic isoenzyme analysis has been used for investigating predator–prey interactions among many different invertebrate taxa, and in a wide range of ecological situations. The predators studied have included phytoseiid mites in orchards, anthocorid and mirid bugs in orchards, notonectid bugs in fresh water, and an Antarctic gamasid mite; laboratory studies have included predation by carabid beetles on lepidopteran larvae (Lövei, 1986).

The prey identified by electrophoretic analysis of gut content of field-collected predators in these studies are listed in Tables 19.1–19.4.

These studies highlight the particular strengths of electrophoresis as a means of analysing predator gut content; they involve mites or very small insects (Murray and Solomon, 1978; van der Geest and Overmeer,

Table 19.1 Diet analysis of predatory mites in orchards. (From Murray and Solomon, 1978; Solomon *et al.*, 1985; van der Geest and Overmeer, 1985; Dicke and de Jong, 1986, 1988.)

Predator	*Prey*
Typhlodromus pyri (Scheuten)	*Panonychus ulmi* (Koch)
	Tetranychus urticae (Koch)
	Aculus schlechtendali (Nal.)
Amblyseius fallacis (Garman)	*T. urticae*
Amblyseius potentillae (Garman)	*T. urticae*
Amblyseius finlandicus (Oud.)	*P. ulmi*
	T. urticae
	Aculus fockeui (Nal. & Tr.)

Table 19.2 Diet analysis of predatory insects in orchards. (From Murray and Solomon, 1978; Skinner, 1983; Heitmans *et al.*, 1986.)

Predator	*Prey*
Anthocoris nemoralis (F.) (Hemiptera)	*Rhopalosiphum insertum* (Walker) (Hemiptera)
	Pterocallis alni (De Geer) (Hemiptera)
	Psylla pyricola För (Hemiptera)
	Fenusa dohrnii (Tisch.) (Hymenoptera)
	Psylla mali (Schmid) (Hemiptera)
Anthocoris nemorum (L.) (Hemiptera)	*R. insertum*
Orius minutus (L.) (Hemiptera)	*P. ulmi*
Orius vicinus Ribaut (Hemiptera)	*A. schlechtendali*
Blepharidopterus angulatus (Fall.) (Hemiptera)	*P. alni*
	F. dohrnii

Table 19.3 Diet analysis of predatory aquatic Hemiptera. (From Giller, 1986.)

Predators	Prey
Notonecta glauca L. and *N. viridis* (Hemiptera)	*Cyclops* sp. (Crustacea) *Simocephalus* spp. (Crustacea) Libellulidae (Odonata) *Anopheles* sp. (Diptera) Tanypodinae spp. (Diptera) Ceratopogonidae sp. (Diptera) Haliplidae (Coleoptera) *Hydroporus* spp. (Coleoptera) *Paracymus* spp. (Coleoptera) Dytiscidae sp. (Coleoptera) *Corixa punctata* (Illig.) (Hemiptera) *Sigara lateralis* (Leach) (Hemiptera) *Plea leachi* Machreg. (Hemiptera) *Gerris lacustris* (Linn.) (Hemiptera) *Cloeon dipterum* (L.) (Ephemeroptera) *Leptosceridae* spp. (Trichoptera)

Table 19.4 Diet analysis of a predatory mite in the Antarctic. (From Lister, 1984; Lister *et al.*, 1987.)

Predator	*Prey*
Gamasellus racovitzai (Trouessart)	*Alaskozetes antarcticus* (Michael) (Cryptostigmata) *Cryptopygus antarcticus* (Willem) (Collembola) *Parisotoma octooculata* (Willem) (Collembola)

1985; Lister *et al.*, 1987; Dicke and de Jong, 1986) for which antiserum preparation may be difficult, or they examine a large number of different prey types (Giller, 1986).

19.3 QUANTIFICATION OF ELECTROPHORESIS

19.3.1 Estimation of predation rate from frequency of positive samples

The discussion so far has addressed purely qualitative detection of prey enzymes; the presence in the predator gut of enzymes characteristic of a particular prey species indicates that the predator has recently consumed material originating from one or more individuals of that species. This information can be used to build up a picture of the range of prey species featuring in the diet of a predator species, and also provides a crude indication of the relative frequency with which the different prey species are taken (Dicke and de Jong, 1988).

If some assumptions are made about the number of prey taken per meal, and the intervals between meals, then it is possible to produce an estimate of predation rate based on the proportion of predators with positive traces of prey. Rothschild (1966) and Dempster (1967) devised methods based on results of a serological test for prey detection; these techniques, and some developments from them, are discussed in Sunderland (1988). Lister *et al.* (1987) developed an expression for calculating attack rate by the Antarctic mite *Gamasellus racovitzai* (Trouessart), based on the proportion of samples showing positive traces of prey in an electrophoretic analysis; this approach required an independent estimate of the proportion of predators not searching for prey, based on field observations.

More satisfactory estimates of consumption rates are possible if the concentration of prey enzyme in the predator gut is determined.

19.3.2 Quantification of enzyme in a sample

Subjective judgements of the relative size and intensity of stained bands on a gel can be made by eye, but the use of a scanning densitometer enables a quantitative assessment to be made. The area under a densitometer trace peak, as a measure of the staining intensity of a band, is likely to be proportional to the quantity of the enzyme present in the sample, within limits. Fitzgerald *et al.* (1986) investigated the relationships between staining intensity and sample concentration. They demonstrated the need to conduct preliminary calibration experiments with a range of sample concentrations, since for some enzymes the relationship was linear, and for others it was not, particularly at high sample concentrations or long staining times. An investigation of the possible causes of this non-linearity showed that the diffusion of substrate into the centre of the 3-mm-thick gels was limiting, and at high enzyme concentrations the available substrate was depleted; the use of thinner gels would reduce this problem. Alternatively, samples could be diluted so that enzyme concentration is within the range for which the relationship with staining intensity is reasonably linear.

19.3.3 Quantification of prey material consumed

A prerequisite for using a particular enzyme as a marker for the quantification of consumption is that the concentration of the enzyme should be constant in all individuals of the prey species. This can be determined by assaying a range of weighed individuals. If different developmental stages of the prey species are likely to be accessible to the predator, then it is also necessary to examine the enzyme concentration in representatives of these stages.

Having established that a particular enzyme is an acceptable quantitative marker for prey material, the next stage is to consider interpretation of its concentration in the gut of a predator. The quantity of prey enzymes in the gut of a predator at the time it is assayed is a function of the quantity of prey material consumed, the interval since it was ingested, and the rate at which the enzyme is broken down in the predator gut; this rate or pattern of breakdown may be referred to as the decay curve. Typically, the period of detectability for prey enzymes in the arthropod predator gut is several days. Murray and Solomon (1978) detected prey esterase in the gut of the predatory mite *Typhlodromus pyri* (Scheuten) 30 hours after consumption of a single *Panonychus ulmi* (Koch). Subsequent investigation of the decay curves for this feeding relationship has shown that, at 20°C, *P. ulmi* esterases are detectable for at least 50 hours. One would expect this process to be related to temperature; prey esterases in the gut of *Notonecta glauca* were detectable for at least twice as long at 4°C as at 12°C (Giller, 1984) and for the Antarctic predatory mite *G. racovitzai* feeding on the springtail *Cryptopygus antarcticus* (Willem), Lister *et al.* (1987) calculated a detection time of up to 20 days. The digestion rate of prey material in the predator gut is also likely to be influenced by the quantity of other prey material in the gut at the time of ingestion, or consumed later.

Although these influences on prey enzyme decay rate are being considered here in the context of electrophoretic isoenzyme analysis, they apply equally to the detection of prey remains by serological methods. Sopp and Sunderland (1989) and Symondson and Liddell (1993d) detail decay curves for prey material detected by ELISA.

19.3.4 Quantification of predation rate

In order to assess the impact of a predatory species on the population dynamics of a prey species, the numbers of prey individuals consumed should be determined. It is thus necessary to interpret the measure of intake of prey biomass in terms of numbers of individuals. This may be achieved with accuracy only if the predator consumes a standard quantity of each prey individual on which it feeds.

Lister *et al.* (1987) developed a model for calculating predation rates by the mite *G. racovitzai*, based on the quantity of prey esterase in a marker band measured by scanning densitometry. In order that the expression should give an estimate of predation rate rather than consumption rate, it was necessary to make an assumption of the mean weight of prey individual attacked.

There are several means by which predators may ingest prey material other than by direct predation, and this would lead to an overestimate of predation. A predator may consume another predator which has consumed the prey species being tested for; a predator may scavenge dead or moribund prey individuals; a predator may consume a scavenger. These processes may give rise to 'false positive' results of gut content analysis. These food chain errors also apply to gut content analysis by serological techniques, as reviewed by Sunderland (1988, 1996, this volume).

19.4 CONCLUSIONS

Electrophoretic techniques have been used to analyse the diet of a wide range of predatory arthropods. The sensitivity and specificity are sufficient for many purposes, particularly if concentration gradient gels are used to give sharply focused band patterns. The positive advantage of electrophoretic techniques is that they require very little preliminary work. Unlike methods based on immunological techniques, no preparation of serological reagents is necessary. This avoids the requirement for special facilities for their production (particularly for monoclonal antibodies) and the associated high costs. Additionally, it is a particular advantage in studies involving many potential prey species, or prey species that are very small so that it may be difficult to collect sufficient biomass for injecting into a mammal for antiserum preparation.

An initial laboratory test with starved and fed predators, and potential prey species, allows prey marker bands to be selected, following which samples collected from the field can be analysed. Sample preparation is simple and rapid, involving the homogenisation in buffer of the excised gut or, if this is impractical, the entire predator.

The technique can be used for the quantitative assessment of prey material in the predator gut, following a preliminary investigation to check the relationship of staining intensity (densitometer peak area) with enzyme concentration. If enzyme decay curves are determined, and mean meal size is estimated, then consumption rate and predation rate can be calculated.

Thus, electrophoretic isoenzyme analysis has a continuing role in predation studies, particularly for small prey species, and for predators consuming many species of prey, or as an initial means of narrowing the range of potential prey species before selecting candidates for antiserum preparation.

REFERENCES

Brookes, C.P. and Loxdale, H.D. (1985) A device for simultaneously homogenizing numbers of individual small insects for electrophoresis. *Bulletin of Entomological Research*, **75**, 377–8.

Dempster, J.P. (1967) The control of *Pieris rapae* with DDT. 1. The natural mortality of the young stages of *Pieris*. *Journal of Applied Ecology*, **4**, 485–500.

Dicke, M. and de Jong, M. (1986) Prey preference of predatory mites: electrophoretic analysis of the diet of *Typhlodromus pyri* Scheuten and *Amblyseius finlandicus* (Oudemans), collected in Dutch orchards. *IOBC/WPRS Bulletin*, **9**(4), 62–7.

Dicke, M. and de Jong, M. (1988) Prey preference of the phytoseiid mite *Typhlodromus pyri*. 2. Electrophoretic diet analysis. *Experimental and Applied Acarology*, **4**, 15–25.

Easteal, S. and Boussy, I.A. (1987) A sensitive and efficient isoenzyme technique for small arthropods and other invertebrates. *Bulletin of Entomological Research*, **77**, 407–15.

Fitzgerald, J.D., Solomon, M.G. and Murray, R.A. (1986) The quantitative assessment of arthropod predation rates by electrophoresis. *Annals of Applied Biology*, **109**, 491–8.

Giller, P.S. (1982) The natural diets of waterbugs (Hemiptera-Heteroptera): electrophoresis as a potential method of analysis. *Ecological Entomology*, **7**, 233–7.

Giller, P.S. (1984) Predator gut state and prey detectability using electrophoretic analysis of gut contents. *Ecological Entomology*, **9**, 157–62.

Giller, P.S. (1986) The natural diet of the Notonectidae: field trials using electrophoresis. *Ecological Entomology*, **11**, 163–72.

Grafton-Cardwell, E.E., Eash, J.A. and Granett, J. (1988) Isozyme differentiation of *Tetranychus pacificus* from *T. urticae* and *T. turkestani* (Acari: Tetranychidae), in laboratory and field populations. *Journal of Economic Entomology*, **81**, 770–5.

Greenstone, M.H. (1996) Serological analysis of arthropod predation: past, present and future, in *The Ecology of Agricultural Pests: Biochemical Approaches*, (eds W.O.C. Symondson and J.E. Liddell), Chapman & Hall, London, pp. 265–300.

Hagler, J.R. and Naranjo, S.E. (1996) Using gut content immunoassays to evaluate predaceous biological control agents: a case study, in *The Ecology of Agricultural Pests: Biochemical Approaches*, (eds W.O.C. Symondson and J.E. Liddell), Chapman & Hall, London, pp. 383–99.

Heitmans, W.R.B., Overmeer, W.P.J. and van der Geest, L.P.S. (1986) The role of *Orius vicinus* Ribaut (Heteroptera; Anthocoridae) as a predator of phytophagous and predacious mites in a Dutch orchard. *Journal of Applied Entomology*, **102**, 391–402.

Höller, C. and Braune, H.J. (1988) The use of isoelectric focusing to assess percentage hymenopterous parasitism in aphid populations. *Entomologia Experimentalis et Applicata*, **47**, 105–14.

Lister, A. (1984) Predation in an Antarctic micro-arthropod community, in *Acarology VI*, Vol. 2 (eds D.A. Griffiths and C.E. Bowman), Ellis Horwood, Chichester, pp. 886–92.

Lister, A., Usher, M.B. and Block, W. (1987) Description and quantification of field attack rates by predatory mites: an example using an electrophoresis method with a species of Antarctic mite. *Oecologia*, **72**, 185–91.

Lövei, G.L. (1986) The use of biochemical methods in the study of carabid feeding: the potential of isoenzyme analysis and ELISA, in *Feeding Behaviour and Accessibility of Food for Carabid Beetles*, (eds P.J. den Boer, L. Grum and J. Szyszko), Warsaw Agricultural University Press, pp. 21–7.

Loxdale, H.D. (1994) Isozyme and protein profiles of insects of agricultural and horticultural importance, in *The Identification and Characterisation of Pest Organisms*, (ed. D.L. Hawksworth), CAB International, Wallingford, pp. 337–75.

Loxdale, H.D., Brookes, C.P. and De Barro, P.J. (1996) Application of novel molecular markers (DNA) in agricultural entomology, in *The Ecology of Agricultural Pests: Biochemical Approaches*, (eds W.O.C. Symondson and J.E. Liddell), Chapman & Hall, London, pp. 149–98.

Murray, R.A. and Solomon, M.G. (1978) A rapid technique for analysing diets of invertebrate predators by electrophoresis. *Annals of Applied Biology*, **90**, 7–10.

Murray, R.A. and Solomon, M.G. (1985) A microtechnique for preparing homogenates of biological samples. *Analytical Biochemistry*, **151**, 400–2.

Murray, R.A., Solomon, M.G. and Fitzgerald, J.D. (1989) The use of electrophoresis for determining patterns of predation in arthropods, in *Electrophoretic Studies on Agricultural Pests*, (eds J. den Hollander and H.D. Loxdale), Clarendon Press, Oxford, pp. 467–83.

Neuhoff, V. (1973) Micro-electrophoresis on polyacrylamide gels, in *Micromethods in Molecular Biology*, (ed. V. Neuhoff), Springer Verlag, Berlin, pp. 1–83.

Osakabe, M. (1984) Esterase zymogram of the citrus red mite, *Panonychus citri* (McGregor), on citrus and pear. *Japanese Journal of Entomology and Zoology*, **28**, 1–4.

Rothschild, G. (1966) A study of a natural population of *Conomelus anceps* Germar (Homoptera: Delphacidae) including observations on predation using the precipitin test. *Journal of Animal Ecology*, **35**, 413–34.

Skinner, R.M. (1983) *The biology of apple aphids and their predators*, PhD thesis, University of London, UK.

Solomon, M.G. and Murray, R.A. (1986) Apparatus for electrophoresis in microcapillary polyacrylamide gels. *Journal of Chromatography*, **355**, 281–4.

Solomon, M.G., Murray, R.A., and van der Geest, L.P.S. (1985) Analysis of prey by means of electrophoresis, in *Spider Mites. Their Biology, Natural Enemies and Control; World Crop Pests*, Vol. 1B, (eds W. Helle and M.W. Sabelis), Elsevier, Amsterdam, pp. 171–3.

Sopp, P.I. and Sunderland, K.D. (1989) Some factors affecting the detection period of aphid remains in predators using ELISA. *Entomologia Experimentalis et Applicata*, **51**, 11–20.

Sunderland, K.D. (1988) Quantitative methods for detecting invertebrate predation occurring in the field. *Annals of Applied Biology*, **112**, 201–24.

Sunderland, K.D. (1996) Progress in quantifying predation using antibody techniques, in *The Ecology of Agricultural Pests: Biochemical Approaches*, (eds W.O.C. Symondson and J.E. Liddell), Chapman & Hall, London, pp. 419–55.

Symondson, W.O.C. and Liddell, J.E. (1993a) The detection of predation by *Abax parallelepipedus* and *Pterostichus madidus* (Coleoptera, Carabidae) on mollusca using a quantitative ELISA. *Bulletin of Entomological Research*, **83**, 641–7.

Symondson, W.O.C. and Liddell, J.E. (1993b) The development and characterisation of an anti-hemolymph antiserum for the detection of mollusc remains within carabid beetles. *Biocontrol Science & Technology*, **3**, 261–75.

Symondson, W.O.C. and Liddell, J.E. (1993c) A monoclonal antibody for the detection of arionid slug remains in carabid predators. *Biological Control*, **3**, 207–14.

Symondson, W.O.C. and Liddell, J.E. (1993d) Differential antigen decay rates during digestion of molluscan prey by carabid predators. *Entomologia Experimentalis et Applicata*, **69**, 277–87.

van der Geest, L.P.S. and Overmeer, W.P.J. (1985) Experiences with polyacrylamide gradient gel electrophoresis for the detection of gut contents of phytoseiid mites. *Mededelingen van Faculteit Landbouwwetenschappen Rijksuniversiteit Gent*, **50**, 469–71.

20

Genetics of esterases in laboratory and feral cotton boll weevils, *Anthonomus grandis* Boheman

C.J. Biggers, B.R. Jones, H.R. Bancroft, and L.A. Dickey

ABSTRACT

Comparisons of esterase electrophoretic patterns of adult and larval whole-body squashes, dissected tissues, and frass have been made utilizing native polyacrylamide gel electrophoresis, isoelectric focusing, and pulsed field electrophoresis. Five electrophoretic esterase regions are exhibited in the cotton boll weevil. Two autosomal codominant alleles, designated Est^f and Est^s, have been demonstrated at the Esterase 4 locus by individual pair matings.

Explanations for molecular polymorphism may include and exclude natural selection. Biochemical studies offer no evidence for either of the Esterase 4 alleles having a selective advantage. Samples of feral weevil populations exhibit shifts in allelic frequencies that can be explained only by a selection model. When biochemical and statistical techniques are combined, an argument can be made for each allele having a selective advantage. Environmental parameters cannot be ignored when assessing the adaptive value of gene products demonstrated by biochemical techniques. Emphasis is placed on common errors of biochemical assessment. Possible implications of a recently discovered Esterase 4 allele are also discussed.

20.1 INTRODUCTION

Since the cotton boll weevil (*Anthonomus grandis*, Boheman) entered the United States at Brownsville, Texas in 1892, it has cost US cotton

The Ecology of Agricultural Pests, Edited by W.O.C. Symondson and J.E. Liddell.
Published in 1996 by Chapman & Hall, London. ISBN 0 412 62190 8

growers more than $13 billion (National Cotton Council Bulletin, 1994). To reduce the loss of cotton production, a resolution was adopted by the National Cotton Council in 1958 to eradicate the cotton boll weevil and marked the beginning of the Boll Weevil Eradication Program (BWEP). A federal boll weevil research laboratory was established in 1960 at Mississippi State University, Starkville, MS. Early research centred on: (i) mass rearing for sterilization and release; (ii) the use of pheromone traps; and (iii) the improvement of the Frego bract trait in cotton varieties (Klassen, 1974).

Although the focus of this paper is the biochemical genetics of the cotton boll weevil, the BWEP provided the impetus for most of the early research on the genetics, physiology and ecology of this pest. A need for marking techniques and the development of special mutant strains for the BWEP release programme resulted in a search for morphological markers. More than a dozen mutants have been reported, but many of them have not been maintained. Studies of meiotic and mitotic chromosomes revealed a diploid number of 42 (North *et al.*, 1981). The focus of the karyotype research was genetic manipulation for boll weevil control. The present plans of the BWEP board are to have an eradication programme underway in every cotton-producing state by 1995.

Population genetic studies have relied heavily on the technique of electrophoresis and continue to do so. This review is concerned primarily with allozyme variations in both field populations and laboratory strains with special emphasis on the adaptive value of Esterase 4.

20.2 INHERITANCE

Biggers and Bancroft (1977) were the first to separate in native polyacrylamide gel the esterases of the boll weevil using whole-body squashes. The esterases were stained using alpha-naphthylacetate as a substrate and Fast Blue RR as a stain for the enzyme–substrate complex. Although polymorphism was found in all four esterase regions (Esterase I–IV from the anode), Esterase II was the easiest to score and was selected for further study.

Esterase II, renamed Esterase 4 from the cathode by Terranova (1982), is the primary focus of this review. Two autosomal codominant alleles, designated $Est\text{-}4^{f}$ and $Est\text{-}4^{s}$ have been demonstrated at the Esterase 4 locus by individual pair matings (Biggers and Bancroft, 1977). A third band which is suspected to be another allele ($Est\text{-}4^{e}$) has recently been observed (Figure 20.1). Further studies are underway to determine the frequency and the mechanism of inheritance of the recently discovered esterase.

Biggers and Bancroft (1977) examined feral weevils from two popula-

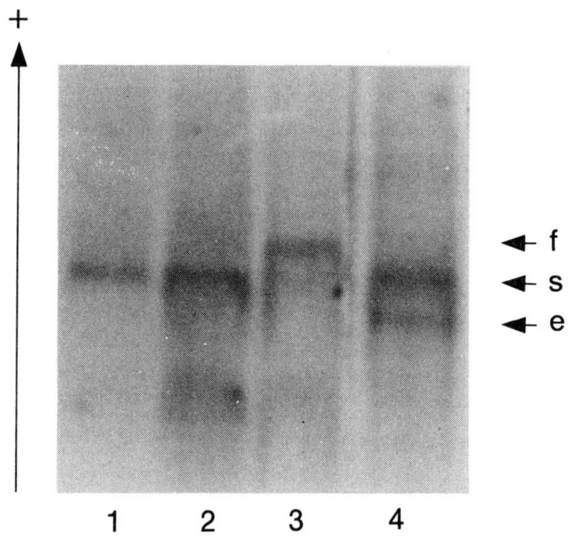

Figure 20.1 Polyacrylamide electropherogram showing *Est-4*^{f}, *Est-4*^{s}, and *Est-4*^{e} typical of individual *Anthonomus grandis* Boh. Lane 1, *Est-4*^{s}; lane 2, *Est-4*^{s}; lane 3, *Est-4*^{f}; and lane 4, *Est-4*^{e}. f, fast; s, slow; e, extra slow.

tions separated by the Mississippi River (Arlington, TN and Mound City, AR) in which the frequencies of the slow isozyme were 0.89 and 0.87, respectively. Terranova (1982) argued, and we agree, that before isoenzymes can be used for establishing standard biochemical profiles on natural populations, an understanding of the genetic basis of each electrophoretic form and the number of alleles segregating at each locus is necessary. The genetic studies reported by Biggers and Bancroft were based on random mating of single pairs followed by electrophoretic analysis of both parents and progeny. Terranova (1982) used a similar method of single pair matings to confirm the work of Biggers and Bancroft and reports the inheritance of four additional autosomal esterase loci (Est 1–3 and 5). Bancroft and Jones (1977) found the same polymorphism for Esterase 4 in frass samples as in whole-body squashes. The use of frass samples allows for single pair matings between weevils of known genotypes instead of random matings, as done by Biggers and Bancroft and Terranova. Frass from weevils of known genotypes can also be used as a standard because only Esterase 4 is detectable in frass. Jones and Bancroft (1986) examined dissected tissues and speculated on the physiological importance of each esterase. They showed that Esterase 3, reported by Terranova (1982) as a single autosomal locus controlled by two co-dominant alleles, was expressed differently in different tissues. They also demonstrated that the Est-3^s isoenzyme was in 80% of the ovarioles and 63% of the fat bodies but rarely found in gut tissue. Conversely, the Est-3^f isoenzyme was much less common in ovarioles (19%) and fat bodies (4.5%) but was expressed in all gut tissue examined. Also, *Est-4*^{f} was present in ovarioles and testes, but *Est-4*^{s} was found in ovarioles but not in testes. These observations suggest that whole-body squashes may not provide a true profile for genetic polymorphism studies.

20.3 RELATIONSHIP BETWEEN EST-4 AND DIAPAUSE

Laboratory weevils have a high frequency for the *Est-4*^{f} allele (0.7) and low frequency of adult reproductive diapause while feral weevils have a low frequency for *Est-4*^{f} and a high frequency of diapause (Biggers and Bancroft, 1977; Jones and Bancroft, 1986). It was further noted that the frequency of *Est-4*^{s} and the ability to diapause are similar. McCoy *et al.* (1968) crossed feral weevils with a high ability to diapause (>35%) with laboratory weevils (Gast strain from Mississippi State University) with a low ability to diapause (<5%). They concluded that diapause was inherited as a single autosomal recessive gene and that the percent of diapause in males was less than in females, but there was no evidence for sex linkage in F_1 or back-cross data.

Lloyd *et al.* (1967) reported that diapause is initiated in the boll weevil when immature stages are subjected to 11 hours of light and night temperatures approach 100°C, or when adults and larvae begin feeding on bolls instead of the flower buds. Earle and Newsom (1964) reported earlier that the larval stages were sensitive to photoperiod but adults were not. They also stated that response to photoperiod was modified by temperature and food. Carter and Phillips (1973) and Phillips and Carter (1978) concluded that changes in diapause incidence seem likely to be either a direct or an indirect response to plant growth substance. A high incidence of diapause is associated with slow fruiting rates, retarded growth, and/or induced maturation of the host plant. They also suggest that the term 'diapause' is used in its broadest sense to include various means of circumventing a wide range of unfavourable conditions. The maturity of the plant may be more useful in inducing diapause than photoperiod (Sterling, 1972). J.R. Phillips (personal communication) states that weevils which enter into mid-season diapause often die before reproducing or feeding.

Diapause in the boll weevil was defined by Brazzel and Newsom (1959) as a cessation of gametogenesis, atrophy of the gonads, increase in fat content, and decrease in respiration. Body fat alone is not a criterion for determining diapause. Phillips (1974) concluded that the weevil has both an aestival diapause (responding to host) and a hibernal diapause (responding to photoperiod) and that the weevil in its northward movement has adapted to photoperiod but retained its ability to respond to the host.

20.4 POPULATION GENETICS

A study was made of a cline to determine if there is any relationship between the frequency of *Est-4* allozymes and various climatic and geographic factors. Weevils were collected from six locations: (1) Brownsville, TX (BR); (2) Corpus Christi, TX (CC); (3) College Station, TX (CS); (4) Curtis, LA (CUR); (5) Choctaw, MS (CH); and (6) Arlington, TN (ARL). Laboratory weevils from Mississippi State University, MS; Baton Rouge, LA; and Florence, SC were also examined. The genotype of the weevils was established for Esterase 4 using a 7.5% running gel and a 3.75% stacking gel to separate the isoenzymes which were stained with alpha-naphthylacetate and Fast Blue RR. The results (Table 20.1 and Figure 20.2) show a steady increase in gene frequency for $Est\text{-}4^s$ and a decline for $Est\text{-}4^f$ from Brownsville, TX to Arlington, TN. Terranova *et al.* (1990) reported a similar increase for $Est\text{-}4^s$ from Cameron, TX to Cleveland County, NC but attributed the changes in allelic frequencies to agronomic, political and economic practices which resulted in a barrier to gene flow. They had no direct evidence of any climatic factor playing a part in the development of the cline and suggested that the differences in the allelic frequencies from Louisiana to North Carolina may need further explanation.

When the genotypic frequencies for the six localities sampled in our study are subjected to correlation analysis against mean January temperature, an interesting pattern emerges. The mean January temperature and the frequency of the homozygous $Est\text{-}4^s$ genotype showed a negative correlation ($r = -0.935$). The frequency of the homozygous *Est-*

Table 20.1 Frequency of Esterase 4 alleles at six collection sites

		Gene frequency	
Collection site	*No. collected*	$Est\text{-}4^s$	$Est\text{-}4^f$
Arlington, TN	115	0.89	0.11
Choctaw, MS	67	0.87	0.13
Curtis, LA	39	0.70	0.30
College Station, TX	105	0.66	0.34
Corpus Christi, TX	81	0.52	0.48
Brownsville, TX	116	0.58	0.42

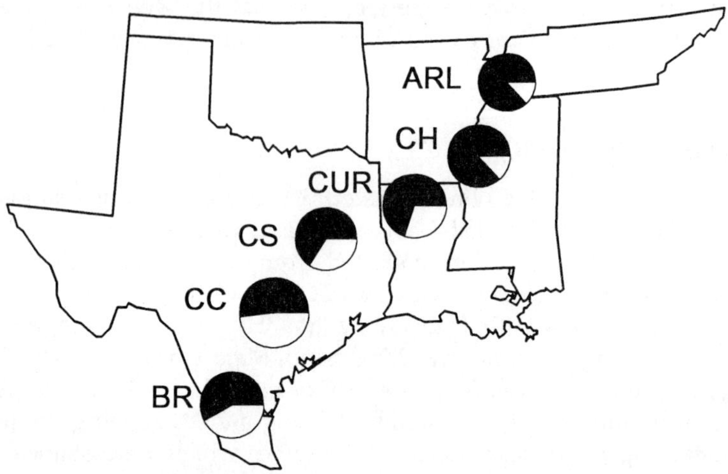

Figure 20.2 A pie-chart representation of the gene frequency of *Est-4*^{s} (shaded area) from Texas to Tennessee. ARL, Arlington, TN; CH, Choctaw, MS; CUR, Curtis, LA; CS, College Station, TX; CC, Corpus Christi, TX; and BR, Brownsville, TX. Frequencies and number of weevils are given in Table 20.1.

4^f genotype and mean January temperature showed a positive correlation (r = 0.903). The frequency of the heterozygous genotype and mean January temperature showed a positive correlation (r = 0.925).

Terranova *et al.* (1991) reported the allelic frequencies of weevils for 14 sites in South Carolina and found no pattern to the distribution of the *Est-4*^{s} allele. However, they claim – if allozyme analyses are reliable to determine the general area from which a weevil may have originated – that four weevils collected in Bladen Co., NC could be traced to the Brownsville, TX population. We do not doubt that an investigator familiar with the diversity in a given population can tell when small samples of weevils are not of that population, but we are sceptical of the claim that the population of origin can be determined for a sample of four weevils. Allelic frequencies based on four individuals are unreliable. For example, the addition of a single allele in a population of four will change the frequency by 0.125 (12.5%).

Another apparent discrepancy in the data of Terranova *et al.* (1991) is the frequency of Esterase 4 alleles from the Gast strain. They report a frequency of *Est-4*^{s} of 0.73, while we have never found it to be above 0.32. We have monitored this population since Biggers and Bancroft (1977) reported that the field-collected and laboratory weevils show a striking contrast in the frequency of the *Est-4*^{s} alleles. In 1977 the slow genotype was predominant in feral weevils, while the fast genotype was predominant in the laboratory-reared populations. The frequency of the *Est-4*^{s} allele in the Gast strain was 0.31 in 1986 and 0.32 in August, 1994. We also examined the frequency of the *Est-4*^{s} allele in the Baton Rouge, LA population (0.42), and the Florence, SC strain (0.18). Terranova *et al.* (1990) reported the frequency of the slow allele in the Gast strain to be 0.73; 0.88 in the Baton Rouge, LA strain; and 0.81 in the Florence, SC strain. A high frequency of *Est-4*^{s} in the Gast strain would not support our hypothesis that *Est-4*^{s} and diapause are related. However, our allelic frequency data (*c.* 0.3), which we have confirmed several times since 1977, has been consistent. Such discrepancy between our data and those reported by Terranova *et al.* cannot be accounted for.

After examining our data we posed the question: will the homozygous *Est-4*^{s} genotype from a strain (Gast) known to be unable to diapause (McCoy *et al.*, 1968) respond differently to diapausing stimuli from Gast homozygous *Est-4*^{f} weevils? To answer this question, individual newly merged weevils were placed in a small holding cup with a single adult food pellet for 3 days. At the end of the 3 days, the frass from each weevil was subjected to electrophoresis to determine the genotype of the weevil. Two lines of weevils were established following this procedure: a homozygous *Est-4*^{s} line and a homozygous *Est-4*^{f} line.

In order to determine the number of eggs per female per day for each genotype under typical laboratory conditions, single pair matings were established by placing a pair of weevils into a capped 30-g plastic cup containing food. Control weevils were held at constant light at 25°C, experimental weevils were held at 11 hours of light at 10°C. The food pellet was replaced on every other day, and the eggs were removed from the old pellet and counted. The results are presented in Table 20.2. Under control conditions there was no difference in egg production between the two genotypes. Diapausing conditions include a lower temperature and shorter day length than the laboratory conditions. Therefore, egg production declines under diapausing conditions for all genotypes (Table 20.2). The *t*-test between weevils of genotype FF and SS under diapausing conditions shows that the SS genotype is reduced significantly more than the homozygous FF genotype ($P < 0.005$). Since both lines were derived from the same laboratory strain, we assume they are genetically similar with the exception of the Est 4 locus. Therefore, we attribute the difference in egg production to the influence of the Esterase 4 genotype.

The experimental weevils were allowed to remain under diapausing stimuli for a further 2 weeks, after which they were examined for gametogenesis and atrophy of gonads. Five of the 75 *Est-4*^{s} females were in the early stages of gonad atrophy but had not entered into firm

Table 20.2 Daily egg production of individuals in diapause and non-diapause conditions (Mean ± SD)

Genotype	*Non-diapause (25°C/constant light)*	*Diapause (10°C/11 h light)*
$Est\text{-}4^f/Est\text{-}4^f$	7.9 ± 0.40 ($n = 20$)	5.0 ± 0.18 ($n = 104$)
$Est\text{-}4^s/Est\text{-}4^s$	7.7 ± 0.47 ($n = 20$)	4.2 ± 0.25 ($n = 75$)

diapause. None of the $Est\text{-}4^f$ weevils showed cessation of gametogenesis. Further work needs to be done in this area.

20.5 ISOELECTRIC FOCUSING AND PULSED FIELD ELECTROPHORESIS

Isoelectric focusing was run using weevils of known genotype as determined by the frass technique described above. Homozygous $Est\text{-}4^s$ and $Est\text{-}4^f$ weevils were used. The pH range of the ampholytes was 3 to 7. Both isoenzymes have an isoelectric point around pH 5, with the faster esterase having a higher isoelectric point than the slow esterase. Electrophoretic and isoelectric focusing data both show a difference in the two isoenzymes. No advantage was found, however, in using pulsed field electrophoresis.

20.6 CONCLUSIONS

1. The ability to determine the genotype from frass samples has allowed us to develop inbred genetic lines in the boll weevil.
2. Two rare alleles of the Esterase 4 locus have been recently reported. The one we report (Fayette County, TN) controls an allozyme which migrates slower than those previously reported while the product found in the Cameron, TX population (Terranova *et al.*, 1990) migrated faster than previously reported for the Esterase 4 locus.
3. Clinal data show that the mean January temperature is correlated negatively with the frequency of the slow allele. We report here that diapause and the slow allele are related in both feral and laboratory populations.

Acknowledgements

The authors would like to thank Dr M.L. Beck of the Department of Biology of The University of Memphis, Memphis, TN for his careful review of the manuscript; Drs G.L. Lentz and P.M. Roberts of the Agricultural Extension Service of the University of Tennessee in Jackson, TN, for providing information and weevils; and Dr Timothy Lockey for his technical assistance.

REFERENCES

Bancroft, H.R. and Jones, B.R. (1977) Genotypes of esterase II determined from frass of *Anthonomus grandis* Boh. (Coleoptera, Curculionidae). *Biochemical Genetics*, **15**, 1175–80.

Biggers, C.J. and Bancroft, H.R. (1977) Esterases of laboratory-reared and field-collected cotton boll weevils, *Anthonomus grandis* Boh.: Polymorphism of adult esterases and formal genetics of esterase II. *Biochemical Genetics*, **15**, 227–33.

Brazzel, J.R. and Newsom, L.D. (1959) Diapause in *Anthonomus grandis* Boh. *Journal of Economic Entomology*, **53**, 603–11.

Carter, F.L. and Phillips, J.R. (1973) Diapause in the boll weevil, *Anthonomus grandis* Boheman, as related to fruiting activity in the cotton plant. *Arkansas Academy of Science, Proceedings*, **27**, 16–20.

Earle, N.W. and Newsom, L.D. (1964) Initiation of diapause in the boll weevil. *Journal of Insect Physiology*, **10**, 131–9.

Jones, B.R. and Bancroft, H.R. (1986) Distribution and probable physiological role of esterases in reproductive, digestive, and fat-body tissues of the adult cotton boll weevil, *Anthonomus grandis* Boh. *Biochemical Genetics*, **24**, 499–508.

Klassen, W. (1971) *Research requirements for boll weevil elimination in the cotton belt. Boll Weevil Suppression, Management, and Elimination Technology*. Proceedings of a Conference, Agriculture Research Service, US Department of Agriculture, pp. 163–6.

Lloyd, E.P., Tingle, F.C. and Gast, R.T. (1967) Environmental stimuli inducing diapause in the boll weevil in North Carolina. *Journal of Economic Entomology*, **58**, 309–12.

McCoy, J.R., Lloyd, E.P. and Bartlett, A.C. (1968) Diapause in crosses of a strain of boll weevils. *Journal of Economic Entomology*, **61**, 163–6.

National Cotton Council (1994) *Boll weevil eradication: a national strategy for success*. Brochure, pp. 1–13.

North, D.T., Leopold, R.A. and Childress, D. (1981) Meiotic and mitotic chromosomes of the cotton boll weevil *Anthonomus grandis grandis* (Coleoptera, Curculionidae). *Canadian Journal of Genetics and Cytology*, **23**, 443–8.

Phillips, J.R. (1974) Diapause as it relates to the boll weevil, *Anthonomus grandis* Boheman. Proceedings of a Conference. Agriculture Research Service, UU Department of Agriculture, pp. 10–11.

Phillips, J.R. and Carter, F.L. (1978) Complexity of diapause in the boll weevil, *Anthonomus grandis* Boheman. *Proceedings of the 12th International Congress for Chronobiology*, Supplement, pp. 621–6.

Sterling, W.L. (1972) Photoperiodic sensitivity in the ontogeny of the boll weevil. *Environmental Entomology*, **165**, 568–71.

Terranova, A.C. (1982) Inheritance of esterases in *Anthonomus grandis grandis* (Coleoptera, Curculionidae). *Review, Entomological Society of America*, **75**, 261–5.

Terranova, A.C., Jones, R.G. and Bartlett, A.C. (1990) The southeastern boll weevil: an allozyme characterization of its population structure. *Southwestern Entomologist*, **15**, 481–96.

Terranova, A.C., Jones, R.G. and A.C. Bartlett. (1991) An allozyme study of the laboratory boll weevil and its influence on some southeastern populations. *Southwestern Entomologist*, **16**, 1–11.

21

Practical applications for techniques to determine the nutritional state and age of field-caught tsetse flies

P.A. Langley

ABSTRACT

The nutritional state and age composition of captured tsetse flies depends on the capture techniques employed. Hence, in order to evaluate biases and to measure seasonal effects upon survival and feeding success in tsetse populations, semi-automated systems have been developed in the laboratory for use on field-collected material. Nutritional state is reflected in the extent of lipid reserves and the amount of residual bloodmeal remaining in the intestine. Lipid content is measured in dried flies by weight differences after chloroform extraction. The residual bloodmeal in the same fly is then measured spectrophotometrically following conversion of the dried abdominal contents to pyridine haemochromagen by first dissolving in sodium hydroxide and ethanol, followed by reacting with pyridine in the presence of sodium dithionite. The absorbance of the pink colour formed is measured at either 558 or 417 nm and is compared with a standard curve prepared with known amounts of haematin.

The chronological ages of individual adult flies can be measured with reference to the degrees of fray on the trailing edge of the wings; six categories are recognized and can provide an estimate of the relative ages of flies of both sexes in a fly population. A highly accurate estimate of the relative ages of females can be obtained by ovarian dissection, since mature eggs from each of the four ovar-

The Ecology of Agricultural Pests, Edited by W.O.C. Symondson and J.E. Liddell.
Published in 1996 by Chapman & Hall, London. ISBN 0 412 62190 8

ioles ovulate singly and sequentially at 9- or 10-day intervals (at 25°C). After the fourth ovulation the presence of relics of the previous oocyte must be identified in order to recognize ovulations 5 to 8. After this, at about 70 days of age, the technique cannot be used alone. However, when used in conjunction with wing fray analysis, it is possible to categorize flies of any age with accuracy. The problems associated with these techniques are that they require the skills of highly trained technicians for their operation and must be performed on freshly caught material.

An alternative method, which is subject to greater inherent variation than that of ovarian categorization is to measure the pteridine content of the head capsule. In many Diptera pteridines accumulate in a predictable manner with time. Storage of desiccated heads in the dark for up to 1 month is followed by extraction into chloroform, separation of the pteridines into an aqueous buffer phase and fluorometric measurement. Standard curves are constructed using laboratory-reared insects of known age. In field experiments in the Ivory Coast and Kenya the pteridine fluorescence technique has shown excellent correlation with wing fray category and with ovarian configuration in the estimation of the chronological age of tsetses.

21.1 INTRODUCTION

Although not classified as conventional agricultural pests, tsetse flies, as vectors of the African trypanosomiases, are responsible for damaging livestock which form an important and integral component of agricultural activity in Africa. They infest approximately 11 million km^2 of sub-Saharan Africa where they transmit sleeping sickness in humans and *nagana* to their livestock.

Two aspects of the behaviour of tsetse are of special interest. First, as adults, both sexes derive all their energy requirements from vertebrate blood. They are active insects and must therefore ingest frequent and large bloodmeals. Second, the females exhibit adenotrophic viviparity, their paired ovaries being reduced to one pair of ovarioles each. Mature oocytes are ovulated one at a time in sequence with an interval of 9–10 days between each ovulation at 25°C. The mature oocyte is fertilized at the moment of ovulation, by spermatozoa stored in the spermathecae for the life of the female, following a single mating early in adult life. The fertilized egg is retained in the oviduct of the female where embryogenesis is completed and the first instar larva hatches after 4–5 days. The larva feeds on a specialized accessory gland fluid produced by the female and completes its larval development within the shelter of the female's abdomen. At the time of birth the third instar larva weighs more than its mother because it carries sufficient food to fuel development and metamorphosis within the puparium over the next 30 days.

Clearly, in studies of tsetse population dynamics – and especially in studies of flight dispersal, feeding success and population growth – it is essential to be able to monitor nutritional reserves and to estimate the age structure of populations. The accuracy of these measurements depends in turn on knowing the biases inherent in the capturing techniques employed.

21.2 SAMPLING METHODS

Adult males live on average only about half as long as females in the field. The assumption is based on the observation that only about half as many males as females are available to a variety of capture techniques, although it is known that equal numbers of adult males and females are recruited to the population. Vale (1974a) revolutionized entrenched ideas concerning the behaviour of tsetse and in particular, the bias in samples caught by the traditional 'fly round' method of sampling populations. In this, a black cloth screen was carried by two field assistants, who stopped periodically and caught tsetse flies which alighted on the screen. Paradoxically the method caught only males and therefore provided at best an 'index' of population density. However, it is now known that neither the upright habit of humans, nor their smell, are attractive except to very hungry flies and that females are less attracted than males in this respect (Vale, 1974a). An alternative method of sampling fly populations was the 'ox fly round'. The presence of the ox altered the bias of the sampling method such that flies of both sexes were caught, though females still were under-represented. However, the composition of the catch depended on whether the ox was stationary or moving. Again, the presence of humans biased the sample compared with that obtained by surrounding an ox or a moving object with invisible electrified nets to intercept flies as they arrived (Vale, 1974a,b).

Sampling methods for tsetse have been simplified by the development of traps which provide both visual and olfactory stimuli that simulate their vertebrate hosts, in order to attract flies from some distance away. This development was reviewed by Wall and Langley (1991). Research continues to improve and refine trapping techniques for tsetse, as well as to explain the biases inherent in such techniques. For instance, it is known that the youngest cohorts of tsetse in a population are not available to capture techniques involving the use of stationary traps or baits, or to host odour alone (Taylor, 1979). Young flies are, however, caught on electrified nets mounted on moving vehicles. Therefore it has been proposed (Langley *et al.*, 1990a) that young adults adopt a watch-and-wait strategy, responding only to movement in their quest for their first bloodmeal. The survival value of this is that under normal circumstances a moving object will signify the arrival of a potential bloodmeal. Hence, young adults with poorly developed flight muscles and only limited energy reserves carried over from the pupa maximize their chances of feeding successfully by responding only to movement. Later, when flight muscles are strong and fuel reserves are greater, the insect can afford to respond to odour attractants in the 'speculative' hunt for bloodmeals.

Argument continues about the biases involved in the use of different capture techniques for tsetse in the field. It is most likely that all the techniques described so far are sampling that proportion of the population which is seeking bloodmeals. Hence, differences in feeding frequency between the sexes will always bias the sample. However, males at least may also visit host animals in their search for females. It has been suggested that females also indulge in mate-seeking (Vale, 1974a), although this seems unlikely, since most females are inseminated at the time of ingestion of their first bloodmeal and thereafter develop strong refractory behaviour towards further mating attempts (Gillott and Langley, 1981).

A sampling method which is independent of the age, sex or nutritional state of a tsetse fly can be used only in certain areas at certain times of the year. It exploits the scototactic behaviour of all tsetse as ambient temperatures rise above 32°C (Huyton and Brady, 1975). Hence, in the Zambezi Valley of Zimbabwe, unbiased samples of flies can be obtained in traps constructed from half-buried oil drums or small thatched enclosures simulating 'refuge' sites into which the flies retreat during the hottest part of the day (Pilson and Pilson, 1967).

Recent developments in the control of tsetse have exploited the behavioural responses of the flies to visual and olfactory stimuli such that insecticide-treated, odour-baited targets of cotton cloth deployed at a density of four per km^2 are at least as effective at causing population suppression as the old techniques of ground-spraying with residual insecticides (Vale *et al.*, 1986). Visually attractive, odour-baited traps are used to monitor fly populations and therefore to monitor the progress of control programmes in terms of a reduction in fly numbers (Wall and Langley, 1991).

Substitution of insect growth regulators (IGRs) for conventional insecticides on odour-baited targets for tsetse control can require estimates to be made of population age as a measure of success of the technique. This is because IGRs effectively sterilize females which do not disappear rapidly from the population, and the population age structure therefore changes as no further young flies are recruited. There are advantages to the use of such control techniques in that if males can be sterilized as well as females, sterilization is superior to killing in terms of the rate and extent of population suppression (Langley and Weidhaas, 1986).

There is some indication that this is feasible using IGRs. Although males are not affected directly, they can transfer sterilizing doses to females during mating and are therefore themselves effectively sterile (Langley *et al.*, 1990b).

21.3 NUTRITIONAL RESERVES

The flight activity of tsetse is fuelled by the mitochondrial oxidation of proline to alanine and the reconstitution of proline from alanine and lipid carbon outside the mitochondria (Bursell *et al.*, 1974). Neutral lipids constitute the major nutritional reserve of the tsetse, the bulk being stored as insoluble triglyceride in the fat body, located mainly in the abdomen (Bursell, 1961). There is little or no capacity for protein storage (Tobe and Langley, 1978) and at times when demands for protein are high, such as in the mobilization of nutrients to feed the growing larva *in utero*, the fly relies on utilizing directly the digestive products of the last bloodmeal ingested before massive larval growth commences combined with lipids derived from the triglyceride store (Cmelik *et al.*, 1969; Langley and Pimley, 1979; Langley *et al.*, 1981). Hence, both in males and females, the extent of the lipid reserves and the amount of blood remaining in the gut are indicators of the nutritional state of the insect.

21.3.1 Measuring nutritional reserves

Flies captured in the field can be stored, desiccated, for long periods before being transferred to the laboratory for processing. After removal of heads, wings and legs, the bodies are dried to constant weight at around 70°C. Following this each fly is subjected to three consecutive daily immersions in 3 ml chloroform. It is then re-dried and re-weighed and its fat content determined by weight difference (Bursell, 1961).

The thorax is now separated from the abdomen and the fat-free dry weight of the thorax is used as an indicator of fly size, provided the fly had already developed its flight muscles before capture (Langley *et al.*, 1990a).

The dried abdomen is subjected to a process which is a modification of techniques developed by Ford *et al.* (1972) and automated by Rogers and Randolph (1978). The abdomen is crushed to a powder in a small sample vial and mixed with 1.5 ml 0.1 M NaOH and 1.0 ml ethanol. This converts haemoglobin and haematin to soluble alkaline haematin and precipitates insoluble matter. After 1 hour, the mixture is re-stirred and centrifuged at 1500 *g* for 10 min. A 2-ml aliquot of the clear supernatant is mixed with 0.5 ml pyridine in the presence of excess sodium dithionite, which gives rise to a pink-coloured compound, pyridine haemochromagen, the intensity of which is proportional to the amount of haematin present in the sample. The optical density of this solution is then read, after appropriate dilution, in a visible-range spectrophotometer at either 417 or 558 nm which are the two peaks of maximum absorption of the compound. Readings are then compared with a standard curve for haematin obtained commercially. Apart from crushing the dried abdomen and adding the sodium dithionite, the other operations are software controlled via IBM PS2 PCs, as are all the computations. The method is more extensively described elsewhere (Langley *et al.*, 1990a).

21.4 AGE DETERMINATION

Estimates of chronological age of both sexes of adult tsetse can be obtained by examination of the trailing edge of the wings. Flies can be assigned to one of six categories based on the degree of wear observed (Figure 21.1). The original experiment (Jackson, 1946) consisted of releasing large numbers of recently hatched adults of *Glossina morsitans* Westwood into the habitat of *G. swynnertoni* Austen. These were recaptured and of course their chronological age was known. However, only males were involved owing to the bias involved in the recapture techniques of the time. The trailing edge of the wing was measured with a planimeter and this measure increased as the degree of wear increased. Jackson found that the mean square root of percentage wear of the

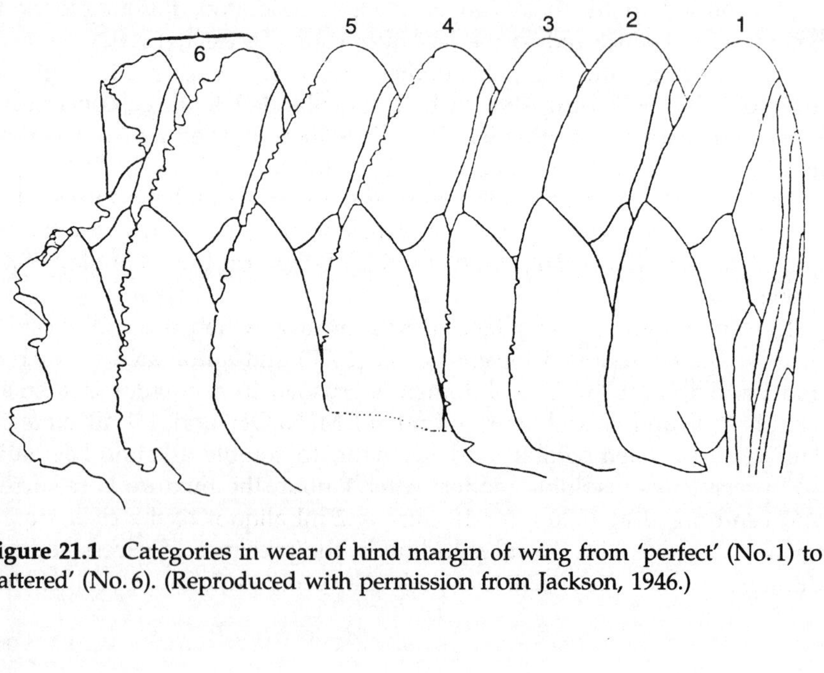

Figure 21.1 Categories in wear of hind margin of wing from 'perfect' (No. 1) to 'tattered' (No. 6). (Reproduced with permission from Jackson, 1946.)

trailing edge of the wing was proportional to chronological age (Figures 21.2 and 21.3). The method does not allow the ages of individuals to be assessed but the age structure of a population can be estimated from such data. There is no reason to believe that the same relationship does not exist for females and so 'wing fray' category can be a useful means of assigning a fly to its correct age in conjunction with other criteria.

For females only, the regular sequence of ovulation from each of four ovarioles in turn, at intervals of 9 or 10 days, independently of ambient temperatures (Williams *et al.*, 1990) [a contention that has recently been seriously questioned (Hargrove, 1994)] permits the estimation of chronological age of individuals accurately, based on ovarian dissection data (Saunders, 1960). It is assumed that ovulation occurs immediately following larviposition. After four ovulations the sequence begins again, but follicular relics of the first four ovulations remain and can be seen by the trained eye. This means that such dissection will allow the number of ovulations that a female has undergone to be detected up to a maximum of eight (approximately 72–80 days). Reference to the developmental state of oocytes in the ovaries at the time of dissection permits an accurate interpolation of age to within a day or two. Any error of judgement can be corrected with reference to the wing fray category of an individual fly.

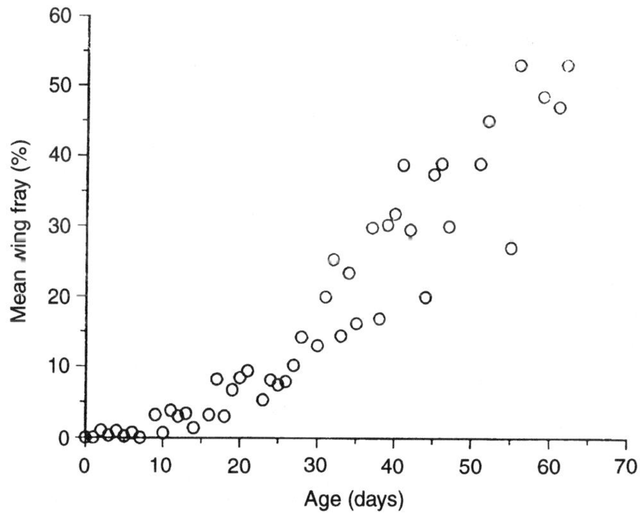

Figure 21.2 Mean wing fray plotted against chronological age of *G. morsitans* males recaptured at intervals after release as newly emerged adults into a habitat of *G. swynnertoni* in Tanzania. (Reproduced with permission from Jackson, 1946.)

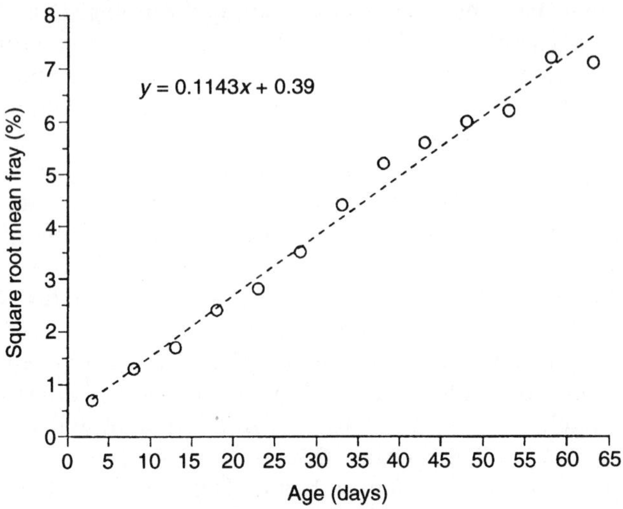

Figure 21.3 Square root mean wing fray plotted against chronological age of *G. morsitans*. (Data as in Fig. 2 of Jackson, 1946.)

21.4.1 Pteridine-based assessment

More recently, the rate of accumulation of pteridine waste products in the head capsule of the tsetse has been shown to be proportional to chronological age, after correcting for the effects of temperature (Lehane and Mail, 1985). The method is less accurate than ovarian dissection (Figure 21.4) for females but has the advantage that it can be used for both sexes. In addition, it is not necessary to have highly trained technical staff on hand to undertake dissections of freshly caught material.

Fly heads can be stored in the dark, desiccated, for up to 12 weeks but thereafter deterioration occurs. The dried head capsule is ground in a glass homogenizer with chloroform: methanol (2:1, v/v), glycine/0.1 M NaOH buffer (pH 10) is added and the whole is vortexed. After centrifugation to separate the two layers, the upper aqueous layer is passed through the flowcell of a Perkin Elmer LS1 fluorimeter fitted with a 340 nm excitation filter and an emission monochromator setting of 450 nm. Fluorescence values are converted to estimates of chronological age on the basis of standard curves determined by linear regression analysis for each sex and species under investigation (Langley *et al.*, 1988).

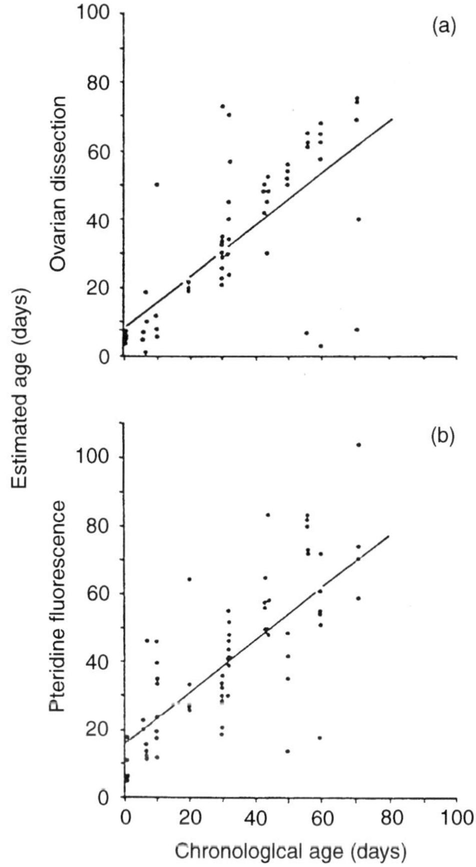

Figure 21.4 Estimates of age of laboratory-reared *Glossina morsitans morsitans* females of known chronological age using (a) the ovarian dissection and (b) the pteridine fluorescence techniques. The lines are drawn from the regression equations: (a) $y = 7.9 + 0.77x$ and (b) $y = 15.9 + 0.77x$. (Reproduced with permission from Langley *et al.*, 1988.)

21.5 PRACTICAL APPLICATIONS

Cycles of feeding activity have been proposed for tsetse in the field based on the measured decline in the amount of residual bloodmeal and the rise and fall of fat reserves between (Randolph and Rogers, 1978; Rogers and Randolph, 1978). Because feeding involves a risk, it has been proposed that tsetse may not begin to seek food until their nutritional reserves fall to pre-determined low levels (Rogers and Randolph, 1986; Randolph *et al.*, 1991, 1992). This concept has been used to support the contention that adult female tsetse may feed only twice or three times during a pregnancy cycle (Rogers and Randolph, 1978, 1986; Randolph and Rogers, 1986), a contention which cannot be supported in terms of the metabolic demands of the fly (Langley and Stafford, 1990).

Under laboratory conditions, normal reproductive effort is sustained only when flies feed four or five times per reproductive cycle (Langley and Stafford, 1990). This is 'explained' by observations of reproductive under-performance by tsetse in the laboratory, related to feeding frequency (Gaston and Randolph, 1993). However, it has been demonstrated that the precise level of nutritional depletion that needs to be reached before a fly will enter a trap depends on the strength of the stimuli received by the fly (Randolph *et al.*, 1989). Hence, flies can be attracted to a potential source of food with higher than expected nutritional reserves. Added to this, wild flies digest bloodmeals even more rapidly than laboratory-reared flies (Langley and Wall, 1990), they must provide fuel for flight activity and their metabolism is temperature-dependent (Bursell and Taylor, 1980). Hence, feeding intervals of 1.9–2.5 days at 25–30°C were calculated in order to keep a female in positive energy balance (Bursell and Taylor, 1980).

Nutritional stress in tsetse results in production of offspring of sub-optimal size (Langley and Pimley, 1975). These are less viable than larger individuals (Phelps and Clarke, 1974). Hence, it seems more logical, even taking into account the risks involved in feeding, that a more 'opportunistic' than 'deterministic' pattern of feeding behaviour should exist among tsetse and that the frequency of feeding in the field probably approaches that measured in the laboratory. Clearly, we have not yet exhausted the value of nutritional analysis of tsetse as a means of interpreting their natural feeding behaviour.

Using thoracic residual dry weight (TRDW) in conjunction with pteridine fluorescence is a means of identifying very young adults (Langley *et al.*, 1990a), the composition of catches of *G. pallidipes* Austen from stationary and mobile baits in Zimbabwe was shown to be markedly different, in that the mobile bait caught a higher proportion of young flies of both sexes than did the stationary bait (Figure 21.5). Ovarian age correlated well with wing fray category as a means of assessing chronological age of *G. tachinoides* Westwood in Ivory Coast (Figure 21.6), while pteridine fluorescence correlated equally well with wing fray index for *G. pallidipes* in Kenya (Figure 21.7).

In view of this background information, pteridine fluorescence and wing fray measurements have been used in conjunction with ovarian dissection data to examine the age structure of tsetse populations. In Ivory Coast, *G. tachinoides* was subjected to control by the use of 20 sterilizing biconical traps, fitted with cotton cones treated with an oil formu-

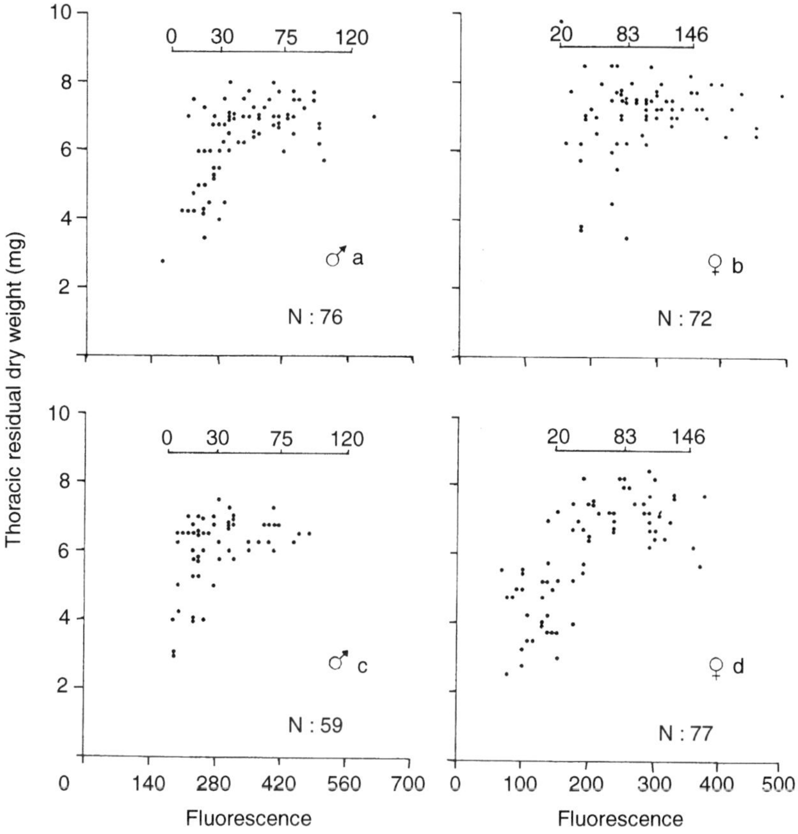

Figure 21.5 Relationship between thoracic residual dry weight and pteridine fluorescence as an indicator of the presence of young flies in catches from a stationary, odour baited (a,b) and a mobile target without odour (c,d). An indication of chronological age in days based on pteridine fluorescence values is given on the horizontal bars above each graph. Age ranges based on regression of fluorescence on chronological age for laboratory-reared flies but using intercepts for field-caught flies. Each point represents one individual fly. N, no. of flies. (Reproduced with permission from Langley *et al.*, 1990a.)

lation of the juvenile hormone mimic pyriproxyfen. The traps were spaced at 300-m intervals along 6 km of the south bank of the Comoe River (Figure 21.8). Flies that enter such traps come into contact with the sterilizing agent as they exit from the top of the trap. A similar control area (separated from the experimental area by 6.5 km of river bank) contained unbaited traps through which flies could pass unheeded.

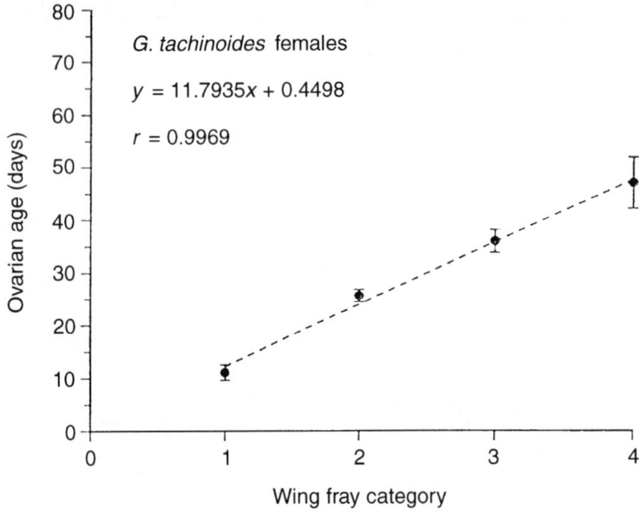

Figure 21.6 Relationship between age as determined by ovarian dissection and wing fray in *G. tachinoides* in Ivory Coast. Vertical bars indicate ± SEM.

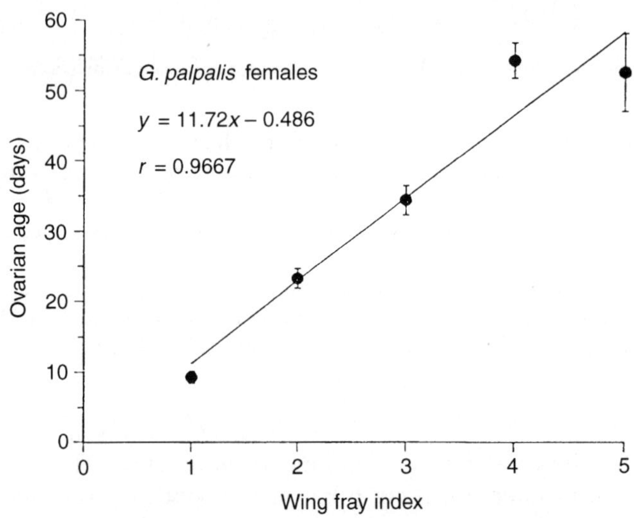

Figure 21.7 Relationship between age as determined by pteridine fluorescence and wing fray in *G. pallidipes* in Kenya. Vertical bars indicate ± SEM.

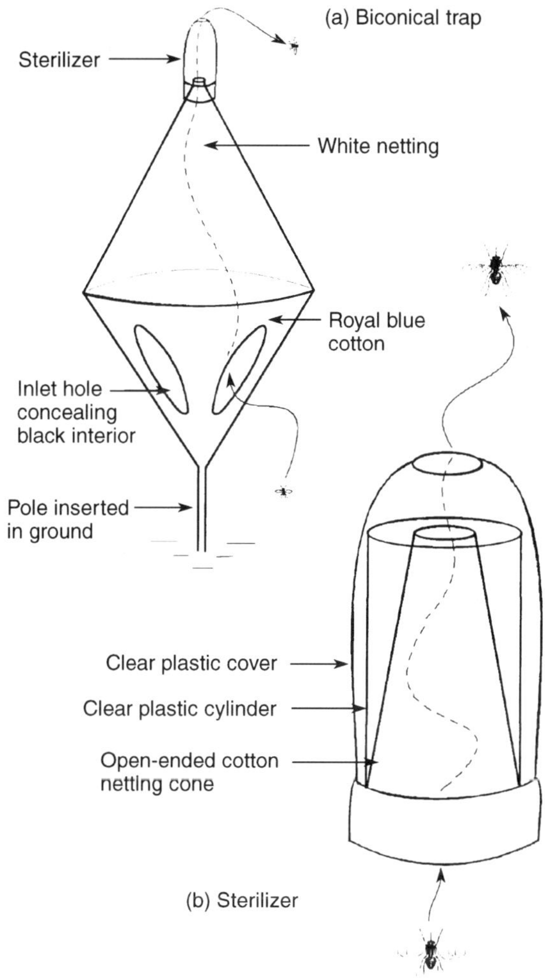

Figure 21.8 (a) Biconical trap fitted with sterilizing cone of cotton cloth treated with oil formulation of the juvenile hormone mimic, pyriproxyfen; (b) detail of sterilizing cone through which trapped flies escape.

Results demonstrated population suppression (Figures 21.9 and 21.10), while the elimination of young flies from the population was indicated by an increase in the ratio of the mean age of flies in the experimental and control areas using wing fray and ovarian dissection (Figures 21.11 and 21.12).

A similar experiment in Kenya (unpublished data) was equally successful against *G. pallidipes* (Figures 21.13–21.15). The fly population

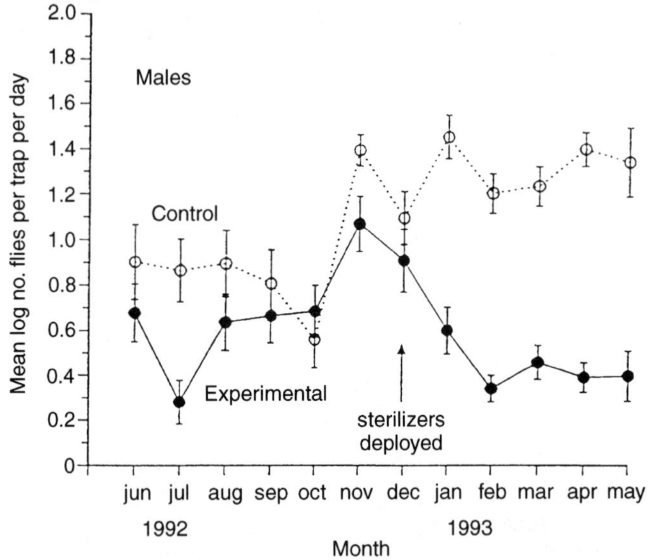

Figure 21.9 Mean $\log_{10}$ numbers of *G. tachinoides* males caught in monitoring traps in experimental and control areas on the Comoe River at Kafolo, Ivory Coast, before and after deployment of sterilizers treated with oil formulation of the juvenile hormone mimic, pyriproxyfen. Vertical bars indicate ± SEM.

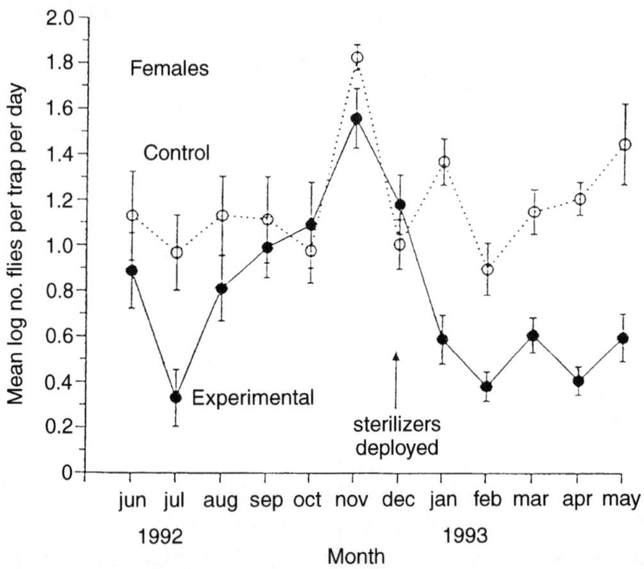

Figure 21.10 Mean $\log_{10}$ numbers of *G. tachinoides* females caught in monitoring traps in experimental and control areas on the Comoe River at Kafolo, Ivory Coast, before and after deployment of sterilizers treated with oil formulation of the juvenile hormone mimic, pyriproxyfen. Vertical bars indicate ± SEM.

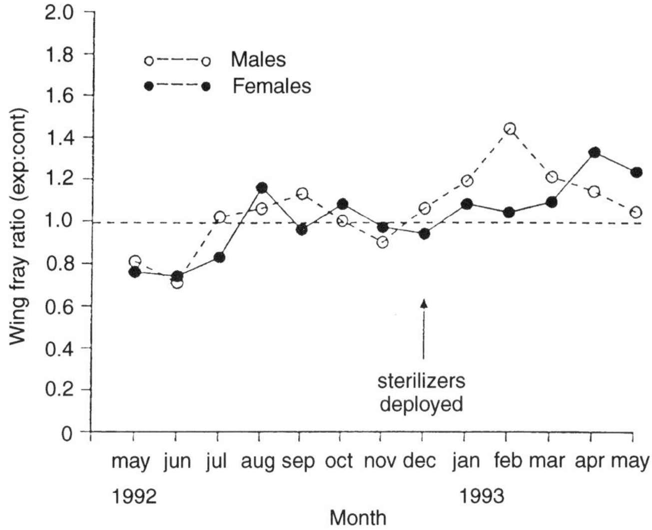

Figure 21.11 Ratio of mean wing fray index (experimental: control) before and after deployment of sterilizers in Ivory Coast in *G. tachinoides*: differences between mean age of experimental and control groups significant at $P < 0.05$ in January and February for males, April and May 1993 for females.

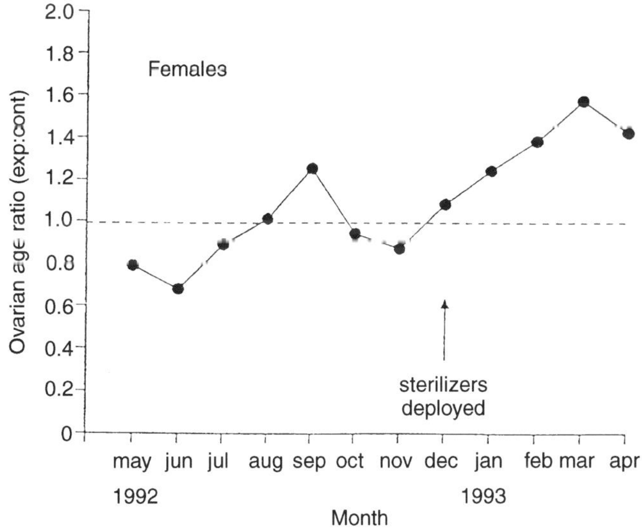

Figure 21.12 Ratio of population age as determined by ovarian dissection in *G. tachinoides* females (experimental: control) before and after deployment of sterilizers in Ivory Coast: differences between experimental and control means significant at $P < 0.05$ from January to April inclusive.

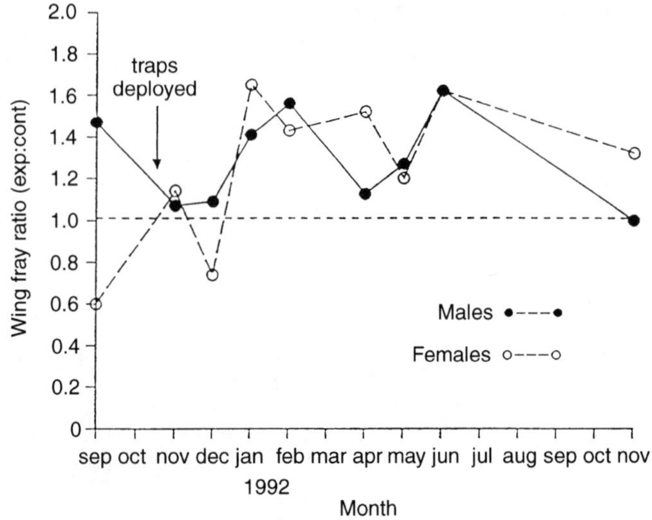

Figure 21.13 Wing fray index ratio for *G. pallidipes* (experimental: control) before and after deployment of sterilizers in Kenya: differences between experimental and control means significant at $P < 0.05$ for January, February, May and June for both sexes.

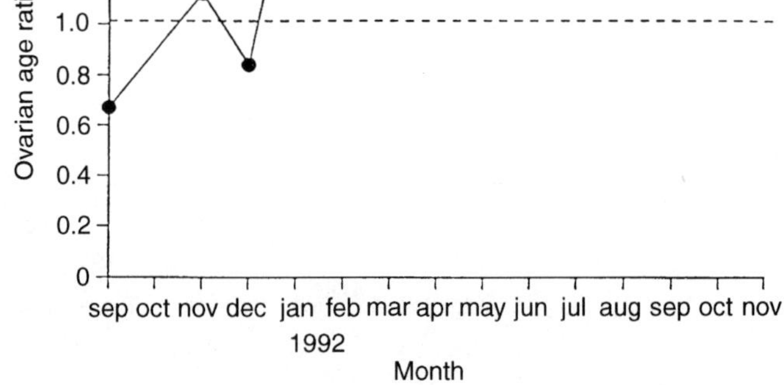

Figure 21.14 Ratio of population age as determined by ovarian dissection in *G. pallidipes* females (experimental: control) before and after deployment of sterilizers in Kenya: differences between experimental and control means significant at $P < 0.05$ for January, February, May, June inclusive.

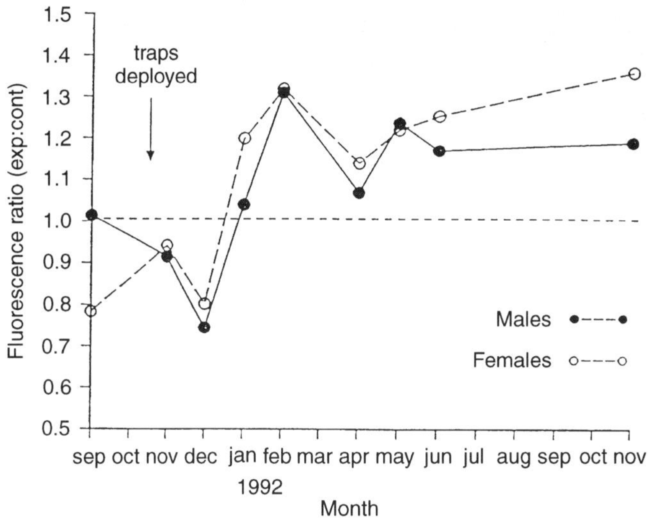

Figure 21.15 Ratio of population age as estimated by pteridine fluorescence in *G. pallidipes* females (experimental: control) before and after deployment of sterilizers in Kenya: differences between experimental and control means significant at $P < 0.05$ between February and November for males; January and November for females.

was suppressed and the incidence of trypanosomiasis in cattle in the experimental area fell to zero within 12 months.

The pteridine fluorescence technique for determining the age of tsetse flies is clearly of great practical value in monitoring the progress of tsetse control programmes involving the use of insect growth regulators.

Further development and a more widespread use of odour-baited, insecticide-treated targets and traps for tsetse control seem inevitable. Therefore, there is an ever-increasing risk of the appearance of either biochemical or behavioural resistance in tsetse populations. These are compelling reasons for using IGRs to effectively sterilize both sexes, rather than relying on conventional insecticides. Hence, the two relatively simple analyses to determine nutritional stage and age in tsetse seem likely to increase in importance as a means of monitoring the progress of tsetse control programmes in the future.

REFERENCES

Bursell, E. (1961) Starvation and desiccation in tsetse flies (*Glossina*). *Entomologica Experimentalis et Applicata*, **4**, 301–10.

Bursell, E. and Taylor, P. (1980) An energy budget for *Glossina* (Diptera: Glossinidae). *Bulletin of Entomological Research*, **70**, 187–96.

Bursell, E., Billing, K.C., Hargrove, J.W., McCabe, C.T. and Slack, E. (1974) Metabolism of the bloodmeal in tsetse flies. *Acta Tropica*, **31**, 297–320.

Cmelik, S.H.W., Bursell, E. and Slack, E. (1969) Composition of the gut contents of third instar tsetse larvae (*Glossina moristans* Westwood). *Comparative Biochemistry and Physiology*, **29**, 447–53.

Ford, J., Maudlin, I. and Humphreys, K.C. (1972) Comparisons between three small collections of *Glossina morsitans morsitans* (Machado) (Diptera: Glossinidae) from the Kilombero River Valley, Tanzania. Part I. Characteristics of flies exhibiting different patterns of behaviour. *Acta Tropica*, **29**, 231–49.

Gaston, K.A. and Randolph, S.E. (1993) Reproductive under-performance of tsetse flies in the laboratory related to feeding frequency. *Physiological Entomology*, **18**, 130-6.

Gillott, C. and Langley, P.A. (1981) The control of receptivity and ovulation in the tsetse fly, *Glossina morsitans. Physiological Entomology*, **6**, 269–81.

Hargrove, J.W. (1994) Reproductive rates of tsetse flies in the field in Zimbabwe. *Physiological Entomology*, **19**, 307–18.

Huyton, P.M. and Brady, J. (1975) Some effects of light and heat on the feeding and resting behaviour of tsetse flies, *Glossina morsitans* Westwood. *Journal of Entomology*, **50**, 123–30.

Jackson, C.H.N. (1946) An artificially isolated generation of tsetse flies (Diptera). *Bulletin of Entomological Research*, **37**, 291–9.

Langley, P.A. and Pimley, R.W. (1975) Quantitative aspects of reproduction and larval nutrition in *Glossina morsitans morsitans* Westw. (Diptera: Glossinidae) fed *in vitro. Bulletin of Entomological Research*, **65**, 129–42.

Langley, P.A. and Pimley, R.W. (1979) Influence of diet on synthesis and utilisation of lipids for reproduction by the tsetse fly *Glossina morsitans. Journal of Insect Physiology*, **25**, 79–85.

Langley, P.A. and Stafford, K. (1990) Feeding frequency in relation to reproduction in *Glossina morsitans morsitans* and *G. pallidipes. Physiological Entomology*, **15**, 415–21.

Langley, P.A. and Wall, R. (1990) The implications of hunger in the tsetse fly, *Glossina pallidipes*, in relation to its availability to trapping techniques. *Journal of Insect Physiology*, **36**, 903–8.

Langley, P. and Weidhaas, D. (1986) Trapping as a means of controlling tsetse, *Glossina* spp. (Diptera: Glossinidae): the relative merits of killing and of sterilization. *Bulletin of Entomological Research*, **76**, 89–95.

Langley, P.A., Bursell, E., Kabayo, J., Pimley, R.W., Trewern, M.A. and Marshall, J. (1981) Haemolymph lipid transport from fat body to uterine gland in pregnant females of *Glossina morsitans. Insect Biochemistry*, **11**, 225–31.

Langley, P.A., Hall, M.J.R., Felton, T. and Ceesay, M. (1988) Determining the age of tsetse flies, *Glossina* spp. (Diptera: Glossinidae): an appraisal of the pteridine fluorescence technique. *Bulletin of Entomological Research*, **78**, 387–95.

Langley, P.A., Hargrove, J.W. and Hall, R.W. (1990a) Maturation of the tsetse fly *Glossina pallidipes* (Diptera: Glossinidae) in relation to trap-orientated behaviour. *Physiological Entomology*, **15**, 179–85.

Langley, P.A., Felton, T., Stafford, K. and Oouchi, H. (1990b) Formulation of pyriproxyfen, a juvenile hormone mimic, for tsetse control. *Medical and Veterinary Entomology*, **4**, 127–33.

Lehane, M.J. and Mail, T.S. (1985) Determining the age of adult male and female *Glossina morsitans morsitans* using a new technique. *Ecological Entomology*, **10**, 219–24.

Phelps, R.J. and Clarke, G.P.Y. (1974) Seasonal elimination of some size classes in males of *Glossina morsitans morsitans* Westw. (Diptera: Glossinidae). *Bulletin of Entomological Research*, **64**, 313–24.

Pilson, R.D. and Pilson, B.M. (1967) Behaviour studies of *Glossina morsitans* Westw. in the field. *Bulletin of Entomological Research*, **57**, 227–57.

Randolph, S.E. and Rogers, D.J. (1978) Feeding cycles and flight activity in field populations of tsetse (Diptera: Glossinidae). *Bulletin of Entomological Research*, **68**, 655–71.

Randolph, S.E. and Rogers, D.J. (1986) The use of discriminant analysis for the estimation of sampling biases for female tsetse. *Ecological Entomology*, **11**, 205–20.

Randolph, S.E., Dransfield, R.D. and Rogers, D.J. (1989) Effect of host odours on trap catch composition of *Glossina pallidipes* in Kenya. *Medical and Veterinary Entomology*, **3**, 297–306.

Randolph, S.R., Rogers, D.J., Dransfield, R.D. and Brightwell, R. (1991) Trap-catches, nutritional condition and the timing of activity of tsetse *Glossina longipennis* (Diptera: Glossinidae). *Bulletin of Entomological Research*, **81**, 455-64.

Randolph, S.E., Williams, B.G., Rogers, D.J. and Connor, H. (1992) Modelling the effect of feeding-related mortality on the feeding strategy of tsetse (Diptera: Glossinidae). *Medical and Veterinary Entomology*, **6**, 231–40.

Rogers, D.J. and Randolph, S.E. (1978) Metabolic strategies of male and female tsetse (Diptera: Glossinidae) in the field. *Bulletin of Entomological Research*, **68**, 639–54.

Rogers, D.J. and Randolph, S.E. (1986) Detection of activity cycles of tsetse from capture–recapture data. *Ecological Entomology*, **11**, 95–109.

Saunders, D.S. (1960) The ovulation cycle in *Glossina morsitans* Westwood (Diptera: Muscidae) and a possible method of age determination for female tsetse flies by the examination of their ovaries. *Transactions of the Royal Entomological Society of London*, **112**, 221–38.

Taylor, P. (1979) The construction of a life-table for *Glossina morsitans* Westwood (Diptera: Glossinidae) from seasonal age-measurements of a wild population. *Bulletin of Entomological Research*, **69**, 553–60.

Tobe, S.S. and Langley, P.A. (1978) Reproductive physiology of *Glossina*. *Annual Review of Entomology*, **23**, 283–307.

Vale, G.A. (1974a) New field methods for studying the responses of tsetse flies (Diptera: Glossinidae) to hosts. *Bulletin of Entomological Research*, **64**, 199–208.

Vale, G.A. (1974b) The responses of tsetse flies (Diptera: Glossinidae) to mobile and stationary baits. *Bulletin of Entomological Research*, **64**, 545–88.

Vale, G.A., Hargrove, J.W., Cockbill, G.F. and Phelps, R.J. (1986) Field trials of baits to control populations of *Glossina morsitans morsitans* Westwood and *G. pallidipes* Austen (Diptera: Glossinidae). *Bulletin of Entomological Research*, **76**, 179–93.

Wall, R. and Langley, P.A. (1991) From behaviour to control: the development of trap and target techniques for tsetse fly population management. *Agricultural Zoology Reviews*, **4**, 137–59.

Williams, B.C., Dransfield, R.D. and Brightwell, R. (1990) Tsetse fly (Diptera: Glossinidae) population dynamics and the estimation of mortality rates from life-table data. *Bulletin of Entomological Research*, **80**, 479–85.

Index

Numbers in **bold** refer to figures; *italic* to tables.

α-glycerophosphate dehydrogenase 100
Abax parallelepipedus *277*, 325, 328–31
Acanthospermum hispidum 377
Acari 9
Acarina *272, 275*
Acetylcholinesterases 100
  assay of 37
  resistance mechanism 43
Aculus
  fockeui 461
  schlechtendali 461
Acyrthosiphon *275*
  pisum 155, 179, 201, *275*
  spartii *272*
Adenylate cyclase, in *Liriomyza* 12
Aedes
  aegypti 40, 179
  albopictus 59, 157–9, 161, 244–5
  contans *273*
  dorsalis *274*
  scutellaris 58, 163
African armyworm 14
Agelaius phoeniceus 139
Ageniaspis fuscicollis 9
Agglutination immunoassays 280, 287
Agonum dorsale 439
Agrimonia 11
Agriolimax reticulatus, see Deroceras reticulatum *273*
Agrotis ipsilon *273, 274*
Alarm pheromones 431
Alaskozetes antarcticus 462
Alfalfa 425
Allelle frequencies 150
  in *Yponomeuta padellus* 12, 14, 20–1
Allozymes 8–15, 150–9, 175–7, 183–4, 250
  aphids 158–9, 163, 175
  brown planthopper 137
  moths 176–7
  mosquitos 156–7
  Yponomeuta 17, 23
  weevils 51–3, 407
Amblyospora 407
Amblyseius
  fallacis 461
  finlandicus 461
  potentillae 461
Amelanchier lamarckii 17
American eels 139
Amphorophora idaei 159
Anagrus epos 9
Anaphes 181
Anastrepha fraterculus 11
Anax parthenope 181
Anguilla rostrata 139
Annelida 11
Anopheles 462
  albimanus 156
  arabiensis 34, 41–3, **157**, 179
  DDT resistance 39–43
  dirus 34, 163
  gambiae 32–6, 39–45, 156, 179, *273*
  infection, species identification, field detection 31–49
  melas 41–3, 156
  merus 156

Anopheles contd
quadriannulatus 156
quadrimaculatus 272
Anthicid 375
Anthocorid bugs *277*, 375, 425, 433, 461
Anthocoris
nemoralis 461
nemorum 461
Anthomus grandis 469–78
Antibody
choice of, for predator gut analysis 349–50
combinatorial infection 361
engineering 337–8, 351–8, 423
gene library construction 351–8
gene library selection (panning) **356**
generation of diversity 351
genetics 350–1, **352**
semi-synthetic gene libraries 361
structure 350–1, **352**
see also Antisera; Immunoassays; Monoclonal antibodies
Antigen purification 270
Antigenic determinants (epitopes) 268
Antisera 323, 385
anti-slug 327–9
predator gut analysis 349, 422
production of 267–9
purification 270–1
Ants 178, 375, 426, 432
AP-PCR, *see* PCR
Aphelinidae 431
Aphelinus asychis 432, 434
Aphids
dispersal of by predators 421
esterase dot-blot **169**
fungus infection of 150
grain 163, 164, **167**
lacewing control 433
parasitism in 150
parasitoids of 154, 156, 180, 232, 430–1, 434
see also Aphidius
phylogenetic tree of holometabolous insects **155**
plant resistance to 209
see also Brevicoryne brassica
predation of 439–40
predator gut analysis of *275*, 341, 423, 424
RAPD-PCR 179–81
rDNA in phylogenetic analysis 152, 157–8,
scavenging of 425–6
wasteful killing of 435
see also Individual species
Aphidius 154–6
colemani 180
ervi 154, 156, 430–1
matricariae 180
microliphii 154, 156
picipes 180
rhopalosiphi 232
smithi 154, 156
urticae 154, 156
Aphis
fabae *276*
gossypii 433
Apis
mellifera 139, **168**, 177, 178, 241, 249, 411
see also Bees
mellifera capensis 164, 165
mellifera scutellata 165
Apodemus sylvaticus *273*
Apple 16, 17
Arachnida *272*, *273*
Arachnocoris 425
Araneae *272*, *273*
Arion 93–131
ater 97–101, 110–13, 115–25, *277*, 324, 329, 337
ater rufus 94, 97–8, 101–7, 111–13, 115–17, 121–5
empiricorum 98–9, 108, 111–3
flagellus 95, 99–107, 110–15, 117, 121–5
hortensis 328
intermedius 324, 334, 337, 362
lusitanicus 94–125
subfuscus 99, 113, 337
sulcatus 108, 111–13, 122
see also Slugs
Armyworm 163
Artioposthia triangulata 338
Arytaina
genistae *272*
spartiophila *272*
Asciodema obsoletum *272*
Ascites 336
see also Monoclonal antibodies

Asellus *272*
meridianus *274*
Aspergillus 251
Augochlorella striata 163

Bacillus thuringiensis 74, 427, 429–30
Bactrocera
cucumis 246, 247–8
cucurbitae 246
dorsalis 246
passiflorae 246–8
tryoni 246, 247–8
xanthodes 246, 247–8
Bathyplectes
annurus 64
curculionis 63–4
stenostigma 64
Beans 403
Bednets, insecticide impregnated 32, 39
Bees 163, 165
RAPDs 180
see also Apis
Bembidion lampros *276*
Bemesia tabaci 180, 233, *276*, 383, 385–93
Bicyclus anynana 163
Bioassays, insecticide resistance 81–7
Biotypes
alfalfa weevils 53–71
aphids 233
electrophoretic analysis 10
host races 16
Bird cherry 17
Black fly 54, 243
Blackthorn 17
Blepharidopterus angulatus 461
Blood meals 421
of tsetse flies 486
Boll Weevil Eradication Program 470
Bollworm 301–21, 375
Bombyx mori 165
Bottlenecks, *see* Population bottlenecks
Brassica
fruticulosa 199, 201–6
napus 208
oleracea 199, 201–6
spinescens 199, 201–6
wild, resistance mechanisms to cabbage aphids 199, 201–6
Brevicoryne brassica 199, 201–8

Broom beetle 292
Bushcricket 164
Butterflies 163, 174

Cabbage looper 386
Cabbageworm 302
Cadra figulilella 59
Calosoma sayi 425
Campoletis sonorensis 66
Cannibalism 371, 377
Canonical discriminant analysis 158
Cantharid larvae 433–4
Capitella capitata 11
Carabid beetles *275*
hyperpredation by 433
lepidopteran larvae predation by 461
pre-oral digestion by 433–4
prey size preference 437
scavengers 425–6, 430
slug predation by 323–6, 328–36
wasteful killing and partial consumption 435
see also Individual species
Carabus violaceous 325
Carbamate resistance 40
Carrion 275
Catalase 11
Caterpillars 421, 425, 438, 439
parasitism 428–9
Catochrysops strabo 377
Cecidomyiid larvae 439
Celastraeceae 11, 17
Cell cloning 336
see also Monoclonal antibodies
Cell fusion 336
see also Monoclonal antibodies
Cellulose acetate 459
Centipedes 433
Ceratitis capitata 163, 165, 176, 246
Ceratopogonidae 462
Ceratovacuna nekoashi *183*
Cereals 135, 429
Chain shuffling, recombinant antibodies 360
Cheiracanthium
inclusum 375
inornatum 367–9, 371–7
Chemiluminescence immunodot assay 305, 312, 314
Chesias legatella 272
Chilo 273

Chilopoda *275, 276*
Chironomus 274
Choristoneura fumiferana 272
Chromosome analysis 156
Chrysanthemum 12
Chrysomelid beetles 55, 163, 435
Chrysoperla carnea 277, 411
Chrysopidae 375, 425, 440
Cicadas 176
Cicindela 174
Clinal variation 11
Cloeon dipterum 462
Cloning
  hybridomas 279
  antibody genes 359
Coccinella
  7-punctata 421
  septempunctata 275, 411
Coccinellidae 180, 341, 375, 421, 425, 440
Coenagrionidae *276*
Coleomegilla maculata 275
Coleoptera 155, 174, *272, 276*, 367
  endosymbionts of 59
  lectins as insecticides 201
  rDNA gene sequences 152, 153
  see also Carabid beetles and Individual species
Collembola *275*
Collops vittatus 277, 383, 388, 389, 392
Combinatorial antibody gene libraries 341, 347–66
  see also Antibody engineering
Combinatorial infection 361
Complementarity Determining Regions (CDRs) of antibodies 350–1
Copepods 438
Coquillettia insignis 276
Cordylochernes scorpioides 165
Corixa punctata 462
Corn borers 401–17
Cotesia marginiventris 301, 303, 306
Cotton 75
Cotton bollworm 301–21
Crataegus 18–22
Crenobia alpine 274
Cross-reaction tests, of antisera 327
Crustacea *276*
Cryptopygus antarcticus 462, 464
Cucumber 12
Culex
  pipiens 58, 83
  quinquefasciatus 170
  tarsalis 276
Cultivation methods, effects on predation 323
Cyclodienes 38
  resistance, of *Drosophila melanogaster* 82, 85–7
Cyclops 462
Cytochrome C 105

Dactylis glomerulata 181
Damselflies 425
Danaus plexipus 15, 174
Daphnia 274
Datura metel 378
DDT 85
  resistance, in *Anopheles* 39–43
  WHO discriminating dosage 41
Delphacidae, *see* Planthoppers
Denaturing gradient gel electrophoesis (DGGE) 137
Dendroctonus frontalis 275
Dendrograms 150, 152–3, 173
Densitometry 457
Dermaptera 425
Deroceras
  agreste 118
  caruanae 94, 324, 337
  reticulatum 94, *277*, 324, 328, 334, 337–9, 423
  see also Slugs
Diaeretiella rapae 180
Diamond black moth 74
Dieldrin resistance 86–7
Diglyphus begini 181
Dinucleotide repeats, mosquito blood meal DNA 39
Diptera 11, 83, 85, 243, 244, *275*, 425
  endosymbionts of 59
  rDNA gene sequences 153, 154
  see also Anopheles; Glossinia
Dipteran larvae 435
Direct squashes, to detect insecticide resistance 81–2
Discus rotundatus 338
Distance-Wagner method 173
Diuraphis noxia 179, 180
DNA
  amplification fingerprinting (DAF) 105

analysis methods, of pest species 233–5
see also Specific techniques
dot blots, 35–6, 38
fingerprinting *40*, 137–8, 159–70, **166,** 183–4, 234
hybridization, phylogeny 171
mitochondrial 154, 173–8, 184, 217–19, *235*, 243
alfalfa weevil 51–2, 54–8
desert locust 213, 218–28
population genetics 80–1
planthoppers 138–9, 141–3
nuclear 51, 139–3, 153, 170, 217–28, 235, 245
RAPD, *see* RAPD-PCR
RFLP 144, 152, 154, 156, 159, 161–2, 173, 177, 458
alfalfa weevils 58
Apis mellifera **168**
fruit flies 247, 250
mtDNA 80
REN 85, 86
ribosomal DNA analysis 31, 40, 42, 156, 235, 243–4, 247–8
Anopheles 34–5
Diptera 153–4
phylogenetic analysis 151
population structure and dynamics 157–9
sequencing of microsporidia 406–7, 411
tandem arrays 35
species-specific probes 34
sequence analysis 51, 53–8, 215, 217–18
software 55
sequence evolution **234**
techniques, in predation analysis 341, 458
technology, to detect insecticide resistance 82–6
Dot blot assays 38, **169**
Dragonflies 181, 433
nuclear genes 224
Drosophila 38, 83, 85, 154, 156, *275*
melanogaster 82, 85, 86, 141, 154, 156
mercatorum 139
pipientis 59
planitibia 141
pseudoobscura 23
simulans 63
yakuba 54, 154, 176, 177, 243
Dryinid parasitoids 434
Dryocosmus kuriphilus 274
Dugesia
polychroa 274
tigrina 273
Dytiscidae 462

Earthworm 327, 339
Earwigs 426
Ecological genetics 73–92
Ecological models 122
Ectoparasitoids, *see* Parasitoids
Elateridae *275*, 434
Electromorphs 8, 33
Electronic recording of behaviour, aphid feeding 200, 202–4
Electrophoresis
agarose 157, 162, 239
cellulose acetate 8–10, 76, 459
characterization of microsporidia 406–7
DGGE 241
enzyme 7–29, 76, 79–81, 100–4, 168, 170, 424, 469–73
heliothine pests 76, 79–80, 81
isoelectric focusing 8, 469, 476
predator gut contents 384
predator-prey interactions 457–68
PAGE 8, 83, 241, 304, **310–11,** 457, 459, 469–73
polyacrylamide, *see* PAGE
pulsed field 406, 469, 476
quantification of 462
SDS-PAGE 205–6, 406
SSCP 241
starch gel 8, 33–4
TGGE 241
ELISA
general techniques and applications 280–2, 285–7, 421–3, 458
Helicoverpa armigera 367, 370–5
heliothines 78
malaria parasites 38
Microplitis croceipes 304, 306
microsporidia 407–9
pink bollworm eggs 383, 387–90
slug predation analysis 323, 326, 327–30, 334–41, 362
whitefly eggs 383, 387–90

EMS mutagenesis 86
Enchytraeidae *275*
Endives 12
Endoparasitoids, *see* Parasitoids
Endosymbionts
  of alfalfa weevil, see *Wolbachia*
  of mosquitoes 58
  of moths 59
Entomogenous fungus 150, 425, 427–8, 429, 430
Entomomophthoraceous fungi 429
Enzyme analysis, *see* Specific enzymes; Electrophoresis
  of scavenging 424
  of weevils 469–78
Enzyme polymorphism, in aphids and Hymenoptera 149
Eogammarus confervicolous 144
Ephemeroptera *274*
Ephestia
  cautella 59
  kuehniella 164
Ephydra riparia *274*
Epiphyas postvittana 273
Epitopes (antigenic determinants) 268, 349
Eriborus terebrans 403
Eroischia brassicae *272*
Esterases 38, 167, 457, 460
  Anopheles 33, 37, 43
  detect insecticide resistance 81–2
  genetics of, cotton boll weevils 469–78
  scavenging 424
  Spodoptera frugiperda 15
  see also Electrophoresis, enzyme
Euphydryas editha, hexokinase 15
Eurygaster integriceps *273*
Euxoa ochrogaster *273*
External transcribed spacers 151

F-statistics 150
Fab antibody fragments 358–9
Fenusa dohrnii 461
Fenvalerate 83
Field sampling
  H. armigera eggs 369–70
  light traps 39–40
  pitfall traps 329–31
  randomized block design 331
  tsetse flies 481–3, **491**
  whitefly in cotton 386–7
Fleas 152
Forficula auricularia *275*
Formicidae 425
Formicomus 367, 368, 371, 372–4
Fragaria 11
Fruit flies 16, 231, *240*, 245–6
  PCR primers for **240**
  see also Drosophila
F_{ST} 14–15, 22
Fungi *275*, 429
  entomogenous 425, 427–8, 429, 430
Fungicides 426

GABA receptor assay 37–8
Galanthus nivalis (snowdrop) lectins as insecticides 201, 210
Gall midges 440
Galleria mellonella 411
Gamasellus racovitzai 462, 463, 464
Gammarus *274*
Gastropoda *273*
  see also Mollusca; Slugs
Gene analysis, *see* Phylogenetic analysis and DNA
Gene flow 13–24, 31, 150, 173–4
  brown planthopper 133–48
  desert locusts 222–3
  heliothine pests 74, 79–81
  insecticide resistance 73, 79–81
  measurement of 13–15
  N_eM 14–16, 22
  practical examples 16–24
  slugs 105–14
  sources of bias 15–16
Gene frequency 14–16
  arionid slugs *106*
  Yponomeuta *20–1*
Gene trees
  alfalfa weevils **57**
  holometabolous insects **155**
  locusts **220, 221**
  planthoppers **143**
Genetic drift 15, 159
  see also Gene flow
Genetic markers, methods **235**
Genetically engineered crops, insect resistant brassicas 200, 201, 209
Genomic subtractive hybridization technique 170
Geocoris 375, 383, 389, 392
  pallens 389

punctipes 277, 389
Geomalacus maculosus 96–7, 107–9, 114
Gerris lacustris 462
Geum 11
Giemsa staining technique 59, 61
Glossina
genetic variation 165
morsitans 484–7
pallidipes 488, 491, 494–5
palpalis 488, 490
swynnertoni 484–5
tachinoides 488, 490, 492–3
Glucose-6-phosphate
dehydrogenase 76
in *Liriomyza* 12
Glucosephosphate isomerase 11
Glutamate-oxaloacetate transaminase 33, 100
Glutathione S-transferase, GST
in insecticide resistance 42–4
measurement of 37–8
substrates of 42
Gonomelus anceps 273
Graphognathus leucoloma 232
Grasshopper 180
Greenbug 158, 159, **160**, 175, 179
Gut analysis of slug predators 323–45

Haemocyanin, slug 327
Haemolymph, slug 327–9, 331
Haliplidae 462
Hardy-Weinberg Law 119, 120, 121, 150, 162
Harpalus pennsylvanicus 275
Harvestmen 425, 433
Hawthorn 16, 17
Helicoverpa 170
armigera 74–87, 367–81
assulta 378
detection of parasitism in 301–21
microsporidia infection of 409
peltigera 377
punctigera 75–7, 79–80
zea 75, *275*, *276*, *277*, 375, 423
Heliothine pests
gene flow 79–81
identification, antibodies 75
identification, isozymes 75–7
identification, mtDNA 80–1
insecticide resistance of 74
species identification 75–9
Heliothis virescens 15, 66, 75, 83, 84, 170, *277*, 375
Helix aspersa 338
Helper phage 355
Hemiptera *275*, *276*
Heterocordylus tibialis *272*
Heteroptera 367, 421, 435, 439
Hexokinase 15
Hippodamia convergens 383, 389, 392, 393
Holometabolous insects 152
Homoptera 174
electrophoretic analysis of 9
endosymbionts of 59
Honeybees 139
RFLP analysis 58
see also Apis mellifera
Hopperburn in rice 135
Hornet 153
Horseshoe crabs 139
Host races 16
Houseflies, insecticide cross-resistance 39
Hoverfly 433, 435
Hybridoma 304
production 271, **269**, **278**–9
technology 323, 347, 385, 423
Hydractinia symbiolongicarpus 241
Hydroporus 462
Hylemya 425
Hymenoptera 174, 301, 303
electrophoretic analysis of 9
endosymbionts of 59
rDNA gene sequences 152, 153
Hypera
brunneipennis, 53
see also Weevils, alfalfa
postica 51–71, *275*
punctata 53, 55, 57
Hyperpredation 432–3

Immunization 267
Immunoassays
choice of antibody 349–50
malaria parasites 38
microsporidia 407–9
predator gut analysis 279–91, 347, 348, 392–5
to detect insecticide resistance 81–2
see also Under specific assay type
Immunoblot assay 60–2, 304, 306–7

Immunocytochemistry, of brassica lectin 208
Immunodiffusion 280
Immunodot assay 282–3, 285–6
  Microplitis croceipes 301, 304–6, 307–18
  Myzus persicae **169**
Immunoelectrophoresis 270, 284–5
Immunosquash technique, for heliothine species identification 76, 77–9
Inbreeding coefficient, F_{ST} 14–15, 22, 150
Indophenol oxidase 100
Inopus rubriceps 275
Inquilinism 178
Insect
  activity-density 333
  blood meal analysis 32
  growth regulators 482
  resistant crops, brassicas 201–10
Insecticide, demeton-S-methyl 426
Insecticide resistance
  bioassays of 36, 81–2
  biochemical assays 36–7
  detection of 36–9
  enzyme electrophoresis 81–2
  genes associated with 82, 167
  genes, detection of 81–2
  heliothine pests 73–92
  immunoassays 81–2
  isozyme/allozyme markers 150
  malaria control 39–45
  molecular analysis of 37–8
  see also Individual insecticides
Insecticide susceptibility status 39
Internal transcribed spacers 151
Intraspecific differentiation 11
Isocitrate dehydrogenase 12, 76
Isoelectric focussing 8, 304, **309,** 459
Isopoda *275*
Isozyme analysis 33–4, 149, 233, 457–65
  species identification 75–7, 150–1
Ixodes ricinus 274

"*kdr*"-type mechanism 37–9

Lacewings 433
Lactuca virosa 204
Ladybirds 421, 433, 437
Lampides boeticus 377
Lasioglossum marginatum 165
Leafhopper phylogeny 156
Leafminer fly 11, 12
Leatherjackets 341, 434
Lectins
  in brassicas 201–10
  insecticidal properties 201, 205–8
  in snowdrop 201
Leersia hexandra 136–7, 142, 144
Legumes 51, 75
Lehmannia marginata 338
Lepidoptera, 9–10, 15, 23, 74, 174, 302, 461
  endosymbionts of 59
  lectins as insecticides 201
  see also Individual species
Leplothorax acervorum 178
Leptinotarsa decemlineata 175
LepTon™ Test Kit, of *Helicoverpa armigera* 77–8
Leptosceridae 462
Lettuce 12
  pest resistance in 204
Leucine aminopeptidase 11, 100
Light traps 40, 75–6
Limax
  ater 97
    see also Arion ater
  pseudoflavus 118
  rufus 97
  see also Arion rufus
  see also Slugs
Limulus 139
Linyphiidae *276*
Liptocorisa oratorius 273
Liriomyza 9, 12
  bryoniae 12
  huidobrensis 12
  sativae 12
  strigata 12
  trifolii 12
Listronotus bonariensis 180, 231–63
Lochea 96, 112–13
Locust 213–29
Locusta migratoria 154
Log-dosage probit lines 40
Lopidea nigridea 276
Lucilia cuprina 83, 85
Lumbricidae *275*
Lutzomyia
  spinicrassa 179
  youngi 179

Lydella thompsoni 403, 411
Lygus
  hesperus *276*, 383, 389–90, 392, 423
  lineolaris *275*
Lymantria dispar 176, 177, *276*
Lysiphlebus testaceipes 180

Macrocentrus grandii 402, 403, 411–14
Macrohaltica jamaicensis 163
Macrosiphum euphorbiae 426
Magicicada *154*
  septendecim 176
  tredecim 176
Maize 375
Malacosoma
  americanum 23
  disstria 272
Malaria 31–49, 156
  control programme 39
  detection of 38
  parasites 31
Malate dehydrogenase 100
Malic enzyme 12
Mannose phosphate isomerase 100
Mansonia dyari *276*
Mecoptera, rDNA gene sequences 153
Medfly 176
Megachile rotundata 165
Mendelian markers 104
Menochilus sexmaculatus 367, 368, 370, 371, 373–**5, 377**
Metaldehyde 325
Methiocarb 324
Metopolophium dirhodum 164, 232, 249
Mice, as pests 427
Microctonus
  aethiopoides 51, 64–7
  colesi 64
  croceipes 301–21
  hyperodae 249, 252
Microsatellite DNA analysis 31, 34, 170–1, **172,** 215, 216
Microsporidium 401–17, 428
  see also Nosema
Microtonus hyperodae 249
Milacidae 94, 337–8
  see also Tandonia
Milichiid flies 425
Minisatellites 161, 164
Mirid bugs 425, 426, 433, 439, 461
Mites 437, 461, 462
  electrophoretic analysis of 9
Mitochondrial DNA (mtDNA), *see* DNA, mitochondrial
Molecular clocks 151, 152, 224
Molecular polymorphism, of cotton boll weevils 469
Molluscs 174
  monoclonal antibodies 423
  predation, *see* Slug predation analysis
  see also Slugs; Individual species
Monoclonal antibodies
  Helicoverpa armigera 76, 77, 370
  malaria parasites 38
  Microplitis croceipes 301, 303–4, 306–18
  predator gut analysis in general 287, 347, 349–50, 383–5, 458
  production techniques **269**, 271, **278**–9
  quantifying predation 422–3
  scavenging 424
  slugs 323, 336–41, 349
  whitefly and bollworm 391–3
  see also Antibody, engineering and immunoassays
Monoxygenases 38
Moribund prey in predation analysis 427, 428
Mosquitoes 38–41, 156, 157, 170, 179
  blood meal, DNA 39, 421
  endosymbionts 58
  see also Anopheles
Moths 59, 170, 177
  see also Individual species
Mountain ash 17
mtDNA, *see* Mitochondrial DNA (Mt DNA)
Musca 434
  domestica 84
  vetustissima *276*
Mutagenesis, of recombinant antibodies 360
Myelomas 271, 278, 336
  see also Monoclonal antibodies
Myrmica 171
Mytilis edulis 174
Myzus persicae 167, **169**, 170

Nabis 375, 388, 389–90, 425, 433
  alternatus 388, 390
Nagana 480
Naphthyl acetate 37, 43

Nasonia wasp 59
Nasonovia ribis-nigri 204
N_eM 14–16, 22
Nematodes 124, *275*, 325, 427
Nephotettix conciteps 201
Nettle 205
Neuroptera, rDNA gene sequences 153
New Zealand, introduced pests 231–63
New Zealand flatworms 338
Nezara viridula *275*
Nicrophorus tomentosus 181
Nilaparvata
  bakeri 133–48
  lugens 133–48, 201
  maeander 133–48
  muiri 133–48
Noctuidae 84, *275*, 425
Non-transcribed (Intragenic) spacers 151
Nosema
  apis 409
  bombycis 407, 409
  furnacalis 401–2, 405–6
  host range 412–3
  immunoassay of 407–9
  locustae 409, 411
  non target effects of 409–12
  meligethi 412
  pyrausta 401–17
Notonecta
  glauca 462, 464
  undulata 411
  viridis 462
Notoxus calceratus 375
Nuclear DNA, see DNA
Nuclear polyhedrosis viruses 428
Nuclear RNA, *see* RNA
Nymphalidae 431

Octanol dehydrogenase 33, 156
Odonata *276*
Oligochaeta *274*
Oncopeltus fasciatus *275*
Onions 403
Operational taxonomic units (OTU's) 98
Operophtera brumata *273*
Orchard pests 461
Organophosphate resistance 38, 40, 85
Orius
  (*Dimorphella*) *maxidentex* 375
  insidiosus, pirate bug 383, 386, 390–3
  minutus 461
  tantillus 367, 368, 370, 371, 373–5, 377
  tristicolor *277*, 383, 389, 391, 392
  vicinus 461
Orthetrum coerulescens 181
Orthoptera 174
  see also Locust
Orthotylus
  adenocarpi 272
  virescens *272*
Orygia pseudotsugata *275*
Oryza sativa 135
  see also Rice
Ostrinia
  furnacalis 401–2, 405–6
  nubilalis *275*, 401–17
Ouchterlony test 280
Oxidases in insecticide resistance 83

p-nitrophenyl acetate 37
PAGE, *see* Electrophoresis
Panning, of phage display antibodies **356,** 357
Panonychus ulmi 464
Parabemisia myricae 180
Paracymus *462*
Parasite infection, in *Anopheles* 32
Parasitic nematode, of slugs 325
Parasitic wasps, *see* Individual species
Parasitism 170, 301, 302, 429–30
Parasitoids 9, 170, 301, 302, 428, 429
  of alfalfa weevils 51–71
  of aphid 180, 232, 249
  of corn borer 401–17
  eggs 428
  host-feeding by 431, 433
  of hymenoptera 63–4
  of *Liriomyza* 13
Pardosa 328
  ramulosa *274*
  sternalis *275*
Paristoma octooculata 462
Pathogenesis related proteins, in plants 208
PCR 138, 151, 152, 235–6
  of alfalfa weevils 51
  allele-specific techniques 87

arbitrarily primed PCR (AP-PCR) 105, 181
assays of *Anopheles* 34–5
based detection of *Wolbachia* 60–2
blood meal identification 291
to detect parasite infection 39
mismatch methods 85
of mtDNA, to detect insecticide resistance 80
RAPD, *see* RAPD-PCR
of rDNA 31, 34–5, 40, 42, 156
RFLP **168**, 215, 217–18
sample preparation 236–8
SSCP 215, 216, 218
techniques 238–42
vector-pathogen diagnosis 291
with universal primers 242–9
Pea aphid 201
Pectinophora gossypiella 277, 383, 385–93, 423
Pelicopoda *273*
Pentatomid bugs 421
Peppers 403
Peptidase 100
Periplaneta americana 277
Permethrin 41
Pest resistance of crops 200–1
Pesticide resistance, *see* Insecticide resistance
Phaenocora typhlops 274
Phage display 351–60
Phagemids 357
Phalangium opilio 277, 327
Phasmarhabditis hermaphrodite 325
Pheromone traps 75, 76, 84
Phidippus audax 276
Philoscia muscorum 273
Phlebotomus papatasi 34
Phoma lingam 208
Phosphoglucomutase 10–1, 100, 156
Phosphogluconate dehydrogenase 76
Phylogenetic analysis 109, 149, 181
aphids 233
biochemical techniques for 215–18
brown planthopper 133–48
desert locusts **220, 221**
genetic distance 152
holometabolous insects 155
maximum parsimony method (PAUP) 55, 57, 242
neighbour-joining method (MEGA) 55
phylogenetic inference package (PHYLIP) 109, 143, 219–21, 242
slugs 93–131
weevils 51–71
see also Population genetics; Gene flow; Gene Trees
Phytodecta olivacea 272
Phytonomus
brunneipennis 53
murina 53
posticus 53
transsylvanicus 53
variabilis 53
Phytoseiid mites 433, 440, 461
Pices *275*
Pieris rapae 272, 274, 276
Pigeonpea 367–81
Piperonyl butoxide 83
Pirate bug 383, 386, 390–1, 392, 393
Pitfall traps 329, 330–1
Plagiognathus politus 425
Planaria torva 274
Planthoppers 133–48, 201, 434
Plasmids 357
Plasmodium
falciparum 32, 38
vivax 32, 38
Plea laechi 462
Plecoptera *274*
Plum 17
Plutella xylostella 74, 276
Podisus maculiventris 275, 276, 277
Poecilimon veluchianus 164
Poecilus cupreus 276
Polistes annularis 171, **172**
Polyacrylamide electrophoresis, *see* Electrophoresis
Polycelis felina 274
Polyclonal antisera, *see* Antisera
Polymerase chain reaction, *see* PCR
Polyphagus predators 330, 334
Polytene chromosomes 32–4
Population bottlenecks 13, 16, 17, 19
cicadas 176
locusts 222
New Zealand possum 254
slugs 104, 120
Population dynamics 13–24, 149
slugs 325, 336
tsetse flies 479–97
Population genetics 7–29, 150, 173, 245

Population genetics *contd*
cotton boll weevils 470, 473–6
desert locust 213–29
DNA fingerprinting 159–70
heliothine pests 73–92
RAPDs 249
slugs 93–131
see also Phylogenetic analysis; Gene flow
Population models, refugial 224–6, 228
Possum 232, 254–6
PCR primers for **240**
Potato 403, 426
Precipitin tests 279–80, 326, 421–2
Predation analysis 419–55
antibody techniques 419–55
electrophoretic techniques 457–68
DNA techniques 291–2
food chain errors 432–3
foraging behaviour 440
of *Helicoverpa* 367–81
quantitation of 462–5
models of 374, 377
of pink bollworm 383 –99
rates of 419, 420, 422, 441
secondary 432
serological analysis of 265–300
applications 283–93
quantitation 289–91
techniques 265–83
of slugs 323–45
of whitefly 383–99
Predator gut content analysis 170
carabids 323–45
electrophoresis 384
field applications 292–3
gut dissection 421
immunoassay of 270, 384, 392–5
see also Specific assays
meal size 340, 393–4, 423–4
quantification 329
radioactive markers 384
serological test summary **272–7**
see also Prey consumption analysis; Predation analysis
Predator–prey encounter rates 437–41
Predatory mites 424
Prey consumption analysis
alternative prey 339–40
antigen decay rates 336, 339–40
detection periods 339–40, 422
electrophoresis 421
moribund prey 427–31
of predatory aquatic Hemiptera 462
pre-oral digestion (extra-oral, extra-intestinal) 433
prey size choice 336, 436–7
quantification of 421–4
radiotracers 421
rates of 419–21, 422, 426, 441, 463–5
scavenging 420, 424–7, 441
slugs 329–36
Prey digestion rates 323, 330, 336, 339–40, 370–1, 393–4
Prey kairomone 440
Prey marking
antibody painting 427
rare elements 427
Prey wounding, wasteful killing 435
Private allelles 16, 22
Propoxur 37
Prosapia bicinta 155
Proteinase inhibitors, induction of 200–1
Protozoa, infection of caterpillars 430
Prunus
cerasifera 17, 18, 22
domestica 17
mahaleb 17
spinosa 17, 19
Pseudoscorpions 165, 440
Psylla
mali 461
pyricola 461
Pterocallis alni 461
Pterostichus *274*
madidus *277*, 328, 330–1, 426
melanarius *276*, 323, 325, 330–5, 338–9
Pyrazophos 426
Pyrethroid resistance 38–40
in heliothine pests 82, 83, 85
Pyrethroid-impregnated bednets 39
Pyrrhosoma nymphuls *273*

Radioimmunoassay 283
Ranatra montezuma *276*
Rape 208, 331
RAPD-PCR (Random amplified polymorphic DNA-PCR) 82, 138, 215, 241–2, 249–55, 458
aphids 178–84
arionid slugs 105–14

baboons 254
possums 254–6
weevils 251–3
Recombinant antibody technology 324, 347–66, **354**
antibody affinities 360
advantages in pest-predator relationships 361, 363
in medical therapeutics 360–1
in predation studies 341
Red-winged blackbirds 139
REN methods 82
Resistance genes 84
Rhagoletis pomonella 16, *276*
Rhopalosiphum
insertum 461
maidis 158, 245
padi 175, 232, *273*, *276*
Rice 205
pests of 135, 201
see Planthoppers
grassy stunt virus 135
ragged stunt virus 135
Rickettsia, see *Wolbachia*
RNA
nuclear 153
ribosomal 34, 105, 155, 243
in phylogenetic analysis 151–3, 167
Rosaceae 11
Rubus 11
Russian wheat aphid 251

Safflower 377
Salicaceae 11
Sandfly 179
Sarothamnus scoparius 272
Sawfly larvae 428, 429
Scathophagid flies 435
Scavenger beetles 425
Scavenging 420, 424–7, 441
scFv antibody fragments 358–9
Schistocerca gregaria 213–29
Schizaphis graminum 158, 159, **160,** 175, 179, 233, 245, 421, 432, 434
scn genes 235, 245
Scorpion flies 152, 425
SDS-PAGE, *see* Electrophoresis
Sesamia inferens *273*
Shrimp 152
Sibling species 9–13, 45, 74
Sigara lateralis 462
Silkworm 165
Silphid beetles 427, 434
Simocephalus 462
Simulium damnosum 34, 163, *274*
Sinea confusa 389
Single strand conformation polymorphism (SSCP) 137
Siphonaptera, rDNA gene sequences 153
Sitobion avenae 163, 164, **167**, *276*, 423, 426, 181–3, **182**
Sitona regensteinensis *273*
Skylarks 427
Sleeping sickness 480
Slugs
anatomical characters 97–9
colour varieties 99–103
distribution and ecology 114–18
ecological genetics 121–24
enzyme systems 100–4
evolutionary history 121–4
German 94
haemocyanin 327
haemolymph 327–9, 331
hybrid populations 116–18
metapopulation structure 124
nematodes 124
plagues 94, 95, 116, 124
predation analysis 323–45
recombinant antibodies to 362
reproductive ecology 118–21
self-fertilization 118-23
Spanish 94
taxonomic genetic markers 104–14
see also Individual species
Snowdrop lectins, as insecticides 201, 210
Soldier fly larvae 435
Solid-phase assays 280–2,
see also ELISA; Immunodot assay
Sorbus aucuparia 17–21
Sorex *273*
Sorghum 135, 158, **160,** 367–81, 403
Southern blot 208–9
Species identification 32–6
Sphecius speciosus 165
Spider mites 9
Spiders 328, 421, 424, 433, 436–7, 440
as predators of *Helicoverpa* 367–9, 371, **372**, 375–7
rDNA gene sequences 152
see also Individual species

Spindle tree 17
Spittlebug 155
Spodoptera
  exempta 15
  frugiperda 15, 163, 303
Springtails 439–40
Squash blot system 35–6
Staphylinids
  beetles *272, 273,* 425, 430, 434–5, 438, 439
  larvae 434, 435
Stigmaeid mites 433
Stigmella
  aurella 11
  basiguttella 10–11
  fragariella 11
  gei 11
  nitens 11
  splendidissimella 11
Sugarcane 135
Sunflowers 75, 375
Superoxide dismutase, in *Anopheles* 33
Sympatric speciation 11, 16
Syrphid larvae 341

Tachinidae 432
Tandem repeats 160–2, 235
Tandonia
  budapestensis 337
  sowerbyi 337
Tanypodinae 462
Taxonomic markers 104–14, 151–7
Tenebrio molitor 411
Tephritidae, *see* Fruit flies
Termites 171
Tetranychid mites 435
Tetranychus
  pacificus 9
  urticae 461
Tetrastichus incertus 64
Thrips 9
Tiger beetles 174
Tiger mosquito 244–5
Tipula 341
Tobacco budworm 66
Tomato 12
Tortricidae 275
Toxorhynchites rutilus 276
Transgenic crops
  for insect resistance 200, 201
  tobacco plants, snowdrop lectin in 210
Trialeurodes abutilonea 180
Tribolium
  castaneum 411
  confusum 59
Trichocorixa reticulata 274
Trichogramma 180, 184
  electrophoretic analysis of 9
  evanescens 180, 411
  nubilale 411
Trichoptera *274*
Trichosurus vulpecula 232, 254–6
Tricladida *273*
Tricoplusia ni 386
Trioxys pallidus 181
Triticum aestivum 205
Trypanosomiases 480
Tsetse flies 479–97
  age determination 484–6
    pteridine fluorescence 486, 488, **489**, 495
  field sampling methods 481–3
  genetic variation 165
  nutritional reserves of 483–4
  see also Glossina
Tubifex tubifex 274
Typhlodromus pyri 461, 464
Tyria jacobaeae 273

Urodacus 277
Uroleucon ambrosiae 179
Urtica dioica 205

Vairimorpha 407, 409
Vespa crabro 154
Viral pathogens 427

Wasp 165, 171, **172,** 421, 425, 427
Weevils 425
  alfalfa 51–71
  argentine stem 180, 231–63
  clover leaf 53, 55, 57
  cotton boll 469–78
  diapause 472–3, 475–6
  PCR primers for **240**
  RAPD-PCR 251–3
  white-fringed 232, *240,* 251–3
  see also Individual species
Wheat 158, **160**, 426
  chitin-binding lectins 205
Wheatgerm agglutinin, insecticidal properties 201

Whitefly 180, 233, 383, 385–93
 see also Individual species
WHO test kits, in tests of insecticide susceptibility 40
Willow 17
Wireworm larvae 434
Wolbachia
 detection of 59–62
 endosymbionts of alfalfa weevils 51–71
 endosymbionts of *Drosophila* 63
Woodlice 423, 432

X-chromosome, banding patterns 33

Yponomeuta
 cagnagellus 17
 evonymellus 17
 malinellus 17
 padellus 10–24
 rorellus 17

Zelus 389
 exsanguis 272
Zootermopsis nevadensis 173

Systematics Association Publications

1. Bibliography of key works for the identification of the British fauna and flora, *3rd edition* (1967)†
   Edited by G. J. Kerrich, R. D. Meikle, and N. Tebble
2. Function and taxonomic importance (1959)†
   Edited by A. J. Cain
3. The species concept in palaeontology (1956)†
   Edited by P. C. Sylvester-Bradley
4. Taxonomy and geography (1962)†
   Edited by D. Nichols
5. Speciation in the sea (1963)†
   Edited by J. P. Harding and N. Tebble
6. Phenetic and phylogenetic classification (1964)†
   Edited by V. H. Heywood and J. McNeill
7. Aspects of Tethyan biogeography (1967)†
   Edited by C. G. Adams and D. V. Ager
8. The soil ecosystem (1969)†
   Edited by H. Sheals
9. Organisms and continents through time (1973)†
   Edited by N. F. Hughes
10. Cladistics: a practical course in systematics (1992)
    P. L. Forey, C. J. Humphries, I. J. Kitching, R. W. Scotland, D. J. Siebert, and D. M. Williams

†Published by the Association (out of print)

Systematics Association Special Volumes

1. The new systematics (1940)
   Edited by J. S. Huxley (Reprinted 1971)
2. Chemotaxonomy and serotaxonomy (1968)*
   Edited by J. G. Hawkes
3. Data processing biology and geology (1971)*
   Edited by J. L. Cutbill
4. Scanning electron microscopy (1971)*
   Edited by V. H. Heywood
   Out of print
5. Taxonomy and ecology (1973)*
   Edited by V. H. Heywood
6. The changing flora and fauna of Britain (1974)*
   Edited by D. L. Hawksworth
   Out of print

7. Biological identification with computers (1975)*
   Edited by R. J. Pankhurst
8. Lichenology: progress and problems (1976)*
   Edited by D. H. Brown, D. L. Hawksworth, and R. H. Bailey
9. Key works to the fauna and flora of the British Isles and north-western Europe, 4th edition (1978)*
   Edited by G. J. Kerrich, D. L. Hawksworth, and R. W. Sims
10. Modern approaches to the taxonomy of red and brown algae (1978)
    Edited by D. E. G. Irvine and J. H. Price
11. Biology and systematics of colonial organisms (1979)*
    Edited by G. Larwood and B. R. Rosen
12. The origin of major invertebrate groups (1979)*
    Edited by M. R. House
13. Advances in bryozoology (1979)*
    Edited y G. P. Larwood and M. B. Abbot
14. Bryophyte systematics (1979)*
    Edited by G. C. S. Clarke and J. G. Duckett
15. The terrestrial environment and the origin of land vertebrates (1980)
    Edited by A. L. Panchen
16. Chemosystematics: principles and practice (1980)*
    Edited by F. A. Bisby, J. G. Vaughan, and C. A. Wright
17. The shore environment: methods and ecosystems (2 Volumes) (1980)*
    Edited by J. H. Price, D. E. G. Irvine and W. F. Farnham
18. The Ammonoidea (1981)*
    Edited by M. R. House and J. R. Senior
19. Biosystematics of social insects (1981)*
    Edited by P. E. House and J.-L. Clément
20. Genome evolution (1982)*
    Edited by G. A. Dover and R. B. Flavell
21. Problems of phylogenetic reconstruction (1982)*
    Edited by K. A. Joysey and A. E. Friday
22. Concepts in nematode systematics (1983)*
    Edited by A. R. Stone, H. M. Platt, and L. F. Khalil
23. Evolution, time and space: the emergence of the biosphere (1983)*
    Edited by R. W. Sims, J. H. Price, and P. E. S. Whalley
24. Protein polymorphism: adaptive and taxonomic significance (1983)*
    Edited by G. S. Oxford and D. Rollinson
25. Current concepts in plant taxonomy (1983)*
    Edited by V. H. Heywood and D. M. Moore
26. Databases in systematics (1984)*
    Edited by R. Allkin and F. A. Bisby
27. Systematics of the green algae (1984)*
    Edited by D. E. G. Irvine and D. M. John
28. The origins and relationships of lower invertebrates (1985)‡
    Edited by S. Conway Morris, J. D. George, R. Gibson, and H. M. Platt
29. Infraspecific classification of wild and cultivated plants (1986)‡
    Edited by B. T. Styles

30. Biomineralization in lower plants and animals (1986)‡
    Edited by B. S. C. Leadbeater and R. Riding
31. Systematic and taxonomic approaches in palaeobotany (1986)‡
    Edited by R. A. Spicer and B. A. Thomas
32. Coevolution and systematics (1986)‡
    Edited by A. R. Stone and D. L. Hawksworth
33. Key works to the fauna and flora of the British Isles and north-western Europe, 5th edition (1988)‡
    Edited by R. W. Sims, P. Freeman, and D. L. Hawksworth
34. Extinction and survival in the fossil record (1988)‡
    Edited by G. P. Larwood
35. The phylogeny and classification of the tetrapods (2 Volumes) (1988)‡
    Edited by M. J. Benton
36. Prospects in systematics (1988)‡
    Edited by D. L. Hawksworth
37. Biosystematics of haematophagous insects (1988)‡
    Edited by M. W. Service
38. The chromophyte algae: problems and perspective (1989)‡
    Edited by J. C. Green, B. S. C. Leadbeater, and W. L. Diver
39. Electrophoretic studies on agricultural pests (1989)‡
    Edited by Hugh D. Loxdale and J. den Hollander
40. Evolution, systematics, and fossil history of the Hamamelidae (2 Volumes) 1989)‡
    Edited by Peter R. Crane and Stephen Blackmore
41. Scanning electron microscopy in taxonomy and functional morphology (1990)‡
    Edited by D. Claugher
42. Major evolutionary radiations (1990)‡
    Edited by P. D. Taylor and G. P. Larwood
43. Tropical lichens: their systematics, conservation, and ecology (1991)‡
    Edited by G. J. Galloway
44. Pollen and spores: patterns of diversification (1991)‡
    Edited by S. Blackmore and S. H. Barnes
45. The biology of free-living heterotrophic flagellates (1991)‡
    Edited by D. J. Patterson and J. Larsen
46. Plant-animal interactions in the marine benthos (1992)‡
    Edited by D. M. Johns, S. J. Hawkins, and J. H. Price
47. The Ammonoidea: environment, ecology, and evolutionary change (1993)‡
    Edited by M. R. House
48. Designs for a global plant species information system‡
    Edited by F. A. Bisby, G. F. Russell, and R. J. Pankhurst
49. Plant galls: organisms, interactions, populations‡
    Edited by Michèle A. J. Williams
50. Systematics and conservation evaluation‡
    Edited by P. L. Forey, C. J. Humphries, and R. I. Vane-Wright
51. The Haptophyte algae‡
    Edited by J. C. Green and B. S. C. Leadbeater

52. Models in phylogeny reconstruction‡
    Edited by R. Scotland, D. J. Siebert and D. M. Williams
53. The Ecology of Agricultural Pests: biochemical approaches**
    *Edited by W. O. C. Symondson and J. E. Liddell**

*Published by Academic Press for the Systematics Association
†Published by the Palaeontological Association in conjunction with Systematics Association
‡Published by the Oxford University Press for the Systematics Association
**Published by Chapman & Hall for the Systematics Association